Technical
Calculus
with
Analytic
Geometry

Mathematics Editor: Heather Bennett
Production: Mary Forkner, Publication Alternatives
Text and Cover Designer: Al Burkhardt
Copy Editor: Linda Thompson
Illustrator: Mary Burkhardt
Photographs: Natural Photo/Robert Vernon Wilson © pp. 1,
66, 182, 246, 286, 340, and 374; Tim Yates, p. 33; Image Bank
West/Grafton Marshall Smith, p. 81; NASA, p. 136
Cover Photograph: Image Bank West/Gabe Palmer
Typesetter: Jonathan Peck Typographers, Ltd.

Printed in the United States of America

1 2 3 4 5 6 7 8 9 10—87 86 85 84 83

Library of Congress Cataloging in Publication Data
Gersting, Judith L.
    Technical calculus with analytic geometry.

    Includes index.
    1. Calculus. 2. Geometry, Analytic. I. Title.
QA303.G43 1984   515'.15   83-14654

ISBN 0-534-02893-4

Technical Calculus with Analytic Geometry

Judith L. Gersting

Indiana University–Purdue University at Indianapolis

Wadsworth Publishing Company

Belmont, California

A Division of Wadsworth, Inc.

Mathematics Editor: Heather Bennett
Production: Mary Forkner, Publication Alternatives
Text and Cover Designer: Al Burkhardt
Copy Editor: Linda Thompson
Illustrator: Mary Burkhardt
Photographs: Natural Photo/Robert Vernon Wilson © pp. 1,
66, 182, 246, 286, 340, and 374; Tim Yates, p. 33; Image Bank
West/Grafton Marshall Smith, p. 81; NASA, p. 136
Cover Photograph: Image Bank West/Gabe Palmer
Typesetter: Jonathan Peck Typographers, Ltd.

Printed in the United States of America

1 2 3 4 5 6 7 8 9 10—87 86 85 84 83

Library of Congress Cataloging in Publication Data
Gersting, Judith L.
   Technical calculus with analytic geometry.

   Includes index.
   1. Calculus. 2. Geometry, Analytic. I. Title.
QA303.G43 1984   515'.15   83-14654

ISBN 0-534-02893-4

Technical
Calculus
with
Analytic
Geometry

Contents

To John, my favorite engineer

Preface to the Instructor

This text is intended for a two-semester course in calculus for technology students. The prerequisites for the course are college algebra and trigonometry. The expected student audience has influenced the choice of topics covered, the style of presentation, and the types of examples and exercises. In my teaching experience I have found technology students to be strongly career oriented; they sincerely work hard to understand material that they feel is pertinent to their goals. This text has been written to attract students' interest by providing motivating examples, by giving them an intuitive understanding of the concepts behind what they are doing, and by providing much opportunity to gain proficiency in techniques and skills.

Topic Coverage

The topics in the table of contents are quite standard, but there are some variations from other books as to where or how a particular topic is covered. A brief but thorough review of the concept of a function is done in the first sections. Polar coordinates, partial derivatives, and double integrals are presented where they are natural extensions of rectangular coordinate work, but they can be left for later or omitted entirely if the instructor desires. The definite integral as the limit of a sum is stressed (in an intuitive way) so that the student gets a good idea of how to use the definite integral in a variety of situations. There is an emphasis on matching an integration problem with the pattern for a given integration rule, and a section is included on partial fractions as an integration technique.

Writing Style

Numerous features are built into the book in an effort to make it a useful learning tool for students. First and foremost, the writing style has been kept clear and readable, informal and conversational. I have tried to write as if I were giving a lecture in class, avoiding the stuffiness that can all too easily slip into written material and cause the students to have to ask for an interpretation of what they've read. The book talks *with* students and acts as their aid and advocate in conquering the calculus.

Worked Examples

Examples are abundant. Each example shows the complete solution method without skipping any intermediate steps. Occasionally alternate solutions to the same problem are given. When a new technique is being demonstrated, a reason is given for each step in the solution process.

Applications

Many examples and exercises are drawn from various fields of technology, as well as such fields as ecology, economics, and physiology. The value of student motivation generated in this manner can be great. I remember a student in my early teaching career who came to ask me about an example that he did not understand. The example was phrased in terms of electronics technology, and the student said, "I want to be an electronics technician, so I figure I really need to understand this example." As teachers, we can capitalize on this sort of interest.

Practice Problems

Practice problems appear in the body of the text in each section of the book. These problems are relatively easy and are intended to be worked by the student as soon as they are encountered. Each new type of example in the text is followed by a practice problem that allows the student to gain immediate reinforcement in applying the problem-solving technique illustrated by the example. Answers to all practice problems are given at the back of the book, many with worked-out solution steps. This sets the method in the student's mind so that when the exercises at the end of the section are worked, the student does not have to begin by searching for an appropriate example and wondering if he or she really understands it once it is located. Students find the practice problems to be extremely helpful.

Learning Aids

Other learning aids are incorporated in the book. Many **Words of Advice** are scattered throughout. These messages from a wise old owl caution the student against common errors. Complex **problem-solving processes** are broken down into a series of step-by-step tasks. Numerous exercises appear at the end of each section. The exercises, for the most part, are paired into **odd-even problems** that are very similar. Answers to all odd exercises are given at the back of the book, some with **expanded solutions.** A **Status Check** at the end of each chapter gives the chapter objectives to help the students review (and to aid the instructor in making out exams). The chapter

Status Check is immediately followed by a section of **additional review exercises.** A **glossary,** with definitions and page references, is included at the back of the book.

Notes on Calculators

The book assumes that the student has a scientific calculator available to use. There are several consequences of this assumption. First, and perhaps most obvious, is that no trigonometric or logarithmic tables appear in the book. Second, the use of calculators allows examples and exercises to be a bit more realistic, rather than being contrived in such a way as to produce integer answers. (In computing answers I have used exact values, such as $\sqrt{3}$, throughout a computation and converted them at the end to decimal equivalents when required. I have used no standard rule, however, for the number of decimal places of accuracy and have simply rounded off answers as seemed appropriate. Thus students should be aware that their answers may differ slightly in numerical value if they have carried more decimal places.) Third, the most interesting consequence of having the calculator available is that it is used to suggest patterns and illustrate concepts such as limit, slope, and convergence of series. A calculator is also useful in approximation techniques for integration and solution of differential equations.

Virtually all examples and exercises use SI (metric) units. An appendix of common metric units and their notation is included at the back of the book. Other appendices give common geometric formulas and a table of integrals.

My thanks to the many reviewers whose helpful suggestions smoothed out the rough spots. In particular, Mr. Blin Scatterday of the Univesity of Akron and Mr. Henry Davison of St. Petersburg Junior College helped in reading proof and checking answers, and I especially appreciate their work. Mr. Scatterday has also prepared a solutions manual for all the even-numbered exercises.

Miss Linda Knabe worked wonders in creating typescript out of handwritten manuscript, and I thank her for her always cheerful help. I would also like to express my appreciation to Rich Jones of Wadsworth, who originally thought of this project and had the patience to see me through it. Finally, my special thanks to my family for their support and encouragement, including their willingness to endure my long periods of reclusion in the den.

Judith L. Gersting

Acknowledgments

My thanks also go to the following individuals who gave of their time to provide valuable comments.

J. Richard Garnham
Rochester Institute
of Technology

Carole E. Goodson
University of Houston,
Central Campus

Henry E. Horwitz
Dutchess Community College

Michael Iannone
Trenton State College

Michael R. John
Wentworth Institute
of Technology

Gerald J. Kirwin
University of New Haven

David Legg
Indiana University–
Purdue University

Wilbur R. LePage
Syracuse University

George R. Mach
California Polytechnic State
University, San Luis Obispo

Charles D. Reinauer
San Jacinto College

V. Merriline Smith
California State Polytechnic
University, Pomona

Thomas J. Stark
Cincinnati Technical College

Margaret Taber
Purdue University

W. Thurmon Whitley
University of New Haven

Don W. Williams
Brazosport College

Note to the Student

You should be aware of several features of this book. As you read, you will find many **practice problems** in the text. These problems are relatively easy, and you will find it helpful to work them as soon as you encounter them. Answers are given at the back of the book. Get your paper and pencil ready now.

There is also a wise old owl who will give you **Words of Advice** throughout; the owl is cautioning you against common errors that students make, so take the advice to heart.

At the end of each chapter there is a section called **Status Check.** This section will help you review all the things you should have learned how to do in that chapter. There are many new definitions to learn in this book, and a **glossary** at the back of the book can help you review or find the meaning of a term you've forgotten.

Many examples and exercises are drawn from various fields of technology, as well as such fields as science, engineering, business, and physiology. Of course, you are not an expert in all of these areas, but each problem explains everything you need to know in order to work it. Don't hesitate to tackle a problem just because it seems to relate to an area unfamiliar to you.

This book assumes that you have a scientific calculator (one with trigonometric and logarithmic functions) available. Be sure that you have read the instructions for your calculator thoroughly and understand how to operate it.

Now, good luck. And remember that in mathematics, practice does not guarantee success, but lack of practice almost always guarantees lack of success. Don't get too far behind, and be sure to ask your instructor about things you don't understand.

Judith L. Gersting

Technical
Calculus
with
Analytic
Geometry

chapter 1

Functions and Graphs

A rectangular coordinate system allows us to locate points on the plane, just as latitude and longitude allow us to locate ourselves physically on the globe.

1-1 Functions

In the physical world, certain quantities have values that do not change, but remain fixed. Such quantities are called **constants**. The number π, which arises in geometry, is a constant. The force of gravitational attraction at a certain point on the earth's surface is a constant. The boiling point of water under standard atmospheric pressure is a constant.

However, many quantities of interest in science and technology do not have fixed values, but instead have values that can change, or vary. Such quantities are called **variables**.

EXAMPLE 1. The ends of a metallic conductor can be connected to a battery. Different batteries may supply different voltages to the conductor. If we use the letter V to represent the voltage supplied to the conductor, V is a variable because it can take on different values depending on whether we use a 6-volt battery, a 12-volt battery, and so on. Applying a voltage to the conductor will cause an electric current to flow along the conductor. Different voltages, in turn, cause different amounts of electric current. If we use I to represent current, then I is also a variable. ■

Variable quantities arising within the same physical situation are often related to one another. Scientists and engineers look for relationships between variables to help explain why things happen the way they do.

EXAMPLE 2.   Copper wire is a metallic conductor. In a fixed length of copper wire of a certain thickness, the following measurements of applied voltages V (in volts) and resulting current I (in amperes), were taken:

V	1	2	4	6	8	10
I	1.6	3.2	6.4	9.6	12.8	16

What is the relationship between V and I? This is probably not obvious, but you can use your calculator to confirm quickly that in each case, if V is divided by I, the result is 0.625. Expressed as an equation, $V/I = 0.625$, or $V = 0.625I$. ■

Repeated experiments and observations, such as those of Example 2, would show us that for any given metallic conductor,

$$\frac{V}{I} = k$$

where k is a constant. This relationship between voltage and current in a metallic conductor is known as *Ohm's law*. Of course, it is not really a "law," it is simply a mathematical relationship that seems to explain a physical phenomenon.

If we know the relationship $V = 0.625I$ of Example 2, we are then able to compute, for that particular conductor, the value of V for any value of I.

Practice 1.   In the piece of copper wire of Example 2, a current of 21 A was measured. What was the applied voltage?

Suppose two variables are related in such a way that a value for the first variable determines a single, unique value for the second variable. Then the relationship is called a **function**, and the second variable is said to be a function of the first variable. Because the value of the second variable depends upon the value of the first variable, the second variable is called the **dependent variable** and the first variable is the **independent variable**.

EXAMPLE 3.   In the equation $V = 0.625I$, V is a function of I; V is the dependent variable and I is the independent variable. ■

EXAMPLE 4. Write the area A of a circle as a function of its radius r.

We are to express a relationship between A and r, where r is the independent variable and A is the dependent variable. When we are done we should be able to calculate a unique value of A for each value of r.

The equation $A = \pi r^2$ expresses this function. If we are given $r = 2.0$, for example, the corresponding unique value for A is $\pi(2.0)^2 = 4\pi$. We can approximate the value of 4π by 12.566. If $r = 1.2$, we can then compute $A = \pi(1.2)^2$, which is approximately 4.524. ■

Practice 2. Write the circumference C of a circle as a function of its radius r.

Although we will usually describe a function by means of an equation, a function does not have to be given by an equation and it need not be mathematical in nature.

EXAMPLE 5. The independent variable in this function represents students in your calculus class. The dependent variable represents left shoes. The function associates with every student that student's left shoe. This relationship, which cannot readily be expressed in equation form, is nevertheless a function, because once the value of the independent variable is known (that is, a particular student is selected), then the value of the dependent variable is known—namely, it is that particular student's left shoe. Note that this example fails if there are barefoot students in the room (because then there is no value of the dependent variable for a value of the independent variable) or if there are any students in the room wearing two left shoes (because this causes a value of the independent variable for which there is not a unique value of the dependent variable). ■

Just as there are functions not given by equations, there are also equations that do not represent functions.

EXAMPLE 6. The equation $y^2 = x^2$ does not represent a function. The variable y is not a function of x because for $x = 2$, there are two values of y (2 and -2) that make the equation true. Knowing the value of x does not give us a unique value for y. Similarly, x is not a function of y. ■

Letters near the end of the alphabet $(x, y, z, w, \ldots)$ are often used to represent variables, and letters near the front of the alphabet $(a, b, c, \ldots)$ stand for constants whose value we may not know. Therefore, in the equation $y = ax + b$, we would assume x and y to be variables and a and b to be constants.

If y is a function of x, that fact is often written as $y = f(x)$. Then we write $y = f(a)$ to denote the value of y when the variable x has value a.

EXAMPLE 7.   Let $y = 3x + 2$, or $y = f(x)$, where $f(x) = 3x + 2$. The notation $f(x) = 3x + 2$ emphasizes that the function is really a set of instructions telling us how to compute the value of the dependent variable given the value of the independent variable. For the function $f(x) = 3x + 2$, the instructions are to multiply the given value, whatever it may be, by 3 and then add 2 to the product. Therefore,

$$f(2) = 3 \cdot 2 + 2 = 6 + 2 = 8$$
$$f(5) = 3 \cdot 5 + 2 = 15 + 2 = 17$$
$$f(-6) = 3(-6) + 2 = -18 + 2 = -16$$
$$f(\Box) = 3 \cdot \Box + 2$$
$$f(T) = 3 \cdot T + 2$$
$$f(a) = 3 \cdot a + 2$$
$$f(f(a)) = 3 \cdot f(a) + 2 = 3(3 \cdot a + 2) + 2 = 9a + 6 + 2 = 9a + 8 \quad \blacksquare$$

As in Example 7, two common ways to define a function of x exist. We might say that y as a function of x is given by the equation $y = 3x + 2$, or we might say, let $f(x) = 3x + 2$. In either case we have given the same set of instructions for computing the dependent value from the independent value. Other variable names can also be used in functions. We could say, let P be a function of t given by $P = 12t^2 + 7$. In this function t is the independent variable and P is the dependent variable. The equation $g(z) = (2z - 1)/z^2$ defines a function named g with independent variable z.

Word of Advice

Once a function $f(x)$ has been defined, you can compute $f(a)$ by substituting a for x in the definition of the function.

Practice 3.   Let $h(t) = 2t^2 - 1$. Find $h(3)$, $h(-1)$, $h(a)$, and $h(h(2))$.

New functions can be created by merging other functions together. Suppose y is a function of u, say $y = f(u)$, and u in turn is a function of x, say $u = g(x)$. Then y is also a function of x, given by $y = f(g(x))$. This new function is called the **composition** of the two original functions.

EXAMPLE 8. Let $y = u^2$ and $u = 4x + 1$. Then $y = u^2 = (4x + 1)^2$ $= 16x^2 + 8x + 1$. Thus $y = 16x^2 + 8x + 1$, so it is clear that y is a function of x. If we leave y in the form $y = (4x + 1)^2$, it is easier to see that y is the composition of the function that says to multiply by 4 and add 1 with the function that says to square. ∎

Practice 4. Let $y = 2z$ and $z = x^3$. Write y as a function of x.

A function may be defined by several equations, each giving the instructions for computing the dependent variable for different sets of values of the independent variable.

EXAMPLE 9. Let $g(t)$ be a function defined by

$$g(t) = \begin{cases} t + 2 & \text{for } t \geq 0 \\ 3 & \text{for } t < 0 \end{cases}$$

Then $g(5) = 5 + 2 = 7$ (using the first equation), and $g(-5) = 3$ (using the second equation). For each value of t, the independent variable, there is a single value of $g(t)$, so $g(t)$ is a function. Note that if $g(t)$ were defined by

$$g(t) = \begin{cases} t + 2 & \text{for } t \geq 0 \\ 3 & \text{for } t \leq 0 \end{cases}$$

then $g(t)$ would no longer be a function, because $g(0)$ could be either 2 (from the first equation) or 3 (from the second equation). ∎

If a function is defined by an equation, we may have to exclude some values of the independent variable. The independent variable cannot be allowed to take on values that result in division by zero. Also, unless we are specifically dealing with a context in which imaginary or complex numbers have meaning, we cannot allow the independent variable to have values that result in the square root of a negative number. (Complex numbers will not be used in this book until Chapter 11.) If the independent variable represents a physical quantity, then certain values may have to be excluded because they do not make sense for that physical quantity.

EXAMPLE 10. For $f(x) = 1/(x + 2)$, $x = -2$ is an excluded value, because this value results in division by zero. <u>The function makes sense for any other value of x.</u> For $g(x) = \sqrt{4 - x^2}$, to avoid imaginary numbers we must exclude values of x for which $4 - x^2 < 0$, or $4 < x^2$. Therefore, we must exclude all values of x for which $x > 2$ or $x < -2$. Put another way, the only allowable values of x are those for which $-2 \leq x \leq 2$. ■

EXAMPLE 11. The deflection y of a cantilevered beam of length L is given as a function of x, the distance from the fixed end of the beam, by the equation $y = k(x^4 - 4Lx^3 + 6L^2x^2)$, where k is a constant. Find the excluded values of x.

Here the form of the equation itself does not lead to excluded values of x. However, x represents a distance, so $x \geq 0$. Also, x is a distance along a beam whose total length is L. Therefore, $x \leq L$. Allowable values are thus $0 \leq x \leq L$. ■

Practice 5. For $y = \sqrt{x}/(x - 3)$, what are the allowable values of x?

Exercises / Section 1-1

Exercises 1–14: Write the function.

1. Write the area A of a square as a function of the length s of a side.

2. Write the volume V of a cube as a function of the length s of an edge.

3. Write the area A of a triangle with base of length 10 as a function of the altitude h.

4. Write the area A of a semicircle as a function of the radius r.

5. Write the temperature F in degrees Fahrenheit as a function of the temperature C in degrees Celsius.

6. Write the temperature C in degrees Celsius as a function of the temperature F in degrees Fahrenheit.

7. In a square piece of cardboard 136 cm per side, equal squares with sides of length x cm are cut from the four corners. The sides are then folded up to make an open box. Write the volume V of the box as a function of x. What are the allowable values of x?

8. A closed cylindrical can has a surface area of 2000 cm². Write the height h of the cylinder as a function of the radius r of the base.

9. A piece of land consists of a rectangle with an equilateral triangle on one end (see Figure 1-1). Write the perimeter P of the entire piece of land as a function of W if the area of the rectangular portion is 240 m².

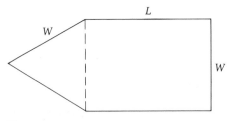

Figure 1-1

10. A piece of wire 70 cm long is cut once; each piece is bent into a circle. Write the total area A of the two circles as a function of C, the circumference of one of the circles.

11. First class postal rates are 20¢ for mail up to and including 1 oz in weight, and an additional 17¢ for each additional ounce or fraction of an ounce. Write the cost P of postage as a function of the weight w for all mail up to 5 oz. (*Hint*: A single equation will not work.)

12. An electric company charges residential rates of 6.1¢ per kWh (kilowatt-hour) used, up to a maximum of 1500 kWh per month. Service above 1500 kWh per month is charged a flat rate of $100 plus 6.8¢ per kWh used in excess of 1500. Write the cost c in dollars for service as a function of p, the total number of kilowatt-hours used per month.

13. The cost c to produce x units of a certain item is given by the equation $c = 600 + 2.8x$. Write x as a function of c.

14. The period T of a simple pendulum as a function of its length L is given by $T = 2\pi\sqrt{L/9.8}$. Write L as a function of T.

Exercises 15–26: Find the values requested.

15. $f(x) = 5x - 4$; $f(-1)$, $f(3)$

16. $f(x) = 2x + 7$; $f(0)$, $f(4)$

17. $g(x) = 3x^2 - 1$; $g(7)$, $g(-6)$

18. $g(x) = 4x^2 - 2x + 5$; $g(4.2)$, $g(-1.8)$

19. $h(z) = \dfrac{z^2}{z - 4}$; $h(a)$, $h(3a)$

20. $M(t) = \dfrac{t + 1}{t^2}$; $M(3w)$, $M(w + 1)$

21. $H(x) = \dfrac{x + 1}{x^2 + 5}$; $H(x^2)$, $H(H(3))$

22. $F(x) = \sqrt{1 - x^2}$; $F(t + 1)$, $F(F(x))$

23. $f(x) = \begin{cases} x + 1 & x \geq 0 \\ x^2 & x < 0 \end{cases}$; $f(3)$, $f(-5)$

24. $g(x) = \begin{cases} x & x \geq 1 \\ -x & 0 < x < 1 \\ 4 & x \leq 0 \end{cases}$; $g(2)$, $g(0.25)$, $g(g(0.5))$

25. The velocity v of a moving object as a function of time t is given by $v = 2t^2 + 14t + 2$. Find the velocity at $t = 4$.

26. The magnetic induction B in a certain coil as a function of distance x along the axis of the coil is given by $B = k_1/2(k_2^2 + x^2)^{3/2}$, where k_1 and k_2 are constants. Find the induction at $x = 18$.

Exercises 27–30: Write y as a function of x.

27. $y = 2w$, $w = x + 1$

28. $y = 3u - 7$, $u = x^2$

29. $y = u^2$, $u = \sqrt{x - 1}$

30. $y = z^2 - 2z$, $z = x^3$

Exercises 31–36: Find the allowable values of the independent variable.

31. $y = \dfrac{2x^2}{x - 1}$

32. $y = \sqrt{1 - x}$

33. $y = \sqrt{t}\sqrt{t + 1}$

34. $y = \sqrt{z^2 - 9}$

35. $y = \dfrac{t - 5}{t^3 - 4t}$

36. $y = \dfrac{3u}{u^2 - u - 6}$

1-2 Types of Functions

Functions are used in all branches of technology to express mathematically the relationships that exist between quantities. A function is a way of building a mathematical model of what actually happens in the physical world. Mathematical functions are of two main types, algebraic and transcendental. An **algebraic function** is one involving only the operations of addition, subtraction, multiplication, division, and taking roots.

EXAMPLE 1. The following are all algebraic functions.

$$f(x) = 3x^2 + 14x - \sqrt{3}$$

$$g(x) = \frac{2x^2}{(x + 1)(x^2 - 4)}$$

$$h(x) = x^3\sqrt{2x^2} + \sqrt{x - 1}$$

$$m(x) = \frac{\sqrt[3]{x^2}}{2\sqrt{x} + 17x^4} \quad \blacksquare$$

Transcendental functions are those functions that are non-algebraic, such as the trigonometric and logarithmic functions.

One kind of algebraic function occurs so frequently that it has a name of its own. A **polynomial** is a function of the form

$$f(x) = a_n x^n + a_{n-1}x^{n-1} + \cdots + a_1 x + a_0$$

where n is a nonnegative integer, called the **degree** of the polynomial, and the a's are constants. In Example 1, $f(x) = 3x^2 + 14x - \sqrt{3}$ is a polynomial of degree 2.

In calculus, it is almost always easier to work with fractional exponents than with radical signs. Remember from algebra that fractional exponents are defined as follows:

$$a^{n/m} = \sqrt[m]{a^n} \quad \text{or} \quad (\sqrt[m]{a})^n$$

Therefore, we will often change $\sqrt{x + 1}$ to $(x + 1)^{1/2}$ and $\sqrt[3]{x^2}$ to $x^{2/3}$. Because exponents will be important in simplifying expressions, you should quickly review the laws of exponents.

$$a^n \cdot a^m = a^{n+m}$$

$$\frac{a^n}{a^m} = a^{n-m}$$

$$(a^n)^m = a^{nm}$$

$$a^0 = 1$$

$$a^{-n} = \frac{1}{a^n}$$

$$(ab)^n = a^n b^n$$

$$\left(\frac{a}{b}\right)^n = \frac{a^n}{b^n}$$

EXAMPLE 2. Simplify the expression for $f(x)$.

$$f(x) = \frac{(x - 2)^3}{[(x - 2)^3]^2}$$

Using the laws of exponents,

$$f(x) = \frac{(x - 2)^3}{[(x - 2)^3]^2} = \frac{(x - 2)^3}{(x - 2)^6}$$

$$= (x - 2)^{3-6} = (x - 2)^{-3}, \quad \text{or} \quad \frac{1}{(x - 2)^3} \quad \blacksquare$$

EXAMPLE 3. Simplify

$$g(x) = \sqrt{3 - x}\,(3 - x)^{1/3}$$

Changing the radical to exponential form, we get

$$g(x) = (3 - x)^{1/2}(3 - x)^{1/3} = (3 - x)^{(1/2)+(1/3)} = (3 - x)^{5/6} \quad \blacksquare$$

EXAMPLE 4. Simplify

$$h(x) = (x^2)^{1/4} + x^{-3/2}$$

We first use a law of exponents to rewrite $h(x)$ as

$$h(x) = x^{1/2} + x^{-3/2}$$

Now an x appears in each term; we factor out x to the greatest negative power to which it appears.

$$h(x) = x^{-3/2}[x^2 + 1]$$

Note that the square brackets contain no fractional powers. $\quad \blacksquare$

EXAMPLE 5. Simplify

$$f(x) = (2x^2 + 1)(3x - 1)^{-3/2} - x^2(3x - 1)^{-1/2}$$

We factor out $(3x - 1)^{-3/2}$ (again, the greatest negative power) and get

$$f(x) = (3x - 1)^{-3/2}[2x^2 + 1 - x^2(3x - 1)]$$

As in Example 4, no fractional powers appear within the square brackets. Thus

$$f(x) = (3x - 1)^{-3/2}[2x^2 + 1 - 3x^3 + x^2]$$
$$= (3x - 1)^{-3/2}[-3x^3 + 3x^2 + 1]$$
$$= \frac{-3x^3 + 3x^2 + 1}{(3x + 1)^{3/2}} \quad \blacksquare$$

Simplification problems like that of Example 5 occur frequently in calculus.

Word of Advice

Fluency in applying the laws of exponents
will stand you in good stead in Chapter 4 and
indeed in much of the rest of this book.
Review the laws of exponents again.

Practice 1.   Simplify $f(x) = (3x - 4)^{1/2} + (2x + 2)(3x - 4)^{-1/2}$.

Many of the formulas and functions used in technology involve proportion, or variation. If $y = kx$, where k is a constant, then y is said to be **directly proportional to** x, or **vary directly as** x. In direct proportion, if x increases in size, so does y; if x decreases, so does y. If $y = k/x$, where k is a constant, then y is said to be **inversely proportional to** x, or **vary inversely as** x. In inverse proportion, if x increases in size, y decreases; if x decreases, y increases.

Direct and inverse proportions can both appear in the same equation. If $y = kx_1x_2/x_3$, for example, then y varies **jointly** as x_1 and x_2 and inversely as x_3.

EXAMPLE 6.   The equation for the area of a circle is $A = \pi r^2$. The area varies directly as the square of the radius.   ■

EXAMPLE 7.   The voltage V produced by a certain thermocouple is directly proportional to the square of the temperature T. If a thermocouple produces $200\ V$ at a temperature of $80°C$, write V as a function of T.

The function has the form $V = kT^2$, where k is a constant. We can compute the value of k from the information given.

$$k = \frac{V}{T^2} = \frac{200}{(80)^2} = \frac{200}{6400} = \frac{1}{32}$$

Therefore, the function is $V = (1/32)T^2$.   ■

Practice 2.   The illuminance E of a light source is inversely proportional to the square of the distance d from the source. If one light source produces an illuminance of 58 lx at a distance of 14.3 m, find E as a function of d.

The *absolute value function* is useful when we want to discuss the size of a number, independent of its sign. We can think of the **absolute value** of a number x, written $|x|$, as the distance of x from zero. Therefore, $|3| = 3$ and $|-3| = 3$. When the number x is positive, $|x|$ is the number itself. When the number x is negative, $|x|$ is $-x$; if this is confusing, note that $|-3| = 3 = -(-3)$. To put it another way,

$$|x| = \begin{cases} x & \text{if} \quad x \geq 0 \\ -x & \text{if} \quad x < 0 \end{cases}$$

EXAMPLE 8. If $f(x) = |x|$, then $f(2) = |2| = 2$ and $f(-7) = |-7| = 7$. ■

Practice 3. If $f(x) = |x|$, what is $f(4)$? $f(-6)$?

In a function $y = f(x)$, there is one dependent variable, y, whose value is determined by the value of the single independent variable x. In this case, f is a function of one variable, meaning one independent variable. But it is possible for the value of a dependent variable to be determined by the values of two or more independent variables. Then we have a **function of several variables**. The notation $z = f(x, y)$ indicates that the dependent variable z takes its value from the values of the independent variables x and y.

EXAMPLE 9. The equation $z = x^2 + y^2$ defines z as a function of x and y. When $x = 1$ and $y = 2$, $z = 1^2 + 2^2 = 5$. ■

EXAMPLE 10. Let $f(x, y) = 2x - 3y + x^2$. Then $f(0, 3) = 2 \cdot 0 - 3 \cdot 3 + 0^2 = -9$, because we evaluate the function with 0 for x and 3 for y. Also, $f(2x, x + 1) = 2(2x) - 3(x + 1) + (2x)^2 = 4x^2 + x - 3$. Here we have substituted $2x$ for x and $x + 1$ for y in the equation defining the function. ■

Practice 4. If $g(x, y) = xy + 2x$, what is $g(2, 4)$?

Just as in equations of one variable, we may have to exclude certain values of the independent variables. Again, we cannot divide by zero and we do not allow square roots of negative numbers, and there may be physical constraints on the values the independent variables can take on.

EXAMPLE 11. Write the area A of a triangle as a function of the altitude a and base b.

The familiar formula $A = \frac{1}{2}ab$ is the function we want. The value of A depends upon the values of both a and b. The values of a and b are independent of each other, and we need both values to compute A. Because a and b represent lengths, $a > 0$ and $b > 0$. ∎

The **trigonometric functions** are examples of transcendental functions. In trigonometric functions, the independent variable is an angle. An angle is generated by rotating a half-line about its fixed endpoint (which becomes the **vertex** of the angle) from an initial

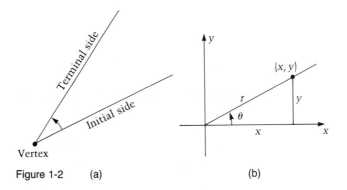

Figure 1-2   (a)   (b)

position to a terminal position (see Figure 1-2(a)). If an angle is in **standard position** on a rectangular coordinate system, its vertex is at the origin, and its initial side is along the positive x-axis (see Figure 1-2(b)). If the angle is measured counterclockwise from its initial side to its terminal side, it is a positive angle; if it is measured clockwise, it is a negative angle. A point (x, y) on the terminal side of an angle has coordinates x and y; a third number associated with (x, y) is the distance r of the point from the origin (Figure 1-2(b)). This distance, which is always positive, is called the **radius vector**. From the Pythagorean theorem, $x^2 + y^2 = r^2$.

If θ is an angle in standard position and a point (x, y) with radius vector r on the terminal side of θ is chosen, then the trigonometric functions of θ are the six ratios

$$\sin \theta = \frac{y}{r} \qquad \csc \theta = \frac{r}{y}$$

$$\cos \theta = \frac{x}{r} \qquad \sec \theta = \frac{r}{x}$$

$$\tan \theta = \frac{y}{x} \qquad \cot \theta = \frac{x}{y}$$

These ratios are independent of the point (x, y) and are, therefore, functions of the angle θ.

One way to measure angles is to use **degree** measure. One complete revolution equals 360°. Each degree is divided into 60 minutes, where a minute is denoted by ', and each minute is divided into 60 seconds, where a second is denoted by ". Thus 34°28'14" denotes an angle of 34 degrees, 28 minutes, 14 seconds.

Angles can also be measured by **radian** measure. One complete revolution equals 2π radians.* Therefore the relation between degree measure and radian measure is

$$360° = 2\pi \quad \text{(radians)}$$

or

$$1° = \frac{2\pi}{360} \quad \text{(radians)} \approx 0.0175 \quad \text{(radians)}$$

$$1 \quad \text{(radian)} = \frac{360°}{2\pi} \approx 57.3° \qquad (\approx \text{ denotes is "approximately equal to")}$$

In calculus, angle measures are almost always given in radians. No units are written with radian measure, and the size of the angle is simply expressed as a number. We may, therefore, see sin 2 (meaning the sine function of an angle of 2 radians), cos $\pi/3$, and so on. Trigonometric functions of angles are easy to find using a scientific calculator.

Word of Advice

If you use a calculator to find trigonometric functions of angles given in radian measure, be sure that your calculator is set in radian (rad) mode.

Practice 5. Use your calculator to find sin 3, cos(−1.2), sin($\pi/4$), and tan(−0.3).

Exercises / Section **1-2**

Exercises 1–8: Simplify the expression.

1. $f(x) = (4x^2 - 2)^{1/2}(4x^2 - 2)^{3/2}$

2. $f(x) = \dfrac{(2x - 7)^3}{(2x - 7)^{1/2}}$

3. $g(x) = (2x - 1)(x^2 + 2)^{1/2} + (x^2 + 2)^{3/2}$

4. $g(x) = (2x - 5)^{-3/4}(x^2 + 1) + x^2(2x - 5)^{1/4}$

5. $h(x) = \sqrt{x + 2}\,(x + 2)^{3/2}$

6. $h(x) = \dfrac{\sqrt[4]{x + 2}}{\sqrt{(x + 2)^3}}$

7. $f(x) = \dfrac{(1 - x)^{-1/2}(2x + 4) - 3(1 - x)^{1/2}}{(1 - x)^{1/2}}$

8. $f(x) = \dfrac{(3x + 2)(2x + 1)^2 - 6x(2x + 1)}{(2x + 1)^4}$

*Formally, a radian is defined to be the measure of an angle with vertex at the center of a circle that subtends an arc on the circle equal in length to the radius of the circle.

Exercises 9–20: Write the function.

9. The pressure P of an ideal gas under constant volume varies directly as the temperature T. Write P as a function of T if $P = 800$ Pa (pascals) when $T = 45°C$.

10. The period P of a pendulum is directly proportional to the square root of its length L. If a pendulum of length 2 m has a period of 2.8 s, write P as a function of L.

11. The time t required to travel a fixed distance varies inversely as the rate r of travel. Write t as a function of r if it takes 1.5 h to travel the distance at the rate of 72 km/h.

12. The frequency f (in hertz) of an electric oscillation in a certain circuit with a capacitor is inversely proportional to the square root of the capacitance C (in farads). If $f = 780$ Hz when $C = 40$ F, write f as a function of C.

13. The crushing load C of a pillar varies directly as the fourth power of its radius r and inversely as the square of its length L. Write an expression for C as a function of r and L.

14. The force F between two electrostatic charges q_1 and q_2 varies jointly as the two charges and inversely as the square of the distance d between the charges (Coulomb's law). Write an expression for F as a function of q_1, q_2, and d.

15. Write the volume V of a rectangular solid as a function of its length L, width W, and height H.

16. Write the surface area A of a rectangular solid as a function of its length L, width W, and height H.

17. Write the volume V of a right circular cylinder as a function of the height h and the radius r of the base.

18. Write the surface area A of a tin can as a function of the height h and the radius r of the base.

19. Write the perimeter P of a rectangle as a function of the length L and width W.

20. Write the amount A in a savings account after t years as a function of the principal P and the interest rate r; assume interest is compounded annually.

Exercises 21–30: Find the value requested.

21. $f(x, y) = 3x + 4y;\ f(2, -3)$

22. $f(x, y) = x^2 - 5xy;\ f(1, 2)$

23. $g(x, y) = \dfrac{xy}{(1 - x)};\ g(-2, 2)$

24. $g(x, y) = \sqrt{x^2 + 2y};\ g(1, 3)$

25. $h(x, y, z) = x^2 + y^2 + z^2 + 4;\ h(1, 1, 2)$

26. $h(x, y, z) = (x - 2)(y + 4)z;\ h(0, 4, 3)$

27. $g(x, y) = x^2 + 2x + y;\ g(2x, y + 1)$

28. $g(x, y) = \dfrac{xy}{(1 + x)(1 - y)};\ g(x^2, x + 1)$

29. $h(x, y) = 2x - \dfrac{x^2 - y^2}{2y};\ h(x^2, y^2) - h(x, y)$

30. $h(x, y) = x^2 + xy;\ h(x + 1, y + 1) - h(x + 2, x)$

Exercises 31–34: Find the allowable values of the independent variables.

31. $z = \dfrac{2x}{y + 1}$

32. $z = \dfrac{\sqrt{1 - y^2}}{x}$

33. $f(x, y) = \dfrac{xy}{x^2 - y^2}$

34. $f(x, y) = \dfrac{3y^2}{x - xy}$

1·3 Graphing

We have represented functions in an algebraic form by using equations. It is also helpful to be able to represent functions in a visual, or geometric, form. **Analytic geometry** allows us to consider relationships between variables from both an algebraic and a geometric point of view.

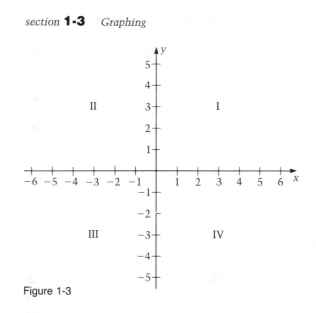

Figure 1-3

In order to visualize relationships, we use a **rectangular coordinate system**. This consists of two perpendicular lines, called axes, which are marked much like a ruler (see Figure 1-3). The horizontal axis, or **x-axis**, measures positive values to the right and negative values to the left; the vertical axis, or **y-axis**, measures positive values upward and negative values downward. Both axes extend infinitely far in each direction. The point where the axes cross, which is the zero value on both scales, is called the **origin** of the coordinate system. The coordinate system divides the plane into four sections, called **quadrants**. The quadrants are numbered counterclockwise with Roman numerals, as in Figure 1-3.

Once we have set up a coordinate system, we can locate a point on the plane by giving its distance and direction away from each of the coordinate axes. Thus, in Figure 1-4, the point shown is 2 units to the right of the y-axis and 3 units below the x-axis. Note that this point is the intersection of the vertical line through the 2 on the x-axis and the horizontal line through the −3 on the y-axis. The point is said to have an **x-coordinate** of 2 and a **y-coordinate** of −3. We can represent the point by the pair of numbers (2, −3). The order of the two numbers here is important, and (2, −3) is called an **ordered pair** of numbers.

For any point with coordinates specified by an ordered pair of numbers, the first coordinate is always the x-coordinate, also called the **abscissa** of the point, and the second coordinate is always the y-coordinate, also called the **ordinate** of the point. (Recall that a rectangular coordinate system is used to define the trigonometric functions of an angle.)

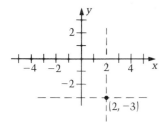

Figure 1-4

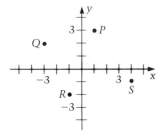

Figure 1-5

EXAMPLE 1. In Figure 1-5, the points shown have coordinates $P = (1, 3)$, $Q = (−3, 2)$, $R = (−1, −2)$, and $S = (4, −1)$. ∎

The scales on the axes show the **integers**, numbers such as 0, 1, 2, ... and $-1, -2, -3, \ldots$. Other points on the axes represent the rest of the **real numbers**. Real numbers are those that can be represented as decimals. A real number that can be represented as a terminating or repeating decimal is called a **rational number**. The numbers $3 = 3.0$, $\frac{1}{4} = 0.25$, and $\frac{2}{3} = 0.666\ldots$ are all rational. (Any rational number can be obtained by dividing an integer by a nonzero integer.) Real numbers with decimal representations that do not repeat and do not terminate are called **irrational numbers**. The numbers $\sqrt{2}$, $\sqrt{7}$, and π, for example, are all irrational.

By means of a coordinate system, any point on the plane has an ordered pair of real numbers, its coordinates, associated with it. Also, any ordered pair of real numbers has a point on the plane associated with it. The **graph** of a point is a mark on a coordinate system corresponding to the point's coordinates. The process of marking the point on the plane is called **graphing**, or **plotting**, the point.

EXAMPLE 2. Locate all the points (x, y) with $x > 2$ and $y \le 1$.

On a coordinate system, the points with abscissas, or x-values, equal to 2 form a vertical line cutting the x-axis at $x = 2$. Those points with abscissas greater than 2 lie to the right of this line. The line itself is not included, because we want no points with x-values of 2. The points with ordinates, or y-values, equal to 1 form a horizontal line cutting the y-axis at $y = 1$. Those points with ordinates less than or equal to 1 lie below or on this line. The shaded region in Figure 1-6 is the set of points we want. ■

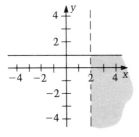

Figure 1-6

Practice 1. Locate all points (x, y) with $-1 \le x < 1$.

Given an equation involving x and y, the **graph of the equation** is the collection of graphs of all points (x, y) whose coordinates satisfy the equation (make the equation true). We can often plot one point by selecting a value for *x* and computing a corresponding value for *y*. Usually an infinite number of points have coordinates that satisfy the equation, and we cannot plot them all. We plot enough points to determine a pattern and then connect these points to complete the pattern.

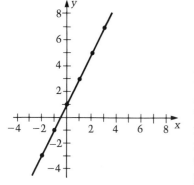

Figure 1-7

EXAMPLE 3. Graph the equation $y = 2x + 1$.

We can make a table of values for *x* and corresponding values for *y*. If we choose an *x*-value of 0, then $y = 2 \cdot 0 + 1 = 1$. Therefore, the ordered pair (0, 1) satisfies the equation and represents a point on the graph. If we choose an *x*-value of 1, then $y = 2 \cdot 1 + 1 = 3$, and the point (1, 3) lies on the graph. Figure 1-7 shows a table of values for *x* and *y*. We have chosen the *x*-values arbitrarily; once an *x*-value is chosen, however, the corresponding *y*-value is specified by the equation. Notice that we chose some negative values, as well as positive values, for *x*. If we carefully plot these points on a rectangular coordinate system, they all seem to lie on a straight line. We fill in the pattern by drawing a straight line. This straight line is the visual form of the equation $y = 2x + 1$.

x	0	1	2	3	−1	−2
y	1	3	5	7	−1	−3	∎

In Example 3, it was easy to do the calculations because only integer arithmetic was involved. In general, use of a calculator will simplify and speed up the sometimes tedious process of finding points that satisfy an equation.

Practice 2. Graph the equation $y = -3.4x + 2.6$. (Use your calculator.)

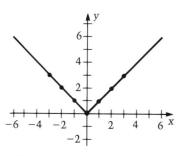

Figure 1-8

EXAMPLE 4. Graph the equation $y = |x|$.

We recall the absolute value function from the previous section and compute some ordered pairs that satisfy the equation. Plotting these points, our graph appears to be two straight lines in a V-shape (Figure 1-8).

x	0	1	2	3	−1	−2	−3
y	0	1	2	3	1	2	3	∎

EXAMPLE 5.   Graph the function defined by

$$f(x) = \begin{cases} 1 & x \le 0 \\ 2 & x > 0 \end{cases}$$

This time our vertical axis will be labeled $f(x)$, rather than y. Again, we make a table of values; the only new idea here is that for any x-value, we must decide which instructions are appropriate for finding the corresponding $f(x)$-value. We know that $f(x)$ is 1 for any x-value less than or equal to 0. For positive x-values, the corresponding $f(x)$-value is 2.

x	-3	-2	-1	0	1	2	3
$f(x)$	1	1	1	1	2	2	2

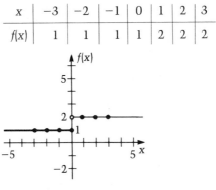

Figure 1-9

Note that in Figure 1-9, the point $(0, 1)$ is shown by a dot, or **closed circle**; $(0, 1)$ is a point on the graph. The point $(0, 2)$ is shown by an **open circle**. This point is not on the graph; it simply denotes the end of the line segment consisting of positive x-values and corresponding $f(x)$-values of 2. The rest of this line segment is part of the graph. ■

From Example 5, we see that we can graph functions that are not given by a single equation.

Practice 3.   Graph the function

$$f(x) = \begin{cases} 0 & x < 1 \\ 1 & x \ge 1 \end{cases}$$

All the graphs we have done so far have involved only straight lines. More complicated equations (more terms, higher powers of x, and so on) have graphs that are not straight lines. The more complicated the equation, the more points of the graph we should plot to make sure that we have captured the correct visual pattern.

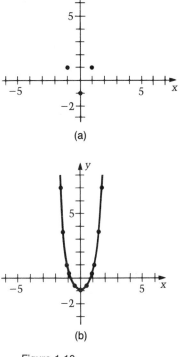

(a)

(b)

Figure 1-10

EXAMPLE 6.   Graph the equation $y = 2x^2 - 1$.

A first attempt might produce the points on the graph shown in Figure 1-10(a).

x	-1	0	1
y	1	-1	1

We might think that the graph resembles that of the absolute value function and is V-shaped. However, if we fill in more points on the graph—including some noninteger values of x between -1 and 1—we see that the curve is more U-shaped. The graph is shown in Figure 1-10(b).

x	-1	0	1	-0.8	-0.4	0.4	0.8	-2	-1.5	1.5	2
y	1	-1	1	0.28	-0.68	-0.68	0.28	7	3.5	3.5	7

∎

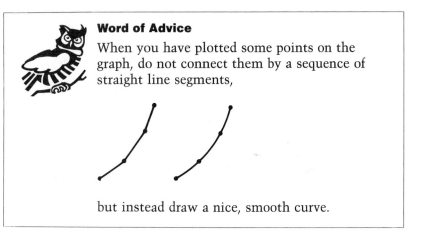

Word of Advice

When you have plotted some points on the graph, do not connect them by a sequence of straight line segments,

but instead draw a nice, smooth curve.

The graph of $y = 2x^2 - 1$ in Figure 1-10(b) is *symmetric about the y-axis*. If we were to draw the figure on a sheet of paper and fold the paper along the y-axis, the two halves would match. The graph of an equation is **symmetric about the y-axis** whenever $-x$ can be substituted for x in the equation with no change, that is, if $(-x, y)$ is on the graph whenever (x, y) is. For $y = 2x^2 - 1$, for example, $2x^2 - 1 = 2(-x)^2 - 1$ because the square of a number equals the square of its negative. A graph is **symmetric about the x-axis** if $-y$ can be substituted for y in the equation with no change, that is, if $(x, -y)$ is on the graph whenever (x, y) is. If we notice before we graph that a graph is symmetric about the x- or y-axis, this cuts down on the amount of work we have to do to draw it.

Practice 4. Graph the equation $y = 2 - x^2$.

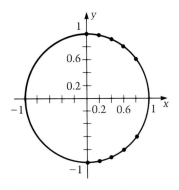

Figure 1-11

EXAMPLE 7. Graph the equation $x^2 + y^2 = 1$.

We note first that the graph of this equation will be symmetric about both the x-axis and the y-axis. In order to create a table of values, we solve this equation for y and get

$$y^2 = 1 - x^2$$
$$y = \pm\sqrt{1 - x^2}$$

The plus-and-minus sign means that there will be two values of y for each value of x. Writing the equation in this form also makes it clear that we are limited in the values we can choose for x. Because we can only graph real values for y, y cannot be the square root of a negative number. Therefore, we must choose values of x that result in nonnegative values of $1 - x^2$, so x cannot be greater than 1. We need not bother computing y for negative values of x; instead we will just make use of the symmetry about the y-axis.

x	0	0.2	0.4	0.6	0.8	1
y	± 1	± 0.98	± 0.92	± 0.8	± 0.6	0

The graph is a circle with center at the origin and radius of 1 unit (Figure 1-11). ■

The equation of Example 7, $x^2 + y^2 = 1$, does not represent a function, because each value of x does not have a single corresponding y-value. Corresponding to the x-value of 0.8, for example, are y-values of 0.6 and -0.6. We can graph equations that are not functions. If there is a unique y-value for every x-value, then a vertical line will intersect the graph only once. If there is more than one y-value for any value of x, there is a vertical line that will intersect the graph more than once. We might call this the graphical test for functions: *If a vertical line can be drawn that intersects the graph of an equation more than once, the equation does not represent a function.*

EXAMPLE 8. Figure 1-12(a) is the graph of an equation that does not represent a function. Figure 1-12(b) is the graph of a function. ■

The graph of the circle $x^2 + y^2 = 1$ illustrates still another type of symmetry, symmetry with respect to the origin. Symmetry about the origin means that for every point on the graph, if a line is drawn

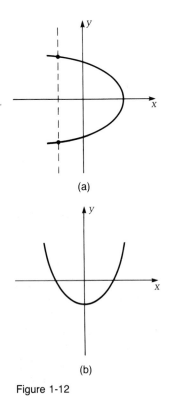

(a)

(b)

Figure 1-12

from the point through the origin, the "opposite" of that point appears on the continuation of the line. A graph is **symmetric about the origin** if $-x$ can be substituted for x and $-y$ can be substituted for y at the same time in the equation with no change—that is, if $(-x, -y)$ is on the graph whenever (x, y) is.

Another aid in graphing some equations is illustrated in the next example.

EXAMPLE 9. Graph the equation $y = 2 - 1/x$.

We must exclude the value $x = 0$, but x can take on values near zero. In fact for values of x close to zero, $1/x$ becomes very large (for example, $1/\frac{1}{10} = 10$, $1/\frac{1}{100} = 100$, and $1/\frac{1}{1000} = 1000$). The closer the value of x to zero, the larger the value of $1/x$. We can see from Figure 1-13 that this affects the shape of the graph near $x = 0$.

x	3	2	1	0.1	0.01	0.001	-3	-2	-1	-0.1	-0.01	-0.001
y	1.67	1.5	1.0	-8	-98	-998	2.3	2.5	3	12	102	1002

The vertical line $x = 0$ is a guideline for the curve, a line to which the curve gets closer and closer. Such guidelines are called **asymptotes**. This graph also has a horizontal asymptote, $y = 2$. For very large positive or negative values of x, $1/x$ is a small number and $y = 2 - 1/x$ is close to 2. ∎

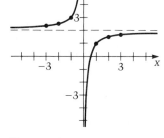

Figure 1-13

Asymptotes often occur in functions that involve quotients of polynomials. Vertical asymptotes may occur at values of x that result in division by zero; horizontal asymptotes can be found by letting x take on very large positive or negative values. If we can identify lines that are asymptotes to a curve, they can help us in drawing the graph of the curve.

Practice 5. Graph the equation $y = 1/x$. Try to identify a horizontal and a vertical asymptote ahead of time.

A function is often described in equation form by $y = f(x)$. It is for this reason that we called the horizontal axis of the coordinate system the x-axis and the vertical axis the y-axis. However, we know that the independent and dependent variables in a function do not have to be named x and y. When a function is graphed, the horizontal axis is used for the independent variable, and the vertical axis for the dependent variable, whatever the names of those variables.

EXAMPLE 10. An electric circuit contains a variable resistor and an 8 Ω resistor in parallel. The total resistance R is given by the equation

$$R = \frac{8R_1}{8 + R_1}$$

where R_1 is the variable resistance in ohms. Graph R (in ohms) as a function of R_1 for values of R_1 between 0.8 Ω and 10 Ω.

Here the horizontal axis is labeled R_1 and the vertical axis is labeled R. We select a number of values for R_1 within the specified range and, using a calculator, compute the corresponding R-values. The endpoints, $R_1 = 0.8$ and $R_1 = 10$, are values we definitely want to include. The graph is shown in Figure 1-14.

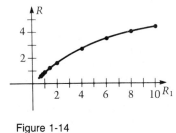

Figure 1-14

R_1	0.8	0.9	1.0	1.5	2.0	4.0	6.0	8.0	10.0
R	0.73	0.81	0.89	1.26	1.6	2.67	3.43	4.0	4.44

Exercises / Section **1-3**

1. Name the coordinates of the points shown:

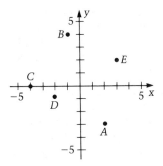

2. On a rectangular coordinate system, plot the points $P = (-2, -3)$, $Q = (0.5, -1.5)$, $R = (-\frac{3}{2}, 4)$, $S = (5, 0)$.

Exercises 3–10: Locate the indicated points on a rectangular coordinate system.

3. All points (x, y) with $x \geq -3$.

4. All points (x, y) with $y > -2$.

5. All points with positive ordinates and negative abscissas.

6. All points (x, y) with ratio y/x negative.

7. All points (x, y) with $x < 2$ and $y < 2$.

8. All points (x, y) with $x \geq -1$ and $y < 1$.

9. All points (x, y) with $|x| \leq 1$.

10. All points (x, y) with $|x| < 2$, $|y| \leq 1$.

Exercises 11–14: Describe the set of points in the shaded region.

11.

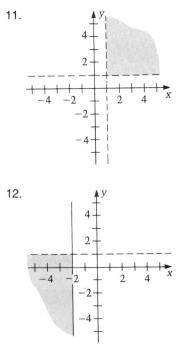

12.

13.

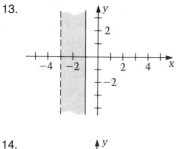

14.

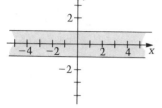

15. The points $(1, 2)$, $(3, 2)$, and $(5, y)$ lie on a straight line. What is the value of y?

16. The points $(-1, -2)$, $(-1, 3)$, $(4, 3)$, and (x, y) form the corners of a rectangle. What are the values of x and y?

Exercises 17–44: Graph the equation.

17. $y = x$

18. $y = -x$

19. $y = 3x - 5$

20. $y = -2x + 4$

21. $y = 6.1 - 3.2x$

22. $y = 0.4 + 0.6x$

23. $y = x^2$

24. $y = -x^2$

25. $y = 3x^2 + 1$

26. $y = 2 - 4x^2$

27. $y = x - 3x^2$

28. $y = 2x^2 - 1.5x$

29. $y = x^2 - 2x + 1$

30. $y = 4x^2 + 0.5x - 2.8$

31. $y = x^3$

32. $y = 3x^3 - x^2$

33. $y = x^{1/3}(1 - x)$

34. $y = x^{4/3} + 4x^{1/3}$

35. $y = \sqrt{x}$

36. $y = \sqrt{1 + x}$

37. $y = \dfrac{1}{x + 3}$

38. $y = \dfrac{1}{x - 1}$

39. $y = \dfrac{1}{x^2}$

40. $y = \dfrac{1}{x^2 - 9}$

41. $y = \begin{cases} -1 & x \le 1 \\ 2 & x > 1 \end{cases}$

42. $y = \begin{cases} 2 & x < 0 \\ x & x \ge 0 \end{cases}$

43. $y^2 = 4x + 1$

44. $x^2 + 4y^2 = 16$

45. In Exercises 17–44, which graphs are symmetric about the y-axis? Which are symmetric about the x-axis?

46. In Exercises 17–44, which graphs do not represent functions?

47. In Exercises 17–44, find any vertical asymptotes.

48. In Exercises 17–44, find any horizontal asymptotes.

49. An electric circuit satisfies the equation $V = IR$, where V is voltage, I is current, and R is resistance. For a circuit with a resistance of 10 Ω, graph V (in volts) as a function of I (in amperes) for $I \ge 0$.

50. A ball is thrown upward from the ground with an initial velocity of 10 m/s. Its distance s above the ground (in meters) at any time t (in seconds) is given by the equation $s = -4.9t^2 + 10t$. Graph s as a function of t for $t \ge 0$. About how long does it take for the ball to hit the ground?

51. The kinetic energy K of an object brought from rest to a velocity v is given by the equation $K = \frac{1}{2}mv^2$, where m is the mass of the object. For an object with a mass of 4 kg, graph K (in joules) as a function of v (in meters per second) for $v \ge 0$.

52. The speed v of exhaust gas escaping from a rocket chamber is given by the equation $v = \sqrt{2(p - 60)/1.3}$, where p is the pressure inside the chamber. Graph v (in meters per second) as a function of p (in pascals) for $p \ge 60$.

53. The volume V of a sphere is given by the equation $V = \frac{4}{3}\pi r^3$, where r is the radius of the sphere. Graph V (in cubic centimeters) as a function of r (in centimeters) for $r \ge 0$.

54. The current I in a vacuum diode is given by the equation $I = 1.8V^{3/2}$, where V is the voltage. Graph I (in amperes) as a function of V (in volts) for $V \ge 0$.

55. In an electroplating laboratory, data is taken on temperature T (in degrees Celsius) of an alloy as a function of time t (in minutes).

t	0	2
T	3	11

Prove that each data point satisfies each of the equations $T = 4t + 3$ and $T = 2t^2 + 3$. Graph each equation on the same coordinate system. If another data point, $(5, 23)$, is obtained, decide by looking at the graph which of the two equations is correct. Estimate the value of T at $t = 4$ min.

56. Data is taken on the intensity I (in candelas) of illumination from a light source as a function of the distance s (in meters) from the source.

s	1	2
I	38	19

Prove that each data point satisfies each of the equations $I = 38/s$ and $3I = 76/s^2 + 38$. Graph each equation on the same coordinate system. If another data point, (4, 14.25), is obtained, decide by looking at the graph which of the two equations is correct. Estimate the value of I at $s = 5$ m.

1-4 Distance Formula and Slope

In order to help us graph equations on a rectangular coordinate system in a systematic way, it will be helpful to be able to compute the distance between two points whose coordinates are known and to measure the amount of slant of a straight line.

For two points on a horizontal line (Figure 1-15(a)), the distance between the two points can be found by subtracting the two x-coordinates. Because distance is positive, we subtract the smaller value from the larger. On the x-axis, larger values are farther to the

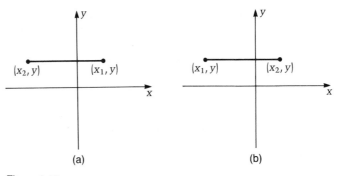

Figure 1-15

right, so for Figure 1-15(a), the horizontal distance is $x_1 - x_2$. However, this is not a good formula to remember. If we label the points as in Figure 1-15(b), then the horizontal distance is $x_2 - x_1$, because the point (x_2, y) is farther right than the point (x_1, y). Either point can be called (x_1, y) or (x_2, y). Therefore, the way to remember horizontal distance is x-right minus x-left, or $x_R - x_L$.

Horizontal distance $= x_R - x_L$

Similarly, the distance between two points on a vertical line is the difference between the two y-coordinates. In the vertical direction, larger values are higher on the graph, so we subtract the lower y-value (y_L) from the upper y-value (y_U).

Vertical distance $= y_U - y_L$

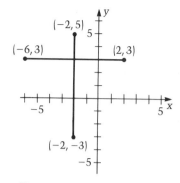

Figure 1-16

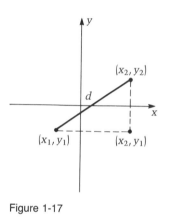

Figure 1-17

EXAMPLE 1. Find the distance between the points $(-6, 3)$ and $(2, 3)$. Find the distance between the points $(-2, -3)$, $(-2, 5)$.

The points $(-6, 3)$ and $(2, 3)$ are on the same horizontal line (Figure 1-16). The distance between the points is $x_R - x_L = 2 - (-6) = 2 + 6 = 8$. The points $(-2, -3)$ and $(-2, 5)$ are on the same vertical line. The distance between the points is $y_U - y_L = 5 - (-3) = 8$. ■

Now let (x_1, y_1) and (x_2, y_2) be any two points in the plane, as in Figure 1-17. To find the distance d between the points, we make d the length of the hypotenuse of a right triangle by inserting the new point (x_2, y_1). Then we know the lengths of the legs of the triangle, because these are horizontal and vertical distances. We use the Pythagorean theorem to find the length d.

$$d = \sqrt{(x_2 - x_1)^2 + (y_2 - y_1)^2} \tag{1}$$

Is this a general formula? Suppose the points (x_1, y_1) and (x_2, y_2) had been reversed. Then

$$d = \sqrt{(x_1 - x_2)^2 + (y_1 - y_2)^2} \tag{2}$$

Because we square $x_2 - x_1$ (or $x_1 - x_2$) and we square $y_2 - y_1$ (or $y_1 - y_2$), the result d is the same whether we use Equation (1) or Equation (2). Therefore, to compute the distance between two points, we use the **distance formula**

$$d = \sqrt{(x_2 - x_1)^2 + (y_2 - y_1)^2}$$

where one point is arbitrarily selected as (x_1, y_1) and the other point is (x_2, y_2).

EXAMPLE 2. Find the distance between $(-2, 1)$ and $(4, 2)$.

Using the distance formula with $(-2, 1) = (x_1, y_1)$ and $(4, 2) = (x_2, y_2)$, we get

$$d = \sqrt{(4 - (-2))^2 + (2 - 1)^2} = \sqrt{36 + 1} = \sqrt{37}$$

If we let $(4, 2) = (x_1, y_1)$ and $(-2, 1) = (x_2, y_2)$, then

$$d = \sqrt{(-2 - 4)^2 + (1 - 2)^2} = \sqrt{37} \quad ■$$

Practice 1. Find the distance between $(1, 3)$ and $(-2, 4)$.

The *slope* of a straight line is an indicator of whether the line slants uphill or downhill as we look from left to right and of the

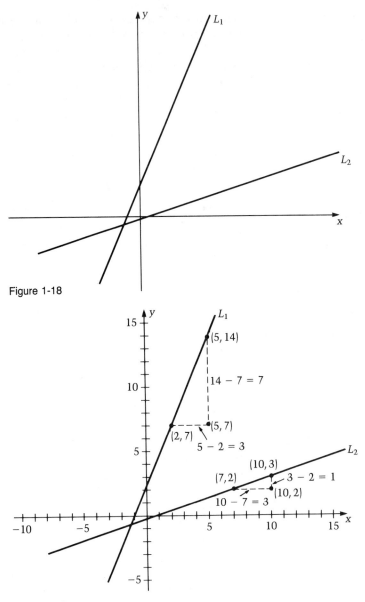

Figure 1-18

Figure 1-19

steepness of the slant. In Figure 1-18, line L_1 has a steep slope, and line L_2 has a shallow slope. The slope of a line can be measured by selecting any two points on the line and computing the ratio of the vertical change between the two points to the horizontal change between the two points.* In Figure 1-19 the horizontal

---

*The slope of a line can also be defined as the tangent of the angle of inclination of the line.

change between the two points on L_1 is 3 units, while the vertical change between the two points is 7 units. The slope of L_1 is the ratio $\frac{7}{3}$. But for the two points on L_2, while the horizontal change between the two points is also 3 units, the vertical change is only 1 unit. The slope of L_2 is only $\frac{1}{3}$.

The **slope** m of a line connecting the two points (x_1, y_1) and (x_2, y_2) is given by the equation

$$m = \frac{y_2 - y_1}{x_2 - x_1}$$

Again, it is arbitrary which of the two points is designated as (x_1, y_1) and which as (x_2, y_2). If we were to change the labeling, we would just change the sign of both the numerator and the denominator in computing m, so the value of m itself would be unchanged.

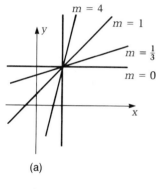

Figure 1-20

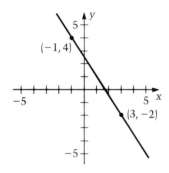

(a)

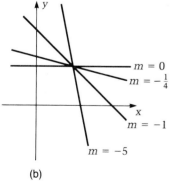

(b)

Figure 1-21

EXAMPLE 3. Find the slope of the line connecting the points $(-1, 4)$ and $(3, -2)$.

Letting $(-1, 4) = (x_1, y_1)$ and $(3, -2) = (x_2, y_2)$, we find

$$m = \frac{-2 - 4}{3 - (-1)} = \frac{-6}{4} = -\frac{3}{2}$$

Letting $(3, -2) = (x_1, y_1)$ and $(-1, 4) = (x_2, y_2)$, we have

$$m = \frac{4 - (-2)}{-1 - 3} = \frac{6}{-4} = -\frac{3}{2}$$

In either case the slope is the same. The negative value for the slope means that the line slants downhill from left to right (see Figure 1-20). ■

From the formula for slope, it is clear that a horizontal line has a slope of zero because the two y-values are the same. As the lines slant more and more steeply uphill, the slopes have larger and larger positive values (Figure 1-21(a)). A vertical line, however, has no slope at all, because the formula for m in this case would involve division by zero. As we move from a horizontal line through lines slanting more and more steeply downhill (Figure 1-21(b)), the slopes are negative numbers with larger and larger absolute values.

Word of Advice

To compute the slope of a line between two points, subtract the *y*-value of one point from the *y*-value of the other. Subtract the *x*-values *in the same order.* Finally, divide the *y*-difference by the *x*-difference.

Practice 2. Find the slope of the line through the points $(-1, -2)$ and $(4, -6)$.

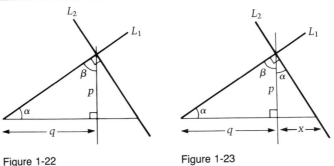

Figure 1-22   Figure 1-23

Two parallel lines clearly have equal slopes. A relationship between the slopes of perpendicular lines also exists. In Figure 1-22, L_1 and L_2 are perpendicular lines. The slope m_1 of L_1 is given by $m_1 = p/q$, where p is the vertical distance and q is the horizontal distance shown. From Figure 1-23, the slope m_2 of L_2 is given by $m_2 = -p/x$, where x is the horizontal distance shown. From the small right triangle, $x/p = \tan \alpha$, or $x = p \tan \alpha$. From the larger right triangle, $\tan \alpha = p/q$. Therefore, $x = p^2/q$. Thus

$$m_2 = \frac{-p}{x} = \frac{-p}{p^2/q} = \frac{-q}{p}$$

Therefore,

$$m_1 m_2 = \left(\frac{p}{q}\right)\left(\frac{-q}{p}\right) = -1$$

In general, when two lines are perpendicular, the product of their slopes is -1. Also, any two lines whose slopes have this relationship are perpendicular.

Slopes	*Lines*
$m_1 = m_2$	Parallel
$m_1 m_2 = -1$	Perpendicular

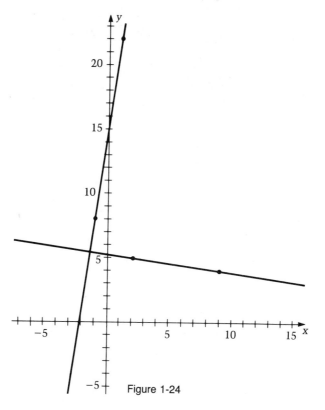

Figure 1-24

EXAMPLE 4. Prove that the line through the points (1, 22) and (−1, 8) is perpendicular to the line through the points (9, 4) and (2, 5).

A sketch of the two lines (Figure 1-24) shows that they seem to be perpendicular. However, while a sketch could probably reveal if two lines were obviously not close to being perpendicular, a drawing is not sufficient to prove that two lines are perpendicular. In general, while drawings can guide our thinking and indicate general shapes, they are too inaccurate to be used as a proof.

Instead, we compute the slopes of the two lines. For the line through (1, 22) and (−1, 8),

$$m_1 = \frac{22 - 8}{1 - (-1)} = \frac{14}{2} = 7$$

For the line through (9, 4) and (2, 5),

$$m_2 = \frac{4 - 5}{9 - 2} = -\frac{1}{7}$$

Because $m_1 \cdot m_2 = 7(-\frac{1}{7}) = -1$, the two lines are perpendicular. ■

Practice 3. Prove that the line through the points (6, 15) and (2, −3) is parallel to the line through the points (3, 9) and (1, 0).

Exercises / Section **1-4**

Exercises 1–10: Find the distance between the two points and the slope of a line connecting the two points.

1. $(2, 0), (3, 4)$
2. $(3, 1), (4, 7)$
3. $(4, -2), (-3, 0)$
4. $(-3, 1), (-6, 2)$
5. $(-4, -2), (3, 7)$
6. $(2, 5), (0, -1)$
7. $(-3, -6), (-1, -4)$
8. $(-2, -1), (-5, -4)$
9. $(1.2, -6.3), (-3.8, 2.8)$
10. $(-3.1, -2.6), (1.5, -0.4)$

11. Prove that the line through the points $(8, 3)$ and $(3, 1)$ is parallel to the line through the points $(0, -8)$ and $(5, -6)$.

12. Prove that the line through the points $(2, -11)$ and $(-2, -6)$ is parallel to the line through the points $(6, 2)$ and $(10, -3)$.

13. Prove that the line through the points $(0, 6)$ and $(-1, 2)$ is perpendicular to the line through the points $(-4, 5)$ and $(4, 3)$.

14. Prove that the line through the points $(1, 2)$ and $(-3, -4)$ is perpendicular to the line through the points $(-1, 6)$ and $(2, 4)$.

Exercises 15–18: Find *a*.

15. The distance between $(2, 1)$ and $(a, -2)$ is 5.

16. The slope of the line through $(2, a)$ and $(3, 6)$ is 10.

17. The line through the points $(-1, a)$ and $(4, -a)$ is parallel to the line through the points $(12, 6)$ and $(2, -6)$.

18. The line through the points $(5, 15)$ and $(2, -3)$ is perpendicular to the line through the points $(a, 5)$ and $(-a, 6)$.

19. Prove that the points $(-4, -1)$, $(-1, 1)$, and $(-10, -5)$ all lie on the same straight line.

20. Prove that the points $(17, 3)$, $(9, -3)$, and $(4, -7)$ do not lie on the same straight line.

21. Prove that the points $(1, 3)$, $(9, -1)$, and $(3, 5)$ determine a right triangle.

22. Find the area of the triangle of Exercise 21.

23. Prove that the points $(-4, 3)$, $(-3, -5)$, and $(4, -1)$ determine an isosceles triangle.

24. Prove that the points $(-3, -4)$, $(8, 6)$, $(-1, 3)$, and $(6, -1)$ are vertices of a parallelogram.

25. Prove that the points $(-3, 5)$, $(-1, -3)$, $(3, -2)$, and $(1, 6)$ are vertices of a rectangle.

26. Find the area of the rectangle of Exercise 25.

STATUS CHECK

Now that you are at the end of Chapter 1, you should be able to:

section **1-1**   Write an equation for one quantity as a function of another, given information that relates the two quantities.

Evaluate a given function at a specific value for the independent variable.

Write the composition of two functions given by equations as a single equation.

Find the allowable values of the independent variable of a function.

section **1-2**   Simplify algebraic expressions by using laws of exponents.

Write an equation for a function that involves direct or inverse proportion.

Write an equation for one quantity as a function of two or more other quantities, given information that relates the quantities.

Evaluate a given function of several variables at specific values for the independent variables.

Find the allowable values of the independent variables of a function of several variables.

section **1-3** Name the coordinates of a point shown on a rectangular coordinate system.

Plot a point on a rectangular coordinate system, given its coordinates.

Locate all points with a given description on a rectangular coordinate system.

Describe the set of points shown in a given region of a rectangular coordinate system.

Sketch the graph of an equation on a rectangular coordinate system.

section **1-4** Given the coordinates of two points, find the distance between them and the slope of a line connecting them.

Prove that a line connecting two given points is parallel to or perpendicular to a line connecting two other given points.

1-5 More Exercises for Chapter **1**

Exercises 1–2: Simplify the expression.

1. $f(x) = \sqrt{x^2 + 1}\sqrt[4]{x^2 + 1}$

2. $g(x) = \dfrac{(3x - 4)^{-1/2}(2x) - (3x - 4)^{1/2}}{(3x - 4)^{1/2}}$

Exercises 3–6: Find the indicated values of the function.

3. $f(x) = 2x^2 + 17x - 9;\ f(-2),\ f(f(4))$

4. $H(x) = \dfrac{x + 1}{(x - 1)^2};\ H(3),\ H(x + 1)$

5. $f(x, y) = 3x^2 - 2xy + 3y;\ f(-1, 4)$

6. $g(x, y) = 2x + \sqrt{y^2 + 1};\ g(3x, y^2)$

Exercises 7–10: Find the allowable values of the independent variables.

7. $y = \sqrt{9 - x^2}$

8. $y = \dfrac{\sqrt{x}}{x^2 - 4}$

9. $z = \dfrac{\sqrt{1 - y}}{x - y}$

10. $z = \dfrac{xy^2}{(x^2 + 4)(y + 1)}$

Exercises 11–12: Find the distance between the two points and the slope of the line connecting the two points.

11. $(4, -2),\ (-3, 1)$ 12. $(-6, -1),\ (0, -4)$

13. Write the surface area A of a cube as a function of the volume V of the cube.

14. Write the amount A in a savings account that started with \$50 and paid 12% interest compounded annually as a function of the number of years t since the account was opened.

15. Total waste water W from an industrial cleaning process is directly proportional to the number of hours t of operation of the facility. Write W as a function of t if $W = 138,000$ L when $t = 5$ h.

16. The electric resistance R of a piece of wire is inversely proportional to its cross-sectional area A. Write R as a function of A if $R = 0.7\ \Omega$ when $A = 0.003$ cm^2.

17. Write the altitude a of a triangle as a function of the base b and area A.

18. Write the length d of the diagonal of a rectangle as a function of the width W and length L.

19. In a rectangular area of constant width 5 units, graph area as a function of length.

20. Graph the surface area A of a cube as a function of length s of a side of the cube.

21. At a certain temperature, an ideal gas satisfies the equation $PV = 28$, where P is pressure (in pascals) and V is volume (in liters). Graph P as a function of V for $V > 0$.

22. The force F (in newtons) between two asteroids due to gravitational attraction is given by the equation $F = 37/r^2$, where r is the distance (in kilometers) between the asteroids. Graph F as a function of r for $r > 0$.

23. The density D of a particular water pollutant varies with the temperature T of a stream by the equation $D = 2T^2 - 15T + 100$. Graph D (in units per cubic centimeter) as a function of T (in degrees Celsius) for $0 \leq T \leq 10$.

24. The frequency f (in hertz) of electric oscillation in one electric circuit is given by the equation $f = 147/\sqrt{C}$, where C is the capacitance (in farads). Graph F as a function of C for $C > 0$.

25. The kinetic energy E of a rotating flywheel (in joules) as a function of angular velocity w (in radians per second) is given by the equation $E = 0.05w^2$. Sketch the graph of E as a function of w for $0 \leq w \leq 50$.

chapter **2**

Straight Lines and Conic Sections

Conic sections are found in a surprising number of places. Here a bridge shows a series of halves of ellipses.

2-1 **Why Analytic Geometry**

We said in Chapter 1 that analytic geometry allows us to consider relationships from both an algebraic and a geometric point of view. Much of our work in this chapter will involve translating back and forth between algebraic and geometric, or visual, information. Thus, given an equation, we want to graph it; our translation task here is:

$$\text{algebraic} \xrightarrow{\text{graph the equation}} \text{geometric}$$

Also, given some geometric information—that is, information about a graph, if not the actual graph—we want to find the equation. Our translation task here is:

$$\text{geometric} \xrightarrow{\text{find the equation}} \text{algebraic}$$

However, we also mentioned that graphical information may not be very accurate, so it is reasonable to ask why we want to graph equations. The answer is that while graphs cannot answer questions for us in as precise a way as equations can, visual information often has a greater impact and can be more easily absorbed than algebraic information.

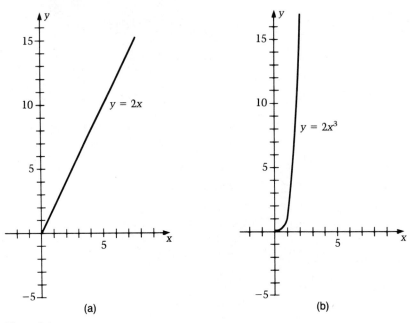

Figure 2-1

EXAMPLE 1. These two tables represent points that satisfy the equations $y = 2x$ and $y = 2x^3$, respectively.

x	0	1	2	3	5
y	0	2	4	6	10

x	0	1	2	3	5
y	0	2	16	54	250

If we examine the two tables carefully, we can see that the value of y increases with increasing values of x for each equation and that the y-values increase much more rapidly for $y = 2x^3$ than for $y = 2x$. However, this same information is conveyed by just a glance at the graphs of the two equations for $x \geq 0$ (see Figure 2-1). On the other hand, if we have only the graphs of the two curves and not the equations, we could not be certain from the graph alone that $(3, 6)$ is a point on the first curve; perhaps it is actually $(3, 5.9)$ or $(3.1, 6)$. This example shows that both kinds of information, algebraic and geometric, are useful. ■

We have already graphed equations by simply plotting a large number of points. We next classify certain types of equations so that we can tell by looking at the form of the equation what kind of picture to expect. For example, in the simplest case, if we can recognize that a particular equation represents a straight line, then we need to plot only two points in order to graph the equation.

2-2 The Straight Line

We first find the equation of a straight line based on geometric information known about the line. Thus our task is:

$$\text{geometric} \xrightarrow{\text{find the equation}} \text{algebraic}$$

What geometric information specifies a particular line? For example, a point on a line and the slope of the line determine a unique line.

Suppose that a line goes through the point (x_1, y_1) and has slope m. The algebraic equation of the line will contain the variables x and y so that any point (x, y) whose coordinates satisfy the equation will lie on the line. In Figure 2-2, the fixed point (x_1, y_1) is shown, along with an arbitrary point (x, y) on the line. Note that x and y are variables, while x_1 and y_1 are fixed, known values.

From Figure 2-2,

$$m = \frac{y - y_1}{x - x_1}$$

or

$$y - y_1 = m(x - x_1) \qquad (1)$$

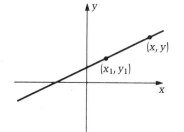

Figure 2-2

Equation (1) is called the **point-slope form** of the equation of a straight line. When we know a point on a line and the slope of the line, we can write the equation of the line by using (1).

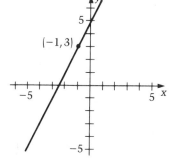

Figure 2-3

EXAMPLE 1. Write the equation of the line passing through $(-1, 3)$ with slope 2.

From Equation (1), with $(x_1, y_1) = (-1, 3)$ and $m = 2$, we can write

$$y - 3 = 2[x - (-1)]$$

which can also be written in the form

$$2x - y + 5 = 0$$

As a check, note that $(-1, 3)$ satisfies this equation. The graph is shown in Figure 2-3. ■

Practice 1. Write the equation of the line passing through $(2, 5)$ with slope -4.

Two points, (x_1, y_1) and (x_2, y_2), also determine a line. Given this information about a line, we can still use Equation (1); we first compute m from the two given points.

EXAMPLE 2. Write the equation of the line that passes through $(-2, 4)$ and $(-4, -5)$.

First we find the slope.

$$m = \frac{4 - (-5)}{-2 - (-4)} = \frac{9}{2}$$

Next we use Equation (1) with either of the two known points as (x_1, y_1).

$$y - 4 = \tfrac{9}{2}[x - (-2)]$$

or

$$2y - 8 = 9(x + 2)$$

Finally,

$$9x - 2y + 26 = 0 \quad \blacksquare$$

Practice 2. Write the equation of the line through the points $(-7, 2)$ and $(3, -1)$.

A special form of Equation (1) occurs when the known point is the **y-intercept** of the line, the point where the line crosses the y-axis. Such a point has coordinates $(0, b)$, and Equation (1) becomes

$$y - b = m(x - 0)$$

or

$$\boxed{y = mx + b} \qquad (2)$$

Equation (2) is called the **slope-intercept form** of the equation of a straight line.

Both Equations (1) and (2) require that we know the slope of the line. A vertical line does not have any slope, but it easy to write the equation of a vertical line. In Figure 2-4, every point on the line has an x-coordinate of 3. All points on the line satisfy the equation $x = 3$, and no points not on the line satisfy this equation. Therefore, $x = 3$ is the equation of the line. Vertical lines in general have equations of the form

$$\boxed{x = k}$$

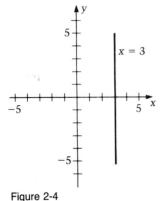

Figure 2-4

Following the same reasoning, horizontal lines have equations of the form

$$y = b$$

This form can also be obtained from Equation (2) by setting $m = 0$; a horizontal line has slope equal to zero.

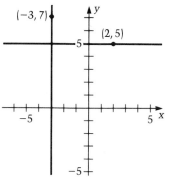

Figure 2-5

EXAMPLE 3. The horizontal line through the point (2, 5) has the equation $y = 5$. The vertical line through the point $(-3, 7)$ has the equation $x = -3$ (see Figure 2-5). ■

When we have simplified the equations in these examples, we have arrived at equations of the general form

$$Ax + By + C = 0 \tag{3}$$

Equations of this form always represent straight lines and are called **linear equations**.

We are now ready to take up our other task with respect to straight lines:

$$\text{algebraic} \xrightarrow{\text{graph the equation}} \text{geometric}$$

Any equation of the form of Equation (3) (or one which can be reduced to this form) will have a straight line as a graph. Two special cases occur. If $A = 0$, then Equation (3) can be written in the form $y = b$, and the graph will be a horizontal line. If $B = 0$, then Equation (3) can be written in the form $x = k$, and the graph will be a vertical line. These cases are easy to graph. In the general case, it is often convenient to graph the straight line by finding the two intercepts, those points for which $x = 0$ and $y = 0$.

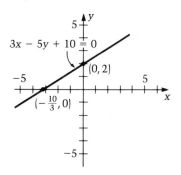

Figure 2-6

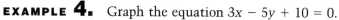

EXAMPLE 4. Graph the equation $3x - 5y + 10 = 0$.

The y-intercept occurs when $x = 0$. Letting $x = 0$ in the equation of the line and solving for y, we get $y = 2$, so the y-intercept is (0, 2). The x-intercept occurs when $y = 0$. Letting $y = 0$ in the equation of the line and solving for x, we get $x = -\frac{10}{3}$; thus the x-intercept is $(-\frac{10}{3}, 0)$. By plotting these two points and connecting them with a straight line, we get the graph of the equation (Figure 2-6). ■

Further information about a line can be obtained by taking the general form $Ax + By + C = 0$ and solving for y. This puts the equation in slope-intercept form, $y = mx + b$. In this form, the coefficient of x is the slope of the line.

EXAMPLE 5. The straight line of Example 4 has the equation $3x - 5y + 10 = 0$. We could find the slope of the line by using the two intercept points we found in Example 4. But we can also solve the equation for y and read the slope directly from the result.

$$3x - 5y + 10 = 0$$
$$-5y = -3x - 10$$
$$y = \tfrac{3}{5}x + 2$$

The slope is 3/5. ■

Practice 3. What is the slope of the line $12x + 4y - 7 = 0$? Graph the line.

There are many relationships between variables arising in different fields of technology that can be expressed as linear equations.

EXAMPLE 6. The length of a spring varies linearly with the amount of force applied to stretch the spring. When 2 N of force are applied, the length of a particular spring is 0.8 m; when 5 N of force are applied, the length of the spring is 1.4 m. Write the equation relating length s and force f. What is the natural length of the spring?

Choosing f as the independent variable and s as the dependent variable, the slope-intercept form becomes

$$s = mf + b \tag{4}$$

The two points (2, 0.8) and (5, 1.4) satisfy the equation. Substituting these values into Equation (4) results in the system of equations

$$0.8 = 2m + b$$
$$1.4 = 5m + b$$

We solve this system of equations and obtain $m = 0.2$ and $b = 0.4$. Equation (4) then becomes

$$s = 0.2f + 0.4 \tag{5}$$

which is the equation relating s and f. The natural length of the spring occurs when the applied force f is zero. From Equation (5), when $f = 0$, $s(0) = 0.4$. The natural length of the spring is 0.4 m. ■

Exercises / Section **2-2**

Exercises 1–28: Write the equation of the line described.

1. Passes through $(2, 0)$ with slope 3.

2. Passes through $(0, -4)$ with slope -2.

3. Passes through $(-1, -4)$ with slope $\frac{1}{2}$.

4. Passes through $(-2, 5)$ with slope $\frac{-2}{3}$.

5. Passes through $(5, 1)$ and $(-2, 3)$.

6. Passes through $(1, -3)$ and $(4, 7)$.

7. Passes through $(3, -4)$ and $(-8, -6)$.

8. Passes through $(-5, -1)$ and $(1, 6)$.

9. Has x-intercept $(4, 0)$ and y-intercept $(0, -6)$.

10. Has x-intercept $(-3, 0)$ and y-intercept $(0, -1)$.

11. Has slope $\frac{1}{3}$ and y-intercept $(0, 2)$.

12. Has slope $\frac{-5}{8}$ and y-intercept $(0, -3)$.

13. Passes through $(1, 4)$ and is parallel to the y-axis.

14. Passes through $(-2, -3)$ and is parallel to the x-axis.

15. Passes through $(5, 2)$ and is parallel to the line $2x - 3y + 7 = 0$.

16. Passes through $(-1, -6)$ and is parallel to the line $x + 4y + 7 = 0$.

17. Passes through $(1, -4)$ and is perpendicular to the line $3x - 6y + 2 = 0$.

18. Passes through $(-5, 1)$ and is perpendicular to the line $5x + 2y - 4 = 0$.

19. Passes through $(-2, 3)$ and is parallel to the line connecting $(-8, -4)$ and $(1, 5)$.

20. Passes through $(5, -6)$ and is parallel to the line connecting $(-7, 4)$ and $(3, -1)$.

21. Passes through $(-1, 4)$ and is perpendicular to the line connecting $(-1, -6)$ and $(2, 3)$.

22. Passes through $(4, -2)$ and is perpendicular to the line connecting $(-5, 4)$ and $(7, -2)$.

23. The perpendicular bisector of the line connecting $(-1, 6)$ and $(3, -4)$.

24. The perpendicular bisector of the line connecting $(2, 5)$ and $(9, 8)$.

25. Is parallel to the line $3x - 5y + 2 = 0$ and passes through the point of intersection of the lines $x + 4y - 22 = 0$ and $3x + y = 0$.

26. Is perpendicular to the line $x + 4y + 7 = 0$ and passes through the point of intersection of the lines $2x + 2y + 3 = 0$ and $5x - 4y - 6 = 0$.

27. Passes through $(-6, -2)$ and has x-intercept equal to twice its y-intercept.

28. Passes through $(7, 5)$ and has y-intercept equal to three times its x-intercept.

Exercises 29–32: Find the slope of the line and sketch its graph.

29. $2x - 3y + 5 = 0$ 30. $x - 6y + 4 = 0$

31. $6x + 3y - 12 = 0$ 32. $2x + 5y - 8 = 0$

Exercises 33–34: Prove that the given lines are parallel.

33. $4x - 8y + 17 = 0$ and $5x - 10y - 8 = 0$

34. $2x + 3y - 4 = 0$ and $8x + 12y + 3 = 0$

Exercises 35–36: Prove that the given lines are perpendicular.

35. $x - 2y + 3 = 0$ and $4x + 2y - 7 = 0$

36. $5x - 8y - 7 = 0$ and $16x + 10y + 5 = 0$

37. The number of revolutions per minute (rpm) produced by a small motor varies linearly with the current (in milliamps) supplied to the motor. When 0 mA of current are supplied, the motor produces 0 rpm. When 20 mA of current are supplied, the motor produces 1375 rpm. Write the equation relating revolutions per minute r and current i.

38. A jogger moves toward a fixed point at a constant velocity; the distance of the jogger from the point varies linearly with the time spent jogging. When the jogger starts out, the fixed point is 30 km away; 1 h later the fixed point is 20.4 km away. Write the equation relating distance d of the jogger from the point and time t.

39. The voltage produced by a temperature transducer varies linearly with the temperature. When the temperature is 30°C, the transducer reads 0.3 V; when the temperature is 90°C, the

transducer reads 4.8 V. Write the equation relating voltage V and temperature T. What is the voltage when $T = 70°C$?

40. The voltage produced by a pressure transducer varies linearly with the atmospheric pressure. When the pressure is 740 mm of mercury, the transducer reads 1.1 V; when the pressure is 770 mm of mercury, the transducer reads 3.2 V. Write the equation relating voltage V and pressure p. What is the voltage when $p = 760$ mm of mercury?

41. For the pressure transducer of Exercise 40, a voltage reading of 2.8 V was taken. What was the approximate corresponding pressure?

42. A technician read the pressure transducer of Exercise 40 and recorded a voltage of 2.3 V

when the atmospheric pressure was 750 mm of mercury. Was the transducer broken?

43. The cost for a certain company to manufacture semiconductors is $.16 per unit. In addition, there is a fixed overhead cost of $3400 per day. Write the equation relating cost C per day and the number x of units produced per day.

44. Probability of onset of a particular genetic disease in an at-risk individual varies linearly with the age of the individual. When the age of the individual is 27, the probability is 27%; when the age is 60, the probability is 38%. Write the equation relating probability p and age a. What is the probability of onset at birth?

2-3 The Parabola

When we developed the equation for a straight line, we began with a typical point (x, y) on the line. The line itself could be thought of as the path traced by this typical point as it moved in a certain way. In this and the next two sections, we discuss other curves that can be described as paths traced by a point moving in some specified manner. There are four types of such curves; a sample of each type is shown in Figure 2-7.

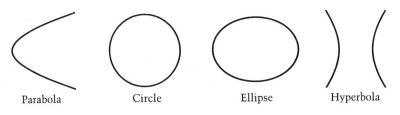

Parabola   Circle   Ellipse   Hyperbola

Figure 2-7

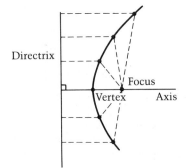

Figure 2-8

A **parabola** is the path traced by a point that is always equally distant from a fixed point, called the **focus**, and a fixed line, called the **directrix**. The type of curve formed appears in Figure 2-8, which shows the moving point stopped in a number of positions. Figure 2-8 also shows the directrix and the focus. The line through the focus perpendicular to the directrix is the **axis** of the parabola, and the point of the parabola on the axis is the **vertex**. The vertex is the "nose," or turning point, of the parabola.

We first develop the equation for a special case of the parabola, where the origin of the coordinate system is the vertex of the parabola and the x-axis of the coordinate system coincides with the

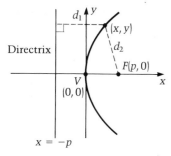

Figure 2-9

axis of the parabola. Figure 2-9 shows the coordinate system placed on the parabola, with the focus located at $(p, 0)$. Because the vertex is halfway between the focus and the directrix, the equation of the directrix is $x = -p$. For the typical point (x, y), the distance d_2 in Figure 2-9 must be the same as the distance d_1. The distance d_1 is a horizontal distance, so we use $x_R - x_L$:

$$d_1 = x - (-p) = x + p$$

To find the distance d_2 between two points, we use the distance formula:

$$d_2 = \sqrt{(x - p)^2 + (y - 0)^2}$$

Because $d_2 = d_1$, we have

$$\sqrt{(x - p)^2 + y^2} = x + p$$

Squaring both sides of the equation and then simplifying, we get

$$(x - p)^2 + y^2 = (x + p)^2$$
$$x^2 - 2px + p^2 + y^2 = x^2 + 2px + p^2$$
$$y^2 = 4px$$

Thus $y^2 = 4px$ is the equation of the parabola in Figure 2-9. We can substitute $-y$ for y in the equation with no change; this indicates that the curve is symmetric about the x-axis, as Figure 2-9 illustrates.

There are three other ways to place the vertex of the parabola at the origin of the coordinate system and the axis along one of the coordinate axes. In each case, we can derive the equation in a manner similar to what we just did. Figure 2-10 summarizes the four cases. In Figure 2-10(a) and (c), p is a positive number, but in

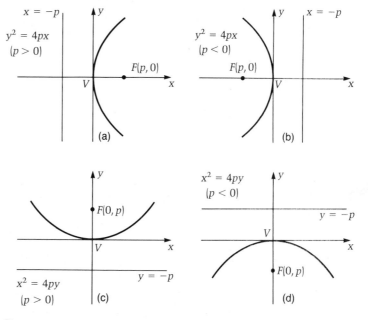

Figure 2-10

Figure 2-10(b) and (d), p is a negative number. The form of the equation in Figure 2-10(c) and (d) indicates symmetry of the curve about the y-axis. In all four cases, the equation contains one squared variable and the other variable to the first power.

We are now able to accomplish

$$\text{algebraic} \xrightarrow{\text{graph the equation}} \text{geometric}$$

for any of these four special cases of parabolas. A look at the form of the equation tells us the general shape of the graph, and we can then plot some points for more exact information. As an additional guide, it may help to note that the smaller the numerical value of p, the more sharply the parabola bends back.

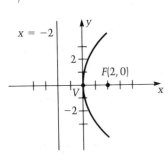

Figure 2-11

EXAMPLE 1.   Sketch the graph of the equation $y^2 = 8x$. Give the coordinates of the focus and the equation of the directrix.

The form of the equation is $y^2 = 4px$, with $4p = 8$, or $p = 2$. This corresponds to Figure 2-10(a), so we have a parabola with vertex at the origin, opening to the right. The focus is $(2, 0)$ and the equation of the directrix is $x = -2$. The curve is shown in Figure 2-11   ■

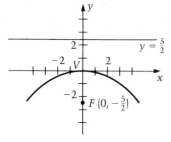

Figure 2-12

EXAMPLE 2.   Sketch the graph of the equation $x^2 = -10y$. Give the coordinates of the focus and the equation of the directrix.

Here the form of the equation is $x^2 = 4py$, with $4p = -10$, or $p = \frac{-5}{2}$. This represents a parabola with vertex at the origin, opening downward. The focus is $(0, \frac{-5}{2})$ and the directrix is the line $y = \frac{5}{2}$. The curve is shown in Figure 2-12.   ■

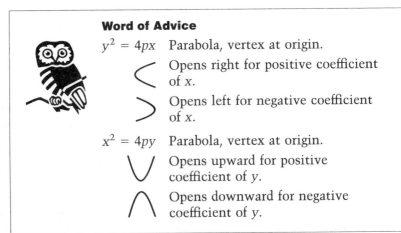

Word of Advice

$y^2 = 4px$ Parabola, vertex at origin.

Opens right for positive coefficient of x.

Opens left for negative coefficient of x.

$x^2 = 4py$ Parabola, vertex at origin.

Opens upward for positive coefficient of y.

Opens downward for negative coefficient of y.

Practice 1. Sketch the graph of the function $x^2 = 12y$. Give the coordinates of the focus and the equation of the directrix.

How do we find the equation of a parabola given some geometric information? This is our other task, namely,

$$\text{geometric} \xrightarrow{\text{find the equation}} \text{algebraic}$$

Because of the definition of a parabola, the geometric information we need to specify a particular parabola must tell us, directly or indirectly, the focus and directrix of the parabola. In the four cases where the vertex is at the origin and the directrix is horizontal or vertical, we simply determine p and use whichever of the four equations is appropriate.

EXAMPLE 3. Find the equation of the parabola with focus $(-5, 0)$ and directrix $x = 5$.

From the information given, the parabola has its vertex at the origin and opens to the left, with $p = -5$. The equation is $y^2 = -20x$. ■

Practice 2. Find the equation of the parabola with focus $(0, 4)$ and directrix $y = -4$.

Exercises / Section **2-3**

Exercises 1–14: Sketch the graph of the equation. Find the coordinates of the focus and the equation of the directrix.

1. $y^2 = 4x$
2. $y^2 = 16x$
3. $y^2 = -12x$
4. $y^2 = -100x$
5. $x^2 = 8y$
6. $x^2 = 20y$
7. $x^2 = -16y$
8. $x^2 = -32y$
9. $y^2 = 3x$
10. $y^2 = -7x$
11. $x^2 = -y$
12. $x^2 = 2y$
13. $2x = 3y^2$
14. $5y = 6x^2$

Exercises 15–24: Write the equation of the parabola described.

15. Focus $(0, -2)$, directrix $y = 2$.

16. Focus $(3, 0)$, directrix $x = -3$.
17. Focus $(0, 5)$, directrix $y = -5$.
18. Focus $(-1, 0)$, directrix $x = 1$.
19. Vertex $(0, 0)$, directrix $y = -3$.
20. Vertex $(0, 0)$, directrix $x = \frac{2}{3}$.
21. Vertex $(0, 0)$, focus $(0, -1)$.
22. Vertex $(0, 0)$, focus $(\frac{7}{4}, 0)$.
23. Vertex $(0, 0)$, axis $x = 0$, passes through $(3, 6)$.
24. Vertex $(0, 0)$, axis $y = 0$, passes through $(-1, -1)$.

25. Spotlights are made with parabolic reflectors at the back, because if a light source is placed

at the focus of a parabolic reflector, the reflected rays of light are all parallel to the axis of the parabola. If a parabolic reflector has the equation $y^2 = 20x$, how far from the vertex of the reflector should the light source be placed (see Figure 2-13)?

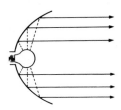

Figure 2-13

26. If a parabolic reflector measures 30 cm across at the opening and is 9 cm deep, where is the focal point?

27. A city decides to build an arch in the shape of a parabola. The arch is 280 m high, and the feet of the arch are 120 m apart at ground level. What is the height of the arch 30 m away from the center point? (*Hint:* Put the origin of a coordinate system at the vertex and find the equation of the parabola.)

28. A cable hangs from a suspension bridge in the approximate shape of a parabola. The supporting towers are 100 m apart and 30 m high. The lowest point of the cable is 6 m above the highway. Find the height of the cable above the highway at a point 20 m away from the low point.

29. The current I (in amperes) in a certain electric circuit as a function of time t (in seconds) is given by the equation $I = 1.08t^2$, $0 \le t < 6$. Sketch the graph of I as a function of t.

30. The period T (in seconds) of a simple pendulum as a function of the length L (in meters) is given by the equation $T = 2\sqrt{L}$. Sketch the graph of T as a function of L for $0 \le L \le 20$.

31. A stone is thrown horizontally from the top of a bridge 30 m high. If a coordinate system is placed with its origin at the starting point, then the path of the stone is given by the equation $x^2 = -68y$. Approximately how far does the stone travel horizontally before it hits the water?

32. Soybeans fed with a special nutrient solution were predicted to gain new growth g (in centimeters) as a function of time t (in days) by a parabolic equation of the shape shown in Figure 2-14. The amount of new growth after 3 days was 4.8 cm. What new growth should be expected after 5 days?

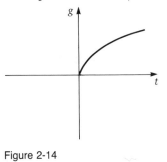

Figure 2-14

2-4   The Ellipse and Circle

In this section we define a curve called an *ellipse*. A circle is a special case of an ellipse, so we talk about circles here also.

An **ellipse** is the path traced by a point which moves so that the sum of its distances from two fixed points, called **foci** or **focal points**, is a constant. The type of curve formed is given in Figure 2-15, with the moving point shown in several positions. You could generate an ellipse by taking a length of string, pinning its ends down with two thumbtacks, and—using your pencil to hold the string taut—moving the pencil along the string.

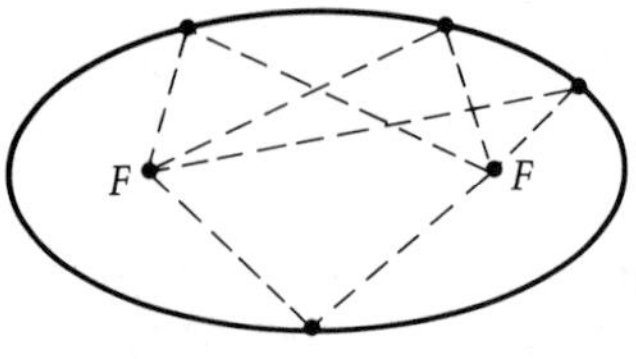

We first develop the equation for a special case of the ellipse, where the origin of the coordinate system is the center of the ellipse

Figure 2-15

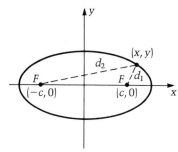

Figure 2-16

and the focal points are on the x-axis. Figure 2-16 shows the coordinate system placed on the ellipse, with the foci located at $(c, 0)$ and $(-c, 0)$. The equation has a simpler form if we let $2a$ represent the fixed sum of the two distances from the foci. From the definition of an ellipse, $d_1 + d_2 = 2a$. Using the distance formula, this means

$$\sqrt{(x - c)^2 + (y - 0)^2} + \sqrt{(x + c)^2 + (y - 0)^2} = 2a$$

This equation can be simplified most easily if we move one radical term to the other side before we square both sides of the equation. We simplify as follows:

$$\sqrt{(x - c)^2 + y^2} = 2a - \sqrt{(x + c)^2 + y^2}$$
$$(x - c)^2 + y^2 = (2a - \sqrt{(x + c)^2 + y^2})^2$$
$$x^2 - 2cx + c^2 + y^2 = 4a^2 - 4a\sqrt{(x + c)^2 + y^2} + (x + c)^2 + y^2$$
$$x^2 - 2cx + c^2 + y^2 = 4a^2 - 4a\sqrt{(x + c)^2 + y^2} + x^2 + 2cx + c^2 + y^2$$
$$4a\sqrt{(x + c)^2 + y^2} = 4a^2 + 4cx$$
$$(a\sqrt{(x + c)^2 + y^2})^2 = (a^2 + cx)^2$$
$$a^2(x^2 + 2cx + c^2 + y^2) = a^4 + 2cxa^2 + c^2x^2$$
$$(a^2 - c^2)x^2 + a^2y^2 = a^2(a^2 - c^2)$$

Now we introduce a new quantity b, defined by $b^2 = a^2 - c^2$. (We shall soon see that this substitution equation has some geometric significance, as well as helping to simplify the form of the equation of the ellipse.) The equation becomes

$$b^2x^2 + a^2y^2 = a^2b^2$$

or, dividing both sides by a^2b^2,

$$\frac{x^2}{a^2} + \frac{y^2}{b^2} = 1 \tag{1}$$

Equation (1) is the equation of the ellipse of Figure 2-16. The x-intercepts are $(a, 0)$ and $(-a, 0)$, and the y-intercepts are $(0, b)$ and $(0, -b)$. These points are shown in Figure 2-17. Now consider the point $(0, b)$. Because it is a point on the ellipse, the sum of its distances from $(-c, 0)$ and $(c, 0)$ must equal $2a$. But $(0, b)$ is equidistant from $(-c, 0)$ and $(c, 0)$. Therefore, the distance from $(0, b)$ to $(c, 0)$ is a. This is shown in Figure 2-17, where we see that we have a right triangle with legs of length b and c and hypotenuse of length a. Thus

$$a^2 = b^2 + c^2$$

which agrees with our substitution equation.

Notice from the above equation that for $c > 0$, $a^2 > b^2$, so a is greater than b. The line segment with endpoints $(-a, 0)$ and $(a, 0)$ is called the **major axis** of the ellipse. Each half of the major axis, from

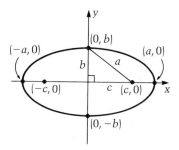

Figure 2-17

$(0, 0)$ to $(a, 0)$ or from $(-a, 0)$ to $(0, 0)$, is a **semimajor axis**. The ends of the major axis, $(-a, 0)$ and $(a, 0)$, are the **vertices** of the ellipse. The line segment from $(0, b)$ to $(0, -b)$ is the **minor axis** of the ellipse, and each half is a **semiminor axis**. Because $a > b$, the major axis is longer than the minor axis.

The height of the ellipse at the focal point $(c, 0)$ can be found by substituting c for x in Equation (1) and solving for y.

$$\frac{c^2}{a^2} + \frac{y^2}{b^2} = 1$$

$$y^2 = b^2\left(\frac{1 - a^2}{a^2}\right) = b^2\left(\frac{a^2 - c^2}{a^2}\right) = b^2\frac{b^2}{a^2}$$

$$y = \pm\frac{b^2}{a}$$

A line from one side of the ellipse to the other that is perpendicular to the major axis and goes through a focus is called a **latus rectum**; its length is $2y = 2b^2/a$.

If we were to place the foci of an ellipse on the y-axis at points $(0, c)$ and $(0, -c)$, the same sort of reasoning as above leads to the equation

$$\frac{y^2}{a^2} + \frac{x^2}{b^2} = 1 \tag{2}$$

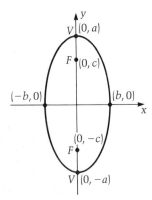

Figure 2-18

This ellipse (Figure 2-18) has its major axis along the y-axis.

In both Equations (1) and (2), the form of the equation indicates that the curve is symmetric about the x-axis, about the y-axis, and about the origin, as Figures 2-16 and 2-18 show. In each equation, both variables are squared, and their terms have positive signs. In fact, the only difference between Equations (1) and (2) is whether the larger number (a^2) appears beneath x^2 or beneath y^2. This determines whether the major axis is horizontal or vertical.

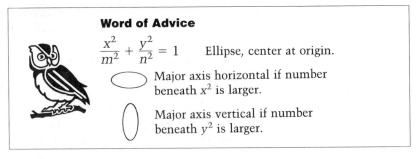

Word of Advice

$$\frac{x^2}{m^2} + \frac{y^2}{n^2} = 1 \qquad \text{Ellipse, center at origin.}$$

Major axis horizontal if number beneath x^2 is larger.

Major axis vertical if number beneath y^2 is larger.

For ellipses with centers at the origin and horizontal or vertical major axes, the task

$$\text{algebraic } \xrightarrow{\text{graph the equation}} \text{geometric}$$

is easy. We divide through, if necessary, to put the equation in the form of (1) or (2). This will tell us whether the major axis is horizontal or vertical. Plotting the four intercepts gives an indication of the shape of the graph.

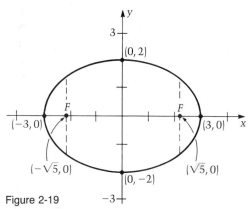

Figure 2-19

EXAMPLE 1. Graph the equation $\frac{x^2}{9} + \frac{y^2}{4} = 1$. Give the coordinates of the vertices and the foci.

This is the equation of an ellipse; because the larger value is beneath x^2, the major axis is horizontal. From $a^2 = 9$, we see that $a = 3$; the vertices are $(3, 0)$ and $(-3, 0)$. We also know that $b^2 = 4$, so the y-intercepts of the curve are $(0, 2)$ and $(0, -2)$. Using the equation $a^2 = b^2 + c^2$, we solve for c and get $c = \sqrt{5}$. The focal points are thus $(\sqrt{5}, 0)$ and $(-\sqrt{5}, 0)$. The length of the latus rectum is $2b^2/a = \frac{8}{3}$; this is an additional aid in sketching the graph. The graph is shown in Figure 2-19. ■

EXAMPLE 2. Graph the equation $16x^2 + 9y^2 = 144$. Give the coordinates of the vertices and the foci.

From the form of the equation (both variables squared and all the terms positive), we know that the equation represents an ellipse; to put it in the form of Equation (1) or (2), we divide through by 144. The result is

$$\frac{x^2}{9} + \frac{y^2}{16} = 1$$

Thus the major axis is vertical; $a = 4$ and $b = 3$. The vertices are $(0, 4)$ and $(0, -4)$. From $a^2 = b^2 + c^2$, $c = \sqrt{7}$. The foci are $(0, \sqrt{7})$, $(0, -\sqrt{7})$. Figure 2-20 shows the graph. ■

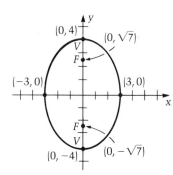

Figure 2-20

Practice 1. Sketch the graph of the equation $\frac{x^2}{1} + \frac{y^2}{4} = 1$. Give the coordinates of the vertices and foci.

Now suppose we let the two focal points of an ellipse move together until they converge to a single point. What does the definition of the ellipse become? We now have "the path traced by a point which moves so that the sum of its distances from a fixed point is a constant." This is the same as "the path traced by a point which moves so that its distance from a fixed point is a constant." And this is just the definition of a **circle**, where the fixed point is the **center** of the circle and the constant distance is the **radius** of the circle. This is why a circle is a special case of an ellipse.

What is the equation of a circle with center at the origin? If we look at Figure 2-16, we have moved the focal points to $(0, 0)$, so that $c = 0$. From the equation $a^2 = b^2 + c^2$, $a^2 = b^2$ when $c = 0$. Equation (1) (and also Equation (2)) becomes

$$\frac{x^2}{a^2} + \frac{y^2}{a^2} = 1$$

or

$$x^2 + y^2 = a^2 \qquad\qquad (3)$$

Equation (3) is the equation of a circle with center at the origin and radius a.

EXAMPLE 3. Graph the equation $x^2 + y^2 = 16$.

This is a circle with center at the origin and radius 4. Figure 2-21 shows the graph. ■

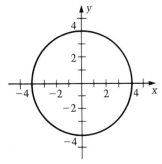

Figure 2-21

The opposite task for the ellipse, namely,

geometric $\xrightarrow{\text{find the equation}}$ algebraic

is also of interest to us. Given enough geometric information, we want to write the equation of the ellipse. For the case where we know that the center of the ellipse is at the origin of the coordinate system and the major axis is horizontal or vertical, the equation will be in the form of Equation (1) or Equation (2). We need enough additional information to be able to determine a and b and to know whether Equation (1) or Equation (2) should be used.

EXAMPLE 4. Find the equation of the ellipse with foci $(\pm 4, 0)$ and vertices $(\pm 5, 0)$.

From the information given, the center of the ellipse is at the origin and the major axis is horizontal, so Equation (1) should be used. We also know that $a = 5$ and $c = 4$. From the relationship $a^2 = b^2 + c^2$, we solve for b and get $b = 3$. The equation is

$$\frac{x^2}{25} + \frac{y^2}{9} = 1$$

The ellipse appears in Figure 2-22. ■

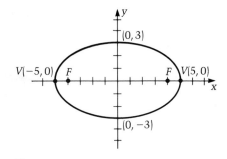

Figure 2-22

EXAMPLE 5. The equation of the circle with center at the origin and radius 7 is $x^2 + y^2 = 49$. ■

Practice 2. Find the equation of the ellipse with center at the origin, focus at $(0, 3)$, and semiminor axis of length 2.

Exercises / Section **2-4**

Exercises 1–16: Sketch the graph of the equation. Give the coordinates of the vertices and foci.

1. $\dfrac{x^2}{25} + \dfrac{y^2}{16} = 1$

2. $\dfrac{x^2}{100} + \dfrac{y^2}{9} = 1$

3. $\dfrac{x^2}{16} + \dfrac{y^2}{36} = 1$

4. $\dfrac{x^2}{25} + \dfrac{y^2}{49} = 1$

5. $\dfrac{x^2}{12} + \dfrac{y^2}{9} = 1$

6. $\dfrac{x^2}{10} + \dfrac{y^2}{3} = 1$

7. $\dfrac{x^2}{8} + \dfrac{y^2}{16} = 1$

8. $\dfrac{x^2}{3} + \dfrac{y^2}{5} = 1$

9. $x^2 + 4y^2 = 4$

10. $2x^2 + 49y^2 = 98$

11. $25x^2 + 4y^2 = 100$

12. $9x^2 + 7y^2 = 63$

13. $4x^2 + 9y^2 = 1$

14. $x^2 + 5y^2 = 1$

15. $2x^2 + y^2 = 1$

16. $5x^2 + 4y^2 = 1$

Exercises 17–18: Sketch the graph.

17. $x^2 + y^2 = 9$

18. $\dfrac{x^2}{25} + \dfrac{y^2}{25} - 1 = 0$

Exercises 19–28: Write the equation of the ellipse described.

19. Foci $(\pm 3, 0)$, vertices $(\pm 6, 0)$.

20. Foci $(0, \pm 5)$, vertices $(0, \pm 8)$.

21. Vertices $(0, \pm 7)$, minor axis of length 4.

22. Center at the origin, major axis horizontal of length 6, minor axis vertical of length 2.

23. Foci $(\pm 10, 0)$, major axis of length 28.

24. Foci $(0, \pm 6)$, minor axis of length 5.

25. Center at the origin, vertex at $(5, 0)$, goes through $(3, \frac{8}{5})$.

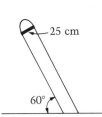

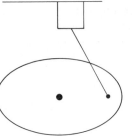

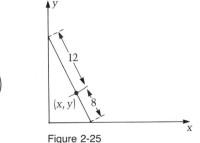

Figure 2-23   Figure 2-24   Figure 2-25

26. Center at the origin, vertex at $(0, -4)$, goes through $(1, 4\sqrt{6/7})$.

27. Center at the origin, passes through $(1, 3)$ and $(2, 2)$.

28. Center at the origin, passes through $(-4, 2)$ and $(2, 3)$.

29. Find the equation of the circle with center at the origin and radius 2.

30. Find the equation of the circle with center at the origin and radius 5.

31. The arch of a bridge has the shape of half an ellipse. If the span of the arch is 40 m and its height at the highest point is 15 m, find the height of the arch at a point 10 m from the center.

32. A reflecting pool has been designed in the shape of an ellipse. If the ellipse is to have a maximum length of 30 m and a maximum width of 8 m, what will the width be at a point 4 m from the end?

33. A spotlight throws a beam of light that is 25 cm in diameter. If the beam hits the stage floor at an angle of 60° with the horizontal, find an equation for the elliptical pool of light on the stage floor (Figure 2-23).

34. An elliptical plate rotates on a shaft through its center. A pin 1 cm from the end of the ellipse is free to rotate and is connected by a shaft to a piston (see Figure 2-24). In a coordinate system with the origin at the center of the ellipse, the ellipse, as shown in Figure 2-24, has the equation $2x^2 + 5y^2 = 115$. Find the stroke of the piston (the maximum up and down distance it moves) as the ellipse revolves.

35. A communications satellite travels in an elliptical orbit, with the center of the earth at one focal point. If the minimum distance of the satellite above the earth is 725 km, the maximum distance is 2900 km, and the radius of the earth is 6440 km, find the equation of the orbit. (Assume the origin is at the center of the ellipse and the major axis is along the x-axis.)

36. A 20-cm ruler is notched at a point 12 cm from one end. If the ruler is moved so that its two ends stay on two adjacent edges of a sheet of paper, find the equation of the curve traced by a pen placed at the notch. (*Hint:* Put the origin of the coordinate system at the corner of the paper and construct similar triangles. See Figure 2-25.)

2-5 The Hyperbola

A **hyperbola** is the path traced by a point which moves so that the difference of its distances from two fixed points, called **foci** or **focal points**, is a constant. We first develop the equation for a special case of the hyperbola, where the focal points are on the x-axis and the origin of the coordinate system is midway between them. We let the

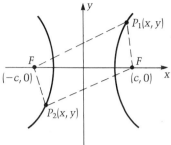

Figure 2-26

constant difference be $2a$. In Figure 2-26, the focal points are at $(c, 0)$ and $(-c, 0)$. Points on the right branch of the hyperbola, such as P_1, are closer to the right focal point and satisfy the equation

$$\sqrt{(x + c)^2 + (y - 0)^2} - \sqrt{(x - c)^2 + (y - 0)^2} = 2a \qquad (1)$$

Points on the left branch of the hyperbola, such as P_2, are closer to the left focal point and satisfy the equation

$$\sqrt{(x - c)^2 + (y - 0)^2} - \sqrt{(x + c)^2 + (y - 0)^2} = 2a \qquad (2)$$

We can simplify Equations (1) and (2) by a procedure similar to the one we used for the ellipse. Because the equation is squared (twice) during this simplification process, the result of simplifying either (1) or (2) is the same equation, namely,

$$\frac{x^2}{a^2} - \frac{y^2}{b^2} = 1 \qquad (3)$$

As in the case of the ellipse, a new variable b was introduced to simplify the form of the equation. But here the equation relating a, b, and c is

$$c^2 = a^2 + b^2$$

Notice that this equation does not guarantee that a is greater than b. We know nothing about the relative size of a and b; we know only that c is greater than either. However, the value b not only simplifies the form of the equation, it also has a geometric interpretation with respect to the graph of the hyperbola. To see this, we solve Equation (3) for y in terms of x.

$$\frac{y^2}{b^2} = \frac{x^2}{a^2} - 1 = \frac{x^2}{a^2} - \frac{a^2 x^2}{a^2 x^2} = \frac{x^2}{a^2}\left(1 - \frac{a^2}{x^2}\right)$$

$$y^2 = \frac{b^2 x^2}{a^2}\left(1 - \frac{a^2}{x^2}\right)$$

$$y = \pm \frac{bx}{a}\sqrt{1 - \frac{a^2}{x^2}} \qquad (4)$$

Now look at this last equation. For large positive or negative values of x, a^2/x^2 is close to zero and y is close to $\pm bx/a$. The larger the numerical value of x, the closer Equation (4) is to

$$y = \pm\frac{bx}{a} \qquad (5)$$

Equation (5) is really two equations, one for the line $y = bx/a$ (through the origin) and one for the line $y = -bx/a$ (through the origin). These two lines are *asymptotes* for the hyperbola, guidelines to which the curve gets closer for larger and larger positive or negative values of x.

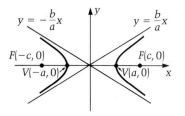

Figure 2-27

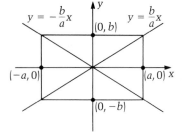

Figure 2-28

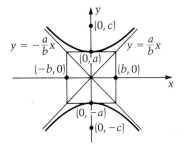

Figure 2-29

Figure 2-27 is a graph of the hyperbola

$$\frac{x^2}{a^2} - \frac{y^2}{b^2} = 1$$

with the asymptotes shown. From the equation of the hyperbola, the x-intercepts are $(a, 0)$ and $(-a, 0)$; these are the **vertices** of the hyperbola. There are no y-intercepts for the hyperbola; setting $x = 0$ in the equation of the hyperbola results in a negative value for y^2. In fact, from Equation (4) we can see that only points with values of x such that $|x| \geq a$ will appear on the graph, as otherwise y involves taking the square root of a negative number.

Graphing the hyperbola is easy if we plot the vertices and draw in the asymptotes. From the equation

$$\frac{x^2}{a^2} - \frac{y^2}{b^2} = 1$$

we can read the values of a and b. The vertices are plotted at $(\pm a, 0)$. The line segment with the two vertices as endpoints is called the **transverse axis** of the hyperbola. Next we plot the points $(0, \pm b)$ (these are not points on the curve, remember), and draw a rectangle with these four points at the centers of the four sides. The diagonals of this rectangle are the lines $y = \pm bx/a$ (see Figure 2-28). The line segment with the points $(0, -b)$ and $(0, b)$ as endpoints is called the **conjugate axis** of the hyperbola.

Now suppose that the hyperbola has a vertical transverse axis, as in Figure 2-29. By reasoning similar to the previous case, we find the equation of the hyperbola to be

$$\frac{y^2}{a^2} - \frac{x^2}{b^2} = 1 \qquad (6)$$

with asymptotes

$$y = \pm \frac{a}{b} x$$

In both Equations (3) and (6), we see that the curve is symmetric about the x-axis, about the y-axis, and about the origin. In each equation, both variables are squared, but they have opposite signs. It is the positive term that indicates the direction of the transverse axis and the placement of the vertices.

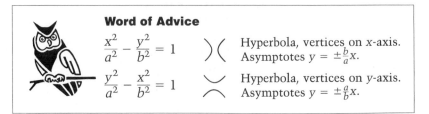

Word of Advice

$\dfrac{x^2}{a^2} - \dfrac{y^2}{b^2} = 1$)(Hyperbola, vertices on x-axis.
Asymptotes $y = \pm \frac{b}{a} x$.

$\dfrac{y^2}{a^2} - \dfrac{x^2}{b^2} = 1$ $\smile$ $\frown$ Hyperbola, vertices on y-axis.
Asymptotes $y = \pm \frac{a}{b} x$.

For hyperbolas with centers at the origin and horizontal or vertical transverse axes, we know how to do

$$\text{algebraic} \xrightarrow{\text{graph the equation}} \text{geometric}$$

We need only to recognize whether we have Equation (3) or Equation (6).

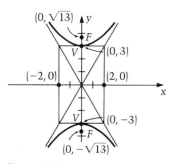

Figure 2-30

EXAMPLE 1.   Graph the equation $\frac{y^2}{9} - \frac{x^2}{4} = 1$. Find the coordinates of the vertices and foci.

Because the y^2-term is positive, the transverse axis runs along the y-axis. The vertices are $(0, 3)$ and $(0, -3)$. The value of b is 2, so we plot the points $(2, 0)$ and $(-2, 0)$ in order to draw the rectangle whose diagonals are the asymptotes. From the equation $c^2 = a^2 + b^2$, the value of c^2 is 13 and the foci are $(0, \sqrt{13})$ and $(0, -\sqrt{13})$. The curve is shown in Figure 2-30. ■

EXAMPLE 2.   Graph the equation $x^2 - 2y^2 - 8 = 0$. Find the coordinates of the vertices and foci.

By writing the equation in the form

$$\frac{x^2}{8} - \frac{y^2}{4} = 1$$

we can see that the transverse axis is horizontal, with vertices at $(\pm\sqrt{8}, 0)$. The conjugate axis runs from $(0, -2)$ to $(0, 2)$. Because $c^2 = a^2 + b^2$, $c^2 = 12$ and the foci are $(\sqrt{12}, 0)$ and $(-\sqrt{12}, 0)$. The graph is shown in Figure 2-31. ■

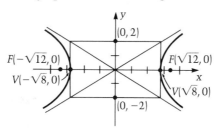

Figure 2-31

Practice 1.   Sketch the graph of the equation $\frac{x^2}{9} - \frac{y^2}{16} = 1$. Give the coordinates of the vertices and foci.

Finally, we want to be able to write the equation of a hyperbola given some geometric information, namely,

$$\text{geometric} \xrightarrow{\text{find the equation}} \text{algebraic}$$

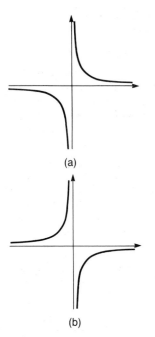

Figure 2-32

For the case where the center of the hyperbola is at the origin and the foci are on one of the coordinate axes, we can write the equation if we can determine a and b and whether the transverse axis is horizontal or vertical.

EXAMPLE 3.   Find the equation of the hyperbola with foci $(\pm 4, 0)$ and transverse axis 4 units long.

From the information given, $c = 4$ and $a = 2$. Because $c^2 = a^2 + b^2$, $b^2 = 12$. The transverse axis runs along the x-axis, and the equation is, therefore,

$$\frac{x^2}{4} - \frac{y^2}{12} = 1$$

The hyperbola is shown in Figure 2-32.   ■

Practice 2.   Find the equation of the hyperbola with foci $(0, \pm 6)$ and vertices $(0, \pm 4)$.

(a)

(b)

Figure 2-33

A hyperbola whose equation has a different form occurs when the coordinate axes are asymptotes to the curve, as in Figure 2-33. This type of hyperbola has an equation of the form

$$xy = k$$

where k is a constant.

EXAMPLE 4.   Sketch the graph of the equation $xy = 5$.

Solving the equation for y in terms of x, we get $y = 5/x$. In this form, we see that $x = 0$ is an excluded value, but that $5/x$ grows numerically larger as x gets closer to zero. The line $x = 0$ is an asymptote. Also, for large positive or negative values of x, $5/x$ is close to zero; so the line $y = 0$ is also an asymptote. Finally, y has positive values for positive x-values, and y has negative values for negative x-values. Thus the curve will appear only in the first and third quadrants. After plotting a few points, we get the graph shown in Figure 2-34.   ■

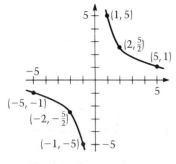

Figure 2-34

Practice 3.   Sketch the graph of the equation $xy = -1$.

Exercises / Section **2-5**

Exercises 1–12: Sketch the graph of the equation. Give the coordinates of the vertices and foci.

1. $\dfrac{x^2}{16} - \dfrac{y^2}{25} = 1$ 2. $\dfrac{x^2}{100} - \dfrac{y^2}{36} = 1$

3. $\dfrac{y^2}{4} - \dfrac{x^2}{1} = 1$ 4. $\dfrac{y^2}{49} - \dfrac{x^2}{16} = 1$

5. $7x^2 - 4y^2 = 28$

6. $12x^2 - 5y^2 = 60$

7. $y^2 - 3x^2 = 3$

8. $4y^2 - 5x^2 = 40$

9. $2x^2 - y^2 + 4 = 0$

10. $y^2 - 3x^2 + 3 = 0$

11. $4x^2 - 16y^2 = 1$

12. $25y^2 - 4x^2 = 1$

Exercises 13–20: Write the equation of the hyperbola described.

13. Foci $(\pm 10, 0)$, vertices $(\pm 7, 0)$.

14. Foci $(0, \pm 4)$, vertices $(0, \pm 3)$.

15. Foci $(\pm 4, 0)$, conjugate axis of length 6.

16. Foci $(0, \pm 15)$, tranverse axis of length 10.

17. Vertices $(0, \pm 6)$, asymptotes $y = \pm 2x$.

18. Foci $(\pm 4, 0)$, asymptotes $y = \pm \frac{2}{3}x$.

19. Vertices $(0, \pm 5)$, passes through $(2, 5\sqrt{13}/3)$.

20. Transverse axis horizontal, center at the origin, passes through $(3, \sqrt{5}/2)$ and $(4, \sqrt{10})$.

Exercises 21–24: Sketch the graph.

21. $xy = 2$ 22. $xy = 6$

23. $xy = -3$ 24. $xy = -4$

25. A precast concrete beam has a cross section in the form of a hyperbola (see Figure 2-35).

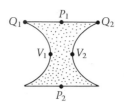

Figure 2-35

Measurements between points are taken as follows:

between V_1 and V_2: 0.08 m

between P_1 and P_2: 0.26 m

between Q_1 and Q_2: 0.30 m

Find the equation of the hyperbola.

26. A radio-transmitting station intermittently sends out pulse signals, which travel directly to a ship at sea and to a second station 250 km from the first. The second station relays these signals. The ship is moving in such a way that the relayed signal is always received by the ship 0.001 s after the original signal. Assume that the speed of transmission of the radio signal is 3×10^5 km/s and find the equation of the ship's path of motion. (*Hint:* Make the two stations the foci. Put the origin of the coordinate system midway between them.)

27. In an optical lens, the power D (in diopters) and the focal length f (in meters) of the lens are related by the equation $Df = 1$. Sketch a graph of D as a function of f.

28. In an ideal gas, when the temperature remains constant, pressure P and volume V are related by the equation $PV = k$. If a certain gas under 15 atmospheres (atm) of pressure occupies 100 L, graph P as a function of V for constant temperature.

2-6 **The Second-Degree Equation**

The circle, ellipse, parabola, and hyperbola were all defined in terms of the path of a point moving according to some fixed rule. However, these four types of curves share two other common features, one geometric and one algebraic. Geometrically, all four curves can be

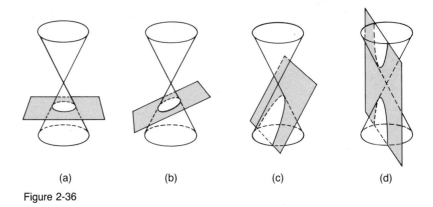

(a)   (b)   (c)   (d)

Figure 2-36

obtained by intersecting a plane with a right circular cone. If the plane is perpendicular to the axis of the cone, the intersection is a circle (Figure 2-36(a)). If the plane is tipped a bit, the intersection is an ellipse (Figure 2-36(b)). If the plane is further tipped, so that it is parallel to the edge of the cone, the intersection is a parabola (Figure 2-36(c)). And, finally, if the plane is tipped to the point where it intersects both halves of the cone, a hyperbola is formed (Figure 2-36(d)). For this reason, the circle, ellipse, parabola, and hyperbola are often called **conic sections**.

Special cases, called **degenerate conic sections**, are obtained by intersecting a plane and a cone. For example, a horizontal plane passed through the point between the two halves of the cone will produce an intersection of a single point.

Algebraically, each of the four curves is a special case of a single equation. The characteristics of the equations for each of the four curves are as follows:

1. Parabola: One variable squared, one linear.
2. Ellipse: Both variables squared, both coefficients positive.
3. Circle: Both variables squared with identical coefficients.
4. Hyperbola: Both variables squared with opposite signs or of the form $xy = k$.

Each of these equations is a special case of the general second-degree equation

$$Ax^2 + Bxy + Cy^2 + Dx + Ey + F = 0 \tag{1}$$

Until now we have dealt only with conic sections where the center (or vertex, in the case of the parabola) is at the origin of the coordinate system and the axes of the curve are horizontal and vertical. The equations of these conic sections are the simplest forms (called **standard forms**) for equations of the conic sections. We can also find equations for conic sections where the axes remain horizontal and vertical, but the center (vertex) is not at the origin. These equations, while involving more terms than the standard forms, will still be instances of Equation (1).

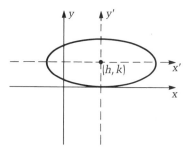

Figure 2-37

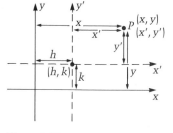

Figure 2-38

How do we find the equation for a conic section with its center or vertex at a point (h, k), for example, where (h, k) is not the origin? One way is to go back to the original definition of the particular conic and derive the equation, much as we did when the center or vertex was at the origin. This process involves a lot of squaring of radicals and algebraic simplification. An easier way to find equations of this type involves introducing a new (temporary) coordinate system. For example, in Figure 2-37, the ellipse shown has its center at (h, k). We want to find its equation in terms of the xy-coordinate system. We introduce a new $x'y'$-coordinate system with its origin at (h, k). We can easily write the equation for the ellipse in terms of x' and y', because this is the case where the center of the ellipse coincides with the origin of the coordinate system. If we convert this equation into one involving x and y rather than x' and y', we are done. We need a relationship between xy-coordinates and $x'y'$-coordinates.

In Figure 2-38, the point P has coordinates (x, y) in one system and (x', y') in the other system. We can see that

$$x = x' + h \qquad y = y' + k$$

or

$$\boxed{x' = x - h \qquad y' = y - k} \tag{2}$$

These equations hold no matter what quadrant (h, k) is in. This process of introducing a new coordinate system parallel to the original one is called **translation of axes**. A translation greatly simplifies the task of writing the equation of a conic section with center (or vertex) not at the origin and horizontal and vertical axes.

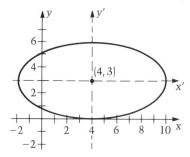

Figure 2-39

EXAMPLE 1. Write the equation of the ellipse with center at $(4, 3)$, a horizontal major axis of length 12, and a minor axis of length 6.

In an $x'y'$-coordinate system with origin at $(4, 3)$, the ellipse has its center at the origin (Figure 2-39). Its equation is

$$\frac{(x')^2}{36} + \frac{(y')^2}{9} = 1$$

Now we make use of Equations (2) with $h = 4$ and $k = 3$.

$$\frac{(x - 4)^2}{36} + \frac{(y - 3)^2}{9} = 1$$

This is the equation of the ellipse in the original xy-coordinate system. After simplifying, this equation becomes

$$x^2 + 4y^2 - 8x - 24y + 16 = 0 \qquad \blacksquare$$

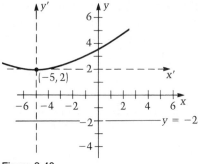

Figure 2-40

EXAMPLE 2. Write the equation of the parabola with vertex at $(-5, 2)$ and directrix $y = -2$.

The $x'y'$-coordinate system will have its origin at $(-5, 2)$. With respect to this system, the parabola has its vertex at the origin (Figure 2-40). The distance between the vertex and the directrix is 4 units, and the parabola opens upward. In the $x'y'$-coordinate system, $p = 4$ and $4p = 16$, so the equation is

$$(x')^2 = 16y'$$

With $(h, k) = (-5, 2)$, we rewrite this equation in the xy-coordinate system and get

$$(x + 5)^2 = 16(y - 2)$$

or

$$x^2 + 10x - 16y + 57 = 0 \quad \blacksquare$$

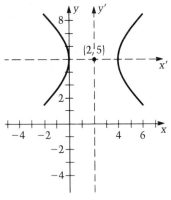

Figure 2-41

EXAMPLE 3. Write the equation of the hyperbola with foci at $(-1, 5)$ and $(5, 5)$ and a transverse axis of length 4 units.

The $x'y'$-coordinate system has its origin at $(2, 5)$ (Figure 2-41). In this hyperbola, $a = 2$ and $c = 3$, so that $b^2 = 5$. The equation of the hyperbola with respect to the $x'y'$-coordinate system is

$$\frac{(x')^2}{4} - \frac{(y')^2}{5} = 1$$

With $(h, k) = (2, 5)$, we rewrite this equation in the xy-coordinate system and get

$$\frac{(x - 2)^2}{4} - \frac{(y - 5)^2}{5} = 1$$

or

$$5x^2 - 4y^2 - 20x + 40y - 100 = 0 \quad \blacksquare$$

Practice 1. Write the equation of the ellipse with center at $(3, -1)$, focus at $(3, 1)$, and vertex at $(3, 2)$.

From the above examples, we note that for any conic section that has been translated from the origin (moved horizontally, vertically, or both), the equation is of the form

$$Ax^2 + Bxy + Cy^2 + Dx + Ey + F = 0 \qquad (1)$$

with $B = 0$. Equations for each type of conic section still have their own characteristics. For example, an equation of the form of (1) with A and C both positive still represents an ellipse, although if $D \neq 0$ or $E \neq 0$, the center is not at the origin. The xy-term ($B \neq 0$) in Equation (1) is introduced by rotating the curve; we have dealt with this only in the case of equations of the form $xy = k$, which represent hyperbolas that have been rotated about the origin. We will not consider any other rotation equations.

The degenerate cases are part of Equation (1) also. To get a single point, let $A = 1$, $B = 0$, $C = 1$, and $D = E = F = 0$ in Equation (1). This results in the equation $x^2 + y^2 = 0$, which initially looks like a circle because it has both variables squared with identical coefficients. A moment's thought, though, tells us that the only solution is the single point $(0, 0)$ (which is a degenerate circle with center at the origin and radius 0).

We have found the equation of a curve, given some geometric information. This involved changing an equation in the $x'y'$-coordinate system to a corresponding equation in x and y. The other half of analytic geometry asks that we start with an equation and graph the curve. For any form of Equation (1) with $B = 0$, we should be able to recognize whether its graph is a circle, ellipse, parabola, or hyperbola. To sketch the graph, we essentially reverse what we have just done: We try to change the xy-equation to a corresponding $x'y'$-equation, where the point (h, k) is chosen to make it easy to graph the $x'y'$-equation on the $x'y'$-coordinate system. Once we have graphed the equation on the $x'y'$-plane, we have also graphed it on the xy-plane.

EXAMPLE 4.   Sketch the graph of the equation

$$x^2 + 4y^2 - 8x - 24y + 16 = 0$$

We recognize from the form of the equation (both variables squared, with positive coefficients) that it represents an ellipse. This particular equation happens to be the one we found in Example 1. In fact, our problem here is the reverse of that in Example 1; this time we want to start with the equation and get to the graph. We can easily graph the equation if we can rewrite it in the form

$$\frac{(x - 4)^2}{36} + \frac{(y - 3)^2}{9} = 1$$

because we recognize that the translation $x' = x - 4$ and $y' = y - 3$ would reduce the equation to the form

$$\frac{(x')^2}{36} + \frac{(y')^2}{9} = 1$$

We graph this on a coordinate system with center $(4, 3)$ and we are done (see Figure 2-39). Therefore, the only hard part is the initial rewriting stage.

From the original equation, we want to introduce terms of the form $(x - h)^2$ and $(y - k)^2$. This can be done by the process of *completing the square*. We first isolate terms involving x and terms involving y (factoring out the coefficient of y^2).

$$x^2 + 4y^2 - 8x - 24y + 16 = 0$$
$$x^2 - 8x \quad\ + 4(y^2 - 6y \quad\ ) = -16$$

We recognize $x^2 - 8x$ as the first two terms in $(x - 4)^2$ and $y^2 - 6y$ as the first two terms in $(y - 3)^2$. We add the missing third term in each case to the left side of the equation; to maintain equality, we must add the same thing to the right side of the equation.

$$x^2 - 8x + 16 + 4(y^2 - 6y + 9) = -16 + 16 + 4 \cdot 9$$
$$(x - 4)^2 + 4(y - 3)^2 = 36$$
$$\frac{(x - 4)^2}{36} + \frac{(y - 3)^2}{9} = 1 \quad \blacksquare$$

Remember that to complete the square of an expression of the form $x^2 - kx$, we add a term of the form $(k/2)^2$—that is, we find half of the coefficient of x and square that, because

$$x^2 - kx + \left(\frac{k}{2}\right)^2 = \left(x - \frac{k}{2}\right)^2$$

Word of Advice

To graph an equation of the form

$$Ax^2 + Cy^2 + Dx + Ey + F = 0$$

we first decide what type of conic it represents, and then we complete the square to rewrite the equation in one of these forms. Parabola:

$$(y - k)^2 = 4p(x - h) \quad \text{or}$$
$$(x - h)^2 = 4p(y - k)$$

Ellipse:

$$\frac{(x-h)^2}{a^2} + \frac{(y-k)^2}{b^2} = 1 \quad \text{or}$$

$$\frac{(y-k)^2}{a^2} + \frac{(x-h)^2}{b^2} = 1$$

Circle:

$$(x-h)^2 + (y-k)^2 = a^2$$

Hyperbola:

$$\frac{(x-h)^2}{a^2} - \frac{(y-k)^2}{b^2} = 1 \quad \text{or}$$

$$\frac{(y-k)^2}{a^2} - \frac{(x-h)^2}{b^2} = 1$$

EXAMPLE 5. Sketch the graph of the equation

$$y^2 - 4y + 6x - 8 = 0$$

The equation represents a parabola, and we complete the square to obtain an equation of the form $(y-k)^2 = 4p(x-h)$.

$$y^2 - 4y \qquad = -6x + 8$$
$$y^2 - 4y + 4 = -6x + 12$$
$$(y-2)^2 = -6(x-2)$$

Letting $x' = x - 2$ and $y' = y - 2$, we get

$$(y')^2 = -6x'$$

This represents a parabola with vertex at $(2, 2)$, opening to the left. The graph is shown in Figure 2-42. ∎

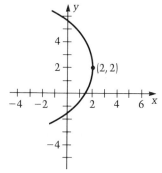

Figure 2-42

Practice 2. Identify the type of conic (parabola, ellipse, circle, or hyperbola), give the coordinates of the center (or vertex in the case of the parabola), and sketch the curve.

$$4x^2 - 5y^2 - 16x - 10y + 91 = 0$$

Conic sections are rather specialized and well-behaved curves with equations that are relatively easy to analyze and graph. Graphing other functions may require much more extensive use of the calculator and plotting of individual points, although an additional tool for analysis will be available after we study the derivative. Conic sections, however, do arise as mathematical models to describe a wide variety of physical situations, from orbits of planets to paths of projectiles and from electromagnetic force fields to motion of atomic particles.

Exercises / Section **2-6**

Exercises 1–16: Write the equation of the curve described.

1. Parabola: vertex $(2, 5)$, directrix $x = -3$.

2. Parabola: vertex $(-2, 1)$, focus $(-5, 1)$.

3. Parabola: focus $(-2, -7)$, directrix $y = -3$.

4. Parabola: focus $(3, -1)$, directrix $y = -7$.

5. Ellipse: center $(2, 6)$, focus $(5, 6)$, vertex $(7, 6)$.

6. Ellipse: foci $(-4, 1)$ and $(-4, -7)$, vertices $(-4, 2)$ and $(-4, -8)$.

7. Ellipse: foci $(5, 3)$ and $(5, -7)$, major axis of length 16.

8. Ellipse: vertices $(-8, 3)$ and $(6, 3)$, minor axis of length 6.

9. Circle: center $(3, 4)$, radius 2.

10. Circle: center $(-1, -1)$, radius $\sqrt{7}$.

11. Circle: center $(-5, -8)$, tangent to the y-axis.

12. Circle: center $(1, -6)$, passes through $(4, -2)$.

13. Hyperbola: foci $(1, -2)$ and $(7, -2)$, vertices $(3, -2)$ and $(5, -2)$.

14. Hyperbola: foci $(-5, 9)$ and $(-5, 1)$, vertices $(-5, 8)$ and $(-5, 2)$.

15. Hyperbola: center $(6, 4)$, focus $(6, -1)$, transverse axis of length 8.

16. Hyperbola: vertex $(-2, -2)$, asymptotes $4x - 3y + 14 = 0$ and $4x + 3y + 26 = 0$.

Exercises 17–32: Identify the type of conic (parabola, ellipse, circle, or hyperbola), give the coordinates of the center (or vertex in the case of the parabola), and sketch the curve.

17. $x^2 - 9y^2 + 8x + 7 = 0$

18. $4x^2 + y^2 - 24x + 4y + 36 = 0$

19. $x^2 + y^2 + 6x - 2y + 3 = 0$

20. $y^2 - 6x + 4y - 14 = 0$

21. $x^2 + y^2 - 2x - 3 = 0$

22. $5x^2 + 9y^2 - 30x - 54y - 54 = 0$

23. $x^2 - 6x - 4y + 9 = 0$

24. $3x^2 - 4y^2 + 12x + 8y - 4 = 0$

25. $2(2x^2 + 8x - 9y) = 11 - 9y^2$

26. $x^2 + 2y + 1 = -4(2x + 1) - y^2$

27. $16x^2 + 9(2y - y^2 + 31) = 96x$

28. $4y^2 = 8y - x + 1$

29. $2(2y^2 - 49x + 18) = 8y - 49x^2 + 179$

30. $2(4x^2 + y^2) + 3y + 16x + 12 = 2(y + 1)^2$

31. $2(2x^2 - 3y + 1) = y^2 + 16x - 9$

32. $(x - 2)^2 - 2y + y^2 + 21 = 6(x - y)$

33. The resistance R (in ohms) of a platinum wire used in a platinum resistance thermometer as a function of temperature T (in degrees Celsius) is given by the equation $R = 11[1 + (4.86 \times 10^{-3})T - (1.05 \times 10^{-5})T^2]$. What type of curve does this equation represent?

34. One leg of a right triangle is 2 units longer than the other. Write the equation relating the length h of the hypotenuse and the length L of the shorter leg. Sketch a graph of the equation.

35. The cross section of a radar dish is part of one branch of a hyperbola. The equation of the hyperbola in a coordinate system with the origin at one vertex is $x^2 - 4y^2 - 24y = 0$. Find the equation of the curve in a coordinate system with the origin at the center of the hyperbola.

36. Measurements of the amount R (in inches) of rainfall at a certain location as a function of the month M of the year indicate a parabolic relationship. From Figure 2-43, write the equation relating R and M.

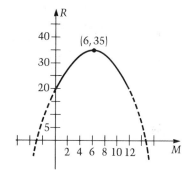

Figure 2-43

STATUS CHECK

Now that you are at the end of Chapter 2, you should be able to:

section **2-2** Write the equation of a straight line, given:

    1. a point on the line and the slope of the line;
    2. two points on the line;
    3. a point on the line and the equation of a parallel line or a perpendicular line.

Find the slope of a line from its equation in the form
    $Ax + By + C = 0$.
Sketch the graph of a line from its equation in the form
    $Ax + By + C = 0$.
Given the equations of two lines, prove that the lines are parallel or perpendicular.
From given information, write a linear equation relating two quantities, and use that equation to find additional information.

section **2-3** Sketch the graph of a parabola from its equation in standard form.
Write the equation of a parabola with vertex at the origin and horizontal or vertical axis, given:

    1. the focus and directrix;
    2. the vertex and directrix;
    3. the vertex and focus.

From given information, write the equation of a parabolic object or a parabolic equation relating two quantities and use that equation to find additional information.

section **2-4** Sketch the graph of an ellipse from its equation in standard form.
Sketch the graph of a circle from its equation in standard form.
Write the equation of an ellipse with center at the origin and major axis horizontal or vertical, given:

    1. the foci and vertices;
    2. the vertices or foci and length of an axis;
    3. two points on the ellipse.

Write the equation of a circle with center at the origin, given its radius.

From given information, write the equation of an elliptical object and use that equation to find additional information.

section **2-5**   Sketch the graph of a hyperbola from its equation in standard form.

Write the equation of a hyperbola with center at the origin and transverse axis horizontal or vertical, given:

1. the foci and vertices;
2. the foci and length of an axis;
3. the vertices or foci and equations of asymptotes;
4. two points on the hyperbola.

Sketch the graph of a hyperbola with equation $xy = k$.

From given information, write the equation of a hyperbolic object or a hyperbolic equation relating two quantities and use that equation to find additional information.

section **2-6**   From given information, write the equation of a conic section translated from the origin.

Given an equation of the form $Ax^2 + Cy^2 + Dx + Ey + F = 0$, identify the type of conic section and sketch the graph.

2-7 More Exercises for Chapter 2

Exercises 1–16: Find the equation of the curve described.

1. Straight line; slope $-\frac{1}{3}$, passes through $(2, 4)$.

2. Straight line; passes through $(-1, -4)$ and $(-3, 7)$.

3. Straight line; parallel to $2x + 3y - 6 = 0$, passes through $(-1, 6)$.

4. Straight line; perpendicular to $x + 3y + 5 = 0$ with y-intercept $(0, 3)$.

5. Parabola; focus $(0, 5)$, directrix $y = -5$.

6. Parabola; vertex $(0, 0)$, axis $x = 0$, passes through $(-\frac{4}{3}, 4)$.

7. Circle; center at the origin, radius = 3.

8. Circle; center at the origin, radius = 12.

9. Ellipse; foci $(\pm 3, 0)$, major axis of length 10.

10. Ellipse; center at the origin, vertex $(0, -4)$, passes through $(1, -2\sqrt{3})$.

11. Hyperbola; foci $(0, \pm 10)$, conjugate axis of length 10.

12. Hyperbola; vertices $(\pm 8, 0)$, asymptotes $y = \pm\frac{1}{2}x$.

13. Parabola; vertex $(4, -3)$, focus $(4, -1)$.

14. Circle; center $(-3, 2)$, tangent to the x-axis.

15. Ellipse; foci $(-7, -3)$ and $(-1, -3)$, vertices $(-9, -3)$ and $(1, -3)$.

16. Hyperbola; center at $(3, 6)$, focus at $(3, 10)$, minor axis of length 4.

Exercises 17–26: Sketch the graph of the equation and find the indicated quantities.

17. $2x - 4y + 1 = 0$; slope.

18. $12x + 3y - 18 = 0$; slope.

19. $y^2 = 12x$; coordinates of focus, equation of directrix.

20. $x^2 = -4y$; coordinates of focus, equation of directrix.

21. $x^2 + y^2 = 25$; radius.

22. $\dfrac{x^2}{4} + \dfrac{y^2}{4} - 1 = 0$; radius.

23. $\dfrac{x^2}{49} + \dfrac{y^2}{81} = 1$; coordinates of vertices and foci.

24. $x^2 + 16y^2 = 16$; coordinates of vertices and foci.

25. $\dfrac{x^2}{49} - \dfrac{y^2}{16} = 1$; coordinates of vertices and foci.

26. $5y^2 - 2x^2 = 10$; coordinates of vertices and foci.

Exercises 27–34: Identify the type of curve (parabola, circle, ellipse, or hyperbola), give the coordinates of the center (or vertex in the case of the parabola) and sketch the curve.

27. $x^2 - 2x - 3y - 8 = 0$

28. $x^2 + y^2 - 2x - 4 = 0$

29. $x^2 - 7y^2 + 6x + 28y - 33 = 0$

30. $2x^2 + 5y^2 - 8x + 10y + 3 = 0$

31. $x^2 + y^2 + 4x + 6y + 6 = 0$

32. $5x^2 - 7y^2 + 30x - 56y - 32 = 0$

33. $16x^2 + 6y^2 + 64x + 12y - 26 = 0$

34. $y^2 - 2x - 8y + 14 = 0$

35. A cable suspended from two supports hangs in (approximately) the shape of a parabola. If the supports are 20 m apart and the cable dips 8 m at its lowest point, find the height of the cable above the lowest point at a distance 2 m from one of the supports.

36. A cross section of the ceiling of a convention center is in the form of half an ellipse. The room is 100 m wide, the walls are 4 m high, and the ceiling is 7 m high in the center of the room. Find the height of a support post located 20 m in from a wall.

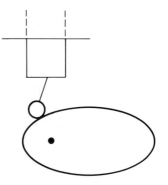

Figure 2-44

37. A brick patio is designed in the shape of an ellipse 8 m long and 6 m wide. The area of the patio (and the number of bricks to order) can be found by using calculus once the equation of the ellipse is known. Write the equation of the ellipse with center at the origin and major axis along the x-axis.

38. An elliptical cam rotates about one of its focal points. One end of a shaft rides along the edge of the ellipse, and the other end is attached to a piston (see Figure 2-44). In a coordinate system with the origin at the center of the ellipse, the ellipse—as shown in Figure 2-44—has the equation $x^2 + 7y^2 = 28$. Find the stroke of the piston (the maximum up-and-down distance it moves) as the ellipse revolves.

39. The earth's orbit is elliptical with the sun at one focal point. If the least distance between the earth and the sun is 1.47×10^8 km and the greatest distance between the earth and the sun is 1.52×10^8 km, write the equation of the orbit. (Assume the origin is at the center of the ellipse and the major axis is along the x-axis.)

chapter **3**

New Coordinate Systems

In the polar coordinate system, it is easy to represent spiral curves. This shell shows a beautiful three-dimensional spiral.

3-1 Graphing in Three Dimensions

We have used a rectangular coordinate system to graph functions of the form $y = f(x)$, functions with one independent variable. However, we know that a function can have several independent variables. In order to graph a function of the form $z = f(x, y)$, we need three coordinate axes, one for each of the independent variables x and y, and one for the dependent variable z. We use a three-dimensional coordinate system, where the three coordinate axes meet at right angles. Figure 3-1 shows the coordinate system we shall use. Remember that we are trying to picture a three-dimensional object on a two-dimensional surface, the page, so a little imagination is required. The coordinate system divides space into eight sections, four above the xy-plane and four below. It may help to picture the first octant of the coordinate system—that section in which x, y, and z all have positive values—as the corner of a room. The xy-plane is the floor of the room, and the xz-plane and zy-plane are the walls.

Points in space are located by coordinates that consist of an ordered triple of numbers (x, y, z), where the x-coordinate measures distance from the zy-plane, the y-coordinate measures distance from

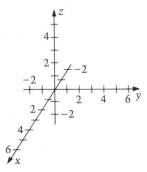

Figure 3-1

66

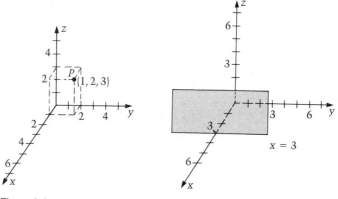

Figure 3-2   Figure 3-3

the xz-plane, and the z-coordinate measures distance from the xy-plane. Figure 3-2 shows the point with coordinates $(1, 2, 3)$.

The simplest curve in two dimensions is the straight line, which has the general equation $Ax + By + C = 0$. In three dimensions, the simplest surface is the plane, which has the general equation

$$Ax + By + Cz + D = 0 \qquad (1)$$

If we are given an equation of this form, how can we graph the plane it represents? There are three possible cases.

1. All but one of A, B, and C are zero.

For example, $2x - 6 = 0$, or $x = 3$, falls into this category. For the equation $x = 3$, in a three-dimensional coordinate system, x has the value 3 for any values of y and z. The graph is a plane parallel to the zy-plane with x-value 3; part of this plane is shown in Figure 3-3.

2. One of A, B, or C is zero.

For example, $3x + 2z = 6$ has this property. For this equation, in a three-dimensional coordinate system, the relationship $3x + 2z = 6$ holds between x and z for every value of y. In particular, when $y = 0$, $3x + 2z = 6$; this is the equation of a straight line in the xz plane and represents the intersection of the surface with the plane $y = 0$. For any plane perpendicular to the y-axis, its intersection with the surface will also have the equation $3x + 2z = 6$. Figure 3-4 illustrates part of the surface in the first octant.

3. All of A, B, and C are nonzero.

Notice that in Case 1, the plane intersects only one of the coordinate axes. In Case 2, the plane intersects two of the coordinate axes. In Case 3, the plane intersects all three coordinate axes. The easiest way to graph the plane is to find those three intersection points and connect them.

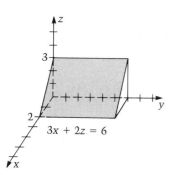

Figure 3-4

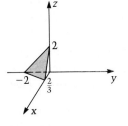

Figure 3-5

Figure 3-6

EXAMPLE 1. Sketch the graph of $2x + 4y + z = 8$.
The three intercept points, determined from the equation, are $(4, 0, 0)$, $(0, 2, 0)$, and $(0, 0, 8)$. These points are plotted and then connected (see Figure 3-5). Notice that, just as in two dimensions, we often show only part of the graph. ■

EXAMPLE 2. Sketch the graph of $3x - y + z = 2$.
The intercepts are $(\frac{2}{3}, 0, 0)$, $(0, -2, 0)$, and $(0, 0, 2)$. The plane is represented in Figure 3-6. ■

Practice 1.   Sketch the graph of $4x + 2y + z - 2 = 0$.

In graphing surfaces, it is helpful to sketch the curve that is the intersection of the surface with the coordinate planes. Such a curve is called a **trace** of the surface. The traces of a plane will always be straight lines, but traces may be more complex curves for more complex surfaces. It can also be helpful to visualize the curve formed by the intersection of the surface with a plane parallel to one of the coordinate planes. Such a curve is called a **section** of the surface.

EXAMPLE 3.   Sketch the graph of $z = 9 - x^2 - y^2$.
The three traces of the surface are the curves formed by the intersection of the surface with the xz-plane ($y = 0$), the zy-plane ($x = 0$), and the xy-plane ($z = 0$). The equations of the traces are found by setting each of the three variables to zero, in turn, in the equation of the surface.

xz-plane:	$z = 9 - x^2$	Parabola, vertex at $x = 0$, $z = 9$, opens downward.
zy-plane:	$z = 9 - y^2$	Parabola, vertex at $y = 0$, $z = 9$, opens downward.
xy-plane:	$x^2 + y^2 = 9$	Circle, center at origin, radius 3.

A section parallel to the xy-plane can be found by setting z equal to a constant k. The equation is

section parallel to xy-plane: $x^2 + y^2 = (9 - k)$

The section is a circle with center at the origin and a radius that decreases for increasing values of k, with the maximum value of k

being 9. Thus horizontal cross sections of the surface are smaller and smaller circles as we move up the z-axis. The surface is called a **circular paraboloid** (see Figure 3-7, below). ∎

EXAMPLE 4. Sketch the graph of $9x^2 - 16y^2 + 4z^2 = 144$. The equations of the traces are

xz-plane:	$9x^2 + 4z^2 = 144$	Ellipse, center at origin, major axis 12 units along z-axis, minor axis 8 units.
zy-plane:	$4z^2 - 16y^2 = 144$	Hyperbola, center at origin, transverse axis 12 units along z-axis.
xy-plane:	$9x^2 - 16y^2 = 144$	Hyperbola, center at origin, transverse axis 8 units along x-axis.

Any section parallel to the xz-plane is an ellipse with major axis along the z-axis; the ellipse grows with larger positive or negative values of y. The surface is called a **hyperboloid of one sheet** (see Figure 3-8). ∎

Three-dimensional surfaces are difficult to draw. Keep in mind that we are graphing only relatively simple, regular surfaces here, and three-dimensional surfaces can be much more complex. Excellent representations of three-dimensional surfaces can be generated by means of computer graphics.

Practice 2. Sketch the graph of $z = x^2 + y^2$.

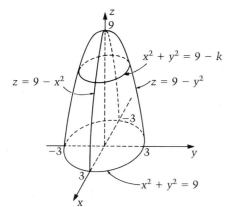

Figure 3-7

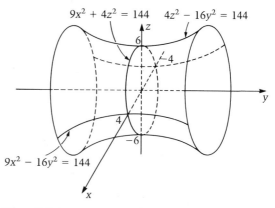

Figure 3-8

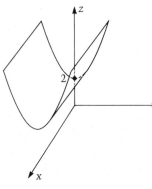

Figure 3-9

A special type of surface occurs when the equation contains only two variables. This means that for any plane perpendicular to the third variable, the equation relating the other two variables still holds. The surface is **cylindrical**.

EXAMPLE 5. Sketch the graph of the surface $y^2 = z - 2$.

Any section perpendicular to the x-axis has the equation $y^2 = z - 2$, a parabola opening upward. Part of the surface appears in Figure 3-9. Of course, we must know that we are working in three dimensions, not two. Merely asking for a sketch of the graph of $y^2 = z - 2$ does not tell us whether a two-dimensional curve or a three-dimensional surface is to be found. ■

Practice 3. Sketch the graph of $x^2 + y^2 = 4$ in three dimensions.

Exercises / Section **3-1**

Exercises 1–18: Sketch the graph in a three-dimensional coordinate system.

1. $y = 2$
2. $z = 4$
3. $2x - 3y = 6$
4. $2y + 5z = 10$
5. $2x + y + 4z = 8$
6. $x - 3y + 6z = 6$
7. $2x + y - 2z + 4 = 0$
8. $3x - 3y - z + 3 = 0$
9. $z = 4 - x^2 - y^2$
10. $x = 2z^2 + y^2$
11. $x^2 + 4y^2 + z^2 = 16$
12. $4x^2 + y^2 + 9z^2 = 36$
13. $4x^2 + 4y^2 - 9z^2 = 36$
14. $4x^2 - y^2 - 25z^2 + 100 = 0$
15. $x^2 + y^2 - 4z^2 = 0$

16. $z = x^2 + 4y^2 + 2$
17. $z^2 + 4y^2 = 4$
18. $x^2 + z = 3$

19. The inlet pipe in a heat-exchanger tank has the equation $y^2 + z^2 = 25$ (in three dimensions). The end of the pipe is cut at an angle, as if a plane were passed through the pipe. Three points of intersection of the plane and cylinder are $(0, 0, 5)$, $(6, 4, -3)$, and $(6, -4, -3)$. Find the equation of the plane.

20. The perimeter P of a rectangle is a function of the length L and the width W. Sketch the graph of P as a function of L and W.

21. The velocity profile for Poiseuille flow (laminar) in a pipe has the equation $x = 9[1 - (y^2 + z^2)]$. Sketch the graph of this surface.

22. A surface is given by the equation $z = y^2 x^2$. What is the shape of a horizontal section of the curve for $z > 0$? For $z = 0$? For $z < 0$? Can you visualize why this surface is called a **saddle**?

3-2 Polar Coordinates

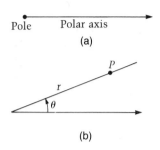

Pole   Polar axis

(a)

(b)

Figure 3-10

All our two-dimensional graphs so far have been drawn on a rectangular coordinate system. This is the coordinate system that is most frequently used, but there are occasions when another coordinate system is more convenient.

In a **polar coordinate system**, we have a fixed point, called the **pole**, and a fixed line, called the **polar axis**, as in Figure 3-10(a). If we put an angle θ with its vertex at the pole and its initial side on the polar axis, then a point P on its terminal side is completely specified by giving the angle θ and the radius vector r (Figure 3-10(b)). The ordered pair (r, θ) is a set of **polar coordinates** for P. For any given (r, θ), there is only one corresponding point in the plane. But, unlike the situation in rectangular coordinates, a single point in the plane has more than one set of polar coordinates.

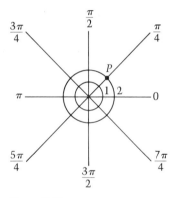

Figure 3-11

EXAMPLE 1. The point $(2, \pi/4)$ shown in Figure 3-11 also has polar coordinates of $(2, 9\pi/4)$, $(2, -7\pi/4)$, $(2, -15\pi/4)$, and so on, because all these angles have the same terminal side as $\pi/4$. It is also customary, in polar coordinates, to allow r to take on negative values. A negative value for r in a set (r, θ) of polar coordinates means that the point, instead of being located on the terminal side of θ, is located $|r|$ units along the extension of the terminal side through the pole. Therefore, point P in Figure 3-11 also has coordinates $(-2, 5\pi/4)$, $(-2, -3\pi/4)$, and so on. ∎

A single point clearly has an infinite number of polar coordinates, although any one set of polar coordinates describes a unique point.

Practice 1. Plot the points $(1, \pi/2)$, $(2, 5\pi/6)$, and $(-1, \pi/3)$ in a polar coordinate system.

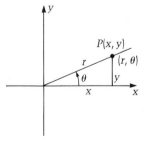

Figure 3-12

Figure 3-12 shows a point P with rectangular coordinates (x, y) and polar coordinates (r, θ). From the figure, we see that

$$r = \sqrt{x^2 + y^2} \qquad \tan \theta = \frac{y}{x} \tag{1}$$

and also that

$$\cos \theta = \frac{x}{r} \qquad \sin \theta = \frac{y}{r}$$

or

$$x = r \cos \theta \qquad y = r \sin \theta \tag{2}$$

Equations (1) are useful when we know the rectangular coordinates of a point and want to find a set of polar coordinates for it. Equations (2) are useful for the opposite case, when we know a set of polar coordinates and want the rectangular coordinates. (Some calculators provide for direct conversion between rectangular and polar coordinates.)

EXAMPLE 2. Find rectangular coordinates for the point with polar coordinates $(3, \pi/6)$.

Using Equations (2),

$$x = 3 \cos \frac{\pi}{6} = 2.598$$

$$y = 3 \sin \frac{\pi}{6} = 1.5$$

Of course, we can use a calculator to find the trigonometric function values required. We would get the same xy-coordinates if we had started with equivalent polar coordinates such as $(3, 13\pi/6)$, or $(-3, 7\pi/6)$. ■

EXAMPLE 3. Find polar coordinates for the point with rectangular coordinates $(1, 2)$.

Using Equations (1),

$$r = \sqrt{x^2 + y^2} = \sqrt{1^2 + 2^2} = \sqrt{5} = 2.24$$

$$\tan \theta = \frac{y}{x} = \frac{2}{1} = 2.0$$

Now how do we find θ, given $\tan \theta$? The TAN^{-1} key, if your calculator has one, will do the job; otherwise use first the INV or ARC key and then the TAN key. The result is

$$\tan \theta = 2.0$$
$$\theta = 1.11$$

Thus $\theta = 1.11$ (in radian measure) is an angle whose tangent is 2.0. One set of polar coordinates for the point is $(2.24, 1.11)$. ■

For any value of the tangent function, your calculator will give you only an angle between $-\pi/2$ and $\pi/2$—that is, a negative angle in quadrant IV if the tangent function is negative, or a positive angle in quadrant I if the tangent function is positive. If the desired angle is not in quadrants IV or I, some adjustments are required.

EXAMPLE 4. Find polar coordinates for the point with rectangular coordinates $(-1, 2)$.

First we find r:

$$r = \sqrt{(-1)^2 + 2^2} = \sqrt{5} = 2.24$$

Next we find $\tan \theta$:

$$\tan \theta = \frac{y}{x} = -2.0$$

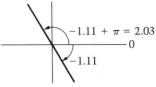

Note that since x is negative and y is positive, the terminal side of θ is in quadrant II. Using a calculator, we find an angle with tangent value of -2.0 to be -1.11, which is a fourth-quadrant angle. The angle in quadrant II with the same tangent value (see Figure 3-13) is $-1.11 + \pi = 2.03$. One set of polar coordinates for the point is $(2.24, 2.03)$. (Another set is $(-2.24, -1.11)$.) ■

Figure 3-13

Practice 2. Find rectangular coordinates for the point with polar coordinates $(-2, 3\pi/4)$.

Practice 3. Find a set of polar coordinates for the point with rectangular coordinates $(-3, 1)$.

Equations (1) and (2) can also be used to change the equation of a curve expressed in rectangular coordinates to one expressed in polar coordinates, and vice versa. Curves that have a fairly simple equation in one system may have a fairly complex equation in the other system.

EXAMPLE 5. Find a polar equation of the circle $x^2 + y^2 = 9$.

Because the circle has its center at the origin and a radius of 3 units, the radius vector is always 3 for any value of θ, and the equation is $r = 3$. However, let us use Equation (2) instead.

$$x^2 + y^2 = 9$$
$$r^2 = 9$$
$$r = \pm 3$$

The equation $r = -3$ is another polar equation for the circle. In either case, the polar form of the equation is simpler than the rectangular form. ■

EXAMPLE **6.** Find the rectangular form of the equation $r = 1 - \cos\theta$.

Using Equations (1) and (2), $r = 1 - \cos\theta$ becomes

$$\sqrt{x^2 + y^2} = 1 - \frac{x}{r} = 1 - \frac{x}{\sqrt{x^2 + y^2}}$$

Multiplying both sides by $\sqrt{x^2 + y^2}$, we get

$$x^2 + y^2 = \sqrt{x^2 + y^2} - x$$
$$x^2 + y^2 + x = \sqrt{x^2 + y^2}$$

or

$$(x^2 + y^2 + x)^2 = x^2 + y^2$$

This is pretty awful, especially if we were to actually try and graph the equation. The original equation, however, is not very difficult to graph in a polar coordinate system, especially with the help of a calculator. ■

Practice 4. Find a polar form of the equation $(x^2 + y^2)^2 = x^2 - y^2$.

Practice 5. Find the rectangular form of the equation $r = 3\cos\theta$.

Exercises / Section **3-2**

Exercises 1–8: Plot the point in a polar coordinate system.

1. $(1, \pi/3)$
2. $(2, 3\pi/4)$
3. $(4, -5\pi/6)$
4. $(2, -7\pi/4)$
5. $(-2, \pi/2)$
6. $(-1, 2\pi/3)$
7. $(-0.8, -\pi)$
8. $(-1.2, -1)$

Exercises 9–16: Find rectangular coordinates for the point whose polar coordinates are given.

9. $(2, \pi/5)$
10. $(3, 7\pi/6)$
11. $(1, -\pi/3)$
12. $(2, -8\pi/5)$
13. $(-2, 4\pi/7)$
14. $(-3, 5\pi/4)$
15. $(-1.6, -2.3)$
16. $(-0.4, -1.2)$

Exercises 17–24: Find a set of polar coordinates for the point whose rectangular coordinates are given.

17. $(2, 5)$
18. $(3, 1.4)$
19. $(2.3, -2)$
20. $(4.2, -1.5)$
21. $(-2, 3)$
22. $(-1.4, 4.6)$
23. $(-0.6, -1.3)$
24. $(-3.1, -0.8)$

Exercises 25–30: Find a polar form of the given equation.

25. $x = -4$
26. $y = 6$
27. $y = 2x$
28. $y = x^2$
29. $x^2 + y^2 = 3y$
30. $(x^2 + y^2)^3 = x^2 y^2$

Exercises 31–38: Find the rectangular form of the given equation.

31. $r = 2 \sin \theta$

32. $r \cos \theta = -1$

33. $r = 5 \cos \theta + 3 \sin \theta$

34. $r^2 = 2 \cos \theta \sin \theta$

35. $r = 1 - \sin \theta$

36. $r = 4(1 + \cos \theta)$

37. $r = \dfrac{9}{5 - 4 \cos \theta}$

38. $r^2(\cos^2\theta + 4 \sin^2\theta) = 16$

39. a. From geometry, we know that the length of an arc cut off on a circle is proportional to the central angle subtended by that arc. Use this fact—along with the formula for the circumference of a circle, $C = 2\pi r$—to prove that the length s of the arc shown in Figure 3-14 is given by the formula $s = r\theta$, where θ is measured in radians.

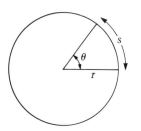

Figure 3-14

b. From geometry, we know that the area of a sector of a circle is proportional to the central angle of the sector. Use this fact—along with the formula for the area of a circle, $A = \pi r^2$—to prove that the area a of the sector shown in Figure 3-15 is given by the formula $a = \frac{1}{2}r^2\theta$, where θ is measured in radians.

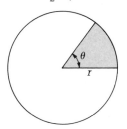

Figure 3-15

40. An audio receiver placed at a fixed point underwater was rotated and the sound levels (in decibels) received from various directions were recorded as follows:

Direction (in radians)	Level (in decibels)
0.0	3.1
0.4	2.6
0.8	1.7
1.2	1.3
1.6	1.1
2.0	0.9
2.4	1.4
2.8	2.1
3.2	1.3
3.6	1.9
4.0	2.4
4.4	1.8
4.8	0.8
5.2	1.7
5.6	2.6
6.0	2.9

Sketch the graph of this data on a polar coordinate system.

41. Optometrists and opthalmologists use the auto-plot visual field test to detect abnormal blind spots in a person's vision. Everyone has a normal blind spot; for the average person, the normal blind spot for the right eye is an ellipse with center (in polar coordinates) at $(15, -0.09)$, a vertical major axis of 8 units, and a horizontal minor axis of 6 units, where the pole is the visual fixed point. Sketch this area on a polar coordinate system.

42. The normal blind spot for the left eye (see Exercise 41) is an ellipse with center (in polar coordinates) at $(15, 3.25)$, a vertical major axis of 8 units, and a horizontal minor axis of 6 units. Sketch this area on a polar coordinate system.

43. An object moves on an *xy*-coordinate system in such a way that at any time t, its position is given by the equations $x = t^2$ and $y = 2t^2 + 1$. Write a single polar equation for the path of the object.

44. The path of a comet has been estimated to have the polar equation $r = 480/(1 + 0.87 \cos \theta)$. Find the equation in rectangular coordinates; what kind of curve is this?

3-3 Polar Graphs

Polar equations for curves are frequently expressed in the form $r = f(\theta)$, with θ the independent variable and r the dependent variable. We can plot the graphs of these equations by selecting values for θ and computing corresponding values for r. A calculator is invaluable in this process. A catalog of some of the more commonplace polar equations gives us an idea of what to expect from the graph.

Form of Equation	Type of Curve
$r = a$	Circle of radius a, center at the pole.
$\theta = a$	Straight line through the pole.
$r = a \sec \theta$	Vertical line.
$r = a \csc \theta$	Horizontal line.
$\left.\begin{array}{l} r = a \sin \theta \\ r = a \cos \theta \end{array}\right\}$	Circle, tangent to pole.
$\left.\begin{array}{l} r = a(1 \pm \sin \theta) \\ r = a(1 \pm \cos \theta) \end{array}\right\}$	**Cardioid**, a heart-shaped curve.
$\left.\begin{array}{l} r = a \pm b \sin \theta \\ r = a \pm b \cos \theta \end{array}\right\}$	**Limaçon**, a generalization of a cardioid, ranging from ◯ to ◯
$\left.\begin{array}{l} r^2 = a^2 \sin 2\theta \\ r^2 = a^2 \cos 2\theta \end{array}\right\}$	**Lemniscate**, a figure-eight shape.
$\left.\begin{array}{l} r = a \sin n\theta \\ r = a \cos n\theta \end{array}\right\}$	**Roses**, with n petals if n is odd, $2n$ petals if n is even.

Symmetry is another aid in graphing polar equations. The graph of an equation in polar coordinates is symmetric about the pole (origin) if $-r$ can replace r in the equation with no change. It is symmetric about the polar axis (x-axis) if $-\theta$ can replace θ in the equation with no change. And it is symmetric about the vertical line through the pole (y-axis) if $\pi - \theta$ can replace θ in the equation with no change.

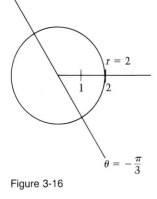

Figure 3-16

EXAMPLE 1. The graphs of the circle $r = 2$ and the line $\theta = -\pi/3$ are shown in Figure 3-16. ■

EXAMPLE 2. The equation $r = 3 \sec \theta$ can be most easily graphed by changing back to rectangular form:

$$r = 3 \sec \theta = \frac{3}{\cos \theta}$$

$$r \cos \theta = 3$$

$$x = 3$$

The graph is a vertical line. ■

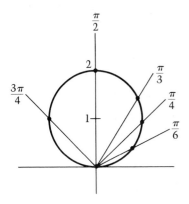

Figure 3-17

EXAMPLE 3. Sketch the graph of the equation $r = 2 \sin \theta$.

We construct a table of values for θ and corresponding values of r.

θ	0	$\pi/6$	$\pi/4$	$\pi/3$	$\pi/2$	$3\pi/4$	π	$5\pi/4$
r	0	1	1.4	1.7	2	1.4	0	-1.4

Plotting these points and connecting them with a smooth line gives the graph in Figure 3-17. Because $\sin(\pi - \theta) = \sin \theta$, the graph is symmetric about the vertical line through the pole. The point $(-1.4, 5\pi/4)$ coincides with $(1.4, \pi/4)$, so we are starting to loop around the circle again, and we do not need to plot any more points. The proof that the graph is actually a circle comes from rewriting the polar equation in rectangular form.

$$r = 2 \sin \theta$$
$$\sqrt{x^2 + y^2} = 2\frac{y}{\sqrt{x^2 + y^2}}$$
$$x^2 + y^2 = 2y$$
$$x^2 + y^2 - 2y = 0$$
$$x^2 + (y - 1)^2 = 1$$

This is a circle with center at $(0, 1)$ and radius 1. ■

Practice 1. Graph the equation $r = 4 \cos \theta$.

EXAMPLE 4. Graph the equation $r = 2(1 - \cos \theta)$.

Plotting the points from the table of values, we get the graph of Figure 3-18, a cardioid. The curve is symmetric about the polar axis because $\cos(-\theta) = \cos \theta$.

θ	0	$\pi/6$	$\pi/4$	$\pi/3$	$\pi/2$	$3\pi/4$	π	$3\pi/2$	$7\pi/4$
r	0	0.3	0.6	1	2	3.4	4	2	0.6

■

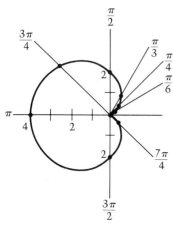

Figure 3-18

Practice 2. Graph the equation $r = 1 + \sin \theta$.

It may be hard to decide which values of θ to use in making a table of data. If the curve seems to be changing shape rapidly, then we need to pick values of θ that are close together, in order not to miss any of the features of the curve; if the curve seems to be changing gradually, then we can let θ jump through a larger interval.

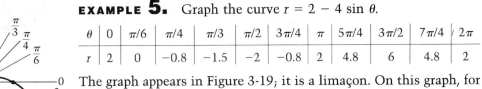

EXAMPLE 5. Graph the curve $r = 2 - 4 \sin \theta$.

θ	0	$\pi/6$	$\pi/4$	$\pi/3$	$\pi/2$	$3\pi/4$	π	$5\pi/4$	$3\pi/2$	$7\pi/4$	2π
r	2	0	-0.8	-1.5	-2	-0.8	2	4.8	6	4.8	2

The graph appears in Figure 3-19; it is a limaçon. On this graph, for example, quite a bit happens between $(-0.8, 3\pi/4)$ and $(2, \pi)$. We may want to find some more values in this range to see that the curve actually goes through the pole and then rises in one of the "shoulders" of the limaçon. ■

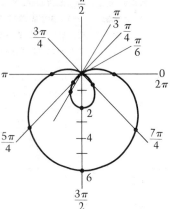

Figure 3-19

EXAMPLE 6. Graph the curve $r^2 = 4 \cos 2\theta$.

For this curve, values of θ for which $\cos 2\theta$ is negative are not allowed because they would produce negative values for r^2, and hence no real values for r.

θ	0	$\pi/6$	$\pi/4$	$\pi/2$	$3\pi/4$	$5\pi/6$	π
r	± 2	± 1.4	0	—	0	± 1.4	± 2

The graph, a lemniscate, is shown in Figure 3-20. It is symmetric about the pole, as well as about the polar axis. ■

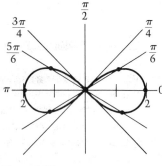

Figure 3-20

Word of Advice

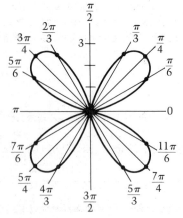

In graphing polar curves, remember that points with negative r-values are not located on the terminal side of θ, but on the extension of the terminal side through the pole.

EXAMPLE 7. Graph the equation $r = 3 \sin 2\theta$.

θ	0	$\pi/6$	$\pi/4$	$\pi/3$	$\pi/2$	$2\pi/3$	$3\pi/4$	$5\pi/6$	π
r	0	2.6	3	2.6	0	-2.6	-3	-2.6	0
θ	$7\pi/6$	$5\pi/4$	$4\pi/3$	$3\pi/2$	$5\pi/3$	$7\pi/4$	$11\pi/6$	2π	
r	2.6	3	2.6	0	-2.6	-3	-2.6	0	

The graph is the rose with four petals (four-leaved rose) shown in Figure 3-21. ■

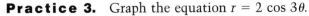

Practice 3. Graph the equation $r = 2 \cos 3\theta$.

Figure 3-21

Exercises / Section 3-3

Exercises 1–26: Graph the given equation on a polar coordinate system.

1. $r = 1$

2. $r = 4$

3. $\theta = \dfrac{\pi}{4}$

4. $\theta = \dfrac{-5\pi}{6}$

5. $r = 2 \csc \theta$

6. $r = 4 \sec \theta$

7. $r = 3 \sin \theta$

8. $r = -2 \cos \theta$

9. $r = 1 + \cos \theta$

10. $r = 2 (1 - \sin \theta)$

11. $r = 1 - 2 \cos \theta$

12. $r = 2 + 3 \sin \theta$

13. $r = 2 - \sin \theta$

14. $r = 3 + 2 \cos \theta$

15. $r^2 = \sin 2\theta$

16. $r^2 = 9 \cos 2\theta$

17. $r = \cos 2\theta$

18. $r = 2 \sin 3\theta$

19. $r = 2 \sin 4\theta$

20. $r = -\cos 3\theta$

21. $r = \dfrac{1}{1 + \sin \theta}$

22. $r = \dfrac{2}{2 - \cos \theta}$

23. $r = 2^\theta$

24. $r = \dfrac{3}{\theta}$

25. $r = 2 \sin \theta \tan \theta$

26. $r^2 = \theta$

27. The shells of many mollusks have a spiral design. The chambered nautilus shell is built around a *logarithmic spiral*, which has the polar equation $r = e^\theta$, where e is a fixed constant that arises in calculus. The value of e is approximately 2.718. Using this value of e, graph the polar equation $r = e^\theta$. (You may be able to compute powers of e directly on your calculator. If so, you will not need to use an approximation for e.)

28. A sewing machine needle for a fancy embroidery stitch traces a curve with rectangular equation

$$3\left(\frac{x^2 - y^2}{x^2 + y^2}\right) = \sqrt{x^2 + y^2}$$

Change this to a polar equation and graph the equation. (*Hint:* Use the trigonometric identity $\cos^2\theta - \sin^2\theta = \cos 2\theta$.)

STATUS CHECK

Now that you are at the end of Chapter 3, you should be able to:

section **3-1**   Sketch the graph of a surface in a three-dimensional coordinate system, given its equation.

section **3-2**   Plot a point on a polar coordinate system, given its polar coordinates.

Given polar coordinates for a point, find its rectangular coordinates.

Given rectangular coordinates for a point, find a set of polar coordinates for the point.

Find a polar form of an equation expressed in rectangular coordinates.

Find the rectangular form of an equation expressed in polar coordinates.

section **3-3**   Graph an equation given in polar coordinates on a polar coordinate system.

3-4 More Exercises for Chapter **3**

Exercises 1–4: Find a polar form of the given equation.

1. $x = -3y$

2. $xy = 4$

3. $x^2 + y^2 = 5x$

4. $x^2 - y^2 = 1$

Exercises 5–8: Find the rectangular form of the given equation.

5. $r = 4$

6. $r = 2 \cos \theta$

7. $r = \dfrac{3}{1 + 4 \cos \theta}$

8. $r = \sin \theta + \cos \theta$

Exercises 9–12: Sketch the graph in a three-dimensional coordinate system.

9. $x + 3y + 2z = 6$

10. $2x - y + 5z = 10$

11. $25x^2 + 4y^2 + 50z^2 = 100$

12. $y = 3x^2 + 3z^2$

Exercises 13–16: Sketch the graph in a polar coordinate system.

13. $r = 2(1 + \cos \theta)$

14. $r^2 = 4 \sin 2\theta$

15. $r = -2 \sin 3\theta$

16. $r = -2 \sin \theta$

chapter **4**

The Derivative

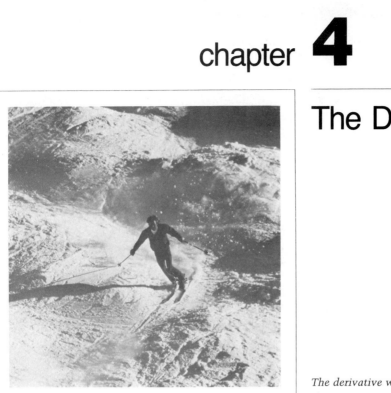

The derivative will allow us to "navigate the slopes."

4-1 Introduction

We have already defined the slope of a straight line. The side view of a ski slope is not likely to be a straight line; it may look much more like Figure 4-1. Yet we still have an intuitive understanding of slope in the sense of "steepness." For example, at points P_1 and P_2 in Figure 4-1, the curve is very flat and the steepness, or slope, is small, while at point Q, the curve is very steep and the slope is large. Mentally, we are attaching tangent lines to the curve at these points and considering the slopes of the tangent lines, as in Figure 4-2. *The*

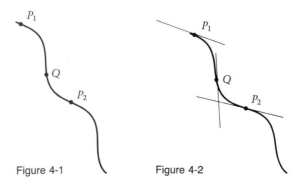

Figure 4-1   Figure 4-2

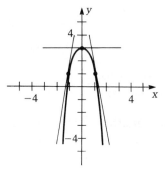

Figure 4-3

slope of a curve at a point is the slope of the tangent line to the curve at that point.

EXAMPLE 1.   Given the function $f(x) = 3 - 2x^2$, we graph it as in Figure 4-3. It appears that the tangent line to $f(x)$ at the point where $x = 0$ is horizontal. Thus the slope of $f(x)$ when $x = 0$ is zero. The slope of $f(x)$ when $x = -1$ is positive, while the slope of $f(x)$ when $x = 1$ is negative.   ∎

Practice 1.   Sketch the graph of the function $f(x) = 3 + x^2$. Is the slope of $f(x)$ when $x = 0$ positive, negative, or zero? What about when $x = -2$? When $x = 2$?

Given a function $f(x)$, the value of the slope of the curve at a specific point depends upon that point, which in turn depends only upon the value of x. *The slope of a function of x is also a function of x, which we will denote by $f'(x)$.*

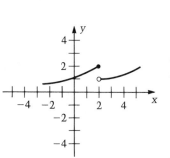

Figure 4-4

EXAMPLE 2.   For the function of Example 1, $f(x) = 3 - 2x^2$, the slope function can be shown to be $f'(x) = -4x$. Then $f'(0) = (-4)(0) = 0$; this agrees with our conclusion of Example 1, where we determined from the sketch that the value of the slope of $f(x)$ when $x = 0$ is zero. Also, $f'(-1) = (-4)(-1) = 4$, so the slope of $f(x)$ when $x = -1$ is positive. Finally $f'(1) = (-4)(1) = -4$, a negative slope.   ∎

Practice 2.   For the function $f(x) = 2x^2 + 5x$, the slope function $f'(x)$ is $f'(x) = 4x + 5$. What is the value of the slope of $f(x)$ when $x = -1$? When $x = 3$?

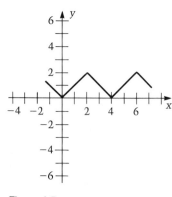

Figure 4-5

Not every function has a slope at each point on the curve. In Figure 4-4, there is a "break" in the curve at $x = 2$. There is no tangent line at this point, so there is no slope to the curve. If the curve of Figure 4-4 is denoted by $f(x)$, then $f'(x)$ would not be defined when $x = 2$. In Figure 4-5, the slope to the curve does not exist where $x = 0$, $x = 2$, $x = 4$, $x = 6$, and so on; the curve in each of these places is too "pointy" to have a tangent line.

When the slope of a function exists at a point, it gives us some information on how the function is changing right around that point. At a point where the slope is 0.33, for example, the tangent line to the curve is nearly horizontal. If we were to move a small amount horizontally, the corresponding vertical change in the curve

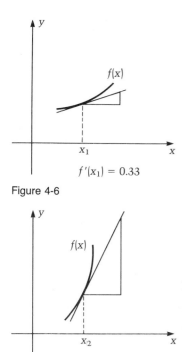

$f(x)$

x_1

$f'(x_1) = 0.33$

Figure 4-6

$f(x)$

x_2

$f'(x_2) = 2$

Figure 4-7

would be small (see Figure 4-6). At a point where the slope is 2, the tangent line to the curve is fairly steep, and the same small change (horizontally) would produce a large vertical change in the curve (see Figure 4-7). The amount that the curve changes vertically with a small horizontal change is called the **rate of change** of the curve. Therefore, the slope function $f'(x)$ is a way to measure the rate of change of $f(x)$ for various values of x.

Whenever one quantity y is a function of another quantity x, we may be interested in the rate at which y changes as x changes. For example, with a moving body whose position changes as a function of time, the rate of change of position with respect to time is the *velocity* of the body. In an electric circuit, where the total amount of charge which has passed a given point changes as a function of time, the rate of change of charge with respect to time is the *current* in the circuit.

In this chapter, we study the *derivative* of a function. In Section 3, we give a formal definition of the derivative. The definition will be mathematical, but it has a wide variety of interpretations and applications. Slope and rate of change are two ways to interpret the derivative.

Exercises / Section **4-1**

Exercises 1–4: By sketching a graph of the function, decide whether the slope is positive, negative, or zero at each value of x.

1. $f(x) = 1.4x^2 + 3.1;$ $x = 0, x = 1.2, x = -2.4$

2. $f(x) = 4.2 - 2x^2;$ $x = 0, x = 3.1, x = -4$

3. $f(x) = 0.5x^2 - 2x;$ $x = 0, x = 2, x = 3$

4. $f(x) = 2 + 2x - x^2;$ $x = 0, x = 1, x = -1.8$

Exercises 5–12: For the function $f(x)$, the slope function $f'(x)$ is given. Find the value of the slope of $f(x)$ at the given value of x.

5. $f(x) = 4x^2 - 2x + 1; f'(x) = 8x - 2; x = 3$

6. $f(x) = 4 - 3x^2; f'(x) = -6x; x = -1$

7. $f(x) = 6 - 2x^3; f'(x) = -6x^2; x = -2$

8. $f(x) = 8x^3 + 2x; f'(x) = 24x^2 + 2; x = 2$

9. $f(x) = 2x^3 - 6x + 4; f'(x) = 6x^2 - 6; x = 0$

10. $f(x) = 3x^3 + 12x^2 - 2; f'(x) = 9x^2 + 24x; x = 1$

11. $f(x) = \sin 2x; f'(x) = 2 \cos 2x; x = \dfrac{\pi}{2}$

12. $f(x) = \cos x; f'(x) = -\sin x; x = \dfrac{\pi}{2}$

Exercises 13–20: Find any values of x for which the slope function is undefined.

13.

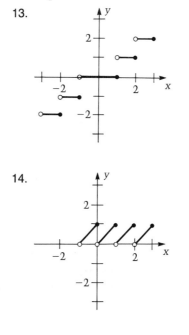

14.

15.   **16.**   **17.**

18.   **19.**   **20.**

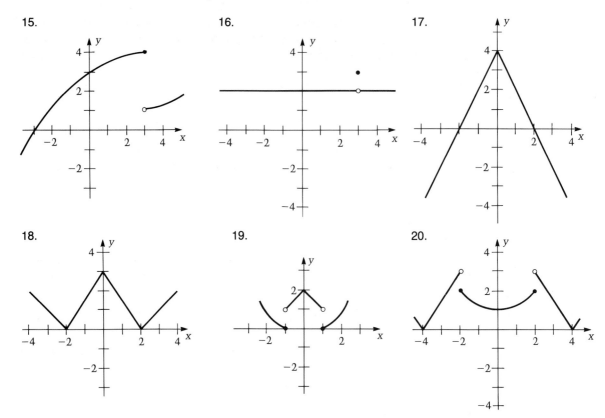

21. In a certain thermocouple, the voltage V produced as a function of temperature T is given by the equation $V = 2.7T + 0.2T^2$. Carefully graph V as a function of T and estimate the value of the slope when $T = 1$. The slope function is $V'(T) = 0.4T + 2.7$. Use this function to find the value of the slope when $T = 1$. Compare the two values.

22. For a certain uniformly loaded beam, the bending moment M as a function of distance x from one end of the beam is given by the equation $M = 2.6x^2 - 1.5x + 0.5$. Carefully graph M as a function of x and estimate the value of the slope when $x = 1$. The slope function is $M'(x) = 5.2x - 1.5$. Use this function to find the value of the slope when $x = 1$. Compare the two values.

4-2 Limits

Before we can define the derivative of a function, we need to understand the idea of *limit*. A limit is a value to which we get closer and closer as we do some operation.

EXAMPLE 1. In the sequence of numbers

$$2.100000, \ 2.010000, \ 2.001000, \ 2.000100, \ 2.000010, \ \ldots$$

the farther we go in the sequence, the closer we get to the value 2. We say that 2 is the limit of the sequence. ∎

In Example 1, we found the limit of a *sequence* of numbers. One way to obtain a sequence of numbers is to evaluate a function $f(x)$ for various values of x.

EXAMPLE 2. The tables below show values of x and corresponding values of $f(x)$ for the function $f(x) = 2x^2 - 3$.

x	0.5000	0.8000	0.9200	0.9900	0.9990
$f(x)$	−2.5000	−1.7200	−1.3072	−1.0398	−1.0040
x	1.2000	1.1000	1.0500	1.0100	1.0010
$f(x)$	−0.1200	−0.5800	−0.7950	−0.9598	−0.9960

In the first table, we have taken values of x that are smaller than 1 but get closer and closer to 1; the limit of the sequence of corresponding function values seems to be −1. In the second table, we have taken values of x that are larger than 1, but get closer and closer to 1; again the limit of the sequence of corresponding function values seems to be −1. When we let x get closer and closer to 1 by using a succession of values both smaller and larger than 1, we say that *x is approaching* 1; we denote this by $x \to 1$. For this function, the limit of $f(x)$ *as* $x \to 1$ is −1. ■

Now we need some notation to describe the type of thing which happened in Example 2. We let a be some number and let $x \to a$. If the values of a function $f(x)$ get very close to some number L as $x \to a$, then L is the **limit** of $f(x)$ *as x approaches a*. We write this

$$\lim_{x \to a} f(x) = L \quad \text{or} \quad \lim_{x \to a} f(x) = L$$

and read: the limit of $f(x)$ as x approaches a is L. Here $x \to a$ means that x gets closer and closer to a through a succession of values both smaller and larger than a, but we do not actually let x equal a. In contrast, the values of $f(x)$ may not only get closer and closer to L, they may equal L.

EXAMPLE 3. Let $f(x)$ be the function $f(x) = 6$. The following table shows that $\lim_{x \to 2} f(x) = 6$.

x	1.900	1.990	1.999	. . .	2.100	2.010	2.001	. . .
$f(x)$	6	6	6		6	6	6	■

EXAMPLE 4.   Find $\lim_{x \to 4}(3x - 4)$.

We evaluate $f(x)$ for the following values of x: 3.900, 3.990, 3.999 and 4.100, 4.010, 4.001. The corresponding values of $f(x)$ are 7.700, 7.970, 7.997 and 8.300, 8.030, 8.003. Since the values of $f(x)$ get closer to 8 as x approaches 4, $\lim_{x \to 4}(3x - 4) = 8$.   ∎

Practice 1.   Find $\lim_{x \to 3}(5 - 3x)$ by evaluating $f(x)$ for the following values of x: 2.500, 2.800, 2.900, 2.990, 2.999 and 3.400, 3.200, 3.110, 3.010, 3.001.

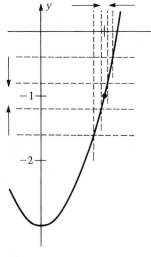

We decided in Example 2 that $\lim_{x \to 1}(2x^2 - 3) = -1$. A graph of the function $f(x) = 2x^2 - 3$ illustrates what is happening as we take this limit. In Figure 4-8, we notice that as the x-values get close to 1 (shown on the horizontal axis), the corresponding $f(x)$-values get close to -1 (shown on the vertical axis). We also notice that $f(1) = -1$. It seems considerably easier to evaluate $\lim_{x \to 1}f(x)$ by simply computing $f(1)$. If all functions behaved as nicely as $f(x) = 2x^2 - 3$, we could always evaluate $\lim_{x \to a}f(x)$ by computing $f(a)$. The function $f(x) = 2x^2 - 3$ is a function that is everywhere **continuous**, meaning that there are no points that are gaps or breaks in the graph of the function. *If a function is continuous at $x = a$, we can always evaluate the limit as $x \to a$ by substituting a for x in the expression for the function.**

Figure 4-8

EXAMPLE 5.   Find $\lim_{x \to 5}(2x + 4)$.

The function $f(x) = 2x + 4$ is a straight line and the graph has no gaps, so it is everywhere continuous. Therefore, $\lim_{x \to 5}(2x + 4) = 2(5) + 4 = 14$.   ∎

Practice 2.   Find $\lim_{x \to -3}(4 - 1.8x)$.

If a function $f(x)$ is not continuous at $x = a$ (is **discontinuous** at $x = a$), then the substitution trick to find $\lim_{x \to a}f(x)$ will not work.

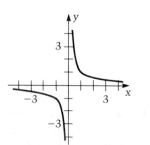

EXAMPLE 6.   Figure 4-9 shows the graph of the function $f(x) = 1/x$. This function is discontinuous at $x = 0$. Also, $\lim_{x \to 0}f(x)$ does not exist; as $x \to 0$, the values of $f(x)$ do not approach a fixed value L. If we try to substitute, we find that $f(0)$ is undefined.   ∎

Figure 4-9

*In fact, we can *define* continuity of $f(x)$ at $x = a$ to mean that $\lim_{x \to a}f(x) = f(a)$.

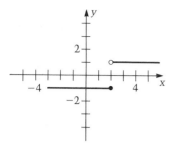

Figure 4-10

EXAMPLE 7. Figure 4-10 shows the graph of the function

$$g(x) = \begin{cases} -1 & \text{for } x \le 2 \\ 1 & \text{for } x > 2 \end{cases}$$

This function is defined to have the value -1 when $x = 2$, but it is discontinuous at $x = 2$. The $\lim_{x \to 2} g(x)$ does not exist; as $x \to 2$, the values of $g(x)$ do not approach a fixed value L. Instead, they take the value -1 for x-values less than 2 and they take the value 1 for x-values greater than 2. If we substitute, we get $g(2) = -1$, but this might mislead us into thinking the limit exists. ■

In both of the functions of Example 6 and Example 7, the limit of the function does not exist as x approaches the point of discontinuity. However, a function may be discontinuous for some value of x and still have a limit as x approaches this value.

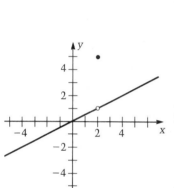
Figure 4-11

EXAMPLE 8. Let $f(x)$ be defined by

$$f(x) = \begin{cases} \frac{1}{2}x & \text{for } x \ne 2 \\ 5 & \text{for } x = 2 \end{cases}$$

Figure 4-11 shows the graph of this function. Note that it is defined, but discontinuous, at $x = 2$. We see from the graph that $\lim_{x \to 2} f(x) = 1$, while $f(2) = 5$. ■

EXAMPLE 9. Let $g(x)$ be defined by

$$g(x) = \begin{cases} \frac{1}{2}x & \text{for } x \ne 2 \\ \text{undefined} & \text{for } x = 2 \end{cases}$$

Figure 4-12 shows the graph of this function. Although $g(2)$ is undefined, it is still true that $\lim_{x \to 2} g(x) = 1$. ■

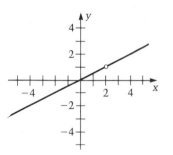

Figure 4-12

We see from these examples that if $x = a$ is a point of discontinuity of a function $f(x)$, there are four possible situations.

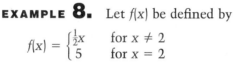

$f(a)$ undefined.	$f(a)$ undefined.	
$\lim_{x \to a} f(x)$ does not exist.	$\lim_{x \to a} f(x)$ exists.	
$f(a)$ defined.	$f(a)$ defined.	But these
$\lim_{x \to a} f(x)$ does not exist.	$\lim_{x \to a} f(x)$ exists.	are not equal.

The point to remember is that $\lim_{x \to a} f(x) = f(a)$ *if and only if $f(x)$ is continuous at $x = a$.*

Points of discontinuity of $f(x)$ will occur at any value of x that results in division by zero (as in Example 6) and can occur when the function is defined by more than one equation (as in Examples 7, 8, and 9). Suppose $f(x)$ is discontinuous at $x = a$, but we want to know whether $\lim_{x \to a} f(x)$ exists and its value if it does exist. We do not necessarily have to resort to graphical methods.

EXAMPLE 10. Find $\lim\limits_{x \to 2} \dfrac{x^2 - 2x}{2x - 4}$.

We note that this function is not continuous at $x = 2$, since substituting 2 for x results in division by zero. However, we can write the function as

$$\frac{x(x - 2)}{2(x - 2)}$$

and divide numerator and denominator by the factor $x - 2$. But wait—are we not dividing by 0, which is illegal? No; we are looking for the limit as $x \to 2$, which means that x does not equal 2 and $x - 2$ does not equal 0. Therefore it is legal to divide numerator and denominator by $x - 2$.

$$\lim_{x \to 2} \frac{x^2 - 2x}{2x - 4} = \lim_{x \to 2} \frac{x(x - 2)}{2(x - 2)} = \lim_{x \to 2} \frac{x}{2} = 1$$

The graph of the original function is shown in Figure 4-13, from which we see that this function is just the function $g(x)$ of Example 9. ∎

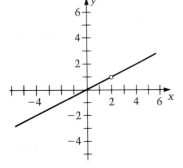

Figure 4-13

EXAMPLE 11. Find $\lim\limits_{x \to -1} \dfrac{x^2 - x - 2}{x + 1}$.

Here the function is discontinuous at $x = -1$, but

$$\lim_{x \to -1} \frac{x^2 - x - 2}{x + 1} = \lim_{x \to -1} \frac{(x + 1)(x - 2)}{x + 1} = \lim_{x \to -1} (x - 2) = -3 \quad ∎$$

Practice 3. Find $\lim\limits_{x \to 3} \dfrac{x^2 + 2x - 15}{3x - 9}$.

Suppose that $g(x)$ and $h(x)$ are two functions of x; suppose also that $g(x)$ has the limit L_1 as $x \to a$ and $h(x)$ has the limit L_2 as $x \to a$. Then the function $f(x) = g(x) + h(x)$ has the limit $L_1 + L_2$ as $x \to a$. In other words, if we want to find the limit of a sum, we can just take the sum of the limits of each term. Similarly, the limit of a product of two functions can be found by applying the limit to each factor, and the limit of a quotient can be found by applying the limit to the numerator and to the denominator.

We can also consider whether a function $f(x)$ has a limit L as the values of x continue to increase, rather than approaching some fixed number a. If this happens, we write

$$\lim_{x \to \infty} f(x) = L$$

(Remember that ∞ does not denote some very big number; $x \to \infty$ means that for any number we think of, x will eventually get bigger than that number.)

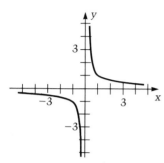

Figure 4-14

EXAMPLE 12. For the function $f(x) = 1/x$, the table of values shows that as $x \to \infty$, $f(x)$ gets closer and closer to 0.

x	10	100	1000	. . .
$f(x)$	0.1	0.01	0.001	. . .

Therefore $\lim_{x \to \infty} 1/x = 0$. This is also clear from the graph of $f(x)$, in Figure 4-14. As we look further to the right on the x-axis, the function values get close to 0. The line $y = 0$ is a horizontal asymptote. ■

EXAMPLE 13. Find $\lim\limits_{x \to \infty} \dfrac{2x^2 - 4}{3x^2 + 2x - 5}$.

Rather than computing values of the function for larger and larger values of x, we again resort to a rewriting trick, this time dividing numerator and denominator by the highest power of x that appears, x^2.

$$\lim_{x \to \infty} \frac{2x^2 - 4}{3x^2 + 2x - 5} = \lim_{x \to \infty} \frac{2 - \dfrac{4}{x^2}}{3 - \dfrac{2}{x} - \dfrac{5}{x^2}}$$

Now we see that as x gets large, $-4/x^2$, $-2/x$, and $-5/x^2$ all approach 0, and only the constant ratio $\frac{2}{3}$ is left. Thus

$$\lim_{x \to \infty} \frac{2x^2 - 4}{3x^2 + 2x - 5} = \lim_{x \to \infty} \frac{2 - \overset{0}{\cancel{\dfrac{4}{x^2}}}}{3 - \underset{0}{\cancel{\dfrac{2}{x}}} - \underset{0}{\cancel{\dfrac{5}{x^2}}}} = \frac{2}{3}$$

If we were to graph

$$y = \frac{2x^2 - 4}{3x^2 + 2x - 5}$$

the line $y = \frac{2}{3}$ would be a horizontal asymptote. ■

Practice 4. Find $\lim\limits_{x \to \infty} \dfrac{4x^2 - 7x + 5}{2x^2 + 4}$.

EXAMPLE 14. The kinetic energy of an object is given by the formula $KE = \frac{1}{2}mv^2$, where m is the mass of the object and v is its velocity. In a 30.0-kg object, the velocity is expected to approach a steady rate of 15.0 m/s. Find the limit of the kinetic energy (in joules).

We substitute 30.0 for m; we want to find $\lim_{v\to 15} \frac{1}{2} \cdot 30v^2 = \lim_{v\to 15} 15v^2$. Because $f(v) = 15v^2$ is a continuous function, we substitute; $\lim_{v\to 15} 15v^2 = 15(15)^2 = 3375$ J. ∎

Exercises / Section 4-2

Exercises 1–6: Evaluate the function for the given values of x and then use your data to estimate the limit.

1. $f(x) = 3x - 7.2$; $x = 1.400, 1.450, 1.490,$ 1.499 and $x = 1.600, 1.550, 1.510, 1.501$; $\lim_{x\to 1.5}(3x - 7.2)$

2. $f(x) = 2 - 4.1x$; $x = 3.500, 3.580, 3.590,$ 3.599 and $x = 3.700, 3.610, 3.604, 3.601$; $\lim_{x\to 3.6}(2 - 4.1x)$

3. $f(x) = \dfrac{x^3 + x^2 - 11x + 10}{x - 2}$; $x = 1.80, 1.90,$ 1.99 and $x = 2.11, 2.01, 2.001$; $\lim_{x\to 2} \dfrac{x^3 + x^2 - 11x + 10}{x - 2}$

4. $f(x) = \dfrac{x^3 + 3x^2 - 11x - 5}{x + 5}$; $x = -5.1, -5.05,$ -5.01 and $x = -4.9, -4.98, -4.99$; $\lim_{x\to -5} \dfrac{x^3 + 3x^2 - 11x - 5}{x + 5}$

5. $f(x) = \dfrac{\sqrt{x} - 7}{2 + 3\sqrt{x}}$; $x = 10, 100, 1000, 10{,}000,$ $100{,}000, 1{,}000{,}000$; $\lim_{x\to\infty} \dfrac{\sqrt{x} - 7}{2 + 3\sqrt{x}}$

6. $f(x) = \dfrac{x^2 - 2x + 5}{2x^3 - 3}$; $x = 10, 100, 1000, 10{,}000$; $\lim_{x\to\infty} \dfrac{x^2 - 2x + 5}{2x^3 - 3}$

Exercises 7–18: Find any values of x for which $f(x)$ is discontinuous. (Drawing graphs may help.)

7. $f(x) = 4x^3 - x$

8. $f(x) = 2x^2 - 5x + \sin x$

9. $f(x) = 2 + \dfrac{1}{x - 4}$

10. $f(x) = \dfrac{3 - 7x}{x - 1}$

11. $f(x) = \begin{cases} x & \text{for } x \geq 2 \\ x^2 & \text{for } x < 2 \end{cases}$

12. $f(x) = \begin{cases} x & \text{for } x \geq 1 \\ x^2 & \text{for } x < 1 \end{cases}$

13. $f(x) = \begin{cases} 3x + 1 & \text{for } x \neq 5 \\ 3 & \text{for } x = 5 \end{cases}$

14. $f(x) = \begin{cases} 2x - 1 & \text{for } x \neq 3 \\ 1 & \text{for } x = 3 \end{cases}$

15. $f(x) = \begin{cases} 4x + 7 & \text{for } x \neq 2 \\ 15 & \text{for } x = 2 \end{cases}$

16. $f(x) = \begin{cases} 5 - 2x & \text{for } x \neq 4 \\ -3 & \text{for } x = 4 \end{cases}$

17. $f(x) = \begin{cases} x & \text{for } x \geq 0 \\ -x & \text{for } x < 0 \end{cases}$

18. $f(x) = \begin{cases} x & \text{for } x \geq 1 \\ -x & \text{for } x < 1 \end{cases}$

Exercises 19–46: Find the limit.

19. $\lim_{x\to 3} (7 - 4x)$

20. $\lim_{x\to -2} (4x + 6)$

21. $\lim_{x\to 5} (3x^2 - 2x + 1)$

22. $\lim_{x\to 2} (2x^2 + 4x - 6)$

23. $\lim_{x\to 0.3} (1 - 2x^2)$

24. $\lim_{x\to 1.7} (3x^3 - 2)$

25. $\lim_{x\to 1.6} \dfrac{2x^2 - 5x + 3}{x - 4}$

26. $\lim_{x\to -3.1} \dfrac{5x^2 - x + 4}{x + 6}$

27. $\lim_{x\to 2} \dfrac{x^2 - 4}{x - 2}$

28. $\lim_{x\to -1} \dfrac{x^2 - 1}{x + 1}$

29. $\lim_{x\to 3} \dfrac{x^2 + x - 12}{x - 3}$

30. $\lim_{x\to 4} \dfrac{x^2 - 2x - 8}{x - 4}$

31. $\lim_{x \to 0} \dfrac{x^2 - 2x}{x}$

32. $\lim_{x \to -2} \dfrac{(x + 1)^2 - 1}{x + 2}$

33. $\lim_{x \to 3} f(x)$, where $f(x) = \begin{cases} x & \text{for } x \neq 3 \\ 5 & \text{for } x = 3 \end{cases}$

34. $\lim_{x \to -1} f(x)$, where $f(x) = \begin{cases} -1 & \text{for } x \neq -1 \\ -3 & \text{for } x = -1 \end{cases}$

35. $\lim_{x \to 0} f(x)$, where $f(x) = \begin{cases} x^2 & \text{for } x > 0 \\ -x & \text{for } x < 0 \end{cases}$

36. $\lim_{x \to 1} f(x)$, where $f(x) = \begin{cases} x & \text{for } x > 1 \\ x^2 & \text{for } x < 1 \end{cases}$

37. $\lim_{x \to -2} (x + 2)\sqrt{x}$ (Be careful— do not substitute.)

38. $\lim_{x \to 1} (x - 1)\sqrt{x - 3}$ (Be careful— do not substitute.)

39. $\lim_{x \to \infty} \dfrac{3}{1 + 2/x}$

40. $\lim_{x \to \infty} \dfrac{2 - 5/x}{4 + 3/x^2}$

41. $\lim_{x \to \infty} \dfrac{3x - 5}{2 - 4x}$

42. $\lim_{x \to \infty} \dfrac{3x^2 + 2x - 7}{x^2 - 4}$

43. $\lim_{x \to \infty} \dfrac{4x^2 + 7x - 10}{8 - 2x^2}$

44. $\lim_{x \to \infty} \dfrac{4 - 8x}{1.5 + 2x}$

45. $\lim_{x \to \infty} \dfrac{x^2 - 3x + 1}{3x^3 - 1}$

46. $\lim_{x \to \infty} \dfrac{2x - 7}{4x^2 - 2x + 1}$

47. In a water purification plant with a biological filter, the ratio r of impurities in the water before filtration to impurities after filtration is given by the equation $r = (x + k)/x$, where k is a constant and x is the rate of flow through the filter. Show that $\lim_{x \to \infty} r(x) = 1$. How effective is the purification process at high rates of flow through the filter?

48. An insect population is controlled by sterilizing a fixed number S of males in each generation. The number N of fertile males in the next generation is given by the equation

$$N = \frac{k}{r/S + r}$$

where k is a constant and r is the ratio of sterilized males to fertile males in the present generation. The ratio r is large when the number of sterilized males far exceeds the number of fertile males in a given generation. Find $\lim_{r \to \infty} N(r)$ and interpret the answer.

49. The cost per unit c (in dollars) to manufacture x units of a product is given by the equation $c = (100 + 5x)/x$. Estimate the cost per unit of large-scale production by computing $\lim_{x \to \infty} c$.

50. The amount E of energy required to refine metal from ore to x percent purity is $E = 10x/(100 - x)$. Show that as $x \to 100$, $E \to \infty$, thus showing that highly refined metal requires a large amount of energy.

51. The cost of water is $100 for each unit or fractional part of a unit used. However, consumers of more than 6 units are classified as industrial users and are charged only $20 per unit. Let $f(x)$ be the cost of x units of water. Graph $f(x)$. Find any values of $x > 0$ for which $f(x)$ is discontinuous.

52. A sheet-metal supplier charges its customers interest at the rate of $1\frac{1}{2}\%$ a month on the first $1000 or less owed plus 1% a month on the amount owed over $1000. Let $f(x)$ be the monthly interest on the amount x owed (in dollars). Graph $f(x)$. Find any values of $x > 0$ for which $f(x)$ is discontinuous.

4-3 The Definition of the Derivative

Calculus is divided into two parts, *differential calculus* and *integral calculus*. We discuss integral calculus in Chapter 6; until then we will be concerned with differential calculus. The process of *differentiating* a function (**differentiation**) is the process of finding its *derivative*. Now that we know how to take limits, we are ready to give a definition of the derivative.

We have seen that a slope function $f'(x)$ can be associated with a function $f(x)$. The slope function has the property that for any

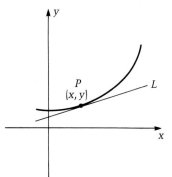

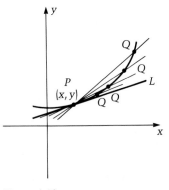

particular value of *x*, such as $x = a$, $f'(a)$ represents the slope of the tangent line to the graph of the function $f(x)$ where $x = a$. Since the slope function $f'(x)$ is one interpretation of the derivative of $f(x)$, we use it to help motivate the actual definition of the derivative.

Suppose that we have a function $f(x)$ and we wish to find a function $f'(x)$ that gives us the value of the slope of $f(x)$ at any point. In Figure 4-15, we show the graph of some general function $f(x)$ with an arbitrary point *P* whose coordinates are (x, y) where $y = f(x)$. We would like an expression in terms of *x* for the slope of the tangent line, *L*, at *P*. In order to find the slope of a line, however, we need the coordinates of two points on the line. Here is where the limit comes in. We let *Q* be a point on the curve that approaches the point *P*. As $Q \to P$ (along the curve), the line through points *P* and *Q*, denoted by *PQ*, approaches the tangent line *L*. Also, the slope of *PQ* approaches the slope of *L* (see Figure 4-16). Thus

$$\text{slope of } L = \lim_{Q \to P} (\text{slope of } PQ)$$

EXAMPLE 1. For the function $f(x) = x^2 + 2$, let us try to find the slope of the tangent line at the point $P = (1, 3)$. (We obtain 3 from computing $f(1) = 1^2 + 2 = 3$.) If we draw a very careful graph (Figure 4-17), we might estimate the slope of the tangent line to be somewhat bigger than 1, perhaps 1.8 or so. Using our limiting scheme, we select three points *Q* approaching *P* on the curve, as shown in Figure 4-18, compute the slopes of the three lines *PQ*, and try to find the limit.

$$P = (1, 3) \qquad Q_1 = (2, 6) \qquad \text{slope } PQ_1 = \frac{6 - 3}{2 - 1} = 3$$

$$P = (1, 3) \qquad Q_2 = (1.5, 4.25) \qquad \text{slope } PQ_2 = \frac{4.25 - 3}{1.5 - 1} = 2.5$$

$$P = (1, 3) \qquad Q_3 = (1.1, 3.21) \qquad \text{slope } PQ_3 = \frac{3.21 - 3}{1.1 - 1} = 2.1$$

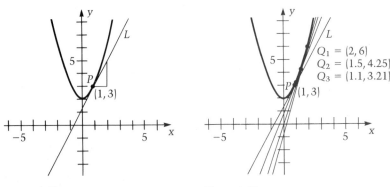

Figure 4-17   Figure 4-18

We now do two more values of Q even closer to P.

$P = (1, 3)$ $Q_4 = (1.01, 3.0201)$ slope $PQ_4 = \dfrac{3.0201 - 3}{1.01 - 1} = 2.01$

$P = (1, 3)$ $Q_5 = (1.001, 3.002001)$ slope $PQ_5 = \dfrac{3.002001 - 3}{1.001 - 1} = 2.001$

Looking at the sequence of slopes, the limit seems to be 2. (Our graphical estimate was evidently a bit low.) ■

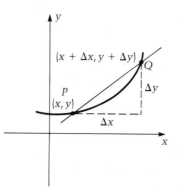

Figure 4-19

Now back to the general problem. Keep in mind that P is a fixed point—although it was arbitrarily chosen—and that Q will be a moving point. Wherever Q happens to be, its coordinates are different from those of P; its x-coordinate differs from that of P by some change (increment) in x. We denote this x-increment by Δx (read: delta x). There is also a y-increment, which we call Δy. These quantities are shown in Figure 4-19. (By the way, Δx is not something called Δ multiplied by something called x; Δx is just an expression for a change in x.) The coordinates of Q are the coordinates of P plus the changes, so $Q = (x + \Delta x, y + \Delta y)$. By definition of the slope of a line, the slope of PQ is $\Delta y/\Delta x$. Therefore,

$$\text{slope of } L = \lim_{Q \to P} \frac{\Delta y}{\Delta x}$$

Another way to describe the fact that $Q \to P$ is to say that $\Delta x \to 0$. We write

$$\text{slope of } L = \lim_{\Delta x \to 0} \frac{\Delta y}{\Delta x}$$

But consulting Figure 4-19 again, we note that Δy represents the difference in height between Q and P; $\Delta y = (y + \Delta y) - y$. Because both Q and P are points on the graph of the function, we know that $y = f(x)$ and that $y + \Delta y = f(x + \Delta x)$. Therefore, $\Delta y = f(x + \Delta x) - f(x)$.

Finally,

$$\text{slope of } L = \lim_{\Delta x \to 0} \frac{f(x + \Delta x) - f(x)}{\Delta x} \tag{1}$$

Using (1), we can solve a problem like Example 1 without doing so much numerical calculation.

EXAMPLE 2.   As in Example 1, we try to find the slope of $f(x) = x^2 + 2$ at the point $(1, 3)$. To compute the expression shown on the right side of Equation (1), there are four steps. We number each one.

1. Find $f(x + \Delta x)$:   Since $f(x) = x^2 + 2$,

$$f(x + \Delta x) = (x + \Delta x)^2 + 2 = x^2 + 2x\Delta x + (\Delta x)^2 + 2$$

2. Subtract $f(x)$ from $f(x + \Delta x)$:

$$f(x + \Delta x) - f(x) = (x^2 + 2x\Delta x + (\Delta x)^2 + 2) - (x^2 + 2)$$
$$= 2x\Delta x + (\Delta x)^2$$

3. Divide by Δx:

$$\frac{f(x + \Delta x) - f(x)}{\Delta x} = \frac{2x\Delta x + (\Delta x)^2}{\Delta x} = 2x + \Delta x$$

4. Take the limit:

$$\lim_{\Delta x \to 0} \frac{f(x + \Delta x) - f(x)}{\Delta x} = \lim_{\Delta x \to 0} (2x + \Delta x) = 2x$$

As a result of the four steps we know that the slope of the tangent line at any point (x, y) is $2x$. Since we are interested in the slope when $x = 1$, we substitute 1 for x in the expression $2x$ and find that the slope at $(1, 3)$ is 2, as we had concluded in Example 1. ∎

It may seem at first that the process we used in Example 2 was at least as hard as what we did in Example 1. The process of Example 2, however, gives us much more information. Suppose, for example, we now decide to find the slope of $f(x) = x^2 + 2$ at the point $(-5, 27)$. Example 1 tells us nothing about this, and we would have to start all over. But Example 2 tells us that the slope *anywhere* on the curve is given by the expression $2x$; when $x = -5$, the slope is $2(-5) = -10$. The whole spirit of mathematics is to solve the general problem whenever possible; then each specific case is easily handled.

The expression in Equation (1) is defined to be the derivative of the function $y = f(x)$, denoted by dy/dx. Thus *the definition of the* ***derivative*** *of a function f(x) is*

$$\boxed{\frac{dy}{dx} = \lim_{\Delta x \to 0} \frac{f(x + \Delta x) - f(x)}{\Delta x}}$$

(This definition is very important, and it might be one of your instructor's favorite exam questions.) There are several notations for the derivative. If $y = f(x)$, then

$$f'(x), \qquad y', \qquad \frac{dy}{dx}, \qquad \frac{df}{dx}, \qquad \frac{d[f(x)]}{dx}$$

all denote the derivative of $f(x)$. The notation dy/dx should be thought of as a single symbol for the derivative, not as d multiplied times y and divided by d multiplied times x. We often call this *the derivative with respect to x*, just to indicate that x is the name of the independent variable. (Notice that $f'(x)$ is the same notation we used in Section 2-1 to denote the slope function. Now we can call the slope function by its formal mathematical name, the derivative.)

Looking at the definition of the derivative, we note again that there is a four-step process involved (sometimes called the *delta process*).

1. Find $f(x + \Delta x)$. This merely involves substituting $x + \Delta x$ for x in the function and then doing any simplifications on the resulting expression.

2. Subtract $f(x)$ from $f(x + \Delta x)$. Whenever you do this subtraction step, some terms should add to zero and drop out.

3. Divide by Δx. Dividing by Δx means creating the fraction $[f(x + \Delta x) - f(x)]/\Delta x$, which has a factor of Δx in the denominator. You should generally be able to factor a Δx out of each term of the numerator and then divide both numerator and denominator by Δx.

4. Find the limit as $\Delta x \to 0$ of $[f(x + \Delta x) - f(x)]/\Delta x$. Remember that $\Delta x \to 0$ means that the values of Δx get closer and closer to 0, but never equal 0. This fact makes dividing numerator and denominator by Δx in Step 3 legal.

EXAMPLE 3.   For the function $f(x) = 1 - x^2$, let us compute the slope at the point $(3, -8)$ by finding the derivative, using the four-step process. (See Figure 4-20.)

1. Find $f(x + \Delta x)$:

$$f(x + \Delta x) = 1 - (x + \Delta x)^2 = 1 - [x^2 + 2x\Delta x + (\Delta x)^2]$$
$$= 1 - x^2 - 2x\Delta x - (\Delta x)^2$$

2. Subtract $f(x)$:

$$f(x + \Delta x) - f(x) = [1 - x^2 - 2x\Delta x - (\Delta x)^2] - (1 - x^2)$$
$$= -2x\Delta x - (\Delta x)^2$$

3. Divide by Δx:

$$\frac{f(x + \Delta x) - f(x)}{\Delta x} = \frac{-2x\Delta x - (\Delta x)^2}{\Delta x} = -2x - \Delta x$$

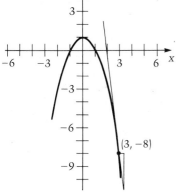

Figure 4-20

4. Take the limit:

$$\lim_{\Delta x \to 0} \frac{f(x + \Delta x) - f(x)}{\Delta x} = \lim_{\Delta x \to 0} (-2x - \Delta x) = -2x$$

The slope at the point $(3, -8)$ is $-2(3) = -6$. ■

Practice 1.   Find the slope of the function $f(x) = x^2$ at the point $(0, 0)$; use the derivative, following the four-step process.

In a functional relationship, the independent variable will not always be named x, and the dependent variable will not always be named y. Whatever the names of the variables, we can always find the derivative of the dependent variable with respect to the independent variable.

EXAMPLE 4.   Given the function $w = 1.2 - 0.3m^2$, find dw/dm.
The independent variable here is m. If we use the notation $w = f(m)$, then the definition of the derivative becomes

$$\frac{dw}{dm} = \lim_{\Delta m \to 0} \frac{f(m + \Delta m) - f(m)}{\Delta m}$$

Using the four-step process,

1. $f(m + \Delta m) = 1.2 - 0.3(m + \Delta m)^2$
$\qquad\qquad\quad = 1.2 - 0.3[m^2 + 2m\Delta m + (\Delta m)^2]$
$\qquad\qquad\quad = 1.2 - 0.3m^2 - 0.6m\Delta m - 0.3(\Delta m)^2$

2. $f(m + \Delta m) - f(m) = [1.2 - 0.3m^2 - 0.6m\Delta m - 0.3(\Delta m)^2]$
$\qquad\qquad\qquad\qquad\qquad - (1.2 - 0.3m^2)$
$\qquad\qquad\qquad\qquad\quad = -0.6m\Delta m - 0.3(\Delta m)^2$

3. $\dfrac{f(m + \Delta m) - f(m)}{\Delta m} = \dfrac{-0.6m\Delta m - 0.3(\Delta m)^2}{\Delta m}$
$\qquad\qquad\qquad\qquad\quad = -0.6m - 0.3\Delta m$

4. $\lim_{\Delta m \to 0} \dfrac{f(m + \Delta m) - f(m)}{\Delta m} = \lim_{\Delta m \to 0} (-0.6m - 0.3\Delta m) = -0.6m$

Thus

$$\frac{dw}{dm} = -0.6m \quad ■$$

Practice 2.   Given the function $s = 3t^2 - 2$, find ds/dt.

Exercises / Section **4-3**

Exercises 1–4: Compute the slope of the tangent line of the function at the given point by three different methods: "graph and guess"; find the limit of a sequence of slopes (use the points whose x-coordinates are given); and use the derivative.

1. $f(x) = 1 + x^2$ at $(2, 5)$; $x_1 = 2.3, x_2 = 2.1, x_3 = 2.01, x_4 = 2.001$

2. $f(x) = x^2 - 4$ at $(-1, -3)$; $x_1 = -1.1, x_2 = -1.05, x_3 = -1.01, x_4 = -1.001$

3. $f(x) = 3x - 1$ at $(1, 2)$; $x_1 = 1.1, x_2 = 1.05, x_3 = 1.01$

4. $f(x) = 5 - 2x$ at $(3, -1)$; $x_1 = 2.9, x_2 = 2.95, x_3 = 2.99$

Exercises 5–10: Compute the slope of the tangent line of the function at the given point by using the derivative.

5. $f(x) = 4 - 3x^2$ at $(1, 1)$

6. $f(x) = 2x^2 - 1$ at $(2, 7)$

7. $f(x) = x^2 + 4x + 4$ at $(0, 4)$

8. $f(x) = -x^2 + 2x - 1$ at $(2, -1)$

9. $f(x) = x^3 + 1$ at $(0.5, 1.125)$

10. $f(x) = 0.5x^3 - 2$ at $(1, -1.5)$

Exercises 11–14: Use the four-step process to find the derivative of the dependent variable with respect to the independent variable.

11. $p = 2q^2 + 7$ 12. $v = 3 - 4w^2$

13. $m = 5 - 2n + 3n^3$ 14. $s = 21 - 7t^2 + t^3$

4-4 General Interpretation of the Derivative

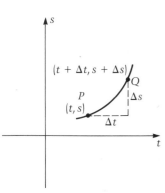

Figure 4-21

If y is a function of x, $y = f(x)$, then the derivative of this function with respect to x, $f'(x)$, is a second function of x. What meaning can we give to this second function? One interpretation, which we have already discussed, is that $f'(x)$ is a function that represents the slope of the tangent line to $f(x)$ at any point.

A second interpretation of the derivative is that $f'(x)$ is a function representing the rate of change of y with respect to x for various values of x. To understand this a little better, suppose we have some object, perhaps a car, moving along a straight line. The distance s of the car away from some fixed point is a function of the time t it has been traveling. Let $s = f(t)$, and suppose that Figure 4-21 represents part of the graph of this function. If P and Q are two points on the curve, then between P and Q there is an increment Δt and a corresponding change in s, Δs. If $\Delta t = 2$ h and $\Delta s = 120$ km, then the car has moved 120 km in 2 h. Its average velocity during this time is

$$\frac{\Delta s}{\Delta t} = \frac{120 \text{ km}}{2 \text{ h}} = 60 \text{ km/h}$$

This represents an *average* rate of change of distance with respect to time across this 2-h interval. We certainly cannot conclude that the car was going exactly 60 km/h at each moment of the entire 2 h. Velocity is a function of time and generally varies from moment to moment. How can we find the velocity at a given instant, such as at the time and position represented by P? If $\Delta s/\Delta t$ represents the average velocity over the interval Δt, then we look at the ratio $\Delta s/\Delta t$

for smaller and smaller values of Δt. As $\Delta t \to 0$, $\Delta s/\Delta t$ approaches the *instantaneous* velocity at time t. Therefore,

$$\lim_{\Delta t \to 0} \frac{\Delta s}{\Delta t} = \text{instantaneous rate of change of } s \text{ with respect to } t$$
$$\text{(instantaneous velocity)}$$

Because $s = f(t)$, we know that $\Delta s = f(t + \Delta t) - f(t)$ and

$$\lim_{\Delta t \to 0} \frac{\Delta s}{\Delta t} = \lim_{\Delta t \to 0} \frac{f(t + \Delta t) - f(t)}{\Delta t} = \frac{ds}{dt}$$

We therefore *find the instantaneous rate of change of s with respect to t at any time t by taking the derivative of s with respect to t.*

EXAMPLE 1.   An object is moving in such a way that its distance (in meters) as a function of time (in seconds) is given by the equation $s = 2 + 3t^2$. Find the instantaneous velocity (in meters per second) of the object when $t = 3$ s.

The expression for the instantaneous velocity at any time t is given by ds/dt. We first compute the derivative, and only then do we consider what happens when $t = 3$. We use the four-step process.

1. $f(t + \Delta t) = 2 + 3(t + \Delta t)^2 = 2 + 3[t^2 + 2t\Delta t + (\Delta t)^2]$
$$= 2 + 3t^2 + 6t\Delta t + 3(\Delta t)^2$$

2. $f(t + \Delta t) - f(t) = [2 + 3t^2 + 6t\Delta t + 3(\Delta t)^2] - (2 + 3t^2)$
$$= 6t\Delta t + 3(\Delta t)^2$$

3. $\dfrac{f(t + \Delta t) - f(t)}{\Delta t} = \dfrac{6t\Delta t + 3(\Delta t)^2}{\Delta t} = 6t + 3\Delta t$

4. $\displaystyle\lim_{\Delta t \to 0} \dfrac{f(t + \Delta t) - f(t)}{\Delta t} = \lim_{\Delta t \to 0} (6t + 3\Delta t) = 6t$

$$\frac{ds}{dt} = 6t$$

The velocity at any time t is $6t$. In particular, when $t = 3$, the velocity is $6 \cdot 3 = 18$ m/s. ∎

Practice 1.   An object is moving in such a way that its distance (in kilometers) as a function of time (in hours) is given by the equation $s = 0.8t^2 - 1$. Find the instantaneous velocity (in kilometers per hour) of the object when $t = 2$ h.

Whenever a quantity y is a function of x, $y = f(x)$, the derivative function $f'(x)$ may be interpreted as the instantaneous rate of change of y with respect to x. From now on, whenever we say *rate of change*, we mean *instantaneous rate of change*. Of course, the dependent and independent variables might not be named y and x.

Solving a problem which says "Find the rate of change of m with respect to n when $n = a$" requires three steps.

1. You must have the function which relates m and n, $m = f(n)$. Often the equation is given as part of the problem. Other times you must derive the equation from information given in the problem; and still other times—for instance if the equation is a well-known formula from geometry—you must supply it. In an actual technological problem, the functional relationship is often taken as the "best guess" that fits observed data about m and n.

2. You then differentiate to find dm/dn, which will be a function of n.

3. Finally, you substitute a for n in dm/dn.

Word of Advice

To find the rate of change of m with respect to n when $n = a$, be sure you differentiate *before* you substitute a for n.

EXAMPLE 2.   The specific heat of a certain gas as a function of temperature is given by the equation $c_p = 2.04 + 0.03T^2$. Find the rate of change of c_p with respect to T (in cal/gm($°$C)^2) when $T = 210°C$.

Here we are given the function, so we may proceed to step 2 and differentiate. We want dc_p/dT. We use the four-step process.

1. $f(T + \Delta T) = 2.04 + 0.03(T + \Delta T)^2 = 2.04 + 0.03T^2$
$$+ 0.06T\Delta T + 0.03(\Delta T)^2$$

2. $f(T + \Delta T) - f(T) = [2.04 + 0.03T^2 + 0.06T\Delta T + 0.03(\Delta T)^2]$
$$- (2.04 + 0.3T^2) = 0.06T\Delta T + 0.03(\Delta T)^2$$

3. $\dfrac{f(T + \Delta T) - f(T)}{\Delta T} = \dfrac{0.06T\Delta T + 0.03(\Delta T)^2}{\Delta T} = 0.06T + 0.03\Delta T$

4. $\lim\limits_{\Delta T \to 0} \dfrac{f(T + \Delta T) - f(T)}{\Delta T} = \lim\limits_{\Delta T \to 0}(0.06T + 0.03\Delta T) = 0.06T$

$$\frac{dc_p}{dT} = 0.06T$$

At $T = 210°C$, the rate of change is $0.06(210) = 12.6$ cal/gm($°$C)^2.   ■

Practice 2.   The total energy W (in joules) supplied to an inductor while the current I (in amperes) increases from 0 to I is given by the equation $W = 1.9I^2$. Find the rate of change of energy with respect to current when the current is 0.8 A.

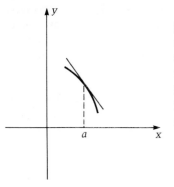

Figure 4-22

Suppose that when we compute dy/dx and evaluate it for some particular value of x, such as $x = a$, we get a negative number. What does a negative rate of change of y with respect to x tell us? If we again think of dy/dx as the slope function, then at $x = a$, the slope of the tangent line is negative. This indicates that at $x = a$, the function $y = f(x)$ is a decreasing function (see Figure 4-22). Therefore, a negative rate of change just means that as x increases, y decreases.

EXAMPLE 3. Water is pumped from a tank in a water purification plant in such a way that the volume of water (in cubic meters) remaining in the tank t hours after the pump is started is given by the equation $W = 2t - 19t^2 + 28.6$. Find the rate of change of W with respect to t when $t = 15$ min.

We first find dW/dt using the four-step process.

1. $f(t + \Delta t) = 2(t + \Delta t) - 19(t + \Delta t)^2 + 28.6$
$$= 2t + 2\Delta t - 19[t^2 + 2t\Delta t + (\Delta t)^2] + 28.6$$
$$= 2t + 2\Delta t - 19t^2 - 38t\Delta t - 19(\Delta t)^2 + 28.6$$

2. $f(t + \Delta t) - f(t) = [2t + 2\Delta t - 19t^2 - 38t\Delta t - 19(\Delta t)^2$
$$+ 28.6] - (2t - 19t^2 + 28.6)$$
$$= 2\Delta t - 38t\Delta t - 19(\Delta t)^2$$

3. $\dfrac{f(t + \Delta t) - f(t)}{\Delta t} = \dfrac{2\Delta t - 38t\Delta t - 19(\Delta t)^2}{\Delta t} = 2 - 38t - 19\Delta t$

4. $\lim\limits_{\Delta t \to 0} \dfrac{f(t + \Delta t) - f(t)}{\Delta t} = \lim\limits_{\Delta t \to 0} (2 - 38t - 19\Delta t) = 2 - 38t$
$$\frac{dW}{dt} = 2 - 38t$$

Because the equation gave us water as a function of time in hours, we must change 15 min to 0.25 h. At $t = 0.25$, the rate of change is $2 - 38(0.25) = -7.5$ m³/h. The negative rate of change shows that around 15 min after the pump is turned on, the amount of water is decreasing, as time increases, at the rate of 7.5 m³/h. ∎

Exercises / Section 4-4

Exercises 1–4: Given the equation for distance s (in meters) as a function of time t (in minutes), find the instantaneous velocity at the time indicated.

1. $s = 3 + 5t$; $t = 1.2$ min

2. $s = 1 + 2t$; $t = 3.4$ min

3. $s = 4 + 2t^3$; $t = 17.2$ min

4. $s = 5t^3 - 2$; $t = 8.5$ min

5. A tortoise and a hare begin a race at time $t = 0$. The distance s_1 (in kilometers) traveled by the tortoise as a function of time (in hours) is given by the equation $s_1 = 2t$. The distance s_2 (in kilometers) traveled by the hare as a function of time (in hours) is given by the equation $s_2 = 10t - t^2$.

a. Find an expression for the instantaneous velocity of the tortoise as a function of time.

Sketch a graph of velocity as a function of time.

b. Find an expression for the instantaneous velocity of the hare as a function of time. Sketch a graph of velocity as a function of time on the same coordinate system as (a).

c. At what time do the two animals have the same velocity?

6. In t minutes after a plane develops engine trouble, it flies a horizontal distance of s kilometers, where $s = 5t - \frac{1}{2}t^2$.

a. Find an expression for the instantaneous horizontal velocity of the plane as a function of the time elapsed after developing engine trouble.

b. If the plane will stall when its horizontal velocity reaches 2 km/min, how many minutes will elapse between the time the engine develops trouble and the time the plane stalls?

7. The temperature T (in degrees Celsius) of a city as a function of the time t (in hours) that has elapsed from a fixed moment is given by the equation $T = -t^2 + 5t + 20$. Find the rate of change of T with respect to t when $t = 2$ h.

8. The cost C (in dollars) to produce x units of a product is given by the equation $C = 250 + 10x - \frac{1}{5}x^2$. Find the rate of change of C with respect to x when $x = 20$.

4-5 Differentiating Polynomials

Good news! You do not have to use the four-step process every time you want to find the derivative of a function!

Now that we have the definition of the derivative well in mind and have seen how the interpretations of the derivative make it a useful idea, we are ready for some shortcuts. Our procedure is to select a general representation of a certain type of function, something to serve as a pattern. We use the four-step process to find the derivative of the pattern. Then, any time we have a function matching the pattern, we know what its derivative looks like. Once again we are doing the general problem once and for all; then we can apply the results to any specific case that fits the pattern.

In this section we solve the general problem of how to differentiate a polynomial. The pattern for a polynomial is a function of the form

$$f(x) = a_n x^n + a_{n-1} x^{n-1} + \cdots + a_1 x + a_0$$

where n is a nonnegative integer and the a's are real numbers. Breaking it down further, any polynomial can be thought of as a sum of three possible types of functions:

1. A constant c.
2. x^n, where n is a positive integer.
3. cx^n, where c is a constant and n is a positive integer.

We will learn how to differentiate each of these three function types and how to differentiate the sum of two functions. Putting all this information together, we will be able to differentiate with ease any polynomial. (Maybe you have already discovered a quick rule that works to differentiate polynomials.)

First let $y = c$, where c represents any constant. Figure 4-23 shows a typical graph of a constant function. This curve has the

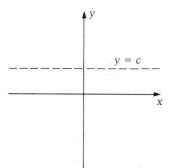

$y = c$

Figure 4-23

same tangent line everywhere, a horizontal line. The slope of a horizontal line is zero. Thinking of the derivative as the slope of the tangent line at any point, we suspect that the derivative of this function will be everywhere zero. We check by using the four-step process on the function $f(x) = c$.

1. $f(x + \Delta x) = c$

2. $f(x + \Delta x) - f(x) = c - c = 0$

3. $\dfrac{f(x + \Delta x) - f(x)}{\Delta x} = \dfrac{0}{\Delta x} = 0$

4. $\lim\limits_{\Delta x \to 0} \dfrac{f(x + \Delta x) - f(x)}{\Delta x} = \lim\limits_{\Delta x \to 0} 0 = 0$

The derivative of any constant function is zero:

$$\frac{dc}{dx} = 0$$

EXAMPLE 1. Find the derivative of $y = 7$.

The function $y = 7$ fits the pattern $y = c$. If we were to use the four-step process on the function $y = 7$, we would get the same result as we did when we let $y = c$, namely, 0. But the nice part is that by using $y = c$ as a pattern for any constant function, we have already done any problem of this type, so we do not have to do it again. We know $dc/dx = 0$, so $d(7)/dx = 0$. ∎

Now we use the four-step process to find the derivative of the function $y = f(x) = x^n$, where n is a positive integer.

1. $f(x + \Delta x) = (x + \Delta x)^n$. If we expand $(x + \Delta x)^n$, we get

$$x^n + nx^{n-1}\Delta x + \frac{n(n-1)}{2} x^{n-2}(\Delta x)^2 + \cdots + (\Delta x)^n$$

2. $f(x + \Delta x) - f(x) = x^n + nx^{n-1}\Delta x + \dfrac{n(n-1)}{2} x^{n-2}(\Delta x)^2$

$+ \cdots + (\Delta x)^n - x^n$

$= nx^{n-1}\Delta x + \dfrac{n(n-1)}{2} x^{n-2}(\Delta x)^2$

$+ \cdots + (\Delta x)^n$

3. $\dfrac{f(x + \Delta x) - f(x)}{\Delta x} = \dfrac{nx^{n-1}\Delta x + \dfrac{n(n-1)}{2}x^{n-2}(\Delta x)^2 + \cdots + (\Delta x)^n}{\Delta x}$

$= nx^{n-1} + \dfrac{n(n-1)}{2} x^{n-2}\Delta x$

$+ \cdots + (\Delta x)^{n-1}$

4. $\lim\limits_{\Delta x \to 0} \dfrac{f(x + \Delta x) - f(x)}{\Delta x} = \lim\limits_{\Delta x \to 0} [nx^{n-1}$

$$+ \frac{n(n-1)}{2} x^{n-2} \Delta x + \cdots + (\Delta x)^{n-1}]$$

$\underbrace{\qquad\qquad\qquad\qquad\qquad\qquad\qquad\qquad\qquad}$

All these terms contain Δx factors, so they all have a limit of 0 as $\Delta x \to 0$.

$$= nx^{n-1}$$

The derivative of x to a positive integral power is the power times x to the one less power:

$$\boxed{\dfrac{d(x^n)}{dx} = nx^{n-1}}$$

EXAMPLE 2. Find the derivative of $y = x^3$.

This function fits the pattern $y = x^n$, so we should multiply the power times x to the one less power.

$$\frac{d(x^3)}{dx} = 3x^{3-1} = 3x^2 \quad \blacksquare$$

EXAMPLE 3. Find the derivative of $y = x^6$.

Using the general rule for the pattern $y = x^n$, we get $d(x^6)/dx$
$= 6x^5$. $\quad \blacksquare$

Practice 1. Find the derivative of $y = x^9$.

Next we want to find the derivative of a function of the form cx^n. This is a constant times a function of x, so we let our pattern be even more general and consider $f(x) = cu(x)$, where $u(x)$ is any function of x. We find the derivative by the four-step process.

1. $f(x + \Delta x) = cu(x + \Delta x)$. We can say nothing more here because we do not know anything about $u(x)$.

2. $f(x + \Delta x) - f(x) = cu(x + \Delta x) - cu(x)$

3. $\dfrac{f(x + \Delta x) - f(x)}{\Delta x} = \dfrac{cu(x + \Delta x) - cu(x)}{\Delta x} = c\left[\dfrac{u(x + \Delta x) - u(x)}{\Delta x}\right]$

4. $\lim\limits_{\Delta x \to 0} \dfrac{f(x + \Delta x) - f(x)}{\Delta x} = \lim\limits_{\Delta x \to 0} c\left[\dfrac{u(x + \Delta x) - u(x)}{\Delta x}\right]$

$$= \lim\limits_{\Delta x \to 0} c \lim\limits_{\Delta x \to 0} \dfrac{u(x + \Delta x) - u(x)}{\Delta x}$$

▲ Applying the limit to each factor of the product.

$$= c\frac{du}{dx}$$

c is a constant, so $\lim_{\Delta x \to 0} c = c$, and $\lim_{\Delta x \to 0}[u(x + \Delta x) - u(x)]/\Delta x$ is the definition of du/dx.

The derivative of a constant times a function of x is the constant times the derivative of the function:

$$\frac{d(cu)}{dx} = c\frac{du}{dx}$$

EXAMPLE 4. Find the derivative of $y = 4x^5$.

This fits the pattern of a constant times a function, so $d(4x^5)/dx = 4\,[d(x^5)/dx] = 4(5x^4) = 20x^4$. Here we have used a pattern within a pattern. The main pattern is "constant times function," and the function is the pattern "*x* to a positive integral power." As we learn more patterns for differentiation, this sort of thing will happen frequently. We must first recognize the main pattern of the problem and follow the rule for it; we then use the proper rules for any subpatterns that occur. ∎

EXAMPLE 5. Find the derivative of $y = 3.7x^{10}$.

$$\frac{d(3.7x^{10})}{dx} = 3.7\frac{d(x^{10})}{dx} = 3.7(10x^9) = 37x^9 \quad ∎$$

Practice 2. Find the derivative of $y = 14x^2$.

EXAMPLE 6. Find the derivative of $y = 3x$.

This is a constant times a function; $d(3x)/dx = 3[d(x)/dx] = 3(1 \cdot x^0) = 3$. Here we have found $d(x)/dx$ by using the rule for *x* to a positive integral power, where the power equals 1. If we graph the function $y = 3x$, we get a straight line with slope everywhere equal to 3, and we see why dy/dx is the constant function 3. ∎

Finally, we want to find the derivative of the sum of two functions, so we let $f(x) = u(x) + v(x)$, where $u(x)$ and $v(x)$ are any functions of *x*. We again use the four-step process.

1. $f(x + \Delta x) = u(x + \Delta x) + v(x + \Delta x)$

2. $f(x + \Delta x) - f(x) = u(x + \Delta x) + v(x + \Delta x) - [u(x) + v(x)]$

$$= u(x + \Delta x) - u(x) + v(x + \Delta x) - v(x)$$

3. $\dfrac{f(x + \Delta x) - f(x)}{\Delta x} = \dfrac{u(x + \Delta x) - u(x)}{\Delta x} + \dfrac{v(x + \Delta x) - v(x)}{\Delta x}$

4. $\displaystyle\lim_{\Delta x \to 0} \dfrac{f(x + \Delta x) - f(x)}{\Delta x} = \lim_{\Delta x \to 0} \dfrac{u(x + \Delta x) - u(x)}{\Delta x}$

$$+ \lim_{\Delta x \to 0} \dfrac{v(x + \Delta x) - v(x)}{\Delta x}$$

$$= \dfrac{du}{dx} + \dfrac{dv}{dx}$$

The derivative of the sum of two functions is the sum of their derivatives:

$$\boxed{\dfrac{d(u + v)}{dx} = \dfrac{du}{dx} + \dfrac{dv}{dx}}$$

EXAMPLE 7. Find the derivative of $y = 9x^3 + 2x$.
The main pattern here is the sum of two functions:

$\dfrac{dy}{dx} = \dfrac{d(9x^3 + 2x)}{dx} = \dfrac{d(9x^3)}{dx} + \dfrac{d(2x)}{dx}$ These are both of the form a constant times a function.

$= 9\dfrac{d(x^3)}{dx} + 2\dfrac{d(x)}{dx}$ These are both of the form x to a positive integral power.

$= 9 \cdot 3x^2 + 2 \cdot x^0$

$= 27x^2 + 2$ ∎

EXAMPLE 8. Find the derivative of $y = 2x^4 - 7x^3 + 5x^2 - 1$.
Remember that $2x^4 - 7x^3 + 5x^2 - 1$ is the same as $2x^4 + (-7)x^3 + 5x^2 + (-1)$, so the main pattern is still a sum.

$\dfrac{d(2x^4 - 7x^3 + 5x^2 - 1)}{dx} = \dfrac{d(2x^4)}{dx} + \dfrac{d(-7x^3)}{dx} + \dfrac{d(5x^2)}{dx} + \dfrac{d(-1)}{dx}$

$= 2 \cdot 4x^3 - 7 \cdot 3x^2 + 5 \cdot 2x + 0$

$= 8x^3 - 21x^2 + 10x$ ∎

With some practice it becomes easy to write the derivative of any polynomial directly.

EXAMPLE 9. Find the derivative of $y = 5x^4 - 6x^3 + 14$.

$\dfrac{dy}{dx} = 20x^3 - 18x^2$ ∎

Practice 3.   Find the derivative of $y = 7x^3 + 2x^2 - 6x + 1$.

We often need to evaluate the derivative for particular values of the independent variable. We use the notation

$$\frac{dy}{dx}\Big|_{x=a}$$

to indicate that we are substituting $x = a$ into the function dy/dx.

EXAMPLE 10.   An object moves in such a way that its distance s (in kilometers) as a function of time t (in hours) is given by the equation $s = 9t^3 + 3t^2 - t$. Find the instantaneous velocity when $t = 2$ h.

We first need to find ds/dt.

$$\frac{ds}{dt} = 27t^2 + 6t - 1$$

Then we substitute $t = 2$ into this expression:

$$\frac{ds}{dt}\Big|_{t=2} = 27 \cdot 4 + 6 \cdot 2 - 1 = 199 \text{ km/h} \quad \blacksquare$$

EXAMPLE 11.   Find the rate of change of the area of a circle with respect to its radius when the radius is 4 cm.

We must supply the equation expressing area A of a circle as a function of radius r. Of course, the formula is $A = \pi r^2$. (Remember that the value $r = 4$ will not concern us until *after* we find dA/dr.) We differentiate and get $dA/dr = 2\pi r$. Evaluating

$$\frac{dA}{dr}\Big|_{r=4}$$

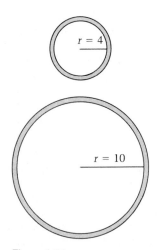

we see that the rate of change of area with respect to radius is 8π cm²/cm for a radius of 4. Note that the rate of change of area with respect to radius when the radius is 10 cm is $2\pi(10) = 20\pi$ cm²/cm. A slight change in the radius of a circle when the radius is 10 cm produces a larger change in area than the same small change in the radius when the radius is 4 cm (see Figure 4-24). The derivative dA/dr is not a constant; rather, it increases as r increases. In fact, $dA/dr = 2\pi r$ is the formula for the circumference of a circle of radius r. Figure 4-24 shows why a small change in the radius of a circle produces a change in area that is numerically about equal to the circumference of the circle, although measured in units of area rather than units of length.   ∎

Figure 4-24

Exercises / Section **4-5**

Exercises 1–18: Find the derivative.

1. $y = -2.8$
2. $y = \frac{4}{3}$
3. $y = x^2$
4. $y = x^{12}$
5. $y = 3.1x^3$
6. $y = -2.4x^2$
7. $y = 4x^5 - 2$
8. $y = 5 - 3x^4$
9. $y = 10x^3 - 4x^2 + 1$
10. $y = 3x^2 + 2x - 5$
11. $y = 4x^4 - 2x^3 + 5x - 17$
12. $y = -5x^4 - 2x^2 + 6x - 1$
13. $y = 1.8x^3 + 2.9x^2 + 7$
14. $y = 3.4x^3 - 1.8x - 2.9$
15. $y = \frac{2}{3}x^{10} - \frac{1}{8}x^3 + \frac{3}{4}x$
16. $y = \frac{4}{5}x^{11} - \frac{2}{3}x^9 + \frac{1}{10}x^2$
17. $y = 8x^5 - \sqrt{7}x$
18. $y = 10x^7 - \sqrt{5}x$

Exercises 19–24: Given the equation for distance s (in kilometers) as a function of time t (in minutes), find the instantaneous velocity at the time indicated.

19. $s = 3t^3 - 14t^2;\ t = 1$ min
20. $s = 29 - \frac{1}{2}t^2;\ t = 2$ min
21. $s = 19 + 0.04t^3;\ t = 6$ min
22. $s = 2t^4 - 5t^2 + 1;\ t = 5$ min
23. $s = 0.5t^6 - 0.01t^3 + 7t;\ t = 1.0$ min
24. $s = 0.8t^5 + 0.03t^2 + 1.9;\ t = 2.0$ min

Exercises 25–30: Find the slope of the tangent line at the point whose x-coordinate is given.

25. $y = 6x^3 - 7x^2 + 4;\ x = 3$
26. $y = 4x^5 + 6x^3 - 5x;\ x = 2$
27. $y = 12 - 8x^5;\ x = 4$
28. $y = 14 + 10x^6;\ x = -3$
29. $y = 4x^5 - 2x^4 + 6x;\ x = -5$
30. $y = 5x^6 + 3x^5 - 8x;\ x = 2$

31. The current I (in amperes) in one circuit as a function of time t (in seconds) is given by the equation $I = 0.06t^2 + 2.1t$. Find the rate of change of current with respect to time at $t = 4$ s.

32. The density D of population of an algae in a lake (in thousands per liter) as a function of water temperature T (in degrees Celsius) is given by the equation $D = 2T^2 - 15T + 100$. Find the rate of change of density with respect to temperature at $T = 10°C$.

33. When a manufacturing company produces no more than 60 of a certain item, the total production cost C (in dollars) is given as a function of the number x of units produced by the equation $C = -0.02x^2 + 1.5x$. Find the rate of change of cost with respect to number of units at $x = 10$.

34. A basic equation in electroplating is

$$m(t) = \frac{MI}{vF}t$$

where $m(t)$ is the mass (in grams) of the substance liberated by time t (in seconds), M is the molecular weight of the substance, v is its valence, F is the Faraday constant, and I is the current in the solution; M, I, v, and F are all constant. Find the rate at which mass is changing with respect to time at $t = 3$ s.

35. The voltage produced by a certain thermocouple as a function of temperature is directly proportional to the square of the temperature. If a thermocouple produces 650 V at a temperature of 100°C, find the rate of change of voltage with respect to temperature when the temperature is 150°C.

36. The kinetic energy of a rotating flywheel is directly proportional to the square of its angular velocity. If a certain flywheel has a kinetic energy of 125 J for an angular velocity of 50 rads/s, find the rate of change of kinetic energy with respect to angular velocity (in joule-seconds) when the angular velocity is 100 rads/s.

37. Under certain conditions, the amount A of insulin secreted by the pancreas into the bloodstream of an individual as a function of the level x of an individual's blood sugar is given by the equation $A = k_1(x^2 - 2k_2x)$, where k_1 and k_2 are constants. Find an expression for the rate of change of A with respect to x.

38. The thrust T of a rocket due to expelled gas at time t after lift-off is $T = -m'(t)v(t)$, where $m(t)$ is the mass of the rocket at time t and $v(t)$ is the velocity of the ejected gases at time t. Find an expression for the thrust as a function of time if $m(t) = 200 - \frac{1}{2}t - \frac{1}{6}t^3$ and $v(t) = 55 - 2t$.

39. A car travels s meters in the t seconds after the brakes are applied, where $s = -9.9t^2 + 79.2t$.
    a. Find an expression for the instantaneous velocity as a function of time.
    b. How many seconds after the brakes are applied will the car stop?
    c. After the brakes are applied, how far will the car travel before it stops?

40. In a bacteria culture, the concentration c of cells at time t is given by the equation $c = 100 + 2t$, and the concentration n of the nutrient on which the cells feed is given by the equa-tion $n = 1000 - 13t - \frac{1}{8}t^2$. The rates of change of c and n are related by the equation $c'(t) + ac(t) = bn'(t)$, where a and b are constants. For $a = \frac{1}{2}$, find the value of b.

41. The reserve funds F held by an insurance company at time t satisfy the equation $F(t) = F(0) + P(t) - C(t)$, where $F(0)$ is the company's reserves when it was founded, $P(t)$ is the amount received in premiums up to time t, and $C(t)$ is the amount paid out in claims up to time t. For $P(t) = 3t - \frac{1}{10}t^2 + \frac{1}{100}t^3$ and $C(t) = 2t - \frac{1}{5}t^3 + \frac{1}{100}t^4$, find $F'(t)$, which tells the rate at which the company's reserves are changing at time t.

42. The revenue R from selling x units of a certain product is $R = 9.8x$. The cost of producing x units of the product is $C = 200 + 3.4x$. Find an expression for the profit P from selling x units, and an expression for the rate of change of P with respect to x (called the *marginal profit*).

4-6   Differentiating Products and Quotients

We can now differentiate polynomial functions without having to resort to the four-step process. In this section we develop the patterns for differentiating functions that are products or quotients of other functions.

In order to deal with a product of functions, we let $f(x) = u(x)v(x)$, where $u(x)$ and $v(x)$ are any functions of x. Applying the four-step process,

1. $f(x + \Delta x) = u(x + \Delta x)v(x + \Delta x)$

2. $f(x + \Delta x) - f(x) = u(x + \Delta x)v(x + \Delta x) - u(x)v(x)$
$$= u(x + \Delta x)v(x + \Delta x) - u(x + \Delta x)v(x)$$
$$+ u(x + \Delta x)v(x) - u(x)v(x)$$

We subtracted and added the same term; this is legal and simplifies things in Step 4.

$$= u(x + \Delta x)[v(x + \Delta x) - v(x)]$$
$$+ v(x)[u(x + \Delta x) - u(x)]$$

3. $\dfrac{f(x + \Delta x) - f(x)}{\Delta x} = u(x + \Delta x)\dfrac{v(x + \Delta x) - v(x)}{\Delta x}$
$$+ v(x)\dfrac{u(x + \Delta x) - u(x)}{\Delta x}$$

4. $\displaystyle\lim_{\Delta x \to 0} \frac{f(x + \Delta x) - f(x)}{\Delta x} = \lim_{\Delta x \to 0} u(x + \Delta x) \lim_{\Delta x \to 0} \frac{v(x + \Delta x) - v(x)}{\Delta x}$

$\displaystyle + \lim_{\Delta x \to 0} v(x) \lim_{\Delta x \to 0} \frac{u(x + \Delta x) - u(x)}{\Delta x}$

$\displaystyle = u(x) \frac{dv}{dx} + v(x) \frac{du}{dx}$ $\quad \begin{array}{l}\lim_{\Delta x \to 0} v(x) \\ = v(x) \text{ because} \\ \Delta x \text{ does not} \\ \text{appear in } v(x).\end{array}$

The derivative of the product of two functions is the first times the derivative of the second plus the second times the derivative of the first:

$$\frac{d(uv)}{dx} = u \frac{dv}{dx} + v \frac{du}{dx}$$

This important rule is called the **product rule**. Say it three times each day before breakfast!

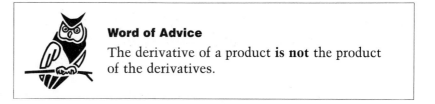

Word of Advice

The derivative of a product **is not** the product of the derivatives.

EXAMPLE 1. Let $y = (2x - 1)(3x^2 + 4x)$. Find dy/dx.

We can do this problem two ways. By multiplying, we see that $y = 6x^3 + 5x^2 - 4x$, so $dy/dx = 18x^2 + 10x - 4$. Using the product rule, the first function is $2x - 1$ and the second is $3x^2 + 4x$, so

$\displaystyle \frac{dy}{dx} = (2x - 1) \frac{d(3x^2 + 4x)}{dx} + (3x^2 + 4x) \frac{d(2x - 1)}{dx}$

$\displaystyle = (2x - 1)(6x + 4) + (3x^2 + 4x)(2)$

$\displaystyle = (12x^2 + 2x - 4) + (6x^2 + 8x)$

$\displaystyle = 18x^2 + 10x - 4 \quad \blacksquare$

Practice 1. Let $y = (3x^2 - 2)(4x + 5)$. Find dy/dx by:
a. multiplying and then differentiating a polynomial;
b. using the product rule.

EXAMPLE **2.**   In a certain gas, the equation relating volume V, pressure P, and temperature T is

$$T = 0.1PV$$

The pressure as a function of time is given by the equation $P = 2.9t^3 - t^2$, and the volume as a function of time is given by the equation $V = 0.7t + 1.8$. Find an expression for the rate of change of T with respect to t.

Here $T = 0.1(2.9t^3 - t^2)(0.7t + 1.8)$. Thus, using first the constant-times-function pattern and then the product pattern,

$$\frac{dT}{dt} = 0.1 \frac{d[(2.9t^3 - t^2)(0.7t + 1.8)]}{dt}$$

$$= 0.1[(2.9t^3 - t^2)(0.7) + (0.7t + 1.8)(8.7t^2 - 2t)]$$

$$= 0.1[2.03t^3 - 0.7t^2 + 6.09t^3 + 14.26t^2 - 3.6t]$$

$$= 0.1(8.12t^3 + 13.56t^2 - 3.6t)$$

$$= 0.812t^3 + 1.356t^2 - 0.36t \quad \blacksquare$$

Next, we consider the quotient of two functions. We let $f(x) = u(x)/v(x)$ and apply the four-step process.

1. $f(x + \Delta x) = \dfrac{u(x + \Delta x)}{v(x + \Delta x)}$

2. $f(x + \Delta x) - f(x) = \dfrac{u(x + \Delta x)}{v(x + \Delta x)} - \dfrac{u(x)}{v(x)}$

$$= \frac{v(x)u(x + \Delta x) - u(x)v(x + \Delta x)}{v(x + \Delta x)v(x)}$$

$$= \frac{v(x)u(x + \Delta x) - v(x)u(x) + v(x)u(x) - u(x)v(x + \Delta x)}{v(x + \Delta x)v(x)}$$

Again, we subtracted and added the same term.

$$= \frac{v(x)[u(x + \Delta x) - u(x)] - u(x)[v(x + \Delta x) - v(x)]}{v(x + \Delta x)v(x)}$$

3. $\dfrac{f(x + \Delta x) - f(x)}{\Delta x} = \dfrac{v(x)\dfrac{u(x + \Delta x) - u(x)}{\Delta x} - u(x)\dfrac{v(x + \Delta x) - v(x)}{\Delta x}}{v(x + \Delta x)v(x)}$

4. $\lim\limits_{\Delta x \to 0} \dfrac{f(x + \Delta x) - f(x)}{\Delta x}$

$$= \frac{\lim\limits_{\Delta x \to 0} v(x) \lim\limits_{\Delta x \to 0} \dfrac{u(x + \Delta x) - u(x)}{\Delta x} - \lim\limits_{\Delta x \to 0} u(x) \lim\limits_{\Delta x \to 0} \dfrac{v(x + \Delta x) - v(x)}{\Delta x}}{\lim\limits_{\Delta x \to 0} v(x + \Delta x)v(x)}$$

$$= \frac{v(x)\dfrac{du}{dx} - u(x)\dfrac{dv}{dx}}{[v(x)]^2}$$

The derivative of a quotient is the denominator times the derivative of the numerator minus the numerator times the derivative of the denominator, all over the denominator squared:

$$\frac{d\left(\dfrac{u}{v}\right)}{dx} = \frac{v\,\dfrac{du}{dx} - u\,\dfrac{dv}{dx}}{v^2}$$

This is called the **quotient rule**—say it five times each day before breakfast!

Unlike the product rule, there is a minus sign involved in the pattern for differentiating a quotient, so be careful to use the rule exactly as stated. (Subtraction is *not* commutative, so $v(du/dx) - u(dv/dx)$ is not the same as $u(dv/dx) - v(du/dx)$.)

EXAMPLE 3. Find the derivative of $y = \dfrac{5x^2 - 2x}{3x + 7}$.

Here y is a quotient of two polynomials, so we apply the quotient rule.

$$\begin{aligned}
\frac{dy}{dx} &= \frac{(3x + 7)\dfrac{d}{dx}(5x^2 - 2x) - (5x^2 - 2x)\dfrac{d}{dx}(3x + 7)}{(3x + 7)^2} \\[2mm]
&= \frac{(3x + 7)(10x - 2) - (5x^2 - 2x)(3)}{(3x + 7)^2} \\[2mm]
&= \frac{(30x^2 + 64x - 14) - (15x^2 - 6x)}{9x^2 + 42x + 49} \\[2mm]
&= \frac{15x^2 + 70x - 14}{9x^2 + 42x + 49} \quad \blacksquare
\end{aligned}$$

Practice 2. Find the derivative of $y = \dfrac{3x - 4}{2x^2}$.

EXAMPLE 4. Find the derivative of $y = \dfrac{(2x + 1)(5x^2 - 4x)}{6x}$.

The basic form of y is that of a quotient, so we use the quotient rule.

$$\frac{dy}{dx} = \frac{(6x)\dfrac{d}{dx}[(2x + 1)(5x^2 - 4x)] - (2x + 1)(5x^2 - 4x)\dfrac{d}{dx}(6x)}{(6x)^2}$$

We now see that in differentiating the numerator of the quotient we must use the product rule.

$$\frac{dy}{dx} = \frac{(6x)[(2x + 1)\frac{d}{dx}(5x^2 - 4x) + (5x^2 - 4x)\frac{d}{dx}(2x + 1)] - (2x + 1)(5x^2 - 4x)\frac{d}{dx}(6x)}{(6x)^2}$$

$$= \frac{(6x)[(2x + 1)(10x - 4) + (5x^2 - 4x)(2)] - (2x + 1)(5x^2 - 4x)(6)}{36x^2}$$

$$= \frac{(6x)(30x^2 - 6x - 4) - (10x^3 - 3x^2 - 4x)(6)}{36x^2}$$

$$= \frac{120x^3 - 18x^2}{36x^2} = \frac{20x - 3}{6}$$

The important thing in a problem of this type is simply to use patience. We analyze the main pattern and apply the appropriate rule. In the course of applying the main rule, we may well have to use other rules on the subpatterns of the problem. In this example, the main pattern is a quotient, but there are polynomials and a product as subpatterns. Writing out each step, without trying to combine several steps into one, reduces the chance of error. ■

Practice 3.   Find the derivative of $y = \dfrac{(2x^2 - 1)(x + 3)}{4x - 2}$.

EXAMPLE 5.   The illuminance E from a light source is given by the equation $E = k/r^2$, where r is the distance from the source and k is a constant. Find an expression for the rate of change of illuminance with respect to distance.

Using the quotient rule as the main pattern, we get

$$\frac{dE}{dr} = \frac{r^2\frac{d(k)}{dr} - k\frac{d(r^2)}{dr}}{(r^2)^2}$$

$$= \frac{r^2 \cdot 0 - k(2r)}{r^4}$$

$$= \frac{-2kr}{r^4} = \frac{-2k}{r^3} \quad ■$$

Exercises / Section **4-6**

Exercises 1–6: Find dy/dx by (a) multiplying and then differentiating; and (b) using the product rule.

1. $y = 3x^2(4x - 7)$

2. $y = (3x + 2)(6x^3)$

3. $y = (4x - 9)(2x + 5)$

4. $y = (14 - 2x)(3x - 2)$

5. $y = (12 - 4x)(2x^2 + 3x)$

6. $y = (4x^2 - 1)(2x + 3)$

Exercises 7–16: Find the derivative.

7. $y = (3x - 4)(2x^5)$

8. $y = (2x^3 - 7)(3x^2)$

9. $y = (5x^2 - 7x)(2x^3 + 5x - 2)$

10. $y = (3x^4 - 2x^2)(4x^2 - x)$

11. $y = \dfrac{2x^2 - 5}{3x}$

12. $y = \dfrac{4x^2}{3x + 2}$

13. $y = \dfrac{(3x - 2)(x + 7)}{3x - 1}$

14. $y = \dfrac{3x^2}{(5x + 7)(2x - 1)}$

15. $y = \dfrac{2}{3x^2 - 7x + 4}$

16. $y = \dfrac{5}{4x^3 - 3x^2}$

Exercises 17–20: Given the equation for distance s (in meters) as a function of time t (in seconds), find the instantaneous velocity at the time indicated.

17. $s = (1.4t^2)(3t + 2);\quad t = 2$ s

18. $s = (2.8t + 7)(0.8t^3);\quad t = 1$ s

19. $s = \dfrac{4.2}{1.8t^2};\quad t = 3$ s

20. $s = \dfrac{3.8t^3}{2t + 7};\quad t = 2$ s

Exercises 21–24: Find the slope of the tangent line at the point whose x-coordinate is given.

21. $y = (3x^3 - 2x^2 + 4x)(2x - 7);\quad x = 1$

22. $y = (3x^2 - 4x + 1)(5x^2 + 2);\quad x = 3$

23. $y = \dfrac{2x^2}{3x + 7};\quad x = 2$

24. $y = \dfrac{(2x - 1)(4x^3)}{5x + 6};\quad x = -1$

25. The current I (in amperes) in a resistor is the voltage V across the resistor divided by the resistance R. If V and R are functions of time t (in seconds) given by $V = 3.4t^3 - 2t$ and $R = 1.9t^2 - 2.6$, find the rate of change of current with respect to time when $t = 4$ s.

26. The volume V (in cubic meters) of a quantity of gas is given by the equation $V = 8.1T/P$, where temperature T and pressure P are functions of time t (in hours) given by $T = 2.7t^3 - 5$ and $P = -2.5t$. Find the rate of change of volume with respect to time when $t = 0.78$ h.

27. The amount W of energy consumed in refining a certain ore to produce metal of x percent purity is given by the equation

$$W = \frac{x^3 + 3x}{100 - x}$$

Find an expression for the rate of change of W with respect to x.

28. After x hours of training, an industrial training program produces a level s of skill in a particular job given by the equation

$$s = \frac{10x^3 - 3x^2 + 4}{x^3 + 1}$$

Find an expression for the rate of change of skill with respect to hours of training.

29. In a model of the muscular system, the length L of a muscle in response to a shock of intensity s is given by the equation

$$L = \frac{K}{(s + A)(s^3 + as^2 + bs + c)}$$

where A, K, a, b, and c are constants. Find an expression for the rate of change of muscle length with respect to shock intensity.

30. Along a certain geologic fault line, one plate of the earth's crust is found to move a distance s (in centimeters) as a function of time t (in years) given by the equation $s = p + t - M/(q + rt^2)$, where P, M, q, and r are constants. Find an expression for the rate of change of s with respect to t.

4-7   Differentiating Powers of Functions

We already have a pattern for differentiating x to a positive integral power. We generalize this in two ways: First, we develop a rule to deal with *any function* of x, instead of just x, to a positive integral power. Then we see that the same rule works where the power is any rational number, not necessarily a positive integer.

We are considering functions of the form $y = u^n$, where u is a function of x and n is a positive integer. (Examples are: $y = (3x - 2)^4$, where $u = 3x - 2$ and $n = 4$; and $y = [3x/(x - 1)]^3$, where $u = 3x/(x - 1)$ and $n = 3$.) Here y is a function of u (so we can talk about dy/du) and u is a function of x (so we can talk about du/dx). But then y (the composition of these two functions) is also a function of x, $y = f(x)$, and we want to find dy/dx. In the definition of the derivative,

$$\frac{dy}{dx} = \lim_{\Delta x \to 0} \frac{f(x + \Delta x) - f(x)}{\Delta x}$$

the expression $f(x + \Delta x) - f(x)$ represents the change Δy corresponding to a change Δx in x. Therefore,

$$\frac{dy}{dx} = \lim_{\Delta x \to 0} \frac{\Delta y}{\Delta x}$$

In the same way,

$$\frac{dy}{du} = \lim_{\Delta u \to 0} \frac{\Delta y}{\Delta u} \quad \text{and} \quad \frac{du}{dx} = \lim_{\Delta x \to 0} \frac{\Delta u}{\Delta x}$$

Using the equation

$$\frac{\Delta y}{\Delta x} = \frac{\Delta y}{\Delta u} \cdot \frac{\Delta u}{\Delta x}$$

we can write

$$\frac{dy}{dx} = \lim_{\Delta x \to 0} \frac{\Delta y}{\Delta x} = \lim_{\Delta x \to 0} \left(\frac{\Delta y}{\Delta u} \cdot \frac{\Delta u}{\Delta x} \right) = \lim_{\Delta x \to 0} \frac{\Delta y}{\Delta u} \lim_{\Delta x \to 0} \frac{\Delta u}{\Delta x}$$

However, because u is a function of x, $\Delta x \to 0$ exactly when $\Delta u \to 0$, so

$$\lim_{\Delta x \to 0} \frac{\Delta y}{\Delta u} \lim_{\Delta x \to 0} \frac{\Delta u}{\Delta x} = \lim_{\Delta u \to 0} \frac{\Delta y}{\Delta u} \lim_{\Delta x \to 0} \frac{\Delta u}{\Delta x} = \frac{dy}{du} \cdot \frac{du}{dx}$$

Thus

$$\boxed{\frac{dy}{dx} = \frac{dy}{du} \cdot \frac{du}{dx}}$$

This equation, known as the **chain rule**, applies whenever we wish to differentiate a composition of functions.

In particular, for $y = u^n$,

$$\frac{dy}{dx} = \frac{d(u^n)}{du} \cdot \frac{du}{dx} = nu^{n-1}\frac{du}{dx}$$

The derivative of a function to a positive integral power is the power, times the function to the one less power, times the derivative of the function:

$$\frac{d(u^n)}{dx} = nu^{n-1}\frac{du}{dx}$$

This is called the **power rule** for differentiation.

EXAMPLE 1. Find y' for $y = (2x^2 + 4)^3$.

This fits the pattern of a function of x to a positive integral power. For this problem, $u = 2x^2 + 4$ and $n = 3$. Following the power rule, we multiply the power times the function to the one less power, times the derivative of the function.

$$\frac{dy}{dx} = 3(2x^2 + 4)^2 \frac{d(2x^2 + 4)}{dx}$$

$$= 3(2x^2 + 4)^2(4x)$$

$$= 12x(2x^2 + 4)^2 \quad \blacksquare$$

EXAMPLE 2. Find $\frac{dy}{dx}$ for $y = (3x - 4)^4$

Here again the power rule applies, with $u = 3x - 4$ and $n = 4$. With practice, we do not need to write u and n, but we still specify them mentally when we apply the power rule.

$$\frac{dy}{dx} = 4(3x - 4)^3 \frac{d}{dx}(3x - 4) = 4(3x - 4)^3(3) = 12(3x - 4)^3 \quad \blacksquare$$

Word of Advice

Whenever the power rule is used to find $d(u^n)/dx$, do not just write down nu^{n-1} and then forget du/dx. Be sure to carry this rule through to its conclusion.

EXAMPLE 3. Find $\frac{dy}{dx}$ for $y = \left(\frac{x - 2}{2x^2 + 1}\right)^5$.

The basic form of y is a function to a power, so we use the power rule.

$$\frac{dy}{dx} = 5\left(\frac{x-2}{2x^2+1}\right)^4 \frac{d}{dx}\left(\frac{x-2}{2x^2+1}\right)$$

Now we must use the quotient rule.

$$= 5\left(\frac{x-2}{2x^2+1}\right)^4 \left[\frac{(2x^2+1)\frac{d}{dx}(x-2) - (x-2)\frac{d}{dx}(2x^2+1)}{(2x^2+1)^2}\right]$$

Finally we differentiate two polynomials.

$$= 5\left(\frac{x-2}{2x^2+1}\right)^4 \left[\frac{(2x^2+1)(1) - (x-2)(4x)}{(2x^2+1)^2}\right]$$

And then we simplify.

$$= \frac{5(x-2)^4(-2x^2+8x+1)}{(2x^2+1)^6}$$

In this example we again get an idea of the "chain reaction" that can occur when we differentiate a complex function. We choose the appropriate pattern for differentiating the problem as a whole, and in the course of applying that pattern, two or three other patterns enter in. ∎

Differentiation of a function often leads to a rather messy expression, which can then be algebraically simplified. Sometimes the simplification is harder than the actual differentiation. (If your answer to a problem does not agree with someone else's answer, it may be your algebra—simplification—rather than your calculus—differentiation—which is at fault.) Most uses of the derivative, as we see in the next chapter, require that the final expression be in factored form.

Practice 1.   Differentiate $y = (3x^2 - 2x + 4)^3$.

Practice 2.   Differentiate $y = (6x + 12)(2x^2 + 7x)^2$ without first expanding and multiplying. (*Hint:* What is the main pattern that applies?)

We developed the power rule for differentiating any function to a positive integral power, and we have no reason to suspect that it will be correct for other powers. Suppose we let the power be 0, and let $y = u^0 = 1$. Because y is a constant function, we know that $dy/dx = 0$. If the power rule for differentiation were applied to $y = u^0$, it would produce $dy/dx = 0 \cdot u^{-1} \cdot (du/dx) = 0$, which is the correct answer. Therefore, the power rule may be used when the power is 0.

Now let $y = u^{-n}$, where n is a positive integer. We can write

$$y = \frac{1}{u^n}$$

and, by using the quotient rule, we see that

$$\frac{dy}{dx} = \frac{u^n \cdot 0 - 1 \cdot nu^{n-1} \cdot \frac{du}{dx}}{(u^n)^2} = (-n)\frac{u^{n-1}}{u^{2n}}\frac{du}{dx} = -nu^{-n-1}\frac{du}{dx}$$

If the power rule for differentiation were applied to $y = u^{-n}$, it would produce $dy/dx = (-n)u^{-n-1}(du/dx)$, which is the correct answer. Therefore, the power rule may be used when the power is a negative integer.

EXAMPLE 4. Differentiate $y = (4x^2 - 3x)^{-2}$
Using the power rule,

$$\frac{dy}{dx} = (-2)(4x^2 - 3x)^{-3}\frac{d}{dx}(4x^2 - 3x)$$

$$= (-2)(4x^2 - 3x)^{-3}(8x - 3)$$

$$= (-16x + 6)(4x^2 - 3x)^{-3} \quad \blacksquare$$

Practice 3. Find $\dfrac{dy}{dx}$ for $y = (2x^3 + 7)^{-5}$.

Sometimes there is a choice of methods which apply to a given problem.

EXAMPLE 5. Find the derivative for $y = \dfrac{1}{x^2}$.
We can use the quotient rule, as follows:

$$\frac{dy}{dx} = \frac{x^2 \cdot 0 - 1(2x)}{x^4} = -\frac{2}{x^3}$$

It is easier, however, to write y in the form $y = x^{-2}$ and then use the power rule.

$$\frac{dy}{dx} = (-2)x^{-3}\frac{d(x)}{dx} = -2x^{-3} = \frac{-2}{x^3} \quad \blacksquare$$

Incidentally, the second approach shown in Example 5 works for any quotient $y = u/v$. The quotient can always be expressed as $y = uv^{-1}$ and treated as a product. (If you have trouble remembering the quotient rule, you may want to try this alternative.)
Our final generalization of the power rule occurs when we

consider a function to a rational number power. We let $y = u^{p/q}$, where p and q are integers, $q \neq 0$, and u is a function of x. If we raise both sides of the equation $y = u^{p/q}$ to the qth power, we get

$$y^q = u^p \tag{1}$$

Now y is some function of x, and each side of Equation (1) consists of a function of x to an integral power. Therefore, the power rule can be applied to each side of the equation. This procedure is a bit different from anything we have done before, because we are not differentiating an equation of the form $y = f(x)$. But it is true that if two functions of x are equal, their derivatives with respect to x will be equal, so we just go ahead and differentiate each side of Equation (1) by the power rule. Recall that for the power rule we multiply the power, times the function to the one less power, times the derivative of the function, where we denote the derivative of the function y by dy/dx and the derivative of the function u by du/dx. (This process is called *implicit differentiation*, and we see it again in Section 4-9.) Differentiating Equation (1), we get

$$qy^{q-1}\frac{dy}{dx} = pu^{p-1}\frac{du}{dx}$$

We want to know what dy/dx is, so we solve this equation for dy/dx:

$$\frac{dy}{dx} = \frac{pu^{p-1}}{qy^{q-1}}\frac{du}{dx}$$

Finally we substitute for y, recalling that $y = u^{p/q}$.

$$\frac{dy}{dx} = \frac{pu^{p-1}}{q(u^{p/q})^{q-1}}\frac{du}{dx}$$

$$= \frac{pu^{p-1}}{qu^{p-p/q}}\frac{du}{dx} \qquad \text{Multiply the denominator exponents.}$$

$$= \frac{p}{q}u^{(p-1)-(p-p/q)}\frac{du}{dx} \qquad \text{Subtract exponents.}$$

$$= \frac{p}{q}u^{p/q-1}\frac{du}{dx} \qquad \text{Simplify.}$$

Now we have learned that for $y = u^{p/q}$,

$$\frac{dy}{dx} = \frac{p}{q}u^{p/q-1}\frac{du}{dx}$$

But this is the power times the function to the one less power times the derivative of the function, which is exactly what the power rule tells us to do. Therefore, *the power rule gives the correct answer and may be applied to differentiate any function of x to any rational power.*

EXAMPLE 6. Differentiate $y = \sqrt{5x + 2}$.

Converting the radical to exponential form, we write $y = (5x + 2)^{1/2}$ and use the power rule.

$$\frac{dy}{dx} = \frac{1}{2}(5x + 2)^{-1/2} \frac{d}{dx}(5x + 2) = \frac{1}{2}(5x + 2)^{-1/2}(5)$$

or, if we want the answer in radical form,

$$\frac{dy}{dx} = \frac{5}{2\sqrt{5x + 2}} \quad \blacksquare$$

EXAMPLE 7. Find the derivative for $y = \dfrac{x^2 - 1}{\sqrt[3]{2x^2 + 5}}$.

We write this in the form $y = (x^2 - 1)(2x^2 + 5)^{-1/3}$ and use the product rule.

$$\frac{dy}{dx} = (x^2 - 1) \frac{d}{dx}(2x^2 + 5)^{-1/3} + (2x^2 + 5)^{-1/3} \frac{d}{dx}(x^2 - 1)$$

$$= (x^2 - 1)\left(-\frac{1}{3}\right)(2x^2 + 5)^{-1/3-1} \frac{d}{dx}(2x^2 + 5)$$

$$\quad + (2x^2 + 5)^{-1/3}(2x)$$

$$= -\frac{1}{3}(x^2 - 1)(2x^2 + 5)^{-4/3}(4x) + (2x^2 + 5)^{-1/3}(2x)$$

To simplify, we factor out $(2x)(2x^2 + 5)^{-4/3}$.

$$\frac{dy}{dx} = 2x(2x^2 + 5)^{-4/3}[-\frac{2}{3}(x^2 - 1) + (2x^2 + 5)^1]$$

$$= 2x(2x^2 + 5)^{-4/3}(-\frac{2}{3}x^2 + \frac{2}{3} + 2x^2 + 5)$$

$$= 2x(2x^2 + 5)^{-1/3}\left(\frac{4x^2 + 17}{3}\right) \quad \blacksquare$$

Most of the time when we have a radical in a calculus problem, we should change it to exponential form at once so that we may deal with a function to a rational power.

Practice 4. Find the derivative for $y = \dfrac{1}{\sqrt{x}}$.

Practice 5. Differentiate $y = x^3(x^2 - 3x)^{1/2}$. (Note the main pattern.)

EXAMPLE 8. The frequency f of vibration of a wire is given by the equation $f = 285\sqrt{T}$, where T is the tension on the wire. Find an expression for the rate of change of f with respect to T.

In exponential form, the equation is $f = 285(T)^{1/2}$, and then

$$\frac{df}{dT} = \frac{285}{2}\,(T)^{-1/2}\,\frac{dT}{dT} = \frac{285}{2}(T)^{-1/2}\quad\blacksquare$$

Recall that an algebraic function is any function that involves only the operations of addition, subtraction, multiplication, division, and taking roots. We now have all the rules we need in order to be able to differentiate any algebraic function.

Also, if we wanted to differentiate, for example, a product of transcendental (nonalgebraic) functions, the main pattern is still a product and we would begin with the product rule. However, in the course of applying the product rule, we would need to be able to differentiate transcendental functions, which we learn to do in Chapter 8.

Exercises / Section **4-7**

Exercises 1–32: Find the derivative.

1. $y = (2x)^3$

2. $y = (x^3 - 4x)^2$

3. $y = (2x^4 - 1.9)^3$

4. $y = (x^4 - 2x^2)^4$

5. $y = \left(\dfrac{2x}{x - 3}\right)^2$

6. $y = \left(\dfrac{x^2}{2x + 1}\right)^2$

7. $y = \dfrac{(2x)\sqrt{x + 1}}{(3x - 2)^2}$

8. $y = \dfrac{\sqrt{x^3 + 4}}{x^2(x - 2)^3}$

9. $y = \dfrac{1}{2x}$

10. $y = \dfrac{1}{x + 2} - x$

11. $y = (2x^3 - 4x + 7)^{-2}$

12. $y = (4x^4 - 3x^3 + 7x^2)^{-3}$

13. $y = \dfrac{2}{(x - 1)^2}$

14. $y = \dfrac{1.9}{(2x + 4)^3}$

15. $y = \sqrt{2x - 8}$

16. $y = \sqrt[3]{x^2 - 1}$

17. $y = x^2\sqrt{x - 1}$

18. $y = \sqrt{x}(2x^2 + 1)^2$

19. $y = \dfrac{\sqrt{x - 1}}{3x^2}$

20. $y = \dfrac{2x^2 - 1}{\sqrt{x}}$

21. $y = \sqrt{\dfrac{2x}{x^2 + 1.8}}$

22. $y = \sqrt{\dfrac{x - 2}{x^2 + 5}}$

23. $y = \sqrt[4]{x}$

24. $y = \sqrt[5]{x - 1}$

25. $y = (x^3 - 4x^2)^{2/3}$

26. $y = (x^2 - 7x)^{3/4}$

27. $y = (x\sqrt{x - 1}\,)^{4/5}$

28. $y = (4x^2\sqrt{x^3}\,)^{1/4}$

29. $y = 2x^{-3} + 7x^{-5}$

30. $y = 8.9x^{-5} + 2.1x^{-4}$

31. $y = \left(\dfrac{x}{\sqrt{x + 4}}\right)^{-2}$

32. $y = \left(\dfrac{\sqrt{2x - 7}}{x^2}\right)^{-1}$

33. Find the x-values of any points on the curve $y = 3x/\sqrt{5x - 1}$ where the tangent line is horizontal.

34. Find the x-values of any points on the curve $y = x^2/\sqrt{2x^2 - 1}$ where the tangent line is horizontal.

35. Distance (in kilometers) as a function of time (in hours) for a particular object is given by the equation $s = (t^3 + 2t)^{2/3}$. Find the instantaneous velocity at $t = 1$ h.

36. Distance (in meters) as a function of time (in seconds) for a particular object is given by the equation $s = (4t^2 + 1.95)^{5/2}$. Find the instantaneous velocity at $t = 1$ s.

37. In a certain electric circuit containing inductance and capacitance, an electric oscillation occurs; the frequency f of oscillation is given by the equation $f = 172/\sqrt{C}$, where C is the capacitance. Find an expression for the rate of change of f with respect to C.

38. If a hole is drilled by a laser, then the depth D of the hole at time t, $t > 1$, is given by the equation $D = k(10 - 2t^{-2} - 5t^{-3})$, where k is a constant that depends on the nature of the material being drilled and the cross-sectional area of the hole. Find an expression for the rate of change of D with respect to t for $t > 1$.

39. Under certain conditions, the pressure P of a gas at time t is given by the equation $P = k(4t^{-3} - t^2)^{-1}$, where k is a constant. Find the derivative of P with respect to t.

40. The force of gravity F exerted by the earth on a certain satellite as a function of time t is given by the equation $F = k(t^3 - 4t^2 + t + 6)^{-2}$, where k is a constant. Find the derivative of F with respect to t.

41. The temperature T (in degrees Celsius) of a certain machine part after the machine has been in operation for t hours is given by the equation $T = 7.9 + 100.8t^{1/5}$. Find an expression for the rate of change of temperature with respect to time.

42. According to relativity theory, if a stick has length L_0 at rest, then its length L if it is moving at velocity v is given by the equation $L = L_0\sqrt{1 - (v/c)^2}$, where c is the speed of light. Find an expression for the rate of change of L with respect to v.

43. The cooling load Q to air condition a building is given by the equation $Q(t) = a[T(t) - 75.6] + M(t)$, where $T(t)$ is the outdoor temperature at time t, $M(t)$ is the interior heating load at time t, and a is a constant. If $T(t) = 5\sqrt{4 - t^2} + 60$ and $M(t) = 10 + 3t^{2/3}$, find dQ/dt.

44. The electric potential V at a point r units along the axis of a uniformly charged circular disk of radius a is given by $V = k(\sqrt{a^2 - r^2} - r)$, where k is a constant. Find dV/dr.

4-8 Higher-Order Derivatives

When we take a function $y = f(x)$ and differentiate it, we get the derivative function $y' = f'(x)$. Occasionally, for reasons that will be clear in a moment, y' is also called the **first derivative** of y with respect to x. Because the derivative y' is also a function of x, it is possible to differentiate it. The resulting function is called the **second derivative** of y with respect to x. Notations for the second derivative are

$$f''(x), \qquad y'', \qquad \frac{d^2y}{dx^2}$$

Of course, we could compute the third derivative $f'''(x)$, the fourth derivative $f^{(4)}(x)$, and so on. Notice that

$$\left(\frac{dy}{dx}\right)^2 \quad \text{and} \quad \frac{d^2y}{dx^2}$$

are not the same; the former denotes the square of the first derivative, while the latter denotes the second derivative.

EXAMPLE 1. Find the second derivative of $y = (2x^3 - 5x)^2$. We find the first derivative by the power rule.

$$\frac{dy}{dx} = 2(2x^3 - 5x)(6x^2 - 5) = (12x^2 - 10)(2x^3 - 5x)$$

To find the second derivative, we differentiate the first derivative by using the product rule.

$$\frac{d^2y}{dx^2} = (12x^2 - 10)(6x^2 - 5) + (2x^3 - 5x)(24x)$$

$$= 120x^4 - 240x^2 + 50 \quad \blacksquare$$

Practice 1.   Find y'' for $y = (2x + 5)^2$.

EXAMPLE 2.   It is easy to see that for any polynomial function, derivatives of a sufficiently high order will eventually produce the zero function. Thus if

$$y = 3x^4 - 5x^3 + 7x^2 - 2x + 4$$

then

$$y' = 12x^3 - 15x^2 + 14x - 2$$
$$y'' = 36x^2 - 30x + 14$$
$$y''' = 72x - 30$$
$$y^{(4)} = 72$$
$$y^{(5)} = 0$$

In fact, if the polynomial is of degree n, then the nth derivative will be a constant function and the $(n + 1)$st derivative will be the zero function. $\blacksquare$

EXAMPLE 3.   Evaluate $\left.\dfrac{d^3y}{dx^3}\right|_{x=0}$ for $y = x^{7/2}$.

Here we must first find the third-derivative function.

$$\frac{dy}{dx} = \frac{7}{2}x^{5/2}$$

$$\frac{d^2y}{dx^2} = \frac{35}{4}x^{3/2}$$

$$\frac{d^3y}{dx^3} = \frac{105}{8}x^{1/2}$$

Then we evaluate $\left.\dfrac{d^3y}{dx^3}\right|_{x=0}$

$$\left.\frac{d^3y}{dx^3}\right|_{x=0} = \frac{105}{8}(0)^{1/2} = 0$$

If we try to evaluate

$$\left.\frac{d^4y}{dx^4}\right|_{x=0}$$

we first find

$$\frac{d^4y}{dx^4} = \frac{105}{16}x^{-1/2} = \frac{105}{16\sqrt{x}}$$

and this function is undefined at $x = 0$. ∎

Practice 2. Evaluate $f''(1)$ for $f(x) = \dfrac{x^2}{2x + 1}$.

For a moving body, **velocity** is the rate of change of position or distance with respect to time. **Acceleration** is the rate of change of velocity with respect to time, so acceleration is the second derivative of distance with respect to time.

EXAMPLE 4. Distance (in meters) as a function of time (in minutes) for a particular object is given by the equation $s = 7t - t^3$. Then velocity v and acceleration a are the functions

$$v = \frac{ds}{dt} = 7 - 3t^2$$

$$a = \frac{d^2s}{dt^2} = -6t$$

The acceleration when $t = 1$ min, for example, is -6 m/min². The negative value indicates that right around $t = 1$ min, the velocity is decreasing and the object is slowing down. ∎

Practice 3. Find the acceleration at $t = 2.1$ min given that distance (in meters) as a function of time (in minutes) is $s = 3.9t^3 - 1.4t$.

Exercises / Section **4-8**

Exercises 1–10: Find the indicated derivative.

1. $y = 3x^4 - 1.8x^2$; y''

2. $y = 2.9x^5 - 18x^3$; y''

3. $y = 9x^4 - 28x^3$; y'''

4. $y = -2.9x^6 + 2.1x^4 - 3x^2$; y'''

5. $y = \dfrac{x^2}{x-1}; \quad \dfrac{d^2y}{dx^2}$

6. $y = \dfrac{2x}{x^2+1}; \quad \dfrac{d^2y}{dx^2}$

7. $y = x\sqrt{x-1}; \quad y''$

8. $y = x^2\sqrt{2x+1}; \quad y''$

9. $y = (2x^2-1)^3; \quad \dfrac{d^3y}{dx^3}$

10. $y = (2-x^4)^3; \quad \dfrac{d^3y}{dx^3}$

Exercises 11–18: Evaluate $f''(1)$.

11. $f(x) = 6x^{14} - 20x^{12} + 13x^8$

12. $f(x) = 8.1x^{12} + 2.9x^9 - 4x^7$

13. $f(x) = \dfrac{4x^2}{3x-7}$

14. $f(x) = \dfrac{2x+5}{x^3}$

15. $f(x) = (x^2+5)^{2/3}$

16. $f(x) = (4-x^3)^{3/4}$

17. $f(x) = \sqrt{2x^3+7}$

18. $f(x) = \sqrt{9-x+2x^4}$

Exercises 19–30: Given the equation for distance s (in kilometers) as a function of time t (in hours), find the acceleration at the time indicated.

19. $s = 1.8t^3 - 2.9t^2; \quad t = 2.8$ h

20. $s = 3.7t^4 - 1.6t^2 + 7.2; \quad t = 1.8$ h

21. $s = (2t^2 - 1.8)^3; \quad t = 1.1$ h

22. $s = (3.4t^3 - t)^4; \quad t = 2.3$ h

23. $s = \dfrac{t}{2t^2-1}; \quad t = 3.4$ h

24. $s = \dfrac{3t^3-6}{2t}; \quad t = 2.5$ h

25. $s = \dfrac{2}{t^4}; \quad t = 3$ h

26. $s = \dfrac{1.8}{t^5}; \quad t = 2$ h

27. $s = \sqrt{t^3 - 2.8}; \quad t = 2$ h

28. $s = \sqrt{3.4 - t^4}; \quad t = 1$ h

29. $s = t^2\sqrt{1+t^2}; \quad t = 1$ h

30. $s = (2t+7)\sqrt{t^3-1}; \quad t = 2$ h

31. For $f(x) = 12 + 2x^2 - x^4$, find $f''(k)$ for all values of k for which $f'(k) = 0$.

32. For $f(x) = 3x^4 - 4x^3 + 6$, find $f''(k)$ for all values of k for which $f'(k) = 0$.

33. The total amount of charge that has passed a given point in a certain electric circuit in t seconds is given by the equation $q = 2t^3 - t^2$. The voltage V (in volts) induced in an inductor in the circuit is $V = 4(d^2q/dt^2)$. Find the voltage at $t = 3.2$ s.

34. In a model of the motion of a particular object in the presence of frictional resistance, the maximal forward force F (in newtons) as a function of time t (in seconds) is given by the equation $F(t) = 10.2s''(t) + 2s'(t) + 5$, where $s(t)$ is the distance of the object from its position at time $t = 0$. If $s(t) = 5t^{3/5}$, what is the force at $t = 3.7$ s?

35. The thrust T of a rocket due to expelled gas at time t after lift-off is $T = -m'(t)v(t)$, where $m(t)$ is the mass of the rocket at time t and $v(t)$ is the velocity of the ejected gases at time t. Write an expression for the rate of change of thrust with respect to time in terms of derivatives of m and v.

36. The force F at time t needed to move a conveyor belt onto which material is being dropped is $F = [m(t) + M]v'(t) + v(t)m'(t)$, where M is a constant, $v(t)$ is the velocity of the belt at time t, and $m(t)$ is the mass of material on the belt at time t. Write an expression for the rate of change of force with respect to time in terms of derivatives of m and v.

37. The price P of an average single-family home in a certain area at time t after the start of record-keeping is $P = 30{,}000 + 2t + \frac{1}{2}t^2$. Find $P''(t)$, which measures how quickly the inflation rate $P'(t)$ is changing at time t.

38. The electrical force F between two charges m_1 and m_2 at distance r is given by the equation $F = gm_1m_2r^{-2}$, where g is a constant. Find d^2F/dr^2.

4-9 Implicit Differentiation

Until now, we have usually differentiated functions given by equations of the form $y = f(x)$. Such equations are said to be *solved explicitly* for y and represent **explicit functions**. Functions represented by equations in which one variable is not isolated are called **implicit functions**.

EXAMPLE 1. $y = 2x - 8$ is an equation that is solved explicitly for y; this is of the form $y = f(x)$. The equation $3y + 3x = 9x - 24$ describes the same function implicitly. Here we can take the implicit equation and solve it for y to get the explicit form. ■

There are equations describing implicit functions that cannot be solved for an explicit form, and there are equations that appear to describe implicit functions, but do not describe functions at all.

EXAMPLE 2. $xy^3 + 2y + x^2 = 4$ describes an implicit function, but we cannot solve this equation explicitly for y. The equation $x^2 + y^2 + 1 = 0$ does not describe a function at all, because for any real-number value for x, there is no real-number value for y that satisfies the equation. We will assume that all equations we encounter in the rest of this section do describe functions. ■

In an implicit function, y is still a function of x, even though we may not be able to solve explicitly for y. We can still attempt to find dy/dx. We do this by simply differentiating both sides of the equation with respect to x; whenever we encounter y, we treat it as a function of x. This process is called **implicit differentiation**. (We used it in Section 4-7 to verify the power rule for rational powers.)

EXAMPLE 3. Given the equation $y^2 - x = 2x^2$, find dy/dx.

We differentiate both sides of the equation. For the y^2-term, we use the power rule for a function of x (multiply the power, times the function to the one less power, times the derivative of the function); the derivative of the function is just dy/dx, whatever that may later turn out to be.

$$2y\frac{dy}{dx} - 1 = 4x$$

Now we solve for dy/dx:

$$\frac{dy}{dx} = \frac{4x + 1}{2y}$$

Notice that dy/dx is expressed in terms of both y and x. ■

Practice 1.   Given that $y^3 = x^2 + 3x$, find dy/dx.

EXAMPLE 4.   Given that $xy^3 - x^2 + y^5 = 0$, find dy/dx.
Differentiating both sides with respect to x,

$$\frac{d(xy^3)}{dx} - \frac{d(x^2)}{dx} + \frac{d(y^5)}{dx} = 0$$

Using the product rule on the first term and the power rule on the second and third terms (remember that y^3 and y^5 are powers of some function of x), we get

$$x\frac{d(y^3)}{dx} + y^3\frac{d(x)}{dx} - 2x + 5y^4\frac{dy}{dx} = 0$$

or

$$x \cdot 3y^2\frac{dy}{dx} + y^3 - 2x + 5y^4\frac{dy}{dx} = 0$$

Solving for dy/dx,

$$(3xy^2 + 5y^4)\frac{dy}{dx} = 2x - y^3$$

$$\frac{dy}{dx} = \frac{2x - y^3}{3xy^2 + 5y^4} \quad\blacksquare$$

Word of Advice

In a problem like Example 4, do not differentiate y^5 as $5y^4$. This would be the correct answer if you were differentiating with respect to y, but you are differentiating with respect to x, and y is a function of x. You must carry out the power rule fully: *the power times the function to the one less power times the derivative of the function.*

Practice 2.   Given that $y^2 - 2xy + x^3 = 0$, find dy/dx.

EXAMPLE 5.   An object moves in such a way that its distance s (in meters) as a function of time t (in minutes) is given by the equation $s^2 - 2t^2 = 8$. Find the velocity and acceleration when $t = 2$ min and $s = 4$ m.

The velocity is given by ds/dt; we use implicit differentiation, remembering that s is a function of t:

$$2s\frac{ds}{dt} - 4t = 0$$

$$\frac{ds}{dt} = \frac{2t}{s}$$

$$\left.\frac{ds}{dt}\right|_{t=4,\ s=4} = 1 \text{ m/min}$$

The acceleration is given by d^2s/dt^2, so we differentiate ds/dt with respect to t, using the quotient rule.

$$\frac{d^2s}{dt^2} = \frac{s(2) - 2t\,\dfrac{ds}{dt}}{s^2}$$

We evaluate d^2s/dt^2 by substituting the given values for s and t and the known value for ds/dt.

$$\left.\frac{d^2s}{dt^2}\right|_{t=2,\ s=4} = \frac{4(2) - 2(2)(1)}{4^2} = \frac{1}{4} \text{ m/min}^2 \quad \blacksquare$$

Practice 3. If $s^3 + t^2 = 2$, find the velocity and acceleration when $t = 1$ min and $s = 1$ m.

EXAMPLE 6. Find the slope of the line tangent to the curve $x^2\sqrt{y+1} = 3$ at the point $(\sqrt{3}, 0)$.

We find dy/dx by using implicit differentiation and the product rule.

$$x^2 \cdot \frac{1}{2}(y+1)^{-1/2}\left(\frac{dy}{dx}\right) + (y+1)^{1/2} \cdot 2x = 0$$

$$\frac{dy}{dx} = \frac{-2x(y+1)^{1/2}}{\frac{1}{2}x^2(y+1)^{-1/2}}$$

$$= \frac{-4(y+1)}{x}$$

Then we evaluate.

$$\left.\frac{dy}{dx}\right|_{x=\sqrt{3},\ y=0} = \frac{-4}{\sqrt{3}} \quad \blacksquare$$

Practice 4. Find the slope of the line tangent to the curve $xy = 4$ at the point $(2, 2)$.

Exercises / Section **4-9**

Exercises 1–4: Find dy/dx by using implicit differentiation; then solve for y explicitly and find dy/dx. Do your answers agree?

1. $7x - 2y = 10$
2. $3x + 4y = 9$
3. $2x^2 - 2y = 3$
4. $3y + x^3 = 7$

Exercises 5–20: Find dy/dx by using implicit differentiation.

5. $y^3 + 2x^2 = x$
6. $3x^4 - y^2 = x$
7. $y^2 + 2y = x^2$
8. $2x + 5 - y^2 = 3y$
9. $y^2 - xy + x^2 = 5$
10. $y^2 - x(y + 1) = 7$
11. $xy^2 - 2x + y^3 = x^2$
12. $3y^2 - xy^3 = 7x^2$
13. $x^4 + 4x^2y^2 - 5xy^3 + 2x = 0$
14. $2x^3 - x^2y + y^3 - 1 = 0$
15. $y^2 - \dfrac{y}{x - 1} = 4$
16. $xy - \dfrac{y^2}{2x} = 7$
17. $y^2 = (2x + 1)^3$
18. $y^3 = (3 - 7x)^2$
19. $(y^2 + 1)^2 = 3(2x - 9)^2$
20. $(3 - y^2)^2 = 4(6x + 5)^2$

Exercises 21–26: Given the equation relating distance s (in kilometers) and time t (in hours), find the velocity and acceleration at the indicated time and place.

21. $s^2 + 2t = 6$; $t = 1$ h, $s = 2$ km
22. $3t - 8 + s^3 = 0$; $t = 0$ h, $s = 2$ km
23. $t^2 + 2st = 9$; $t = 1$ h, $s = 4$ km
24. $3s - s^2t = 1$; $t = 2$ h, $s = 1$ km
25. $(s^2 - 5)^2 = t^2$; $t = 4$ h, $s = 3$ km
26. $(6 - 2s^2)^2 = 4t$; $t = 1$ h, $s = 2$ km

Exercises 27–30: Find the slope of the tangent line at the given point.

27. $xy^2 + 2y = x^2$; $(2, 1)$
28. $3xy - y^2 = x + 5$; $(-5, 0)$
29. $x\sqrt{y} + x^2 = 0$; $(-1, 1)$
30. $3x^2\sqrt{y} - x = 2$; $(1, 1)$

31. The velocity v of sound waves in hydrogen at constant temperature is given by the equation $v^2 = 15.5P$, where P is the pressure. Find dv/dP by using implicit differentiation.

32. An equation of state for helium is

$$PV = n\left(RT - \frac{aP}{T} + bP\right)$$

where n, R, a, and b are constants, P is the pressure, V is the volume, and T is the temperature. Assuming volume to be constant, find the rate of change of pressure with respect to temperature by using implicit differentiation.

4-10 Partial Derivatives

We know from Chapter 1 that functions may have more than one independent variable.

EXAMPLE 1. Let $z = 9 - x^2 - y^2$. Here the dependent variable, z, is a function of two independent variables, x and y; $z = f(x, y)$. We can graph z as a surface in a three-dimensional coordinate system (see Example 3 of Section 3-1). The result is shown in Figure 4-25. ∎

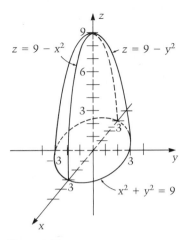

Figure 4-25

In a three-dimensional coordinate system, an equation of the form $x = c$, where c is a constant, represents a plane parallel to the yz-plane. If we let x take the constant value c in an equation of the form $z = f(x, y)$, we then get an expression for z as a function of y only, $z = g(y)$. This function represents the intersection of the surface $z = f(x, y)$ with the plane $x = c$. We used such sections of the surface to help us graph functions of two variables.

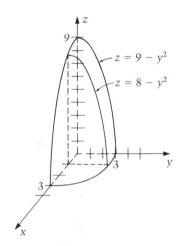

Figure 4-26

EXAMPLE 2. If we let $x = 1$ in the function $z = 9 - x^2 - y^2$ of Example 1, we get the function $z = 8 - y^2$. This parabola is the intersection of the surface with the plane $x = 1$; the first octant is shown in Figure 4-26. If we differentiate z with respect to y in the equation $z = 8 - y^2$, we obtain $-2y$, which is an expression for the slope of the tangent line to the curve $z = 8 - y^2$ at any (y, z) point. Of course we are still in the plane $x = 1$ (see Figure 4-27). Thus the slope of the tangent line to the surface $z = 9 - x^2 - y^2$ in the plane $x = 1$ at the point $(1, 2, 4)$ is

$$-2y\Big|_{(1,\,2,\,4)} = -4 \quad \blacksquare$$

More generally, in any function $z = f(x, y)$, we can simply consider x to have a constant value (we do not actually have to replace x by c). Then we can differentiate with respect to y, holding x constant. The result is called the **partial derivative** of z with respect to y, denoted by

$$\frac{\partial z}{\partial y} \quad \text{or sometimes} \quad f_y(x, y)$$

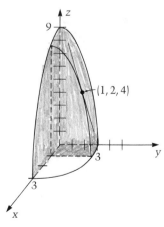

Figure 4-27

Geometrically, $\partial z/\partial y$ represents the slope of the tangent line to the surface in any plane $x = c$. Also, $\partial z/\partial y$ represents the rate of change of z with respect to y while x is constant.

EXAMPLE 3.   Let $z = 3x - 2xy + 3y^2$. Then $\partial z/\partial y$ is obtained by thinking of x as a constant:

$$\frac{\partial z}{\partial y} = -2x + 6y$$

The slope of the tangent line to the surface $z = 3x - 2xy + 3y^2$ in the plane $x = 2$ at the point $(2, 3, 21)$ is

$$-2x + 6y\bigg|_{(2, 3, 21)} = 14 \quad \blacksquare$$

Practice 1.   Let $z = 2xy + 4x^2$. Find $\partial z/\partial y$, and then find the slope of the tangent line to the surface in the plane $x = 1$ at the point $(1, 2, 8)$.

The notation $\partial p/\partial q$ indicates that p is a function of q, as well as of other variables, and that in differentiating, all the other variables are to be considered constant.

EXAMPLE 4.   Let $w = 3uv + v^2 + 2u$. Find $\partial w/\partial u$ and $\partial w/\partial v$.
    We find $\partial w/\partial u$ by thinking of v as constant while we differentiate. Thus $\partial w/\partial u = 3v + 2$. We find $\partial w/\partial v$ by thinking of u as constant while we differentiate. We get $\partial w/\partial u = 3v + 2v$. $\blacksquare$

Practice 2.   Let $r = 2st^2 - t^4$. Find $\partial r/\partial s$ and $\partial r/\partial t$.

EXAMPLE 5.   A magnet placed in a nonuniform magnetic field B will be subjected to a force F given by the expression

$$F = k\frac{\partial B}{\partial y}$$

where k is a constant. If B is given by the expression $B = 1.8x^2y^2$, find the expression for F.
    To find $\partial B/\partial y$, we differentiate $B = 1.8x^2y^2$ with respect to y, while thinking of x as a constant. We get $\partial B/\partial y = 1.8x^2(2y)$, so that $F = 3.6kx^2y$. $\blacksquare$

Exercises / Section **4-10**

Exercises 1–10: Find the indicated partial derivatives.

1. $z = 3xy^2 - 2x^2y$; $\dfrac{\partial z}{\partial x}, \dfrac{\partial z}{\partial y}$

2. $z = 3x\sqrt{y} + x^3$; $\dfrac{\partial z}{\partial x}, \dfrac{\partial z}{\partial y}$

3. $z = \dfrac{x}{1 + y}$; $\dfrac{\partial z}{\partial x}, \dfrac{\partial z}{\partial y}$

4. $z = \dfrac{2x - y}{y^2}$; $\dfrac{\partial z}{\partial x}, \dfrac{\partial z}{\partial y}$

5. $p = (2q^2 - 1)(r + 2)$; $\dfrac{\partial p}{\partial q}, \dfrac{\partial p}{\partial r}$

6. $s = (3t - 4)(2t + u^2)$; $\dfrac{\partial s}{\partial t}, \dfrac{\partial s}{\partial u}$

7. $w = u^2 \sqrt{1 + v}$; $\dfrac{\partial w}{\partial u}, \dfrac{\partial w}{\partial v}$

8. $m = \sqrt{r^2 - 2}(s^2 + 1)$; $\dfrac{\partial m}{\partial r}, \dfrac{\partial m}{\partial s}$

9. $u = (wv + w^3)^2$; $\dfrac{\partial u}{\partial w}, \dfrac{\partial u}{\partial v}$

10. $r = (s - st^2)^4$; $\dfrac{\partial r}{\partial s}, \dfrac{\partial r}{\partial t}$

11. For $z = 3xy^2 - 2x^2$, find the slope of the tangent line to the surface in the plane $x = 2$ at the point $(2, 1, -2)$.

12. For $z = 2y^3 - 3x^2y$, find the slope of the tangent line to the surface in the plane $x = 3$ at the point $(3, 2, 38)$.

13. For $z = x\sqrt{y}$, find the slope of the tangent line to the surface in the plane $y = 4$ at the point $(1, 4, 2)$.

14. For $z = (3x + y)\sqrt{x}$, find the slope of the tangent line to the surface in the plane $y = 2$ at the point $(1, 2, 5)$.

15. For $z = ax^3 + by^3$, show that $x\dfrac{\partial z}{\partial x} + y\dfrac{\partial z}{\partial y} = 3z$.

16. For $z = ax^2 + bxy + cy^2$, show that
$$x\frac{\partial z}{\partial x} + y\frac{\partial z}{\partial y} = 2z.$$

17. For $z = \dfrac{xy}{x^2 + y^2}$, show that $x\dfrac{\partial z}{\partial x} + y\dfrac{\partial z}{\partial y} = 0$.

18. For $z = \sqrt{x^3 - 2y^3}$, show that $x\dfrac{\partial z}{\partial x} + y\dfrac{\partial z}{\partial y} = \dfrac{3}{2}z$.

19. The formula for the volume V of a right circular cylinder of radius r and height h is $V = \pi r^2 h$. Find an expression for the rate of change of volume with respect to radius for constant height.

20. The formula for the volume V of a right circular cone is $V = \frac{1}{3}\pi r^2 h$, where r is the radius of the base and h is the height. Find an expression for the rate of change of volume with respect to height for constant radius.

21. The power dissipation P in an electric circuit with current I and resistance R is given by the equation $P = I^2R$. Find expressions for $\partial P/\partial I$ and $\partial P/\partial R$.

22. The potential V of a conductor is given by the equation $V = Q/C$, where Q is the total charge of the conductor and C is its capacitance. Find expressions for $\partial V/\partial Q$ and $\partial V/\partial C$.

23. The amount of heat Q flowing in one unit of time across a cross-sectional area A at a distance x from the end of an insulated metal bar is given by the equation
$$Q = kA\frac{\partial T}{\partial x}$$
where k is a constant and T, the temperature in the bar, is a function of x and time t. For $T = 3.4xt + 2.1x^2 - t^3$, find an expression for Q.

24. For chemical systems under certain conditions, the heat capacity c at constant volume is given by the equation $c = \partial U/\partial T$ where U, the internal energy of the system, is a function of volume V and temperature T. For $U = V\sqrt{T} + 2VT$, find an expression for c.

25. The speed v of a longitudinal wave in a certain substance is given by the equation
$$v = \sqrt{kV\left(\frac{\partial P}{\partial V}\right)}$$
where k is a constant. If pressure P, volume V, and entropy S are related by the equation
$$VS = P^2$$
find an expression for v.

26. The *isothermal compressibility k* of a gas is defined to be

$$k = -\frac{1}{V}\frac{\partial V}{\partial P}$$

If volume V, temperature T, and pressure P are related by the equation

$$\left(P + \frac{a}{V^2}\right)(V - b) = cT$$

where a, b, and c are constants, find an expression for k.

27. Within a certain temperature range, the growth G (in centimeters) of a particular plant is given by the equation

$$G = kT(W + 0.003T^2)$$

where k is a constant, T is the temperature (in degrees Celsius), and W is the daily amount (in milliliters) of water fed to the plant. Find $\partial G/\partial W$ and $\partial G/\partial T$ when $W = 35$ mL and $T = 12°C$.

28. The illuminance E on a surface s units away from a fluorescent lamp is given by the equation

$$E = \frac{2I}{LS}$$

where L is the length of the lamp and I is its intensity. For a lamp of fixed length, find $\partial E/\partial s$ (in lux per meter) and $\partial E/\partial I$ (in lux per candela) when $s = 12$ m, $L = 1.2$ m, and $I = 4000$ cd.

STATUS CHECK

Now that you are at the end of Chapter 4, you should be able to:

section **4-1**   By sketching the graph of a function, decide whether its slope at a given point is positive, negative, or zero.
Given the slope function $f'(x)$ for a function $f(x)$, find the value of the slope at a given point.

section **4-2**   Find any points of discontinuity of a function.
Find $\lim_{x \to a} f(x)$ for $f(x)$ a polynomial or quotient of polynomials.
Find $\lim_{x \to \infty} f(x)$ for $f(x)$ a quotient of polynomials.

section **4-3**   Use the four-step process to find the derivative of a polynomial function.

section **4-5**   Differentiate a polynomial function.
Given a polynomial expression for distance as a function of time, find the velocity at a specific time.
Find the slope of the tangent line to a polynomial function at a given point.
Given one quantity as a polynomial function of a second quantity, find an expression for the rate of change of the first quantity with respect to the second and be able to evaluate this expression at a given value of the second quantity.

section **4-6**   Differentiate products and quotients of polynomials.
Given an expression for distance as a function of time that is a product or quotient of polynomials, find the velocity at a specific time.
Find the slope of the tangent line at a given point to a function that is a product or quotient of polynomials.

Given an expression for one quantity as a product or quotient of polynomials in a second quantity, find an expression for the rate of change of the first quantity with respect to the second and be able to evaluate this expression at a given value of the second quantity.

section **4-7** Differentiate algebraic functions.

Given distance as an algebraic function of time, find the velocity at a specific time.

Find the slope of the tangent line to an algebraic function at a given point.

Given one quantity as an algebraic function of a second quantity, find an expression for the rate of change of the first quantity with respect to the second and be able to evaluate this expression at a given value of the second quantity.

section **4-8** Given y as an algebraic function of x, find higher-order derivatives of y and evaluate these derivatives for specific values of x.

Given an expression for distance as an algebraic function of time, find the acceleration at a specific time.

section **4-9** Given y as an implicit algebraic function of x, find dy/dx by using implicit differentiation.

Given distance as an implicit function of time, find velocity and acceleration at a specific time and distance.

section **4-10** Given a function of several variables, find the partial derivative with respect to each variable.

Find the slope of the tangent line to a given surface at a fixed point and in a fixed plane.

4-11 More Exercises for Chapter **4**

Exercises 1–14: Find the limit.

1. $\lim\limits_{x \to 2.1} (3x^2 - 4.2)$

2. $\lim\limits_{x \to -1.8} (2 - 1.4x^2)$

3. $\lim\limits_{x \to -2} (2x^2 - 5x + 1)$

4. $\lim\limits_{x \to 3} (4 - 6x + 5x^2)$

5. $\lim\limits_{x \to 2} \dfrac{x^2 + 2x - 8}{x - 2}$

6. $\lim\limits_{x \to -1} \dfrac{3 + 2x - x^2}{1 + x}$

7. $\lim\limits_{x \to 1} f(x)$ where $f(x) = \begin{cases} 2x & \text{for } x \neq 1 \\ 0 & \text{for } x = 1 \end{cases}$

8. $\lim\limits_{x \to 2} f(x)$ where $f(x) = \begin{cases} x - 2 & \text{for } x \neq 2 \\ 2 & \text{for } x = 2 \end{cases}$

9. $\lim\limits_{x \to 1} f(x)$ where $f(x) = \begin{cases} 2x & \text{for } x \leq 1 \\ x + 1 & \text{for } x > 1 \end{cases}$

10. $\lim\limits_{x \to -1} f(x)$ where $f(x) = \begin{cases} -x & \text{for } x < -1 \\ x + 2 & \text{for } x \geq -1 \end{cases}$

11. $\lim\limits_{x \to \infty} \dfrac{2x^2 + 4x - 3}{3x^2 + 1}$

12. $\lim\limits_{x \to \infty} \dfrac{2 - x^2}{3x - 4x^2}$

13. $\lim\limits_{x \to \infty} \dfrac{3x - 1.4}{3.1 - x^2}$

14. $\lim\limits_{x \to \infty} \dfrac{2x - x^2}{2x^3 - 4x}$

Exercises 15–40: Find the derivative dy/dx.

15. $y = 2x^3 - 0.5x^2 - 1.1$

16. $y = x^3 + 0.6x^2 + 0.02$

17. $y = \dfrac{1}{2x}$

18. $y = \dfrac{1}{x + 2}$

19. $y = x^2 - \dfrac{1}{2x}$

20. $y = \dfrac{1}{x + 2} - x$

21. $y = \dfrac{0.2x}{1 + x}$

22. $y = \dfrac{x + 1}{2 - x}$

23. $y = (x - 2)(2x + 4)$

24. $y = (x^2 - 1)(2 - x)$

25. $y = (x^3 - 2x)^2$

26. $y = (x - 0.4x^2)^4$

27. $y = \sqrt{x^2 - 3}$

28. $y = \sqrt{4 - x^2}$

29. $y = x^3(3x^2 - 4)^2$

30. $y = (2 + x^4)^2(x - 2)^3$

31. $y = \sqrt{x}(x - 1)^2$

32. $y = x^2\sqrt{x^2 - 1}$

33. $y = \dfrac{\sqrt{x - 1}}{(x^2 + 4)^3}$

34. $y = \dfrac{(x - 2)^2}{\sqrt{x}}$

35. $y = [\sqrt{x}(x - 1)]^{3/2}$

36. $y = \sqrt{x^3(x^2 - 1)^5}$

37. $xy = 3 - 4y^2$

38. $xy^2 = x^2 + y^2$

39. $y^3 = \sqrt{x^3 - 1} + y$

40. $xy^2 = (1 - y)^3$

Exercises 41–46: Find d²y/dx².

41. $y = 4x^3 - 12x^2 + 5x - 2$

42. $y = -3x^4 + 6x^3 - 7x^2 + 4$

43. $y = \sqrt{x}(2 + x^2)^3$

44. $y = x^2\sqrt{1 - x^3}$

45. $y = \dfrac{x}{1 + x^2}$

46. $y = \dfrac{x^3}{1 - x}$

Exercises 47–50: Find ∂z/∂x and ∂z/∂y.

47. $z = (xy + 1)^2$

48. $z = \sqrt{1 - x^2y}$

49. $z = x^3(1 + y^2)$

50. $z = x^2y^2(1 - x)$

51. Heat loss from a building through windows is a function of the window area. Heat loss in a given building is 0.008°C/h for every square meter of window up to and including 1500 m²; for more than 1500 m² of window, heat loss is a constant 12°C/h. Let $f(x)$ be the heat loss through x square meters of window area. Graph $f(x)$. Find any values of $x > 0$ for which $f(x)$ is discontinuous.

52. A manufacturer's cost to produce x units of a product for $x \le 2000$ is a base cost of $2000 for labor plus $.05 per unit produced for raw mate-rials. For $x > 2000$, labor costs increase by $500 due to overtime pay, but the cost per unit for every unit over 2000 is $.03. Let $f(x)$ be the cost to produce x units. Graph $f(x)$. Find any values of $x > 0$ for which $f(x)$ is discontinuous.

53. The period T of a simple pendulum is given by the formula $T = 2\pi\sqrt{L/g}$ where $g = 9.8$ m/s². Find $\lim_{L \to 4} T$ (in seconds) to two decimal places.

54. The power P in an electric furnace is given by the equation $P = I^2R$, where I is the current and R is the resistance. For a constant resistance of 8 Ω, find $\lim_{I \to 18} P$ (in watts).

55. The electrical force F between two charges m_1 and m_2 at distance r is given by the formula $F = gm_1m_2/r^2$, where g is a constant. Find $\lim_{r \to \infty} F$ and interpret the answer.

56. The quantity q of pollutants in the output of a water purification plant with a biological filter of depth x is given by the formula

$$q = \frac{1}{1 + kx}$$

where k is a constant. Find $\lim_{x \to \infty} q$ and in-terpret the answer.

57. The distance s (in meters) of an object as a function of time t (in seconds) is given by the equation $s = \sqrt{t}(1 + t)$. Find the velocity and acceleration when $t = 3$ s.

58. The distance s (in kilometers) of an object as a function of time t (in hours) is given by the equation $s = \sqrt{1 + t^3}$. Find the velocity and acceleration when $t = 1.4$ h.

59. Find the slope of the tangent line to the curve $y = x(1 + x^2)^2$ at the point $(1, 4)$.

60. Find the slope of the tangent line to the curve $y = \sqrt{1 - x^3}$ at the point $(-2, 3)$.

61. Find the slope of the tangent line to the curve $x^2y = 2$ at the point $(-1, 2)$.

62. Find the slope of the tangent line to the curve $2xy - y^3 = x^3$ at the point $(1, 1)$.

63. Find the slope of the tangent line to the surface $z = 2x^2y - y^3$ in the plane $x = 2$ at the point $(2, 3, -3)$.

64. Find the slope of the tangent line to the surface $z = \sqrt{x}(1 + y^2)$ in the plane $y = 2$ at the point $(4, 2, 10)$.

65. Liquid petroleum expands when heated. A quantity of petroleum occupying 1000 m^3 of volume at 0°C will occupy V cubic meters of volume at temperature T (0°C $\leq T \leq$ 120°C), where $V = 1000 + 0.8994T + (1.396 \times 10^{-3})T^2$. How fast is the volume changing with respect to temperature when the temperature is 30°C?

66. The number s of people infected with a virus t days after the first case is reported is estimated to be given by the formula $s = 40t^2 - 2t^3$ for $0 \leq t \leq 20$ days. Find the rate at which the virus is spreading on the tenth day.

67. The total resistance R of a circuit containing two resistors R_1 and R_2 in parallel is given by the equation $R = R_1R_2/(R_1 + R_2)$. If R_1 and R_2 are functions of time t given by $R_1 = 3t^2 + 1$ and $R_2 = t - 1$, find an expression for the rate of change of R with respect to t.

68. The area A of the pupil of a human eye is a function of the amount r of light which reaches the retina per unit time. If

$$A = \frac{40r^{-0.4} + 23.4}{r^{-0.4} + 3.92}$$

find an expression for the rate of change of A with respect to r.

69. A satellite moving in a circular orbit about the earth experiences a gravitational force f given by the equation $f = mgR^2/x^2$, where m is the mass of the satellite, g is the force of gravity on the earth's surface, R is the radius of the earth, and x is the distance of the satellite above the earth's surface. Find an expression for the rate of change of f with respect to x.

70. A current I in a coil produces a magnetic field; the magnetic induction B at a point x units along the axis of a certain coil of radius r is given by

$$B = \frac{kIr^2}{2(r^2 + x^2)^{3/2}}$$

where k is a constant. Find an expression for the rate of change of B with respect to x (r and I are constant).

71. In a vacuum diode, the current I as a function of voltage V is given by $I = kV^{3/2}$, where k is a constant. Use implicit differentiation to find an expression for the incremental resistance dV/dI.

72. The reaction R to a dosage D of a given drug is $R^2 = D^2(C - 0.38D)$, where C is a constant. Use implicit differentiation to find an expression for the sensitivity to the drug, dR/dD.

73. The wind-chill factor Q used in weather forecasts is given by

$$Q = (10\sqrt{w} + 10.4 - w)(33 - T)$$

where w is the average wind speed in meters per second and T is the temperature in degrees Celsius. Find expressions for $\partial Q/\partial w$ and $\partial Q/\partial T$.

74. An ideal gas satisfies the equation $PV = kT$, where P is pressure, V is volume, T is temperature, and k is a constant. Show that for an ideal gas,

$$\left(\frac{\partial V}{\partial T}\right)\left(\frac{\partial T}{\partial P}\right)\left(\frac{\partial P}{\partial V}\right) = -1$$

chapter  5

Using the Derivative

*Velocity and acceleration involve derivatives
with respect to time of equations of motion.*

5-1   Tangents and Normals

We are now quite familiar with the derivative and know how to differentiate any algebraic function. In this chapter we turn our attention in more detail to the various ways in which derivatives can be used.

One interpretation of the derivative function dy/dx is that it represents the slope of the tangent line to the curve $y = f(x)$ at any point. This makes it easy for us to find the actual equation of the tangent line to the curve $y = f(x)$ at some particular point (x_1, y_1). Evaluating dy/dx at the point (x_1, y_1) gives us the slope of this particular tangent line. Of course, we also know that this tangent line goes through the point (x_1, y_1). We therefore know the slope of the line and a point on the line, so we can use the point-slope form to write the equation of the line.

EXAMPLE 1. Write the equation of the tangent line to the curve $y = 2 + x^2$ at the point $(-1, 3)$.

First we note that $(-1, 3)$ satisfies the equation $y = 2 + x^2$ and is therefore a point on the curve (otherwise the problem does not even make sense). Next we find the derivative dy/dx:

$$\frac{dy}{dx} = \frac{d(2 + x^2)}{dx} = 2x$$

The slope of the tangent line at the point $(-1, 3)$ is

$$\frac{dy}{dx}\bigg|_{(-1,\, 3)} = 2(-1) = -2$$

Using the point-slope form for the equation of a line with point $(-1, 3)$ and slope -2, we get

$$y - 3 = -2(x + 1)$$

or

$$2x + y - 1 = 0$$

From this equation we see that the tangent line has a y-intercept of $(0, 1)$, and this helps us graph the line (see Figure 5-1). ∎

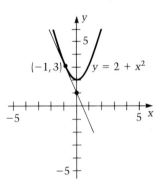

Figure 5-1

Given the function $y = f(x)$, we do three things to get the equation of the tangent line at the point (x_1, y_1).

1. Find dy/dx.
2. Evaluate

$$\frac{dy}{dx}\bigg|_{(x_1,\, y_1)}$$

Call this result m.
3. Use the equation $y - y_1 = m(x - x_1)$ and simplify.

Practice 1. Write the equation of the tangent line to the curve $y = 3x^3 - 1$ at the point $(-1, -4)$.

We can also find the equation of the *normal line* to a curve at a given point. The **normal line** at a point is the line perpendicular to the tangent line at that point. Its slope is, therefore, the negative reciprocal of the slope of the tangent line at the point. Again, we can use the derivative to find the slope of the tangent line and then the slope of the normal line. Knowing the slope of the normal line and a point it goes through, we can write its equation.

EXAMPLE 2. Find the equation of the normal line to the curve $y = 2 + x^2$ at the point $(-1, 3)$.

As in Example 1, $dy/dx = 2x$ and

$$\frac{dy}{dx}\Big|_{(-1,\,3)} = -2$$

This is the slope of the tangent line; the slope of the normal line is the negative reciprocal $\frac{1}{2}$. The equation of the normal line is found by using the point-slope form:

$$y - 3 = \tfrac{1}{2}(x + 1)$$

Simplifying, we get

$$2y - 6 = x + 1$$

or

$$x - 2y + 7 = 0$$

The y-intercept of the normal line is $(0, \frac{7}{2})$, the x-intercept is $(-7, 0)$, and the graph is shown in Figure 5-2. ■

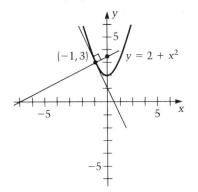

Figure 5-2

Practice 2. Write the equation of the normal line to the curve $y = 3x^3 - 1$ at the point $(-1, -4)$.

EXAMPLE 3. Find equations of the tangent line and the normal line to the circle $x^2 + y^2 = 25$ at the point $(3, 4)$.

Here we use implicit differentiation in order to get dy/dx:

$$2x + 2y \frac{dy}{dx} = 0$$

$$\frac{dy}{dx} = -\frac{x}{y}$$

$$\left. \frac{dy}{dx} \right|_{(3, 4)} = -\frac{3}{4}$$

The equation of the tangent line is

$$y - 4 = -\frac{3}{4}(x - 3)$$

Simplifying, this equation is

$$4y - 16 = -3(x - 3)$$
$$4y - 16 = -3x + 9$$
$$3x + 4y - 25 = 0$$

The slope of the normal line is $\frac{4}{3}$ and the equation of this line is

$$y - 4 = \frac{4}{3}(x - 3)$$
$$3y - 12 = 4(x - 3)$$
$$3y - 12 = 4x - 12$$
$$4x - 3y = 0$$

Notice that the normal line goes through $(0, 0)$, which is the center of the circle. (Recall from geometry that a line perpendicular to a tangent line of a circle goes through the center of the circle.) Figure 5-3 illustrates this situation. ■

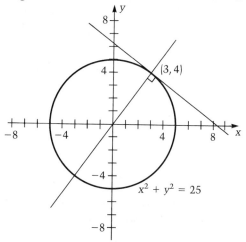

Figure 5-3

Practice 3. Write equations of the tangent line and normal line to the curve $y^2 - 2x - 2y - 9 = 0$ at the point $(-3, 3)$.

EXAMPLE 4. In irrotational, steady fluid flow, streamlines are normal to the curves of equal velocity potential. If $x^2 - 6x + y^2 + 1 = 0$ is an equipotential curve, find the equation of the streamline at the point $(1, 2)$.

We want the normal line to $x^2 - 6x + y^2 + 1 = 0$ at $(1, 2)$. Using implicit differentiation to find dy/dx,

$$2x - 6 + 2y \frac{dy}{dx} = 0$$

$$\frac{dy}{dx} = \frac{3 - x}{y}$$

$$\left.\frac{dy}{dx}\right|_{(1, 2)} = \frac{3 - 1}{2} = 1$$

The slope of the normal line is -1, and its equation is

$$y - 2 = -1(x - 1)$$

or

$$x + y - 3 = 0 \quad \blacksquare$$

Exercises / Section **5-1**

Exercises 1–10: Find equations of the tangent line and normal line to the curve at the given point.

1. $3x - 2y + 4 = 0$; $(2, 5)$

2. $2x + 3y + 1 = 0$; $(-2, 1)$

3. $y = 4 - x + 3x^2$; $(-1, 8)$

4. $y = 3x^2 - 2x + 5$; $(2, 13)$

5. $y = x^4 - 2x^2$; $(2, 8)$

6. $y = 3x^5 - 4x + 2$; $(-1, 3)$

7. $y^2 - 3x^2 + 2x - 3 = 0$; $(1, 2)$

8. $3x^3 - 2y^2 + 3y + 23 = 0$; $(-2, 1)$

9. $2xy + y^2 - 3 = 0$; $(1, 1)$

10. $x^2y - 3y^2 + 10 = 0$; $(-1, 2)$

11. A suspension cable is supported at each end by posts that are 40 m apart; the cable sags 5 m in the middle. Assume that the cable takes the shape of a parabola, and put the origin of the coordinate system at the vertex of the parabola. What is the equation of the cable in this coordinate system? What are the equations of the tangent lines to the cable at the top of each support post (the lines along which tension acts to support the cable)?

12. A rotating grinding wheel is in the shape of a circular disk. The circle itself is given by the equation $x^2 + y^2 = 160$, where the origin of the coordinate system is at the center of the circle. As the wheel rotates, a small piece breaks off at the point $(-4, 12)$ on the coordinate system and flies off tangentially to the wheel. Give the equation for the path of the broken piece.

13. One circle with radius $\sqrt{5}$ has its center at the origin. Find the coordinates of the center of a second circle of radius 5 that does not overlap the first circle and is tangent to the first circle at the point $(1, 2)$. Write the equation of the tangent line to both circles at the point $(1, 2)$.

5-2 Curve Sketching

In Chapter 1 we discussed a number of considerations that can aid us in graphing a function, such as symmetry, asymptotes, and excluded regions. The derivative provides another very important tool to help us graph functions. All these tools are used to give us a general idea of the shape of the function and some key points on the graph. If we want a more detailed and accurate graph, we must plot a number of points; in this case the calculator relieves us of some of the burden of computation. (A computer with a plot program is even better.)

In a function such as the one shown in Figure 5-4, there are three key points where important things happen, *M*, *m*, and *I*, so these are three points to locate on our graph.

M is a **relative maximum point** (higher than any other point around it).

m is a **relative minimum point** (lower than any other point around it).

I is an **inflection point** (more on this later).

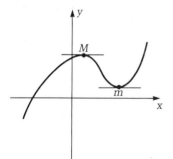

Figure 5-4

While *M* is the highest point in its own neighborhood, there are higher points on the graph; *M* is a relative maximum, but not an absolute maximum. Similarly, *m* is a relative minimum, but not an absolute minimum. The points *M* and *m* are important because they are "turning points" on the graph. In Figure 5-5 we have drawn tangent lines to the curve at these turning points. The tangent lines are horizontal, with a slope of zero. This means that the first derivative of the function is zero at a relative maximum or minimum point. (We assume here that we are dealing with functions whose derivatives exist at all points. In Figure 5-6, for example, there are many relative maximum and minimum points, but tangent lines—and derivatives—are undefined at them.)

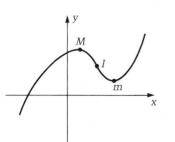

Figure 5-5

Our first step in locating relative maximum or minimum points is to find points on the curve where the first derivative is zero. Such points are called **critical points**. Any relative maximum or minimum points must occur at critical points, but not all critical points are maximum or minimum points. (See Figure 5-7, for example, where there is a critical point but no maximum or minimum.)

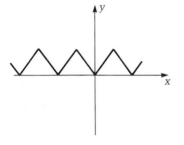

Figure 5-6

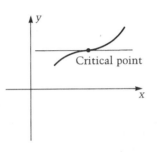

Critical point

Figure 5-7

To find critical points for the function $y = f(x)$, we find dy/dx, set it equal to zero, and solve for x (and the corresponding y-values).

EXAMPLE 1.   Find the critical points for the function $y = \frac{1}{3}x^3 - \frac{1}{2}x^2 - 2x + 2$.

We find $dy/dx = x^2 - x - 2$, set it equal to zero, and solve for x:

$$x^2 - x - 2 = 0$$
$$(x - 2)(x + 1) = 0$$
$$x = 2, -1$$

These are the x-coordinates of the critical points. The corresponding y-coordinates are found by substituting these values for x back into the original equation. The critical points are $(2, -\frac{4}{3})$, $(-1, \frac{19}{6})$ ∎

Practice 1.   Find the critical points for the function $y = x^3 + 3x^2 - 9x - 10$.

Having found the critical points, how do we decide which are maximum, which are minimum, and which are neither? We observe from Figure 5-8 that just to the left of a maximum point, the slope of the tangent line (the derivative) is positive and just to the right of a maximum point, it is negative. There is a change of sign of the first derivative from positive to negative as we go from left to right through a maximum point. Passing left to right through a minimum point, the first derivative changes sign from negative to positive (Figure 5-8). Notice that in Figure 5-7, the first derivative does not change sign, but is positive on both sides of the critical point. Checking for changes of sign of the first derivative around a critical point is called the **first derivative test for maxima and minima**.

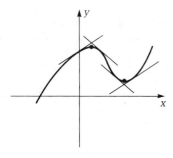

Figure 5-8

EXAMPLE 2.   For the function $y = \frac{1}{3}x^3 - \frac{1}{2}x^2 - 2x + 2$, $dy/dx = (x - 2)(x + 1)$ and the critical points are $(2, -\frac{4}{3})$, $(-1, \frac{19}{6})$. To check for a change of sign of the derivative as we pass left to right through $(2, -\frac{4}{3})$, it is convenient to leave dy/dx in its factored form and consider the signs of each factor for $x < 2$ and $x > 2$. When we consider $x < 2$, we need not substitute an actual value for x; we can simply reason that if x is less than 2 and we subtract 2, the result is negative, so $x - 2 < 0$. Similar reasoning can be used for other values.

$$x < 2: \quad x - 2 < 0, \quad x + 1 > 0, \quad \text{and} \quad (x - 2)(x + 1) < 0$$
$$x > 2: \quad x - 2 > 0, \quad x + 1 > 0, \quad \text{and} \quad (x - 2)(x + 1) > 0$$

The first derivative changes from negative to positive. This means that $(2, -\frac{4}{3})$ is a minimum point. For the critical point $(-1, \frac{19}{6})$,

$$x < -1: \quad x - 2 < 0, \quad x + 1 < 0, \quad \text{and} \quad (x - 2)(x + 1) > 0$$
$$x > -1: \quad x - 2 < 0, \quad x + 1 > 0, \quad \text{and} \quad (x - 2)(x + 1) < 0$$

The first derivative changes from positive to negative, and $(-1, \frac{19}{6})$ is a maximum point.   ■

We begin to see why it is convenient to have the derivative of a function in factored form.

Practice 2.   Use the first derivative test to locate any maximum or minimum points in the function $y = x^3 + 3x^2 - 9x - 10$ of Practice 1.

The second derivative of y, y'', is the derivative of the first derivative and represents the rate of change of the first derivative. When the second derivative is negative, the values of the first derivative are decreasing as we move from left to right on the graph; whenever this happens, the curve itself is *concave downward*, as in Figure 5-9. Note that in Figure 5-9(a), the slope of the tangent line decreases from 1.2 to 0.25, and in Figure 5-9(b), it decreases from -0.2 to -1.0. In particular, if the second derivative is negative at a critical point, that point will be a relative maximum, as in Figure 5-9(c). When the second derivative is positive, the values of the first derivative are increasing as we move from left to right on the graph; whenever this happens, the curve itself is *concave upward*, as in Figure 5-10. In particular, if the second derivative is positive at a critical point, that point will be a relative minimum, as in Figure

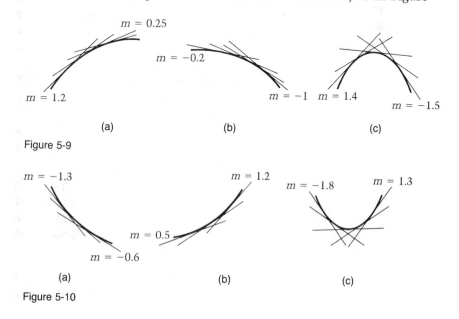

$m = 0.25$

$m = -0.2$

$m = 1.2$

$m = -1 \quad m = 1.4$

$m = -1.5$

(a)   (b)   (c)

Figure 5-9

$m = -1.3$

$m = 1.2$

$m = -1.8$

$m = 1.3$

$m = 0.5$

$m = -0.6$

(a)   (b)   (c)

Figure 5-10

5-10(c). Checking the sign of the second derivative at a critical point is called the **second derivative test for maxima and minima**.

EXAMPLE 3. For the function $y = \frac{1}{3}x^3 - \frac{1}{2}x^2 - 2x + 2$ of Examples 1 and 2,

$$y' = x^2 - x - 2$$
$$y'' = 2x - 1$$

For the critical point $(2, -\frac{4}{3})$,

$$y''\Big|_{x=2} = 3 > 0$$

and $(2, -\frac{4}{3})$ is a minimum point. For the critical point $(-1, \frac{19}{6})$,

$$y''\Big|_{x=-1} = -3 < 0$$

and $(-1, \frac{19}{6})$ is a maximum point. This confirms what we already know from the first derivative test. ∎

The second derivative test is easier to apply than the first derivative test, although it does not always work; if the second derivative is zero at a critical point, then the first derivative test must be used.

Practice 3. For the function $y = x^3 + 3x^2 - 9x - 10$ of Practices 1 and 2, use the second derivative test to determine maxima and minima.

An **inflection point** is a place on the curve where the concavity changes from upward to downward, or vice versa. Thus the second derivative is zero at an inflection point and changes sign around the point.

EXAMPLE 4. For our old friend $y = \frac{1}{3}x^3 - \frac{1}{2}x^2 - 2x + 2$, $y'' = 2x - 1$. The second derivative is zero when $x = \frac{1}{2}$. For $x < \frac{1}{2}$, $y'' < 0$ and the curve is concave downward. For $x > \frac{1}{2}$, $y'' > 0$ and the curve is concave upward. The point $(\frac{1}{2}, \frac{11}{12})$ is an inflection point.

Finally, we sketch a graph of this function (Figure 5-11) by using the fact that $(2, -\frac{4}{3})$ is a minimum point, $(-1, \frac{19}{6})$ is a maximum point, $(\frac{1}{2}, \frac{11}{12})$ is an inflection point, and $(0, 2)$ is an intercept. ∎

Practice 4. Find any inflection points in $y = x^3 + 3x^2 - 9x - 10$ and sketch the graph.

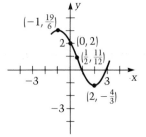

Figure 5-11

The total procedure to sketch the graph of a function $y = f(x)$ consists of the following steps:

1. Find critical points by setting $y' = 0$.
2. Evaluate y'' at these points (second derivative test); if $y'' > 0$, the point is a minimum, if $y'' < 0$, the point is a maximum, and if $y'' = 0$, the test fails.
3. For any critical points where $y'' = 0$, test for changes of sign of y' (first derivative test); if y' changes from positive to negative, the point is a maximum and if y' changes from negative to positive, the point is a minimum.
4. Set $y'' = 0$ and check for changes of sign of y'' around these points. If y'' changes sign, the point is an inflection point.
5. Use maximum and minimum points, inflection points, intercepts, and other aids such as symmetry and asymptotes to sketch the curve; use a calculator to plot a few more points if necessary.

EXAMPLE 5. Sketch the graph of $y = x^5 - 15x^3$.
We first find the derivatives:

$$y' = 5x^4 - 45x^2$$
$$y'' = 20x^3 - 90x$$

To find the critical points, we solve $5x^4 - 45x^2 = 0$.

$$5x^4 - 45x^2 = 0$$
$$5x^2(x^2 - 9) = 0$$
$$x = 0, 3, -3$$

We try the second derivative test:

$$y''\Big|_{x=0} = 20x^3 - 90x\Big|_{x=0} = 0 \qquad \text{Test fails.}$$

$$y''\Big|_{x=3} = 20x^3 - 90x\Big|_{x=3} > 0 \qquad (3, -162) \text{ is a minimum.}$$

$$y''\Big|_{x=-3} = 20x^3 - 90x\Big|_{x=-3} < 0 \qquad (-3, 162) \text{ is a maximum.}$$

Now we use the first derivative test for the critical point $(0, 0)$.

$$x < 0: \quad y' = 5x^2(x^2 - 9) < 0$$
$$x > 0: \quad y' = 5x^2(x^2 - 9) < 0$$

There is no change of sign of the first derivative around $x = 0$; so $(0, 0)$ is neither a maximum nor a minimum.
We look for inflection points by setting $y'' = 0$.

$$20x^3 - 90x = 10x(2x^2 - 9) = 0$$

$$x = 0, \frac{3}{\sqrt{2}}, -\frac{3}{\sqrt{2}}$$

Now we look for changes of sign of y'' around these points.

$$x < 0: \qquad y'' > 0; \qquad x > 0: \qquad y'' < 0$$

$(0, 0)$ is an inflection point.

$$x < \frac{3}{\sqrt{2}}: \qquad y'' < 0; \qquad x > \frac{3}{\sqrt{2}}: \qquad y'' > 0$$

$\left(\frac{3}{\sqrt{2}}, -100\right)$ is an inflection point.

$$x < -\frac{3}{\sqrt{2}}: \quad y'' < 0; \qquad x > -\frac{3}{\sqrt{2}}: \quad y'' > 0$$

$\left(-\frac{3}{\sqrt{2}}, 100\right)$ is an inflection point.

Finally we note that since replacing x by $-x$ and y by $-y$ produces no change in the equation, the graph is symmetric about the origin. Intercepts are $(0, 0)$, $(\sqrt{15}, 0)$, and $(-\sqrt{15}, 0)$. The graph is shown in Figure 5-12. Notice that $(0, 0)$ is an inflection point with a horizontal tangent. ∎

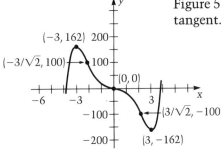

Figure 5-12

Practice 5. Sketch the graph of $y = x^4$.

Practice 6. Sketch the graph of $y = x^4 - 2x^2$.

EXAMPLE 6. Sketch the graph of $y = x^2 + \frac{16}{x^2}$.

The derivatives are

$$y' = 2x - \frac{32}{x^3} = \frac{2x^4 - 32}{x^3}$$

$$y'' = 2 + \frac{96}{x^4}$$

Critical points occur where $y' = 0$, which is where $x = \pm 2$. We note that $y'' > 0$ everywhere, so that $(2, 8)$ and $(-2, 8)$ are minimum points. Furthermore, the curve is everywhere concave upward, and there are no inflection points. Because we can substitute $-x$ for x

with no change in the equation, the graph is symmetric around the y-axis. The function is undefined for $x = 0$, and the line $x = 0$ is a vertical asymptote. Also, as $x \to \infty$, $16/x^2 \to 0$ and $y \to x^2$. This means that as x gets large, the curve approaches the parabola $y = x^2$. The graph is shown in Figure 5-13. ∎

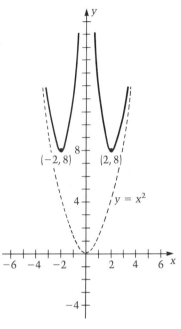

Figure 5-13

Exercises / Section **5-2**

Exercises 1–20: Sketch the graph of the function; indicate any maximum points, minimum points, and inflection points.

1. $y = 3x^2 - 12x$
2. $y = 2 - 8x - 2x^2$

3. $y = 3x^2 - 2x^3$

4. $y = x^3 - 2x^2 - 4x + 1$

5. $y = \frac{1}{3}x^3 - \frac{1}{5}x^5$
6. $y = \frac{1}{4}x^4 + x^3$

7. $y = 7 + 6x^2 - x^4$
8. $y = x^4 - 8x^2 + 3$

9. $y = 8x^3 - 2x^4$
10. $y = x(x - 1)^3$

11. $y = 3x^5 - 10x^3 + 15x$

12. $y = (2x - 1)^5 + 32$

13. $y = \dfrac{x}{x - 4}$
14. $y = \dfrac{x}{x + 1}$

15. $y = \dfrac{3}{x^2 + 1}$
16. $y = \dfrac{2x}{x^2 + 1}$

17. $y = \dfrac{x^2}{x + 1}$
18. $y = \dfrac{1}{x - 2} + \dfrac{x}{4}$

19. $y = \dfrac{1}{x^2 - 1}$
20. $y = \dfrac{x^2 + 1}{x^2 - 1}$

21. Industrial wastes are to be stored in closed cylindrical tanks. Each tank must hold 60 m³ of material. Write a function for the area A of sheet metal needed to construct the tank as a function of the radius r of the cylinder. Sketch a graph of the function.

22. The bending moment M of a certain beam is given by the following equations, where x is the distance from one end of the beam:

$$M = 533x - \frac{1}{12}x^3 \quad \text{for} \quad 0 < x < 10$$

$$M = \frac{-x^3}{12} - 467x + 10000 \quad \text{for} \quad 10 \le x < 20$$

Sketch a graph of M as a function of x.

5-3 Maxima and Minima

Suppose we can express two related quantities x and y in the form of a function $y = f(x)$. Then we can find maximum or minimum values of the function, just as we did in the preceding section where we wanted to graph the function.

EXAMPLE 1. The electrical power P (in watts) produced by a certain source is given by

$$P = 5.12R - \tfrac{8}{3}R^3$$

where R is the resistance (in ohms) in the circuit. For what resistance is the power a maximum? What is the maximum power?

We have an equation giving P as a function of R. To find the value of R that maximizes P, we find dP/dR, set it equal to zero, and solve for R.

$$\frac{dP}{dR} = 5.12 - 8R^2$$

$$5.12 - 8R^2 = 0$$

$$R = \pm\sqrt{\frac{5.12}{8}} = \pm 0.8$$

We do not consider a negative resistance, so we concentrate on the critical value $R = 0.8$. To test whether a maximum value of P occurs here, we use the second derivative test.

$$\frac{d^2P}{dR^2} = -16R$$

The second derivative is negative at $R = 0.8$ (in fact it is negative for any positive value of R). Therefore, P has its maximum value when $R = 0.8 \ \Omega$. That maximum value is found by substituting 0.8 for R in the original equation relating P and R.

$$P = 5.12(0.8) - \tfrac{8}{3}(0.8)^3$$

$$= 4.096 - 1.365$$

$$= 2.731 \text{ W}$$

To sum up, maximum power is 2.731 W, and this maximum occurs for a resistance of 0.8 Ω. ∎

We notice that Example 1 asks two questions: What is the value at which a maximum occurs, and what is the maximum value? In problems like this, the value at which a maximum (or minimum) occurs is the critical value of the independent variable found by setting the first derivative equal to zero, while the maximum (or minimum) value itself refers to the dependent variable and is found by putting the critical value back into the original equation. Many maximum-minimum problems ask only one of the two questions, and it is important in solving the problem to know which question you are trying to answer.

Word of Advice

In a maximum (minimum) problem involving the function $y = f(x)$, the maximizing (minimizing) value x_0 is not the same as the maximum (minimum) value $f(x_0)$.

Practice 1. A manufacturer of oil filters finds that the production cost c as a function of the number x of units produced per month is given by the equation

$$c = 500 - \frac{x}{2} + \frac{1}{1000} x^2$$

What is the minimum monthly production cost?

We may have some quantity y, which we want to maximize or minimize, but we are not directly given an equation for y. In this case, we must find a way, using the information given in the problem, to express y as a function of a single variable x. Once we have $y = f(x)$, we can proceed just as above.

EXAMPLE 2. An apricot grower finds that if no more than 25 trees per half acre are planted, each tree yields a harvest of about 550 apricots. For each tree over 25 per half acre, the yield per tree is reduced by about 15 apricots per tree. How many trees per half acre should be planted to maximize the total yield?

Here the independent variable is the number of trees per half acre, and the dependent variable, which we want to maximize, is the total yield. If there are x trees planted per half acre, the yield per tree is the optimum 550 reduced by 15 for each tree over the 25, or $550 - 15(x - 25)$. The total yield y is the yield per tree times the number of trees, so

$$y = [550 - 15(x - 25)]x$$

or

$$y = 925x - 15x^2$$

To find the value of x for which y is maximum, we set dy/dx equal to zero and solve for x.

$$\frac{dy}{dx} = 925 - 30x$$

$$925 - 30x = 0$$

$$x = 30.83$$

Because $y'' = -30 < 0$, the value of y is a maximum for $x = 30.83$. We clearly cannot deal in fractional parts of trees, so we round the answer to 31 trees per half acre. We could also compute values of y for $x = 30$ and $x = 31$ to determine where the larger yield occurs; it is at $x = 31$. ∎

If a picture can be drawn to illustrate the problem, the picture may make it easier to formulate an equation.

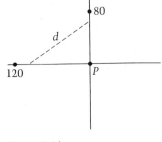

Figure 5-14

EXAMPLE 3. Point P lies at the intersection of two game trails that cross at right angles. A deer is 80 m due north of P and traveling south at the rate of 40 m/min. A mountain lion is 120 m due west of P and traveling east at the rate of 40 m/min. What is the minimum distance between the two animals?

Here we can draw a picture (Figure 5-14). We must be sure to draw a general picture; that is, the animals are moving and do not remain at fixed distances of 80 m and 120 m from P. We want to find the minimum value for d. We can express d as a function of the sides of the triangle, and these in turn are a function of the time t that has elapsed. From Figure 5-15, we get the equation

$$d^2 = (80 - 40t)^2 + (120 - 40t)^2$$
$$= 20{,}800 - 16{,}000t + 3200t^2$$

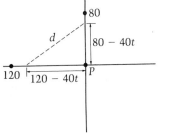

Figure 5-15

The minimum value of d^2 will tell us the minimum value of d, so we let $y = d^2$ and minimize y. This avoids having to differentiate a square root. Thus

$$y = 20{,}800 - 16{,}000t + 3200t^2$$

$$\frac{dy}{dt} = -16{,}000 + 6400t$$

$$-16{,}000 + 6400t = 0$$

$$t = 2.5$$

The second derivative of y is positive everywhere, and so y is a minimum (as is d) when $t = 2.5$ min. The minimum distance is

$$d = \sqrt{[80 - 40(2.5)]^2 + [120 - 40(2.5)]^2}$$
$$= 28.28 \text{ m} ∎$$

EXAMPLE 4. Rectangular sheets of cardboard, which are 48 cm by 90 cm, are to be folded into open cartons by cutting a square out of each corner and folding up the sides. How big should the square be in order to maximize the volume of the carton?

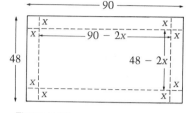

Figure 5-16

Here again we can draw a picture (Figure 5-16), where we let x represent the side of the square. We want to find the value of x that will make the volume V a maximum. The volume of a rectangular solid is the product of length, width, and height. Therefore, from Figure 5-16, we can write the equation for V as follows:

$$V = (90 - 2x)(48 - 2x)x$$
$$= 4(1080 - 69x + x^2)x$$
$$= 4(1080x - 69x^2 + x^3)$$

Now we differentiate:

$$\frac{dV}{dx} = 4(1080 - 138x + 3x^2)$$
$$= 12(360 - 46x + x^2)$$
$$= 12(x - 10)(x - 36)$$

Critical values of x are 10 and 36. However, 36 is too big to fit the physical requirements of the problem, so we discard it as a possible solution. The second derivative of V is

$$\frac{d^2V}{dx^2} = 12(-46 + 2x)$$

and

$$\frac{d^2V}{dx^2}\bigg|_{x=10} < 0$$

Thus by the second derivative test, a maximum volume occurs when the square is 10 cm by 10 cm. ∎

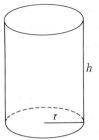

Figure 5-17

EXAMPLE 5.   Closed cylindrical storage tanks, each with a capacity of 16π m^3, are to be constructed for storage of chemical fertilizers. Find the most economical dimensions.

The most economical dimensions are those which require least material; this means that the surface area is to be a minimum. For a cylinder of radius r and height h (Figure 5-17), the total surface area A is given by the equation

$$A = 2\pi r^2 + 2\pi r h$$

Here A has been expressed as a function of the two variables r and h. In order to express A as a function of a single variable, we use the additional information about the volume,

$$V = \pi r^2 h = 16\pi$$

from which

$$h = \frac{16}{r^2}$$

Putting this expression for h into the equation for A, we get

$$A = 2\pi r^2 + 2\pi r\left(\frac{16}{r^2}\right)$$

$$= 2\pi r^2 + \frac{32\pi}{r}$$

This gives us an expression for A as a function of r alone. In order to find the value of r for which A is a minimum, we differentiate.

$$\frac{dA}{dr} = 4\pi r - \frac{32\pi}{r^2} = \frac{4\pi r^3 - 32\pi}{r^2} = \frac{4\pi(r^3 - 8)}{r^2}$$

Setting dA/dr equal to zero,

$$\frac{4\pi(r^3 - 8)}{r^2} = 0$$

The solution to this equation is $r^3 - 8 = 0$, or $r = 2$. Using the second derivative test,

$$\frac{d^2A}{dr^2} = 4\pi + \frac{64\pi}{r^3}$$

which is positive when $r = 2$, and the area is a minimum. Because we want both dimensions of the cylinder, we must compute h as well. From the equation $h = 16/r^2$, the value of h for $r = 2$ is 4. The most economical dimensions are a radius of 2 m and a height of 4 m.   ∎

Practice 2.   A rectangular garden of area 30 m^2 is to be enclosed by fencing. Find the dimensions that will minimize the amount of fencing material needed.

Not all functions have relative maximum or minimum points. Therefore, situations can arise that sound very similar to the problems we have been doing but that do not have solutions.

EXAMPLE 6. A company that manufactures circuit boards for microcomputers estimates that its profit p can be given as a function of the number x of boards produced per week by the equation

$$p = 3x^5 - 10x^3 + 15x$$

How many boards per week should be produced in order to maximize the profit?

Here we already have the equation for p as a function of x, and we proceed by differentiating.

$$\begin{aligned}
\frac{dp}{dx} &= 15x^4 - 30x^2 + 15 \\
&= 15(x^4 - 2x^2 + 1) \\
&= 15(x^2 - 1)^2
\end{aligned}$$

Setting dp/dx equal to zero, we get the equation

$$(x^2 - 1)^2 = 0$$

which has solutions of 1 and -1. We discard -1 as it has no meaning in this problem. Also, we are suspicious that profit can be maximized by producing only one unit per week! Computing d^2p/dx^2, we get

$$\begin{aligned}
\frac{d^2p}{dx^2} &= 60x^3 - 60x \\
&= 60x(x^2 - 1)
\end{aligned}$$

which has the value zero when $x = 1$, and the second derivative test fails. Going back to the first derivative test, we note that dp/dx is never negative, and so there is no change of sign of dp/dx around $x = 1$ and no maximum value of p at $x = 1$. In fact, since dp/dx for $x > 1$ is positive, the graph of the function $p(x)$ is increasing for $x > 1$. Maximum profit is limited only by the number of units that can be produced per week. ■

Exercises / Section **5-3**

1. A ball thrown straight up in the air is at height h units after t seconds, where $h = 5 + 35t - 16t^2$. Find the maximum height the ball reaches.

2. A cell culture grows at a rate of growth r, given by $r = 28t - t^2$. Here r is measured in hundreds per minute and t in minutes. Find the maximum rate of growth.

3. Two light sources, one eight times as bright as the other, are located 40 m apart. At a point between the two sources and x meters from the brighter one, the light intensity I (in candelas) from both sources is given by the equation

$$I = \frac{8}{x^2} + \frac{1}{(40 - x)^2}$$

Find the value of x at which the intensity is a minimum. (*Hint:* If $m^3 = n^3$, then $m = n$.)

4. The heat output H of a heating element is a function of the number of hours t the element has run. The equation giving H as a function of t is $H = 132.24t - 0.034t^3$. Find the value of t at which H is a maximum.

5. The voltage V (in volts) of a certain thermocouple varies with the temperature T (in degrees Celsius) according to the equation $V = \sqrt{0.0002T^4 - T + 125}$. Find the value of T for which V is a minimum.

6. The equation $y = k(16x^4 - 12Lx^3 + L^2x^2)$, where k is a constant, gives the deflection of a beam of length L at a distance x from one end. What value of x results in maximum deflection?

7. What number exceeds its square by the maximum amount?

8. Find two numbers whose sum is 60 and whose product is a maximum.

9. A piece of wire 50 cm long is to be cut in two pieces, with one part bent into a circle and the other into a square. Where should the wire be cut so that the sum of the two enclosed areas is minimal?

10. A piece of wire 76 cm long is to be cut in two pieces, with one part bent into a circle and the other into an equilateral triangle. Where should the wire be cut so that the sum of the two enclosed areas is minimal?

11. A fenced-in rectangular garden of 12 m² is to be built against a house (see Figure 5-18). No fence is needed on the house side. Find the dimensions of the garden in order to minimize the cost of fencing.

12. A rancher wants to construct three identical cattle pens, as in Figure 5-19. A total of 200 m of fencing can be used. Find the dimensions of each individual pen in order to maximize the total area.

13. A river runs along the back of a rectangular lot. No fence is needed along the river. On the front of the lot there is an 18-m opening. Fencing for the front of the lot costs $8.00/m, and along the sides it costs $6.50/m. Find the dimensions of the largest lot that can be fenced for a total of $900.

14. A laboratory facility is to be constructed in a rectangular shape. It must have 60 m² of area, and one wall is to be glass. Cost per running meter of the glass wall is $350; cost per running meter for the other three walls, which are cement block, is $240. What should the dimensions be in order to minimize the cost of the wall construction?

15. A certain brand of stereo speaker sells for $25. The production cost to manufacture x of these speakers per year is $C = 150 + 2.4x + 0.02x^2$. What is the maximum yearly profit from this type of speaker?

16. A computer terminal sells for $2572. The cost to manufacture x terminals per year is $C = 5600 + 940x + 0.8x^2$. What is the maximum yearly profit?

17. A company can sell 3250 semiconductors per month when the price is 15¢. The company estimates that for each penny increase in price, it will sell 90 fewer per month. What price will make the gross income a maximum?

18. A rental agency for an apartment complex rents 250 units when the monthly rent is $350. A survey of tenants reveals that for each $10 increase in rent, four units will become vacant. What rental fee will maximize gross income?

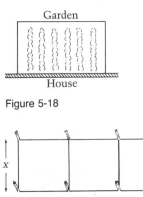

Figure 5-18

Figure 5-19

19. An advertising brochure is to be printed on paper whose total area is 500 cm². There will be margins of 2 cm on each side and 4 cm at top and bottom. Find the dimensions of the paper that will produce a maximum of printed area.

20. Postal regulations require that the sum of the three dimensions of a rectangular package be no more than 120 cm. Find the dimensions of a rectangular package with a square end such that postal requirements are met and the volume is a maximum.

21. A cylindrical boiler with a copper bottom and tin sides will be open at the top. The volume is to be 0.8 m³. If tin costs $35/m² and copper costs $82/m², find the most economical dimensions.

22. A grain silo consists of a cylinder (with a bottom but no top) surmounted by a hemisphere. If a silo is to hold 50π m³, find the most economical dimensions.

23. What are the dimensions of the largest rectangle that can be inscribed in a circle of radius r?

24. A long sheet of aluminum 36 cm wide is to be made into a gutter by folding up the sides at right angles. What size strips should be folded up to maximize the capacity of the gutter?

25. Power cable is to be strung from point A to point B, where B lies 70 m down from a point directly opposite A across a 20-m canyon (see Figure 5-20). It costs three times as much to get cable across the canyon as it does to take it along land. What should be the path of the cable (shown in dotted lines) to minimize the cost?

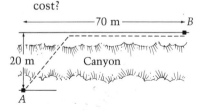

Figure 5-20

26. According to the laws of optics, a ray of light leaving point A and hitting a mirror at point C will be reflected and travel to point B in such a way that angle θ_1 equals angle θ_2 (see Figure 5-21). Show that $\theta_1 = \theta_2$ results in the minimum path for the light ray.

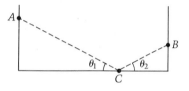

Figure 5-21

5-4 Motion

Another interpretation of the derivative is that it represents rate of change of a dependent variable with respect to an independent variable. Often the independent variable is time, and the derivative then represents time rate of change. Recall that, for a moving object, the rate of change of position with respect to time is the *velocity* of the object, and the rate of change of velocity with respect to time is *acceleration*.

In Chapter 4, we used the derivative to compute velocity and acceleration for bodies moving in a straight line (one-dimensional motion). Given an equation $s = s(t)$ to express distance s along a straight line from some fixed point as a function of time t, we can find velocity (ds/dt) and acceleration (d^2s/dt^2). Motion along a straight line is called **rectilinear motion**. In this section, we study motion in a plane (two-dimensional motion), called **curvilinear motion**.

In order to analyze curvilinear motion, we need the concept of a vector. A **vector quantity** is a quantity with both magnitude and direction. For example, weather reports give wind velocity in terms of both magnitude and direction (i.e., 10 miles per hour from the west). Velocity is a vector quantity; other examples of vector quantities are acceleration and force.* Vector quantities are represented by **vectors**, directed line segments of length proportional to the magnitude of the vector quantity and pointing in the direction of the vector quantity.

Two vectors may be added to find a **resultant**, or **sum**, vector, a single vector that alone has the same effect as the two original vectors. Thus Figure 5-22(a) shows vectors **A** and **B** representing two forces, such as two people pushing upon an object at point *P*. The net effect is the same as the single force represented by **R**. As shown in Figure 5-22(b), **R** can be obtained by making **A** and **B** the sides of a parallelogram, with **R** the diagonal of the parallelogram. (A boldface letter indicates a vector; the magnitude of a vector **A** is denoted by *A*.)

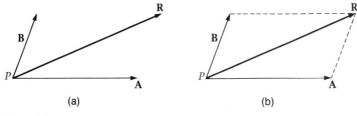

(a)   (b)

Figure 5-22

Not only can we add two vectors to find their resultant using the *parallelogram rule*, we can also do the inverse operation. Given a single vector that is the resultant of two vectors, we can find the two vectors. This process is called *resolving* the vector into its **components**. A single vector, however, can have many sets of components, as shown in Figure 5-23. It is often useful to resolve a vector into its horizontal and vertical components. To do this, we put the vector in standard position on a rectangular coordinate system—that is, we put one end at the origin with the vector pointing away from the origin—and give its direction by means of

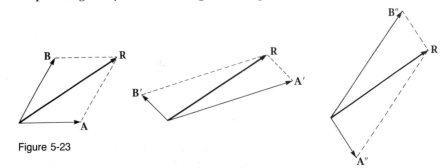

Figure 5-23

*A quantity with magnitude but no direction is called a **scalar quantity**, or **scalar**; examples of scalar quantities are speed, temperature, and mass.

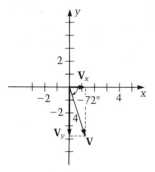

Figure 5-24

the angle it makes with the positive x-axis. Then we just use a little trigonometry.

EXAMPLE 1. Find the horizontal and vertical components of a vector **V** of magnitude 4 units and directed at an angle of $-72°$.

From Figure 5-24, we get the following expressions for the magnitude V_x of the horizontal component $\mathbf{V}_x$ and the magnitude V_y of the vertical component $\mathbf{V}_y$:

$$\frac{V_x}{V} = \cos(-72°)$$

$$V_x = V \cos(-72°) = 4(0.309) = 1.236$$

$$\frac{V_y}{V} = \sin(-72°)$$

$$V_y = V \sin(-72°) = 4(-0.951) = -3.804$$

The negative magnitude of $\mathbf{V}_y$ indicates that $\mathbf{V}_y$ points downward rather than upward. ■

In order to add two vectors more accurately than we can using the graphical parallelogram rule, we resolve each vector into its horizontal and vertical components. We can then accurately add the horizontal components, because they both lie on the x-axis. Similarly, we add the vertical components. This gives us the horizontal and vertical components of the resultant vector, and from these it is easy to find the magnitude and direction of the resultant.

EXAMPLE 2. Find the resultant of vectors **A** and **B**, when **A** is of magnitude 2 and direction $-32°$ and **B** is of magnitude 6 and direction $110°$ (Figure 5-25(a)).

$$A_x = 2 \cos(-32°) = 2(0.848) = 1.696$$
$$A_y = 2 \sin(-32°) = 2(-0.530) = -1.060$$
$$B_x = 6 \cos(110°) = 6(-0.342) = -2.052$$
$$B_y = 6 \sin(110°) = 6(0.940) = 5.640$$
$$R_x = A_x + B_x = 1.696 + (-2.052) = -0.356$$
$$R_y = A_y + B_y = -1.060 + 5.640 = 4.58$$

From Figure 5-25(b), we see that we can find the magnitude of **R** by using the Pythagorean theorem. Thus $R = \sqrt{R_x^2 + R_y^2} = \sqrt{(-0.356)^2 + (4.58)^2} = 4.59$. We can find the direction α by first noting that

$$\tan \alpha = \frac{R_y}{R_x} = \frac{4.58}{-0.356} = -12.865$$

Also, α is a second-quadrant angle (which we know from the signs of R_x and R_y, even without Figure 5-25(b)), and therefore the value of α is 94.4°. The resultant vector has magnitude 4.59 units and direction 94.4°. ■

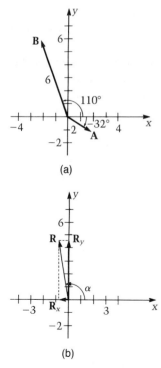

Figure 5-25

Practice 1.   Find the resultant **R** of vectors **A** and **B**, where **A** has magnitude 5 and direction 75° and **B** has magnitude 3 and direction −18°.

One way to describe curvilinear motion of an object is to specify the *x*- and *y*-coordinates of the object as functions of time *t*. This is done by means of a set of **parametric equations** of the form

$$x = x(t)$$
$$y = y(t)$$

where *t*, the independent variable, is called the **parameter**. If we are given such a set of equations it is possible to find the magnitude and direction of the velocity vector at any particular time t_0. To do this, we note that $x = x(t)$ is an expression for the horizontal position of the body as a function of *t*, so that dx/dt is an expression for the magnitude of the horizontal velocity as a function of *t*. If we evaluate dx/dt at $t = t_0$, we have the magnitude, v_x, of the horizontal component of velocity, $\mathbf{v_x}$, at time t_0. By using dy/dt, we can find the magnitude, v_y, of the vertical component of velocity, $\mathbf{v_y}$, in the same way. We then find the resultant velocity vector **v**, just as in Example 2.

EXAMPLE 3.   An object moves according to the set of parametric equations $x = 3t$, $y = 18t − 9t^2$. Find the magnitude and direction of the velocity when $t = 2$.

The object follows the parabolic path shown in Figure 5-26. We compute the magnitudes of the velocity components as follows:

$$v_x = \frac{dx}{dt} = 3 \qquad\qquad v_x\Big|_{t=2} = 3$$

$$v_y = \frac{dy}{dt} = 18 - 18t \qquad v_y\Big|_{t=2} = -18$$

We find the magnitude of the velocity by computing

$$v = \sqrt{3^2 + (-18)^2} = 18.25$$

and the direction α by

$$\tan \alpha = \frac{v_y}{v_x} = \frac{-18}{3} = -6$$

$$\alpha = -80.5° \quad \blacksquare$$

Because acceleration is the time rate of change of velocity, we find the magnitudes of the components of the acceleration vector by

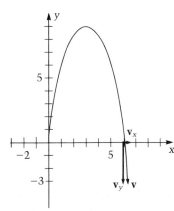

Figure 5-26

differentiating the magnitudes of the velocity components with respect to time.

EXAMPLE 4. A particle moves according to the parametric equations $x = t^3 - 2t^2 + 1$, $y = t^5 - t^7$. Find the velocity and acceleration vectors at $t = 1$.

For the velocity, we differentiate once.

$$v_x = \frac{dx}{dt} = 3t^2 - 4t \qquad\qquad v_x\Big|_{t=1} = -1$$

$$v_y = \frac{dy}{dt} = 5t^4 - 7t^6 \qquad\qquad v_y\Big|_{t=1} = -2$$

$$v = \sqrt{(-1)^2 + (-2)^2} = 2.24$$

$$\tan \alpha = \frac{v_y}{v_x} = \frac{-2}{-1} = 2$$

$$\alpha = 243.4° \quad \text{or} \quad \alpha = -166.6°$$

Observe that α is in the third quadrant because both v_x and v_y are negative. We differentiate again for acceleration.

$$a_x = \frac{dv_x}{dt} = 6t - 4 \qquad\qquad a_x\Big|_{t=1} = 2$$

$$a_y = \frac{dv_y}{dt} = 20t^3 - 42t^5 \qquad\qquad a_y\Big|_{t=1} = -22$$

$$a = \sqrt{2^2 + (-22)^2} = 22.1$$

$$\tan \beta = \frac{a_y}{a_x} = \frac{-22}{2} = -11$$

$$\beta = -84.8°$$

The velocity vector has magnitude 2.24 and direction 243.4°, and the acceleration vector has magnitude 22.1 and direction $-84.8°$. ■

Practice 2. The parametric equations of a moving object are $x = 3t^2 - 1$, $y = 2t^4$. Find the velocity and acceleration vectors at $t = 1$.

Curvilinear motion may also be described by giving the path of the moving object in the form $y = f(x)$. Both y and x are functions of time. We can differentiate both sides of the equation $y = f(x)$ *with respect to time;* this will introduce $v_y = dy/dt$ and $v_x = dx/dt$, the magnitudes of the velocity components.

EXAMPLE 5. A particle moves along the curve $y = x^2 - 4$ in such a way that the magnitude v_x of the horizontal component of velocity is always 3. Find the magnitude and direction of the velocity at the point $(2, 0)$.

We differentiate both sides of $y = x^2 - 4$ with respect to time.

$$\frac{dy}{dt} = 2x\frac{dx}{dt}$$

or

$$v_y = 2xv_x$$

But $v_x = 3$, so we evaluate v_y at $(2, 0)$ and get

$$v_y\bigg|_{(2,0)} = 2(2)(3) = 12$$

The magnitude and direction of **v** are found in the usual way.

$$v = \sqrt{3^2 + 12^2} = 12.4$$

$$\tan \alpha = \frac{12}{3} = 4$$

$$\alpha = 76° \quad \blacksquare$$

Practice 3.   An object moves along the path $y^2 = 4x + 1$ with a vertical component of velocity of constant magnitude 4. Find the velocity vector at the point $(2, 3)$.

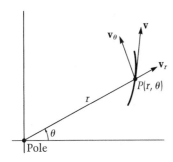

Figure 5-27

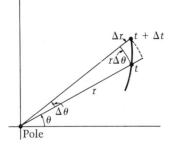

Figure 5-28

Polar coordinates can also be used when describing curvilinear motion. The position of an object in the polar coordinate system is given by specifying the values of r and θ (θ is generally measured in radians). A set of parametric equations for motion in the plane gives both r and θ as functions of time. The velocity of the object at time t_0 is found from perpendicular components. Instead of the horizontal and vertical components used with the rectangular coordinate system, the perpendicular components are taken in the direction of increasing r and in the direction of increasing θ. Figure 5-27 illustrates the velocity vector **v** and its components $\mathbf{v}_r$ and $\mathbf{v}_\theta$ at a particular point along the path of a moving object. Figure 5-28 shows the point at time Δt later. There has been an increment Δr, and an increment $r\Delta\theta$ along the arc which corresponds to the change $\Delta\theta$ (see Exercise 39, Section 3-2). The velocity component $\mathbf{v}_r$ measures the instantaneous time rate of change in the direction of increasing r, so its magnitude is given by the equation

$$v_r = \lim_{t \to 0}\frac{\Delta r}{\Delta t} = \frac{dr}{dt}$$

The velocity component v_θ measures the instantaneous time rate of change along the arc, so its magnitude is given by the equation

$$v_\theta = \lim_{t \to 0} \frac{r\Delta\theta}{\Delta t} = r\frac{d\theta}{dt}$$

To sum up, the velocity components in polar coordinates have magnitudes

$$v_r = \frac{dr}{dt} \qquad v_\theta = r\frac{d\theta}{dt}$$

EXAMPLE 6. The equations of motion for a moving body are $r = 3t^4 - 6t$, $\theta = 2t^2$. Find the magnitude of the velocity vector when $t = 2$.

$$v_r = \frac{dr}{dt} = 12t^3 - 6 \qquad\qquad v_r\Big|_{t=2} = 90$$

$$v_\theta = r\frac{d\theta}{dt} = (3t^4 - 6t)(4t) \qquad v_\theta\Big|_{t=2} = 288$$

$$v = \sqrt{90^2 + 288^2} = 301.7 \quad\blacksquare$$

Practice 4. Parametric equations for a moving object are $r = 2t^3 - t^2$, $\theta = 4t$. Find the magnitude of the velocity vector when $t = 1$.

Exercises / Section **5-4**

Exercises 1–4: Find the resultant (magnitude and direction) of the given vectors **A** and **B**.

1. Magnitude of **A** = 5, direction of **A** = 84°; magnitude of **B** = 3, direction of **B** = −38°.

2. Magnitude of **A** = 6.1, direction of **A** = 78°; magnitude of **B** = 8, direction of **B** = 149°.

3. Magnitude of **A** = 3.7, direction of **A** = 253°; magnitude of **B** = 0.8, direction of **B** = 101°.

4. Magnitude of **A** = 2.5, direction of **A** = −28°; magnitude of **B** = 5.4, direction of **B** = −69°.

Exercises 5–10: For the following parametric equations of a moving object, find the velocity and acceleration vectors at the given value of time.

5. $x = 4t^2 + 3t$, $y = t^5 - 2t^2$; $t = 2$

6. $x = t^4 - 3t^2$, $y = 2 - t^2$; $t = 1$

7. $x = \dfrac{3}{t - 1}$, $y = t^2$; $t = 1.4$

8. $x = \dfrac{4}{2 - 2t^3}$, $y = \dfrac{1}{t^2}$; $t = 2$

9. $x = \sqrt{2 + 3t}$, $y = 2t^3$; $t = 3$

10. $x = \sqrt{t^3}$, $y = 2t^2 - 3$; $t = 1.2$

Exercises 11–14: For the following parametric equations of a moving object, find the magnitude of the velocity vector at the given value of time.

11. $r = t^4 - 3t^3$, $\theta = t^2 - 2t$; $t = 2$

12. $r = 3t - 2t^3$, $\theta = t^4 + t^2$; $t = 3$

13. $r = \dfrac{3}{2 + t}$, $\theta = \dfrac{1}{t^2}$; $t = 1$

14. $r = \sqrt{t^3 - 4}$, $\theta = t^5$; $t = 2$

15. A particle moves along the curve $y = 2x^3$ with a horizontal component of velocity of constant magnitude 2. Find the velocity vector at the point $(1, 2)$.

16. A particle moves along the curve $y = \sqrt{x + 2}$ with a vertical component of velocity of constant magnitude 4. Find the velocity vector at the point $(2, 2)$.

17. Find the acceleration vector at the point $(1, 2)$ for the particle of Exercise 15.

18. Find the acceleration vector at the point $(2, 2)$ for the particle of Exercise 16.

19. A float placed on the surface of a stream to trace its currents moves according to the parametric equations $x = 4t^2 - 5$, $y = 12t + 3$, where distance is in meters and time is in minutes. Find the velocity vector at $t = 3$ min.

20. The electron in a hydrogen atom has an orbit described by the parametric equations

$$x = \frac{2}{\sqrt{t^2 + 4}} \qquad y = \frac{t}{\sqrt{t^2 + 4}}$$

Find the velocity vector at $t = 2$.

21. A moving object follows the curve $y = 6x^2 + 2/x$ in such a way that $v_y = 2x$. Find the velocity vector at the point $(1, 8)$.

22. A cannonball fired over flat ground travels the path $y = x - 0.02x^2$ with a horizontal component of velocity of magnitude $0.2x$. Find the velocity vector when the cannonball hits the ground if distance is in meters and time is in seconds.

5-5 Related Rates

As we saw in the previous section, we are often interested in the situation where a quantity p is a function of time; then dp/dt represents the (instantaneous) rate of change of p with respect to time. Now suppose that there is a second quantity, such as q, which is also a function of time. Then dq/dt represents the time rate of change of q. If, in addition, the quantities p and q are related by some equation, then dp/dt and dq/dt are also related. This relation between the time rates of change of p and q may be found by starting with the equation relating p and q and differentiating both sides (implicitly) with respect to time.

EXAMPLE 1. The voltage V that produces a current i in a wire of radius r is $V = 0.04i/r^2$. Find the rate at which the voltage is increasing if the current is increasing at the rate of 0.03 A/s in a wire of radius 0.12 cm.

We first note that V and i are quantities that vary with respect to time, while the radius r is a constant. Hence we can put the constant value for r directly into the equation and get

$$V = \frac{0.04i}{(0.12)^2} = 2.78i$$

Now we differentiate both sides of this equation with respect to t.

$$\frac{dV}{dt} = 2.78 \frac{di}{dt}$$

Finally we insert the known rate of change, $di/dt = 0.03$.

$$\frac{dV}{dt}\Big|_{di/dt=0.03} = 2.78(0.03) = 0.08$$

The voltage is increasing at the rate of 0.08 V/s. ■

EXAMPLE 2. An object weighing 45 kg on the earth's surface has weight W kilograms at a point r kilometers above the surface, where

$$W = 45\left(1 + \frac{r}{2500}\right)^{-2}$$

Find how fast the weight of the object is changing when it is 525 km above the surface of the earth if its altitude is increasing at the rate of 33 km/s.

Unlike Example 1, r in this problem is a variable that changes with respect to time. We differentiate both sides of the equation with respect to time. The right side of the equation requires the use of the power rule and, because we are differentiating with respect to t rather than r, there is a factor of dr/dt.

$$\frac{dW}{dt} = 45(-2)\left(1 + \frac{r}{2500}\right)^{-3}\left(\frac{1}{2500}\right)\frac{dr}{dt}$$

Substituting $r = 525$ and $dr/dt = 33$, we get

$$\frac{dW}{dt}\Big|_{r=525,\ dr/dt=33} = 45(-2)\left(1 + \frac{525}{2500}\right)^{-3}\left(\frac{1}{2500}\right)(33) = -0.67$$

At a height of 525 km, the weight is decreasing (the negative sign indicates that as t increases, W decreases) at the rate of 0.67 kg/s. ■

Practice 1. The kinetic energy (in joules) of a moving body is given by the equation $KE = \frac{1}{2}mv^2$, where m is the mass of the body and v is its velocity. Suppose a body of mass 15 kg is accelerating at the rate of 4 m/s². How fast is its kinetic energy changing at the moment when the velocity is 45 m/s? (Remember that acceleration is dv/dt.)

Often the equation relating the original quantities p and q must be found; this might require use of some geometric formula or a relationship that can be seen by drawing a picture of the problem. Once we find the equation relating the two quantities, we differentiate with respect to time and substitute the known values, just as in Examples 1 and 2.

EXAMPLE 3. A spherical meteorite burns up as it enters the earth's atmosphere. If the rate at which it burns causes the radius to decrease by 0.3 cm/s, what is the rate of change of volume when the radius is 74 cm?

The related quantities here are the radius of a sphere and the volume of the sphere. The formula for the volume of a sphere in terms of its radius is

$$V = \tfrac{4}{3}\pi r^3$$

We differentiate this equation with respect to time.

$$\frac{dV}{dt} = 4\pi r^2 \frac{dr}{dt}$$

Substituting $r = 74$ and $dr/dt = -0.3$ (the radius is decreasing), we get

$$\frac{dV}{dt}\bigg|_{r=74,\ dr/dt=-0.3} = 4\pi(74)^2(-0.3) = -20{,}644.1$$

The volume is decreasing at the rate of 20,644.1 cm³/s. ∎

EXAMPLE 4. A weather balloon is rising vertically at the rate of 2 m/s. An observer is standing on the ground 100 m away from a point directly below the balloon. At what rate is the distance between the observer and the balloon changing when the balloon is 200 m high?

Figure 5-29 illustrates the situation. Notice that the 100 m along the ground is a fixed distance, while the height x of the balloon and the distance s between the balloon and the observer are variables. In particular, it would be incorrect to label the vertical leg of the triangle with the value 200 because it has that value only for one particular instant, and we must draw our picture for the general case. From Figure 5-29, we see that the relation between s and x is

$$s^2 = x^2 + 100^2 \tag{1}$$

Taking derivatives with respect to time,

$$2s\frac{ds}{dt} = 2x\frac{dx}{dt}$$

or

$$s\frac{ds}{dt} = x\frac{dx}{dt} \tag{2}$$

The known values to substitute into this equation are $x = 200$ and $dx/dt = 2$. This leaves the two unknowns s and ds/dt. However, by going back to Equation (1), we can compute the value of s at the moment when $x = 200$.

$$s^2 = 200^2 + 100^2$$
$$s = \sqrt{50{,}000} = 223.6$$

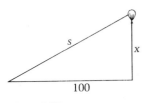

Figure 5-29

Now we have everything we need to solve Equation (2) for ds/dt.

$$(223.6)\frac{ds}{dt} = (200)(2)$$

$$\frac{ds}{dt} = 1.79 \ m/s \quad \blacksquare$$

Word of Advice

When drawing pictures for "related-rates" problems, be careful to label with constant values only those quantities that are constant throughout the problem. Label all quantities that change as variables.

Practice 2. A 6-m ladder leans against a vertical wall. If the bottom of the ladder slides along the floor away from the wall at the rate of 0.3 m/s, how fast is the top of the ladder sliding down the wall when the top of the ladder is 3 m from the floor?

Exercises / Section **5-5**

1. The electric resistance R (in ohms) of a particular resistor is given by $R = 2.000 + 0.002T^2$, where T is the temperature (in degrees Celsius). If the temperature is increasing at the rate of 0.4°C/s, how fast is the resistance changing when the temperature is 120°C?

2. The voltage V (in volts) of a certain thermocouple changes with temperature T (in degrees Celsius) according to the equation $V = 3.100T + 0.004T^2$. If the temperature is increasing at the rate of 2.1°C/min, find how fast the voltage is changing when $T = 90$°C.

3. An ideal gas is expanding adiabatically—with no change of heat—and the relationship between pressure and volume of the gas is $pv^{1.6} = c$, where c is a constant. The volume is increasing at the rate of 0.3 m³/min. What is the rate of change of pressure when the volume is 0.7 m³ and the pressure is 40 kPa?

4. Standard barometric pressure at sea level is 760 mm. (millimeters of mercury). At that pressure the boiling point of water is 100°C. For barometric pressure p between 660 mm. and 860 mm. the boiling point T of water (in degrees Celsius) is approximately

$$T = 100 + 28.012\left(\frac{p}{760} - 1\right)$$

$$- 11.64\left(\frac{p}{760} - 1\right)^2 + 7.1\left(\frac{p}{760} - 1\right)^3$$

Find the rate of change of the boiling point when the pressure is 720 mm and is decreasing at the rate of 9 mm/min.

5. The total resistance R of a circuit containing resistors R_1 and R_2 in parallel is

$$R = \frac{R_1 R_2}{R_1 + R_2}$$

If R_1 is decreasing at the rate of 0.4 Ω/min and R_2 is decreasing at the rate of 0.7 Ω/min, find the rate of change of R when R_1 is 5.00 Ω and R_2 is 7.00 Ω.

6. The equation

$$\frac{1}{s_1} + \frac{1}{s_2} = \frac{1}{f}$$

relates the distance s_1 of an object from a thin lens of focal length f to the distance s_2 of the image from the lens. If an object is moving away from a lens of focal length 15 cm at the rate of 5 cm/min, how fast is its image moving toward the lens when the object is 40 cm from the lens?

7. The velocity v of blood flow in an artery of radius R at a fixed point a distance r units from the center of the artery is given by the equation $v = c(R^2 - r^2)$, where c is a constant. As a result of taking two aspirin, the radius R of the artery increases at the rate of 1.8×10^{-4} cm/min. Assume that c has the value 1, and find the rate of change of v (in centimeters per minute per minute) at the fixed point at the moment when R is 0.03 cm.

8. A spherical tank of radius a contains a volume V of liquid when filled to a maximum depth h, where $V = \frac{1}{3}\pi h^2(3a - h)$. Water is being pumped into a spherical water tower tank 4 m in radius at the rate of 0.6 m³/s. Find how fast the depth of the water is increasing when $h = 1.5$ m.

9. A melting ice cube is losing volume at the rate of 7 cm³/min. How fast is the edge shrinking when it is 3.2 cm in length?

10. A length of pig iron in the shape of a right circular cylinder expands as it is heated. If the length increases at the rate of 0.08 cm/min and the radius increases at the rate of 0.003 cm/min, find the rate of change of volume when the length is 173 cm and the radius is 7 cm.

11. A circular oil slick is spreading from an offshore well. Its radius increases at the rate of 5 km/day. At what rate is its area increasing when its radius is 30 km?

12. A softball diamond is in the shape of a square, with sides of length 18 m. A player is running from first base to second base at a speed of 4 m/s. Find the rate of change of the player's distance from home plate when the player is 5 m from first base.

13. A piston in a cylinder of radius 0.75 m is pulled out at the rate of 0.4 m/s. Find the rate of change of volume contained in the cylinder.

14. The ends of a water trough 2 m long are equilateral triangles with one vertex pointing downward. If water is being pumped into the trough at the rate of 0.8 m³/min, how fast is the water level rising when the water is 0.3 m deep?

15. Sawdust from a milling operation is falling into a conical pile at the rate of 0.2 m³/s. The height of the pile is always three fourths of the radius of the base. How fast is the radius changing when the pile is 1.8 m high?

16. A conical filter funnel is filled at the rate of 300 cm³/min, and liquid drains out of the funnel at the rate of 200 cm³/min. The diameter of the funnel at its open end equals the depth of the funnel. How fast is the liquid in the funnel rising when it is 7 cm deep?

17. A train starts north at 2:00 P.M. traveling at 80 km/h. At 4:00 P.M. another train starts from the same point, traveling west at 95 km/h. How fast are the two trains separating at 5:00 P.M.?

18. A winch on a dock 6 m high is pulling in a rope attached to a boat on the water. The rope is being drawn in at the rate of 2 m/s. How fast is the boat approaching the bottom of the dock when it is 20 m away from the bottom of the dock?

19. A man 2 m tall walks away from a lamp post 5 m high at the rate of 8 km/h. How fast does the end of his shadow move?

20. In Exercise 19 above, how fast does the man's shadow lengthen?

5-6 Differentials

When $y = f(x)$, we have used both $f'(x)$ and dy/dx to denote the derivative of y with respect to x. Thus

$$f'(x) = \frac{dy}{dx}$$

It may have occurred to you that dy/dx looks like a fraction, a quotient composed of something called dy divided by something called dx. In fact, we are now about to define dy and dx in such a way that this interpretation is correct.

For $y = f(x)$, we define dx by

$$dx = \Delta x$$

where Δx is a change or increment in x, and we define dy by

$$dy = f'(x)\,dx \tag{1}$$

The quantity dx is called the **differential of x** and the quantity dy is called the **differential of y**. From Equation (1), it is clear that $f'(x)$ is indeed dy divided by dx, but this still does not give us a very good idea of what the differentials really are. Let us return to a picture we used when developing the definition of the derivative.

In Figure 5-30 we have shown two points on a curve $y = f(x)$. The points are separated by the increments Δx, which can now be called dx, and Δy. The quantity Δy is the change in the height of the curve itself as x changes by a value Δx. In Figure 5-31 we have added the tangent line to the curve at point P. We know that the slope of the tangent line is $f'(x)$ or, in terms of differentials, dy divided by dx. Therefore, the quantity dy is the change in height of the tangent line as x changes by a value Δx. We also see from Figure 5-31 that for small values of Δx, dy can be used to approximate Δy and is often much easier to compute.

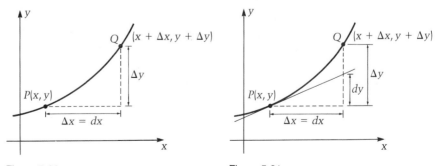

Figure 5-30   Figure 5-31

EXAMPLE 1.   For $y = 4x^2 - 2x + 1$, find the differential dy.

The function here is $f(x) = 4x^2 - 2x + 1$. By taking the derivative, $f'(x) = 8x - 2$. From Equation (1),

$$dy = f'(x)\, dx = (8x - 2)\, dx \quad \blacksquare$$

EXAMPLE 2.   Compare the values of Δy and dy for the function $y = 4x^2 - 2x + 1$ with $x = 2$ and $\Delta x = 0.1$.

The quantity Δy represents the actual change in the value of y as x moves from 2 to 2.1, and can be computed as in the delta process.

$$\begin{aligned}
\Delta y = f(x + \Delta x) - f(x) &= [4(x + \Delta x)^2 - 2(x + \Delta x) + 1] \\
&\quad - (4x^2 - 2x + 1) \\
&= (4x^2 + 8x\Delta x + 4(\Delta x)^2 - 2x - 2\Delta x + 1) \\
&\quad - (4x^2 - 2x + 1) \\
&= 8x\Delta x + 4(\Delta x)^2 - 2\Delta x
\end{aligned}$$

$$\Delta y \Big|_{x=2,\, \Delta x=0.1} = 8(2)(0.1) + 4(0.01) - 2(0.1) = 1.44$$

The differential dy is (from Example 1)

$$dy = (8x - 2)\, dx$$

or, because $dx = \Delta x$,

$$dy = (8x - 2)\Delta x$$

Substituting the values $x = 2$ and $\Delta x = 0.1$, we get

$$dy \Big|_{x=2,\, \Delta x=0.1} = (8(2) - 2)(0.1) = 1.4$$

Here dy is 1.4 and the actual change Δy is 1.44. $\blacksquare$

Again, Δy represents an *actual* change in y for some change Δx in x, and dy *approximates* this change in y when Δx is small.

Practice 1.   For $y = (2x - 1)^2$, find the differential dy.

Practice 2.   Compute Δy and dy for $y = (2x - 1)^2$, with $x = 4$ and $\Delta x = 0.2$.

Suppose a quantity y is computed at a fixed value of x by the function $y = f(x)$. Then an error Δx in the value of x, such as might be encountered due to inaccuracies of measurement, produces a corresponding error Δy in the function value. The error Δy can be approximated by using dy.

EXAMPLE 3. The radius of a spherical ball bearing is estimated to be 0.6 cm $\pm$ 0.002 cm. Estimate the maximum error in computing the volume.

The formula for the volume V of a sphere of radius r is $V = \frac{4}{3}\pi r^3$. The error Δr of 0.002 produces a corresponding error ΔV in the volume. We use dV to approximate ΔV.

$$dV = 4\pi r^2\, dr$$

Substituting $r = 0.6$ and $dr = \Delta r = 0.002$, we get

$$dV\Big|_{r=0.6,\ dr=0.002} = 4\pi(0.6)^2(0.002) = 0.009$$

We estimate the error in volume computation to be within 0.009 cm^3. ∎

Practice 3. The edge of a cube is determined by measurement to be 3.4 cm $\pm$ 0.1 cm. Estimate the maximum error in computing the volume.

The actual error in the measurement of a quantity is called the **absolute error**. Error is often expressed in terms of **relative error**, which is the ratio of the absolute error to the size of the quantity itself; it is usually given as a percentage. The idea behind relative error is that a given error in measuring a small quantity is "worse" than an error of the same size in measuring a large quantity.

EXAMPLE 4. In Example 3, the absolute error in measuring the radius is 0.002 cm. The relative error is the ratio of the absolute error to the size of the radius itself, which is (0.002 cm)/(0.6 cm) = 0.003 = 0.3%. The absolute error in the volume is approximately 0.009 cm^3, while the size of the volume itself is $V = \frac{4}{3}\pi(0.6)^3 = 0.905$ cm^3. Therefore the relative error in measuring the volume is (0.009 cm^3)/(0.905 cm^3) = 0.0099 = 0.99%. ∎

Practice 4. Estimate the relative error in measuring the volume of the cube of Practice 3.

Differentials have been useful in this section to approximate quantities. The differential notation $dy = f'(x)\, dx$ is also used in the next chapter, where we discuss integration.

Exercises / Section **5-6**

Exercises 1–8: Find the differential dy.

1. $y = 2x^6 - 4x^3 + 5x^2$

2. $y = 3x^5 - 2x^2 + 4x$

3. $y = (2x^2 - 1)^{2/3}$ 4. $y = (x - 3x^5)^{3/4}$

5. $y = \dfrac{x^2 - 1}{2x}$ 6. $y = \dfrac{2x + 3}{x^2 - 2}$

7. $y = \sqrt{x^2 - 3}$ 8. $y = \dfrac{1}{\sqrt{2x + 4}}$

Exercises 9–12: Compute Δy and dy for the given values of x and Δx.

9. $y = 2x^2 - 5x + 3$, $x = 1$, $\Delta x = 0.2$

10. $y = 3 + 2x - 4x^2$, $x = 3$, $\Delta x = 0.1$

11. $y = 4x^3 - 2x$, $x = 2$, $\Delta x = 0.1$

12. $y = 2x^4$, $x = 3$, $\Delta x = 0.01$

Exercises 13–16: Compute $f(1.01) - f(1)$, and compare the result with dy for $x = 1$, $\Delta x = 0.01$.

13. $y = (2x - 1)^3$ 14. $y = (4 - x^2)^4$

15. $y = \sqrt{x}(x^2 - 3)$

16. $y = \dfrac{1}{\sqrt{x}} - \dfrac{1}{\sqrt{2x - 1}}$

Exercises 17–22: Solve by using differentials.

17. A circular sheet metal piece was cut with a radius of 12 cm. Later it was learned that the radius was 0.2 cm too large. Estimate the area of metal wasted.

18. A square carpet piece was cut with sides 9 m long. The sides then had to be shortened by 0.02 m. Estimate the area of carpet wasted.

19. The radius of a 4 m high circular cylinder is estimated to be 3 m ± 0.001 m. Estimate the maximum error and the relative error in computing the volume of the cylinder.

20. The radius of a sphere is estimated to be 15 cm ± 0.2 cm. Estimate the maximum error and the relative error in computing the surface area of the sphere.

21. A spherical bead of radius 2 cm is coated with a layer of paint 0.001 cm thick. Estimate the volume of paint used.

22. Extruded metal tubing is drawn into a cylinder 27 cm long with a radius of 2 cm and a thickness of 0.1 cm. Estimate the volume of metal used.

23. According to Boyle's law, the pressure P and the volume V of a gas confined in a closed container are related by the equation $PV = c$ where c is a constant. Show that the differentials dP and dV are related by the equation $P\, dV + V\, dP = 0$.

24. Show that the relative error in computing the volume of a sphere is approximately three times the relative error in measuring the radius.

5-7   Applications of Partial Derivatives

So far in this chapter we have made use of the derivatives of functions of a single variable to solve many types of problems. But we also know (Section 4-10) how to take partial derivatives—that is, derivatives of functions of more than one variable. We can use partial derivatives to solve many of the same types of problems for functions of several variables.

For example, the differential dy of a function $y = f(x)$ was defined to be

$$dy = f'(x)\, dx$$

The analogous idea for a function $z = f(x, y)$ is the **total differential** dz, defined by

$$dz = \frac{\partial f}{\partial x}\, dx + \frac{\partial f}{\partial y}\, dy$$

The quantity dz approximates the change Δz in the function $z = f(x, y)$ for small changes $dx = \Delta x$ and $dy = \Delta y$ in x and y. The total differential can be extended to functions of three or more variables.

EXAMPLE 1. A right circular cylinder has a radius of 2 cm ± 0.1 cm and a height of 10 cm ± 0.2 cm. Estimate the maximum error in computing the volume.

The formula for the volume V of a right circular cylinder is $V = \pi r^2 h$. The two partial derivatives we need are

$$\frac{\partial V}{\partial r} = 2\pi rh$$

$$\frac{\partial V}{\partial h} = \pi r^2$$

The total differential dV is

$$dV = \frac{\partial V}{\partial r}\, dr + \frac{\partial V}{\partial h}\, dh$$

$$= (2\pi rh)\, dr + (\pi r^2)\, dh$$

We substitute $r = 2$, $dr = 0.1$, $h = 10$, and $dh = 0.2$ and get

$$2\pi(2)(10)(0.1) + \pi(2)^2(0.2) = 13.823$$

The maximum error in computing the volume is approximately 13.823 cm^3. ■

Practice 1. Paper for a rectangular poster was cut 35 cm wide and 55 cm long. Later it was trimmed by 0.2 cm in width and 0.6 cm in length. Estimate the area of paper trimmed off.

Partial derivatives also aid in locating relative maximum or minimum points. In Figure 5-32, M is a relative maximum point on the graph of some function $z = f(x, y)$. At a relative maximum point (or a relative minimum point), every tangent line to the surface is horizontal. In particular, at a maximum or minimum point,

$$\frac{\partial z}{\partial x} = 0 \quad \text{and} \quad \frac{\partial z}{\partial y} = 0$$

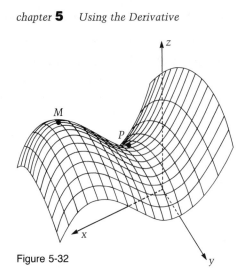

Figure 5-32

Points where $\partial z/\partial x = 0$ and $\partial z/\partial y = 0$ are critical points where maximums or minimums could occur. But, just as in the case of functions of one variable, not every critical point is necessarily a maximum or minimum point. In Figure 5-32, for example, point P is neither a relative maximum nor a relative minimum point, because there are nearby points on the surface that are higher than P and nearby points on the surface that are lower than P. Yet P is a critical point at which $\partial z/\partial x = 0$ and $\partial z/\partial y = 0$. In this section, however, we only consider surfaces whose critical points are relative maximum or minimum points.

EXAMPLE 2. Find any maximum or minimum points for the function $z = x^2 + xy + y^2 + 3x - 3y + 1$.

To find the critical points, we take partial derivatives and set them equal to zero.

$$\frac{\partial z}{\partial x} = 2x + y + 3 = 0$$

$$\frac{\partial z}{\partial y} = x + 2y - 3 = 0$$

This gives us a system of linear equations in x and y. After using your favorite method to solve a system of two equations in two unknowns, you should find that the solution is $x = -3$, $y = 3$. We decide whether the point $(-3, 3, -8)$ is a maximum or minimum by applying a first derivative test. We look at $\partial z/\partial x$ in the plane $y = 3$:

$$\left.\frac{\partial z}{\partial x}\right|_{y=3} = 2x + y + 3 \bigg|_{y=3} = 2x + 6$$

For $x < -3$, $2x + 6 < 0$ and for $x > -3$, $2x + 6 > 0$. The slope of the tangent line to the curve in the plane $y = 3$ goes from negative to positive as x goes through the value -3, and $(-3, 3, -8)$ is a minimum point. ■

Practice 2. Find any maximum or minimum points for the function $z = 6 - x^2 - y^2$.

EXAMPLE 3. Find three positive real numbers whose sum is 100 and whose product is a maximum.

If we call the three numbers q, r, and s, we then have the equation

$$q + r + s = 100 \tag{1}$$

The quantity to be maximized is P, where

$$P = qrs \tag{2}$$

Solving Equation (1) for q and substituting this into Equation (2), we get

$$P = (100 - r - s)rs$$
$$= 100rs - r^2s - rs^2$$

This gives us P as a function of the two variables r and s. In order to find where P is a maximum, we look for critical points. Setting the two partial derivatives equal to zero,

$$\frac{\partial P}{\partial r} = 100s - 2rs - s^2 = 0 \quad \text{or} \quad 100 - 2r - s = 0 \tag{3}$$

$$\frac{\partial P}{\partial s} = 100r - r^2 - 2rs = 0 \quad \text{or} \quad 100 - r - 2s = 0 \tag{4}$$

Note that we have thrown away the solution $s = 0$, $r = 0$. Clearly if s or r is zero, then P is zero and the product is not a maximum.

The system of Equations (3) and (4) can be solved as follows. We first solve Equation (3) for r.

$$r = \frac{100 - s}{2}$$

Then we substitute this expression for r into Equation (4) and solve for s.

$$100 - \left(\frac{100 - s}{2}\right) - 2s = 0$$
$$s = 33\tfrac{1}{3}$$

Then we solve for q and r, each of which also has the value $33\tfrac{1}{3}$.

To confirm that the values $q = r = s = 33\tfrac{1}{3}$ do give a maximum value for P, we do a first derivative test. From Equation 3,

$$\left.\frac{\partial P}{\partial r}\right|_{s=\frac{100}{3}} = 100\left(\frac{100}{3}\right) - 2r\left(\frac{100}{3}\right) - \left(\frac{100}{3}\right)^2$$

$$= \frac{200}{3}\left(\frac{100}{3} - r\right)$$

For $r < 33\tfrac{1}{3}$, this expression is positive, and for $r > 33\tfrac{1}{3}$, this expression is negative, indicating that we are at a maximum point. The numbers are all $33\tfrac{1}{3}$. ■

One final application of a maximum-minimum problem occurs when we try to "curve-fit" data to a straight line. In Chapter 2 we learned how to write the equation of a straight line given two points on the line. Two points determine a unique straight line. Suppose, however, we have a whole collection of data points relating two quantities x and y. Suppose, furthermore, that we have reason to believe that a good mathematical model of the relationship between x and y is that y varies linearly with respect to x. This means that there should be some straight line $y = mx + b$ such that our data points are all fairly close to the line, although perhaps none of them are actually on the line. Thus in Figure 5-33, the given data points suggest that the straight line shown is a good representation of the relation between x and y. If we can find the equation of this line, we can use the equation to predict the value of y for other values of x.

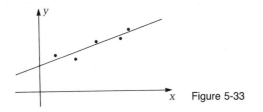

Figure 5-33

Our problem then becomes how to find values of m and b such that $y = mx + b$ is the best possible line to "fit" the given data. We want to choose the line to minimize the distances of the data points from the line. The vertical difference between a data point (x_1, y_1) and the line is found by subtracting the y-value of the line at $x = x_1$ from the y-coordinate of the point, y_1. In order to prevent positive and negative differences from canceling each other, we square this difference. We do this for each data point, add the squares, and then try to minimize the sum. This **method of least squares** is carried out in the next example.

First we will introduce some new notation in order to simplify things. The expression

$$\frac{p_1 + p_2 + \cdots + p_n}{n}$$

adds n p-values and then divides by the number of values. We use $\bar{p}$ to denote this expression; $\bar{p}$ is just the average of the n values of p.

EXAMPLE 4. Let (x_1, y_1), (x_2, y_2) and (x_3, y_3) be three given data points. We want to find m and b such that the line $y = mx + b$ best represents the points.

We find the vertical difference between each point and the line:

$$y_1 - (mx_1 + b)$$
$$y_2 - (mx_2 + b)$$
$$y_3 - (mx_3 + b)$$

We then square each difference and add the results.

$$S = [y_1 - (mx_1 + b)]^2 + [y_2 - (mx_2 + b)]^2$$
$$+ [y_3 - (mx_3 + b)]^2$$
$$= y_1^2 - 2y_1mx_1 - 2by_1 + m^2x_1^2 + 2mx_1b + b^2$$
$$+ y_2^2 - 2y_2mx_2 - 2by_2 + m^2x_2^2 + 2mx_2b + b^2$$
$$+ y_3^2 - 2y_3mx_3 - 2by_3 + m^2x_3^2 + 2mx_3b + b^2$$

Here, S is a function of the two variables m and b and we want to minimize S; so we look for critical points.

$$\frac{\partial S}{\partial m} = -2y_1x_1 + 2mx_1^2 + 2x_1b - 2y_2x_2 + 2mx_2^2$$
$$+ 2x_2b - 2y_3x_3 + 2mx_3^2 + 2x_3b$$

$$\frac{\partial S}{\partial b} = -2y_1 + 2mx_1 + 2b - 2y_2 + 2mx_2 + 2b$$
$$- 2y_3 + 2mx_3 + 2b$$

If we set $\partial S/\partial b = 0$ and solve for b, we get

$$b = \frac{(y_1 + y_2 + y_3) - m(x_1 + x_2 + x_3)}{3}$$

or, using our new notation,

$$b = \bar{y} - m\bar{x} \tag{5}$$

If we set $\partial S/\partial m = 0$, then

$$m(x_1^2 + x_2^2 + x_3^2) = (x_1y_1 + x_2y_2 + x_3y_3) - b(x_1 + x_2 + x_3)$$

Dividing through by 3 and using the new notation,

$$m\overline{x^2} = \overline{xy} - b\bar{x}$$

Substituting from Equation (5) into this equation results in

$$m\overline{x^2} = \overline{xy} - (\bar{y} - m\bar{x})\bar{x}$$

or, solving for m,

$$m = \frac{\overline{xy} - \bar{y}\bar{x}}{\overline{x^2} - \bar{x}^2} \tag{6}$$

Equations (5) and (6) give expressions for b and m, which may be evaluated once we know the actual data points. ∎

Although Equations (5) and (6) in Example 4 were developed for only three data points, the expressions for m and b are valid for any number of points. Thus the least-squares line $y = mx + b$ is found by using the formulas

$$b = \bar{y} - m\bar{x}$$

$$m = \frac{\overline{xy} - \bar{y}\bar{x}}{\overline{x^2} - \bar{x}^2}$$

EXAMPLE 5. The following data relates bacteria concentration per cubic centimeter of pond water to the temperature of the water. Use the method of least squares to express concentration as a linear function of temperature.

Concentration (in thousands/cm^3)	160	172	180	192
Temperature (in °C)	25	25.2	25.5	26

Letting x represent temperature, the independent variable, and y represent concentration, the dependent variable, we construct a table to help us compute the values we need.

x	y	xy	x^2
25	160	4000	625
25.2	172	4334.4	635.04
25.5	180	4590	650.25
26	192	4992	676
Totals 101.7	704	17916.4	2586.29

$$\bar{x} = \frac{101.7}{4} = 25.425$$

$$\bar{x}^2 = 646.431$$

$$\bar{y} = \frac{704}{4} = 176$$

$$\overline{xy} = \frac{17916.4}{4} = 4479.1$$

$$\overline{x^2} = \frac{2586.29}{4} = 646.573$$

We can compute m,

$$m = \frac{\overline{xy} - \bar{y}\bar{x}}{\overline{x^2} - \bar{x}^2} = \frac{4479.1 - (176)(25.425)}{646.573 - 646.431} = 30.28$$

and then b,

$$b = \bar{y} - m\bar{x} = 176 - (30.28)(25.425) = -593.87$$

The equation is thus

$$y = 30.28x - 593.87 \quad \blacksquare$$

Practice 3. Use the method of least squares to express y as a linear function of x for the given data.

x	1	3	5	7	9
y	30	32	33	35	36

Exercises / Section 5-7

Exercises 1–4: Find the total differential.

1. $z = x^3 - 2x^2 + 3y^3$
2. $z = 2x^2 - 7x + 4y^2$
3. $z = x^3 - 3xy + x^2y^2$
4. $z = x\sqrt{1 + y}$

Exercises 5–8: Find any maximum or minimum points for the given function.

5. $z = 2x + 4y - x^2 - y^2 - 3$
6. $z = x^2 + 2xy + 3y^2$
7. $z = x^2 + xy + y^2 - 2x + 5$
8. $z = x^2 - 2x - y^3$

Exercises 9–10: Use the method of least squares to express y as a linear function of x.

9.

x	10	12	14	16	18
y	65	58	53	47	40

10.

x	40	42	43	44	46	49
y	15	18	20	24	28	33

Exercises 11–16: Solve by using differentials.

11. The power P used by an electric resistor of resistance R is $P = V^2/R$, where V is the voltage across the resistor. Estimate the change in power (in watts) for a change in voltage from 120 V to 121 V and a change in resistance from 18 Ω to 18.1 Ω.

12. A wooden corner brace is made in the shape of a right triangle. The legs of the triangle are measured to be 5 cm ± 0.02 cm and 12 cm ± 0.2 cm. Estimate the maximum error in computing the length of the hypotenuse.

13. The total resistance R for two resistances R_1 and R_2 in parallel is given by

$$\frac{1}{R} = \frac{1}{R_1} + \frac{1}{R_2}$$

If R_1 is measured to be 100 Ω ± 1 Ω and R_2 is measured to be 150 Ω ± 1.8 Ω, estimate the maximum error and the relative error in computing the resistance R.

14. A right circular cylinder with no top or bottom has a radius of 0.8 m ± 0.2 m and a height of 1.3 m ± 0.4 m. Estimate the maximum error and the relative error in computing the surface area.

15. A covered rectangular storage bin has dimensions 40 cm by 60 cm by 18 cm. It is to be completely lined with a layer of waterproofing material 0.15 cm thick. Estimate·the volume of material required.

16. A re-entry vehicle is in the shape of a right circular cone of radius 1.2 m and height 2.8 m. Its outside surface is coated with a layer of a fire-resistant material, which increases the radius by 0.03 m and increases the height by 0.08 m. Estimate the volume of material used.

17. Find three positive real numbers q, r, and s whose sum is 20 such that the product qrs^2 is a maximum.

18. A closed rectangular packing crate is to hold 64 m³. Find the most economical dimensions.

19. A particular photographic image transmitted by satellite consists of a grid of points of varying greyness. The greyness G at any point (x, y) is given by the equation $G = 2x^2 + 0.3xy + y^2$. Find the point of minimum greyness.

20. An open rectangular chest is to be covered on the outside with 4 m² of mahogany paneling. Find the dimensions that will maximize the volume.

21. A company plans to manufacture closed rectangular cases for storing electronics parts. Each case must hold 3200 cm³. The front of the case will be glass, which costs twice as much per square centimeter as the plastic used for the rest of the case. Find, to two decimal places, the most economical dimensions.

22. An open rectangular box is to contain a volume of 400 cm³. The material for the bottom is $.40/cm² and the material for the sides is $.25/cm². Find, to two decimal places, the most economical dimensions.

23. Values of current and voltage in an electrical experiment are shown in the following table. Use the method of least squares to express current i as a linear function of voltage V.

Voltage (in V)	6.1	6.8	7.3	7.9	8.2
Current (in mA)	12.3	12.9	13.4	13.8	14.5

24. The viscosity of liquids decreases as the temperature increases. The following table gives experimental values of viscosity for different temperatures for a certain liquid. Use the method of least squares to express viscosity v as a linear function of temperature T.

Viscosity (in Pa-s)	0.21	0.19	0.18	0.17	0.15
Temp. (in °C)	15	17	20	22	25

STATUS CHECK

Now that you are at the end of Chapter 5, you should be able to:

section **5-1**   Find the equations of the tangent line and normal line to a given curve at a given point.

section **5-2**   Sketch the graph of an algebraic function, indicating maximum points, minimum points, and inflection points.

section **5-3**   Given one quantity as an algebraic function of a second quantity, find the maximum or minimum value of the first quantity.

Given one quantity as an algebraic function of a second quantity, find the value of the second quantity that maximizes or minimizes the first quantity.

From information about two related quantities, write an algebraic expression for one quantity in terms of another in order to maximize or minimize that quantity.

section **5-4**   Find the magnitude and direction of the resultant vector of two given vectors.

Given a set of rectangular parametric equations for the position of a moving object, find its velocity and acceleration vectors at a specific time.

Given a set of polar parametric equations for the position of a moving object, find the magnitude of the velocity vector at a specific time.

Given the equation of the path of a moving object in the form $y = f(x)$, find the velocity and acceleration vectors at a specific point.

section **5-5**   Given information relating two quantities that are functions of time and the time rate of change of one of the quantities, find the time rate of change of the other quantity.

section **5-6**   Find the differential dy from a given function $y = f(x)$.

Evaluate the differential dy for given values of x and dx.

When one quantity is a given function of a second quantity and there is a small error in measuring the second quantity, use differentials to approximate the maximum and relative corresponding errors in computing the first quantity.

section **5-7**   Find the total differential dz from a given function $z = f(x, y)$.

Find maximum or minimum points for functions of two variables.

Given data relating two quantities, use the method of least squares to express one quantity as a linear function of the other.

When one quantity is a given function of several other quantities and there is a small error in measuring these other quantities, use the total differential to approximate the maximum and relative corresponding errors in computing the first quantity.

5-8 More Exercises for Chapter **5**

Exercises 1–4: Find equations of the tangent line and normal line to the curve at the given point.

1. $y = 2x^2 - 5x + 1$; $(1, -2)$

2. $y = x^3 - 2x$; $(2, 4)$

3. $y^2 + 3x^2 - 7 = 0$; $(1, 2)$

4. $y^2 - xy + x^2 - 1 = 0$; $(1, 1)$

Exercises 5–12: Sketch the graph of the function, including any maximum points, minimum points, and inflection points.

5. $y = 3x - 4x^2$

6. $y = 2x^2 - 6x + 1$

7. $y = 2x^3 - 6x + 4$

8. $y = 4 + x - 3x^3$

9. $y = x^4 - 2x^3$

10. $y = 2x^4 - x^2 + 5$

11. $y = \dfrac{x}{x - 1}$

12. $y = \dfrac{1}{4 - x^2}$

Exercises 13–16: For the following equations of motion, find the velocity and acceleration vectors at the given values of time.

13. $x = t^2 - 4t^3$, $y = 2t^2 - 1$, $t = 1$

14. $x = 2t^3 - 4t^2 + 5$, $y = 14 - t^4$, $t = 2$

15. $x = \dfrac{2}{t^2}$, $y = \dfrac{4}{1 - t}$, $t = 2$

16. $x = \sqrt{t}$, $y = t^3 - t^2$, $t = 1$

Exercises 17–18: For the following equations of motion, find the magnitude of the velocity vector at the given value of time.

17. $r = 2t^2 - t + 5$, $\theta = t^3$, $t = 2$

18. $r = \sqrt{t + 3}$, $\theta = \dfrac{4}{1 + t^2}$, $t = 1$

Exercises 19–22: Find the differential.

19. $y = 7x^3 - 3x^2 + 4x - 5$

20. $y = (x^4 - 2x)^{1/4}$

21. $y = \dfrac{x}{x^2 + 5}$

22. $y = \sqrt{4 - 8x^2}$

Exercises 23–24: Compute Δy and dy for the given values of x and Δx.

23. $y = 4 - x^2$, $x = 2$, $\Delta x = 0.1$

24. $y = (4 - 3x)^2$, $x = 1$, $\Delta x = 0.1$

Exercises 25–26: Find the total differential.

25. $z = 3y^4 - 2y^3 + 3x^2$

26. $z = x(1 - y^2)$

Exercises 27–28: Find any maximum or minimum points for the given functions.

27. $z = x^2 - 2xy + 5y^2 + 4$

28. $z = x^2 - 3x + y^2$

Exercises 29–30: Use the method of least squares to express y as a linear function of x.

29.

x	22	25	27	29	30
y	54	50	47	44	42

30.

x	5	10	15	20	25
y	70	71	73	75	77

31. Write the equations of all lines tangent to the ellipse $x^2 - xy + 2y^2 - 4x + 2y + 2 = 0$ and parallel to the line $x - 4y - 2 = 0$.

32. Make the parabola $y = ax^2 + bx + c$ pass through the point $(-1, 12)$ and be tangent to the line $5x - y - 10 = 0$ at the point $(2, 0)$.

33. The strength of a rectangular beam is proportional to the breadth and the square of the depth. What are the dimensions of the strongest beam that can be cut from a circular log of diameter 54 cm?

34. A rectangular poster is to contain 900 cm^2 of printed matter with margins of 10 cm each at top and bottom and 5 cm at each side. Find the overall dimensions to make the poster in order to minimize the material needed.

35. A container for frozen strawberries is to be a rectangular solid with a square bottom. The sides of the carton will be cardboard, but the top and bottom will be aluminum. Cardboard costs k cents per square centimeter and aluminum costs twice that much. The carton must hold 220 cm^3. What are the most economical dimensions?

36. Find the dimensions of the largest rectangular parking area that can be placed on a right triangular lot, as shown in Figure 5-34.

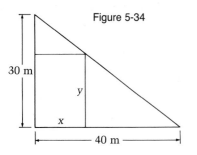

Figure 5-34

37. The cooling load to air-condition a house at time t is f units, where

$$f(t) = 3[T(t) - 75] + M(t)$$

Here, $T(t)$ is the outside temperature and $M(t)$ is the internal heating. For $0 \le t \le 2$ h, $T(t) = 75 + 2t - \frac{5}{3}t^3$ and $M(t) = 5t^2 - 6t$; find the maximum cooling load during this time period.

38. The distance D in kilometers between a space vehicle and a meteor is $D = (10 - t)^2 + (20 - 3t)^2$ for $0 \le t \le 10$ h. Find the closest the meteor comes to the space vehicle during this time period.

39. In studies of epidemics, an inflection point of the function describing the number of individuals infected indicates the point in time when the epidemic passes its most virulent stage. If the number I of individuals infected t days after the start of the epidemic is given by $I = 500t + 900t^2 - 100t^3$, find when the inflection point occurs.

40. The velocity (in centimeters per second) of an object as a function of time t (in seconds) is given by the equation

$$v = \frac{3c}{2}t - \frac{3c}{4}t^2$$

where c is the speed of light. If the object at rest has length 2 cm, then according to relativity theory, the length L (in centimeters) at time t is

$$L = 2\sqrt{1 - \left(\frac{v}{c}\right)^2}$$

Find the minimal length of the object.

41. A particle moves along the curve $y = x^2 + 2x$ with a vertical component of velocity of constant magnitude 12. Find the velocity vector at the point $(2, 8)$.

42. A particle moves along the curve $y^2 = 4x$ with a horizontal component of velocity of constant magnitude 2. Find the velocity vector at the point $(4, 4)$.

43. The power P (in watts) in an electric circuit is given by the equation $P = Ri^2$, where R is the resistance (in ohms) and i is the current (in amperes). If $R = 80\ \Omega$ and the current is increasing at the rate of 0.3 A/min, how fast is the power changing when $i = 4$ A?

44. According to Coulomb's law, the force of repulsion F (in newtons) between two positive charges of q_1 and q_2 coulombs is $F = (9 \times 10^9)q_1q_2/s^2$ where s is the distance (in meters) between the charges. If two charges of 8×10^{-6} and 5×10^{-6} coulombs are separating at the rate of 0.3 m/s, find the rate at which the force is decreasing when the charges are 2.3 m apart.

45. A circular metal plate in a piece of machinery contracts from cooling when the macine is turned off. The area decreases at the rate of 1.1 cm²/h. Find the rate at which the radius is decreasing when the radius is 0.6 cm.

46. Water is being pumped into a vertical cylindrical tank at the rate of 0.8 m³/min. If the radius of the tank is 1.3 m, how fast is the surface rising?

47. A jogger leaves point P traveling west at 24 km/h. One-half hour later, another jogger leaves P traveling south at 18 km/h. One-half hour after that, how fast are the two joggers separating?

48. A lightpost stands 7 m from a house and 5 m off to the side of the path from the house to the street. If a woman walks along the path at the rate of 2 m/s, how fast does her shadow move along the wall of the house when she is 3 m away from the house?

49. The resistance R (in ohms) of a certain resistor varies with the temperature T (in degrees Celsius) according to the formula $R = 3.00 + 0.02T - 0.003T^2$. Use the differential to estimate the change in R as T changes from 80°C to 81°C.

50. The curved surface of a circular cylindrical drum of radius 0.2 m and height 1.1 m is to be lined with an insulating material 0.01 m thick. Use the differential to estimate the amount of insulating material required.

51. The pressure P, volume V, and temperature T of a certain quantity of gas are related by the equation $PV = kT$, where k is a constant. Use differentials to estimate the change in pressure for a change in volume from 800 to 825 units and a change in temperature from 70 to 69 units.

52. A rectangular block of stone is measured to have dimensions 0.5 m, 0.8 m, and 1.2 m, with a maximum error in any dimension of 0.03 m. Use differentials to estimate the maximum error in computing the volume.

53. A closed rectangular box uses material for the bottom and top that is twice as expensive per square centimeter as material for the sides. If the box is to hold 1000 cm^3, find—to two decimal places—the most economical dimensions.

54. Find three positive real numbers p, q, and r such that $pqr = 27$ and the sum $p + q + r$ is a minimum.

chapter 6

Integration

Movies achieve their effect by slicing an action into small time frames and smoothing the frames together by speeding them up; integration works by slicing a quantity into small elements and smoothing the elements together by taking a limit.

6-1 The Indefinite Integral

Until now, we have been concerned with the process of differentiation. Given a function, we have tried to find its derivative (or its differential, which is just about the same thing). Now we want to consider the reverse process. Suppose we are given an expression which is the derivative (or differential) of some function; can we find the original function? This reverse process is called **integration**, or **antidifferentiation**. The result of the process is called the **integral**, or **antiderivative**, of the given expression.

EXAMPLE 1. Find an antiderivative of $18x^2$.

We want to find a function $F(x)$ such that $dF(x)/dx = 18x^2$. Recall that the power rule for differentiation reduces the power by one, so $F(x)$ must have involved x^3. But differentiating x^3 would also have introduced a factor of 3 into the coefficient, so we should consider how 18 can be written with a factor of 3. Thus if we write $18x^2$ as $6(3x^2)$, we see that $3x^2$ comes from differentiating x^3 and that the 6 is just a constant multiple. We conclude that $F(x) = 6x^3$. To check we compute $dF(x)/dx$.

$$\frac{d(6x^3)}{dx} = 6(3x^2) = 18x^2$$

Therefore, $6x^3$ is an antiderivative of $18x^2$. ∎

Practice 1. Find an antiderivative of $12x^3$.

In Example 1, we were asked to find *an* antiderivative of $18x^2$, rather than *the* antiderivative. All the following functions are antiderivatives of $18x^2$:

$$6x^3, \qquad 6x^3 + 3, \qquad 6x^3 - 17, \qquad 6x^3 + \frac{\pi}{2}$$

Once an antiderivative function is found, any constant can be added to the function, because the derivative of a constant is zero. For example,

$$\frac{d}{dx}(6x^3 - 17) = 18x^2$$

Now suppose that a function $F(x)$ is an antiderivative of a function $f(x)$. Then, as we have just observed, $F(x) + C$, where C is any constant, is also an antiderivative of $f(x)$. This means that

$$\frac{d(F(x) + C)}{dx} = f(x)$$

or

$$d(F(x) + C) = f(x)\ dx \tag{1}$$

We now write Equation (1) in a different way, introducing some new notation:

$$F(x) + C = \int f(x)\ dx \tag{2}$$

The symbol $\int$ in Equation (2) is called an **integral sign** and is used to denote integration, the process of finding antiderivatives. The function behind the integral sign, $f(x)$, is called the **integrand**. The entire right side of Equation (2) is called the **indefinite integral** of $f(x)$. The word *indefinite* refers to the presence of the arbitrary constant C, known as the **constant of integration**. On the right side of Equation (2), notice the presence of the integral sign to the left of $f(x)$ and the dx to the right of $f(x)$. We can think of $\int \quad dx$ as left and right parentheses framing $f(x)$. The integral sign is never found without a corresponding dx (or dy, or whatever the variable name is).

EXAMPLE 2. Integrate: $\int 18x^2\ dx$.

Here we want the indefinite integral of $18x^2$. This simply involves finding an antiderivative of $18x^2$ and adding the constant of integration. This problem is essentially the one in Example 1, and

$$\int 18x^2\ dx = 6x^3 + C \quad \blacksquare$$

To emphasize again what the integral notation means, note that

$$\int f(x) \, dx = F(x) + C \quad \text{means} \quad dF(x) = f(x) \, dx$$

Performing an integration is rather like factoring. When you factor an expression, you make an educated guess what the factors are, but the proof that your guess is correct involves multiplying and getting the original expression. When you take an indefinite integral, you make an educated guess at what the answer is, but the verification that your answer is correct involves differentiating it and getting the original function. *You can always check your answer when taking an indefinite integral.*

EXAMPLE 3.   Verify that $\int (2x^2 - 4)^2 4x \, dx = \dfrac{(2x^2 - 4)^3}{3} + C$.

We differentiate $(2x^2 - 4)^3/3 + C$ using the power rule.

$$d\left(\frac{(2x^2 - 4)^3}{3} + C\right) = \frac{1}{3} \, 3(2x^2 - 4)^2 4x \, dx = (2x^2 - 4)^2 4x \, dx$$

This result is the expression behind the integral sign. This verifies that the given answer is correct.   ∎

In order to improve our educated guessing when taking the indefinite integral of a function, we develop some integration rules. Let c be a constant and u be some function of x. Then

$$\int c \, du = c \int du = cu + C$$ (Rule 1)

To check: $d(cu + C) = c \, du$ because a constant factor can always be pulled out of the derivative. This integration rule says that *a constant factor can be passed through an integral sign.*

For u and v both functions of x,

$$\int (du + dv) = \int du + \int dv = u + v + C$$ (Rule 2)

To check: $d(u + v + C) = du + dv$ because the derivative of a sum is the sum of the derivatives. This integration rule says that *the integral of a sum is the sum of the integrals.*

Finally, for u a function of x and $n \neq -1$,

$$\int u^n \, du = \frac{u^{n+1}}{n+1} + C \qquad \text{(Rule 3)}$$

To check:

$$d\left(\frac{u^{n+1}}{n+1} + C\right) = \left(\frac{1}{n+1}\right)(n+1)u^{(n+1)-1}\frac{du}{dx} \, dx = u^n \, du$$

because of the power rule for differentiation. This integration rule says that *the integral of u to a power times du is u to the one more power divided by that power.* The case where $n = -1$ must be excluded, because application of Rule 3 for this case would result in division by zero.

EXAMPLE 4. Integrate: $\int 18x^2 \, dx$.

This is the problem we solved in Examples 1 and 2, but this time we apply the integration rules.

$$\int 18x^2 \, dx \overset{\text{(Rule 1)}}{=} 18 \int x^2 \, dx \overset{\text{(Rule 3)}}{=} 18\frac{x^3}{3} + C = 6x^3 + C$$

In applying Rule 3, $u = x$ and $n = 2$. The expression $\int x^2 \, dx$ matched the form $\int u^n \, du$ needed to apply the rule. ∎

EXAMPLE 5. Integrate: $\int (2x^2 - 5x + 4) \, dx$.

Using the integration rules,

$$\int (2x^2 - 5x + 4) \, dx \overset{\text{(Rule 2)}}{=} \int 2x^2 \, dx + \int -5x \, dx + \int 4 \, dx$$

$$\overset{\text{(Rule 1)}}{=} 2\int x^2 \, dx - 5\int x \, dx + 4\int dx$$

$$\overset{\text{(Rule 3)}}{=} 2\frac{x^3}{3} - \frac{5x^2}{2} + 4x + C$$

Notice that although we do several integrations, we incorporate all the constants of integration into one at the end. ∎

EXAMPLE 6. Integrate: $\int \left(\sqrt{x} - \frac{2}{x^3} \right) dx.$

$$\int \left(\sqrt{x} - \frac{2}{x^3} \right) dx = \int (x^{1/2} - 2x^{-3}) \, dx$$

(Rule 2)
$\downarrow$
$$= \int x^{1/2} \, dx + \int -2x^{-3} \, dx$$

(Rule 1)
$\downarrow$
$$= \int x^{1/2} \, dx - 2 \int x^{-3} \, dx$$

(Rule 3)
$\downarrow$
$$= \frac{x^{3/2}}{\frac{3}{2}} - 2 \frac{x^{-2}}{-2} + C$$

$$= \frac{2}{3} x^{3/2} + x^{-2} + C \quad \blacksquare$$

Practice 2. Integrate: $\int (2 - 3x^2) \, dx.$

Practice 3. Integrate: $\int \frac{1}{\sqrt{x}} \, dx.$

Word of Advice

Don't forget the C at the end of every indefinite integral problem.

In order to apply Integration Rule 3 (the *power rule*), you must have something of the exact form $\int u^n \, du$. In trying to apply Rule 3 to a given integration problem, you must identify u and n and then make sure that the rest of the expression is exactly du.

EXAMPLE 7. Integrate: $\int (2x^2 - 4)^2 4x \, dx.$

This is the problem in Example 3, so we already know the answer. If we try to fit the expression to the form shown in Rule 3, we see that $2x^2 - 4$ is raised to the power 2, and we try letting $u = 2x^2 - 4$, with $n = 2$. For $u = 2x^2 - 4$, $du = 4x \, dx$, and this factor is indeed present. By the power rule, then,

$$\int (2x^2 - 4)^2 4x \, dx = \frac{(2x^2 - 4)^3}{3} + C \quad \blacksquare$$

EXAMPLE 8. Integrate: $\int (x^3 - 1)^3 x^2 \, dx$.

Let us try to fit this into the form needed to apply the power rule, $\int u^n \, du$. If there is more than one expression raised to a power, a good rule of thumb is to let u be the most complex expression. Thus we try $u = x^3 - 1$ and $n = 3$. For $u = x^3 - 1$, $du = 3x^2 \, dx$. We see that we have the factor x^2, but not the 3. We are saved, however, because we miss by only a constant factor, and this can be gotten around by the following trick:

$$\text{(Multiply by } \tfrac{3}{3})$$
$$\downarrow$$
$$\int (x^3 - 1)^3 x^2 \, dx = \int \frac{1}{3} \cdot 3(x^3 - 1)^3 x^2 \, dx$$

$$\text{(Rule 1)} \qquad\qquad\qquad \text{(Rule 3)}$$
$$\downarrow \qquad\qquad\qquad\qquad\qquad \downarrow$$
$$= \frac{1}{3} \int (x^3 - 1)^3 3x^2 \, dx = \frac{1}{3} \frac{(x^3 - 1)^4}{4} + C \quad \blacksquare$$

In Example 8, note how we made use of the ability to move a *constant* through an integral sign. This technique works only when a *constant factor* is all that is missing from du.

EXAMPLE 9. Integrate: $\int \sqrt{x^2 + 2x}\,(x + 1) \, dx$.

We again try to use the power form. Letting $n = \frac{1}{2}$ and $u = x^2 + 2x$, we find that $du = (2x + 2) \, dx$. Our left-over factor of $x + 1$ fails to match du by a constant factor of 2, so we insert 2 (and compensate with a factor of $\frac{1}{2}$ outside the integral sign). Thus

$$\int \sqrt{x^2 + 2x}\,(x + 1) \, dx = \frac{1}{2} \int \sqrt{x^2 + 2x}\,(2)(x + 1) \, dx$$

$$= \frac{1}{2} \int \sqrt{x^2 + 2x}\,(2x + 2) \, dx$$

$$= \frac{1}{2} \frac{(x^2 + 2x)^{3/2}}{\frac{3}{2}} + C$$

$$= \frac{1}{3} (x^2 + 2x)^{3/2} + C$$

Notice that the factor $2x + 2$ had to be present to apply the power rule, but it is part of du and does not appear in the final answer. $\blacksquare$

Using the power rule becomes easier with practice, but at first it is very helpful to write n and u explicitly, and then compute du.

Word of Advice

In using the power rule

$$\int u^n \, du = \frac{u^{n+1}}{n+1} + C$$

once u is chosen, du is completely determined. The expression du must be present in order to apply the power rule, but it does not itself appear in the answer.

Practice 4. Integrate: $\int (x^5 - 2)^2 x^4 \, dx$.

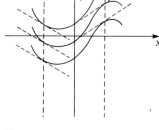

Figure 6-1

The constant of integration, C, which enters into the indefinite integral, has a geometric interpretation. Suppose

$$F(x) + C = \int f(x) \, dx$$

Then $dF(x) = f(x) \, dx$ or $dF(x)/dx = f(x)$. The function $f(x)$ is, therefore, the slope function for $F(x)$. Integrating $\int f(x) \, dx$ involves finding a function, given its slope function. For each slope function, there is a whole family of curves that have the same slope for each value of x (see Figure 6-1). Assigning a particular value to C corresponds to choosing one curve in the family. Knowing a point on the curve determines a value for C.

EXAMPLE 10. Find y as a function of x given that $dy/dx = 2x^3 + 4x$ and that the curve passes through the point $(1, 4)$.

Here $dy = (2x^3 + 4x) \, dx$, and

$$y = \int (2x^3 + 4x) \, dx = \int 2x^3 \, dx + \int 4x \, dx = \frac{2x^4}{4} + \frac{4x^2}{2} + C$$

or

$$y = \frac{x^4}{2} + 2x^2 + C$$

This identifies the family of curves with slope function $2x^3 + 4x$. To pick out that member of the family passing through $(1, 4)$, we substitute 1 and 4 for x and y, respectively, and solve for C.

$$4 = \frac{1}{2} + 2(1) + C$$

$$C = \frac{3}{2}$$

The function is

$$y = \frac{x^4}{2} + 2x^2 + \frac{3}{2} \quad \blacksquare$$

Practice 5. Find y as a function of x given that $dy/dx = \sqrt{x} + 2$ and that the curve passes through the point $(2, 6)$.

The next chapter is devoted to applications of integration. In Chapter 9 we concentrate on developing a whole list of integration rules to supplement the power rule, which is really the only integration form we have at the moment. These additional rules help us integrate many more types of functions. If we are faced with an integral $\int f(x)\,dx$, we try to match it up with one of our integration forms. If we are not successful, it may be that the function $f(x)$ is **integrable** (its antiderivative exists) by some form we do not know, or it may be that $f(x)$ is not integrable at all—that is, that $f(x)$ is not the derivative of any function. There are functions that simply are not the result of differentiating other functions.

Exercises / Section **6-1**

Exercises 1–32: Integrate:

1. $\displaystyle\int x^3\,dx$

2. $\displaystyle\int 2x^5\,dx$

3. $\displaystyle\int -\frac{x^4}{2}\,dx$

4. $\displaystyle\int 4x^7\,dx$

5. $\displaystyle\int x^{2/3}\,dx$

6. $\displaystyle\int 2x^{-3/4}\,dx$

7. $\displaystyle\int \frac{4}{x^2}\,dx$

8. $\displaystyle\int \frac{-2}{\sqrt{x}}\,dx$

9. $\displaystyle\int (5x^5 - 3x^3 + 2x)\,dx$

10. $\displaystyle\int (4 - 6x^2 + x^5)\,dx$

11. $\displaystyle\int (2\sqrt{x} - 3x^4)\,dx$

12. $\displaystyle\int \left(\frac{3}{x^4} - 4x^2 + \frac{2}{\sqrt{x}}\right)\,dx$

13. $\displaystyle\int (2x^2 - 1)^2\,dx$

14. $\displaystyle\int x^4(x^2 + 2)\,dx$

15. $\displaystyle\int \sqrt{x}(x - 3)^2\,dx$

16. $\displaystyle\int \left(\frac{1}{x^2} + \frac{1}{x^3} + \frac{1}{x^4}\right)\,dx$

17. $\displaystyle\int (3x^2 - 1)^3 6x\,dx$

18. $\displaystyle\int (x^4 + 2x)^2(4x^3 + 2)\,dx$

19. $\displaystyle\int (1 - 3x)^4(-3)\,dx$

20. $\displaystyle\int (5x^7 + 2)^2 35x^6\,dx$

21. $\displaystyle\int 2x\sqrt{x^2 + 1}\,dx$

22. $\displaystyle\int (1 + x^3)^{2/3}(3x^2)\,dx$

23. $\displaystyle\int (x^2 + 14)^5 x\,dx$

24. $\displaystyle\int (x^4 - 2)^3 x^3\,dx$

25. $\displaystyle\int \sqrt{1 - 4x}\, dx$

26. $\displaystyle\int \sqrt[4]{2 + x^3}(x^2)\, dx$

27. $\displaystyle\int (4x^2 - 7)^{2/3}4x\, dx$

28. $\displaystyle\int (3x - x^3)^4(1 - x^2)\, dx$

29. $\displaystyle\int \frac{x\, dx}{\sqrt{2 + x^2}}$

30. $\displaystyle\int \frac{(1 + 3x)\, dx}{\sqrt{2x + 3x^2}}$

31. $\displaystyle\int (x^4 - 2x^2 + 5)^4(8x^3 - 8x)\, dx$

32. $\displaystyle\int \frac{6x^2 - 12x}{(x^3 - 3x^2 + 1)^2}\, dx$

Exercises 33–36: Find y as a function of x, given dy/dx and a point on the curve.

33. $\dfrac{dy}{dx} = 4x + 2; \quad (0, 1)$

34. $\dfrac{dy}{dx} = 12x^3; \quad (2, 1)$

35. $\dfrac{dy}{dx} = (2x^3 - 2x)^2(3x^2 - 1); \quad (2, 300)$

36. $\dfrac{dy}{dx} = \sqrt{x^4 - 2x}\,(2x^3 - 1)\, dx; \quad (0, 5)$

37. Find the equation of a curve that has slope $\sqrt{3x - 8}$ and that passes through the point $(8, \frac{5}{3})$.

38. Find the equation of a curve that has slope
$$\sqrt{x^2 + x}\,(4x + 2)$$
and that passes through the point $(2, 0)$.

39. Find the equation of a curve, given that its second derivative is 5, the point $(2, 3)$ lies on the curve, and the slope at that point is 7.

40. Find the equation of a curve, given that its second derivative is $6x^3$, the point $(2, -4)$ lies on the curve, and the slope at that point is 12.

6-2   Area Under a Curve; the Definite Integral

In this section, we talk about areas under curves and limits of sums. This will seem to be a radical departure from the topics of differentiation and integration that have occupied us until now. However, in the next section we come upon the Fundamental Theorem of Calculus, which ties all of these ideas together.

Geometry provides us with formulas for computing the areas of various figures, such as rectangles, triangles, and circles. A more general area problem is to find the area of a figure enclosed by curves whose equations are known. Figure 6-2 shows the area enclosed by the curves $y = x$ and $(y - 5) = -(x - 3)^2$. To illustrate how we might determine such an area, we start with an extremely simple case.

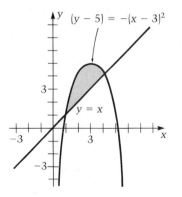

Figure 6-2

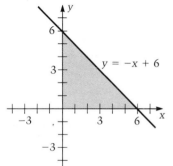

Figure 6-3

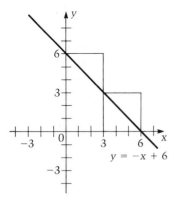

Figure 6-4

EXAMPLE 1. Let us find the area bounded by the curves $y = -x + 6$, $y = 0$, and $x = 0$. This area is shown in Figure 6-3. Of course, the area is easily found from the geometric formula for the area of a triangle, $A = \frac{1}{2}bh$; for this case,

$$A = \tfrac{1}{2}(6)(6) = 18$$

However, we take a different approach to the problem, one that will generalize to figures for which geometric formulas do not exist (such as that shown in Figure 6-2). We begin by breaking the base of the figure into two sections and making each section the base of a rectangle, as shown in Figure 6-4. A first guess (obviously too large) for the area of the triangle is to add the areas of the two rectangles. The height of each rectangle is given by the value of the function $f(x) = -x + 6$ at the left-hand edge of the rectangle. The area of each rectangle is its height times its base. The first approximation, A_2 (the subscript denotes the number of rectangles), to the actual area is given by

$$
\begin{aligned}
A_2 &= f(0)(3) + f(3)(3) \\
&= (-0 + 6)(3) + (-3 + 6)(3) \\
&= 27
\end{aligned}
$$

A better approximation is obtained by taking three rectangles, as in Figure 6-5. This approximation, A_3, is given by

$$
\begin{aligned}
A_3 &= f(0)(2) + f(2)(2) + f(4)(2) \\
&= (-0 + 6)(2) + (-2 + 6)(2) + (-4 + 6)(2) \\
&= 24
\end{aligned}
$$

Using six rectangles, as in Figure 6-6, is even better:

$$
\begin{aligned}
A_6 &= f(0)(1) + f(1)(1) + f(2)(1) + f(3)(1) + f(4)(1) + f(5)(1) \\
&= 6 + 5 + 4 + 3 + 2 + 1 \\
&= 21
\end{aligned}
$$

We claim that if we continue to use more and more rectangles, our approximate values, while always too high, get closer to the right answer. In other words, the sequence of values A_2, A_3, A_6, . . .

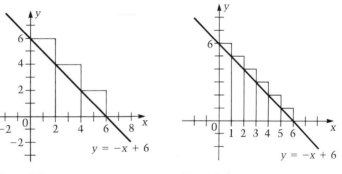

Figure 6-5

Figure 6-6

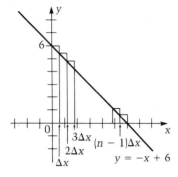

Figure 6-7

approaches the correct value of 18, or has a limit of 18. To see that this is true, we write a general expression for A_n, the sum of the areas of n rectangles. We let Δx denote the width of each rectangle. (Of course, because we divided 6 units into n equal parts, we know that $\Delta x = 6/n$, and we will use this fact later.) The n points along the x-axis where $f(x)$ is to be evaluated are $0, \Delta x, 2\Delta x, \ldots, (n - 1)\Delta x$ (see Figure 6-7). Adding the areas of the n rectangles,

$$
\begin{aligned}
A_n &= f(0)\Delta x + f(\Delta x)\Delta x + f(2\Delta x)\Delta x + \cdots + f((n - 1)\Delta x)\Delta x \\
&= (6)\Delta x + (-\Delta x + 6)\Delta x + (-2\Delta x + 6)\Delta x + \cdots \\
&\quad + [-(n - 1)\Delta x + 6]\Delta x
\end{aligned}
$$

Multiplying, we find there are n terms of the form $6\Delta x$ and $n - 1$ other terms with a common factor of $(-1)(\Delta x)^2$. Therefore,

$$
A_n = [1 + 2 + \cdots + (n - 1)](-1)(\Delta x)^2 + n6\Delta x
$$

The expression $1 + 2 + \cdots + (n - 1)$ equals $(n^2 - n)/2$ (this can be proved by mathematical induction, or by noting that the expression $1 + 2 + \cdots + (n - 1)$ is the first $n - 1$ terms of an arithmetic progression). Using this, we can write

$$
A_n = \frac{(n^2 - n)}{2}(-1)(\Delta x)^2 + n6\Delta x
$$

or, since $\Delta x = 6/n$,

$$
\begin{aligned}
A_n &= \frac{n^2 - n}{2}(-1)\left(\frac{36}{n^2}\right) + n6\left(\frac{6}{n}\right) \\
&= \frac{n^2 - n}{n^2}\left(\frac{-36}{2}\right) + 36 \\
&= \left(1 - \frac{1}{n}\right)(-18) + 36
\end{aligned}
$$

Finally

$$
\lim_{n \to \infty} A_n = \lim_{n \to \infty} \left[(-18)\left(1 - \frac{1}{n}\right) + 36\right] = -18 + 36 = 18
$$

The limiting value is the actual area. ∎

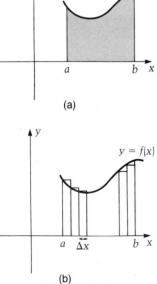

(a)

(b)

Figure 6-8

Suppose, in general, that we have a function $y = f(x)$ that has positive values and we want to find the area A beneath that function and above the x-axis (usually called the *area under the curve*) between $x = a$ and $x = b$, as in Figure 6-8(a). Following the ideas of Example 1, we let A_n denote the sum of the areas of n rectangles, each of width Δx (see Figure 6-8(b)). Then A is given by

$$
A = \lim_{n \to \infty} A_n
$$

Example 1 was unusual in the sense that we were able to find a simple expression for A_n and actually compute the limit to find a numerical answer. Because we shall not often be able to do this, we may have to be content for the moment with approximations.

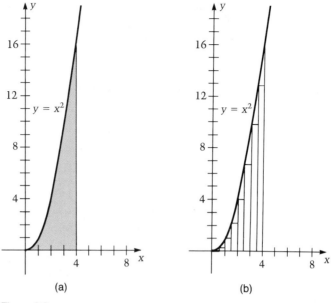

(a)   (b)

Figure 6-9

EXAMPLE 2. Approximate the area under the curve $f(x) = x^2$ from $x = 0$ to $x = 4$ by finding A_8.

The area we want to approximate is shown in Figure 6-9(a) and our approximation is shown in Figure 6-9(b). For eight rectangles, the width Δx of each rectangle is $\frac{4}{8} = 0.5$. Therefore,

$$A_8 = f(0)\Delta x + f(0.5)\Delta x + f(1)\Delta x + f(1.5)\Delta x + f(2)\Delta x + f(2.5)\Delta x$$
$$+ f(3)\Delta x + f(3.5)\Delta x$$
$$= [f(0) + f(0.5) + f(1) + f(1.5) + f(2) + f(2.5) + f(3) + f(3.5)]\Delta x$$
$$= [0^2 + (0.5)^2 + 1^2 + (1.5)^2 + 2^2 + (2.5)^2 + 3^2 + (3.5)^2]0.5$$
$$= 17.5$$

We really do not know at this point whether A_8 is a very good approximation to the actual area; since $A = \lim_{n \to \infty} A_n$, maybe we should have computed A_{16}, or A_{32}! As a matter of fact, the value of A_{16} is 19.375, so we were certainly not very close. By examining Figure 6-9(b), we see that the value of A_n for each n is too small. The value of A_n gets closer to the actual area as n increases. That is why A_{16} in this problem is larger than A_8. The actual area, we now know, is larger than 19.375. ∎

Practice 1. Approximate the area under the curve $f(x) = x + 2$ from $x = 0$ to $x = 3$ by finding A_3; also find A_6.

We have been careful to draw all our rectangles so that each height is given by $f(x)$, where x is the left-hand edge of the rectangle.

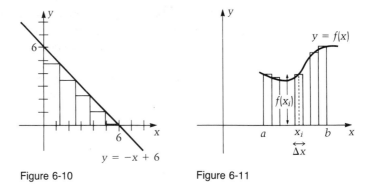

Figure 6-10   Figure 6-11

This is not necessary, however. If we reconsider the problem of Example 1, for instance, we see that we can evaluate the heights of rectangles at the right-hand edge, as in Figure 6-10 and take $\lim_{n\to\infty} A_n$. Here A_n for each n will be smaller than the actual area A, but as the number of rectangles increases, we get closer to A.

In fact, for the area under $f(x)$ from $x = a$ to $x = b$, we can divide the area into n rectangles of width Δx any way we choose, so long as the height of the ith rectangle is given by $f(x_i)$ for some value x_i within the base of the rectangle (see Figure 6-11). This allows us to express A_n as

$$A_n = f(x_1)\Delta x + \cdots + f(x_i)\Delta x + \cdots + f(x_n)\Delta x$$

Using Σ (the capital Greek letter sigma) to denote the process of summing, we write $\sum_{i=1}^{n} f(x_i)\Delta x$ (read: the sum from $i = 1$ to n of $f(x_i)\Delta x$) to stand for $f(x_1)\Delta x + \cdots + f(x_i)\Delta x + \cdots + f(x_n)\Delta x$. In this notation the area A is given by

$$A = \lim_{n\to\infty} A_n = \lim_{n\to\infty} \sum_{i=1}^{n} f(x_i)\Delta x \tag{1}$$

Expressions such as the right side of Equation 1 are the heart of integral calculus, and they appear in a surprising number of applications.*

The right side of Equation (1) is also called the **definite integral of $f(x)$** from a to b and is written $\int_a^b f(x)\,dx$. (We see in the next section why the integral sign has reappeared here in this new context.) Thus

$$\int_a^b f(x)\,dx = \lim_{n\to\infty} \sum_{i=1}^{n} f(x_i)\Delta x \tag{2}$$

The notation in Equation (2) expresses the thought that as n gets large, Δx gets small and is denoted as the differential of x, dx; the summation sign Σ is replaced by $\int$ (a distorted letter S, for summation), and the a and b, called the limits of integration, show the interval from which the typical element can come.

---

*The expression $\sum_{i=1}^{n} f(x_i)\Delta x$ is called a **Riemann sum**, named after the German mathematician G. F. B. Riemann (1826–1866). Another interesting historical fact is that Isaac Newton (English, 1642–1727) and Gottfried Leibniz (German, 1646–1716) independently developed the foundations of calculus in the 1660s and 1670s.

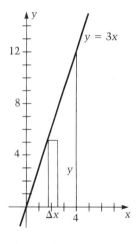

Figure 6-12

EXAMPLE 3. Write the definite integral expression for the area under the curve $y = 3x$ from $x = 0$ to $x = 4$.

The area, together with a typical rectangular element, is shown in Figure 6-12. The area of a typical element is $f(x)\Delta x = y\Delta x$. Summing up the areas of the elements and taking the limit of the sum as the number of elements becomes infinite, we get the expression

$$A = \int_0^4 y \, dx = \int_0^4 3x \, dx$$

The limits of integration denote the interval from 0 through 4, from which the typical element can come in the sense that the typical element can be no farther left than 0 and no farther right than 4.

For this case, we can use geometry to compute the area.

$$A = \frac{1}{2}(4)(12) = 24 \quad \blacksquare$$

Practice 2. Write the definite integral expression for the area under the curve $y = x^2$ from $x = 2$ to $x = 4$.

In order to compute area under a curve, our approach has been to compute the area of a typical element, sum up these areas, and then let the number of elements become infinite. By Equation (2), we are therefore able to write a definite integral expression for the area. However, the definite integral can be used whenever we compute some quantity for a typical element, sum up such quantities, and then let the number of elements become infinite. Further applications of the definite integral, in addition to area, will be explored in Chapter 7. In the meantime, we deal with the actual evaluation of definite integral expressions.

Exercises / Section **6-2**

Exercises 1–4: Approximate the area under the given curve by computing A_n for the indicated value of n. Then use a formula from geometry to compute the actual area under the curve.

1. $f(x) = 3x$ from $x = 0$ to $x = 2$; A_8

2. $f(x) = -2x + 8$ from $x = 1$ to $x = 4$; A_6

3. $f(x) = 2x + 2$ from $x = 1$ to $x = 3$; A_4

4. $f(x) = \sqrt{1 - x^2}$ from $x = 0$ to $x = 1$; A_5

Exercises 5–8: Approximate the area under the given curve by computing A_n for the two indicated values of n.

5. $f(x) = x^3$ from $x = 0$ to $x = 2$; A_4, A_8

6. $f(x) = 2 - x^2$ from $x = 0$ to $x = 1$; A_5, A_{10}

7. $f(x) = \dfrac{1}{x}$ from $x = 2$ to $x = 5$; A_3, A_6

8. $f(x) = 2x^2 - 3x$ from $x = 2$ to $x = 2.5$; A_5, A_{10}

Exercises 9–12: Write the definite integral expression for each quantity.

9. The area under the curve $y = 3x^2 + 1$ from $x = 1$ to $x = 2$.

10. The area under the curve $y = 2x + 5$ from $x = -2$ to $x = 3$.

11. The area under the curve $y = -2x + 6$ from $x = -1$ to $x = 1$.

12. The area under the curve $y = x^3$ from $x = 0$ to $x = 2$.

6-3 The Fundamental Theorem

In this section we learn the Fundamental Theorem of Calculus. Anything with a name as imposing as that must be pretty important, and indeed the Fundamental Theorem (discovered in the seventeenth century) is important for two reasons. First, it allows us to evaluate many definite integrals. But it also ties together the ideas of differentiation and integration with the idea of the limit of a sum.

We suppose that $f(x)$ is a function that has positive values between a and b and that x is some arbitrary value between a and b. We let A denote the area under the curve from a to x (the shaded area in Figure 6-13). As x is increased by a value Δx, there is a corresponding change in area, ΔA. From Figure 6-14 it is clear that $f(x)\Delta x$ (the area of the smaller rectangle) is less than or equal to ΔA (the shaded part), which in turn is less than or equal to $f(x + \Delta x)\Delta x$ (the area of the larger rectangle). Writing this as an inequality,

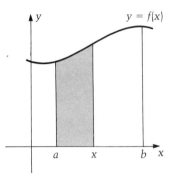

Figure 6-13

$$f(x)\Delta x \le \Delta A \le f(x + \Delta x)\Delta x$$

Dividing by Δx (a positive value), we get

$$f(x) \le \frac{\Delta A}{\Delta x} \le f(x + \Delta x)$$

Now we take the limit as $\Delta x \to 0$:

$$\lim_{\Delta x \to 0} f(x) \le \lim_{\Delta x \to 0} \frac{\Delta A}{\Delta x} \le \lim_{\Delta x \to 0} f(x + \Delta x)$$

or

$$f(x) \le \lim_{\Delta x \to 0} \frac{\Delta A}{\Delta x} \le f(x)$$

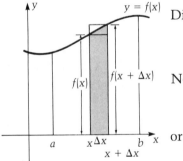

Figure 6-14

which means that

$$\lim_{\Delta x \to 0} \frac{\Delta A}{\Delta x} = f(x) \tag{1}$$

The left side of Equation (1) is the definition of the derivative dA/dx (remember all that stuff back in Chapter 4?), so we can rewrite Equation (1) as

$$\frac{dA}{dx} = f(x)$$

or

$$dA = f(x)\,dx$$

This equation says that the function A is an antiderivative of the function $f(x)$, or, using the indefinite integral notation from Section 6-1,

$$A + C_1 = \int f(x) \, dx \tag{2}$$

But we also know that

$$\int f(x) \, dx = F(x) + C_2 \tag{3}$$

where $F(x)$ is any function such that $dF(x) = f(x) \, dx$ (or $F'(x) = f(x)$). Combining Equations (2) and (3),

$$A = F(x) + C \tag{4}$$

Remember that A represents the area under the curve from a to x, where x is any value between a and b. In particular, for $x = a$, $A = 0$; using this in Equation (4),

$$0 = F(a) + C$$

and, therefore,

$$C = -F(a)$$

Equation (4) then becomes

$$A = F(x) - F(a)$$

Now letting $x = b$,

$$A = F(b) - F(a) \tag{5}$$

Equation (5) gives an expression for the area under the curve $y = f(x)$ from a to b. But we also know that this area is given by the definite integral of $f(x)$ from a to b. Combining this fact with Equation (5), we finally have

$$\int_a^b f(x) \, dx = F(b) - F(a) \tag{6}$$

Although our discussion justifies Equation (6) only for the case of a function $f(x)$ that has positive values between a and b, this is not necessary. Equation (6) is true even if $f(x)$ takes on negative values, and it is true for definite integrals that represent other quantities as well as area. Equation (6) is the Fundamental Theorem.

$$\int_a^b f(x) \, dx = F(b) - F(a), \qquad \text{where } F(x) \text{ is any function such that } F'(x) = f(x)$$

The Fundamental Theorem says that to evaluate a definite integral, we do not have to use limits of sums. If we can find any antiderivative function of the integrand, we need only evaluate that function at two places and subtract the values. The notation $F(x)|_a^b$ is used to denote $F(b) - F(a)$. (Because subtraction is not commutative, we must be sure to compute $F(\text{upper limit}) - F(\text{lower limit})$, and not the other way around.)

EXAMPLE 1.   Evaluate $\int_0^2 x^2 \, dx$.

We first find an antiderivative of x^2, namely, $x^3/3$. Then

$$\int_0^2 x^2 \, dx = \frac{x^3}{3}\bigg|_0^2 = \frac{8}{3} - 0 = \frac{8}{3} \quad \blacksquare$$

EXAMPLE 2.   Evaluate $\int_1^4 (2x + 3) \, dx$.

$$\int_1^4 (2x + 3) \, dx = \frac{2x^2}{2} + 3x \bigg|_1^4 = 16 + 12 - (1 + 3) = 24$$

Note that in the second step, no integral sign appears. The integration (antidifferentiation) process has been completed, and only arithmetic remains. $\blacksquare$

EXAMPLE 3.   Evaluate $\int_1^8 \sqrt[3]{x} \, dx$.

$$\int_1^8 \sqrt[3]{x} \, dx = \int_1^8 x^{1/3} \, dx = \frac{3}{4} x^{4/3} \bigg|_1^8$$

$$= \frac{3}{4} (2)^4 - \frac{3}{4} (1) = 12 - \frac{3}{4} = \frac{45}{4} \quad \blacksquare$$

Practice 1.   Evaluate $\int_0^2 3x \, dx$.

Practice 2.   Evaluate $\int_1^4 (3x^2 - \sqrt{x}) \, dx$.

We have talked about two kinds of integrals in this chapter, indefinite integrals and definite integrals. Let us make sure we understand the distinctions between the two and also how they are related. One obvious distinction is that the indefinite integral, $\int f(x) \, dx$, has no limits of integration, while the definite integral, $\int_a^b f(x) \, dx$, does have limits of integration. The *indefinite integral*

$$\int f(x) \, dx$$

is a *function*, an antiderivative of $f(x)$, and it always has the constant of integration C added at the end. Thus $\int x^2\, dx = x^3/3 + C$. On the other hand, the *definite integral*

$$\int_a^b f(x)\, dx$$

involves a and b, and if a and b are numbers, then $\int_a^b f(x)\, dx$ is a *number*. Thus (see Example 1) $\int_0^2 x^2\, dx = \frac{8}{3}$. But, according to the Fundamental Theorem, we can evaluate $\int_a^b f(x)\, dx$ by first finding an antiderivative of $f(x)$, and $\int f(x)\, dx$ is another name for an antiderivative. We could rewrite the Fundamental Theorem in the form

$$\int_a^b f(x)\, dx = \int f(x)\, dx \Big|_a^b$$

In evaluating $\int_a^b f(x)\, dx$, we also find $\int f(x)\, dx$. (We do not have to include the constant of integration when evaluating a definite integral because it cancels out; by this we mean that $F(x) + C\big|_a^b = F(b) + C - [F(a) + C] = F(b) - F(a)$, so C did not have to appear.)

EXAMPLE 4. Evaluate $\displaystyle\int_{-2}^{1} (x^2 - 1)^3 x\, dx$.

We try to find an antiderivative by the power rule. Letting $n = 3$ and $u = x^2 - 1$, we get $du = 2x\, dx$, and we see that we need only the constant factor of 2. We write

$$\int_{-2}^{1} (x^2 - 1)^3 x\, dx = \frac{1}{2}\int_{-2}^{1} (x^2 - 1)^3 2x\, dx = \frac{1}{2}\frac{(x^2 - 1)^4}{4}\Big|_{-2}^{1}$$

$$= \frac{1}{2}\left[0 - \frac{(3)^4}{4}\right] = -\frac{81}{8}$$

We need not be alarmed that the value of this definite integral is a negative number; remember we are not necessarily computing area, which we expect to be positive.   ∎

EXAMPLE 5. Evaluate $\displaystyle\int_{-1}^{1} \sqrt{2 - x^3}\, x^2\, dx$.

Letting $n = \frac{1}{2}$ and $u = 2 - x^3$, we get $du = -3x^2\, dx$. Thus

$$\int_{-1}^{1} (2 - x^3)^{1/2} x^2\, dx = -\frac{1}{3}\int_{-1}^{1} (2 - x^3)^{1/2}(-3)x^2\, dx$$

$$= -\frac{1}{3}\cdot\frac{2}{3}(2 - x^3)^{3/2}\Big|_{-1}^{1}$$

$$= -\frac{2}{9}\{(2 - 1)^{3/2} - [2 - (-1)]^{3/2}\}$$

$$= -\frac{2}{9}[1 - (3)^{3/2}]$$

$$= \frac{2}{9}(\sqrt{27} - 1)   ∎$$

Word of Advice

In computing $F(b) - F(a)$, be very careful of minus signs. Do not try to combine too many steps at once, or you will lose a minus sign somewhere and get the wrong answer.

Practice 3.   Evaluate $\int_{-1}^{1} \dfrac{x^2}{(x^3 - 2)^3} \, dx$.

EXAMPLE 6.   Find the area under the curve $y = 3x$ from $x = 0$ to $x = 4$.

The area is given by the definite integral $\int_0^4 3x \, dx$ (see Example 3, Section 6-2).

$$\int_0^4 3x \, dx = 3 \int_0^4 x \, dx = 3 \left. \frac{x^2}{2} \right|_0^4 = 24 - 0 = 24$$

This agrees with the value for the area that we previously computed by using geometry.   ∎

Exercises / Section **6-3**

Exercises 1–24: Evaluate:

1. $\displaystyle\int_1^2 x^3 \, dx$

2. $\displaystyle\int_2^5 3x \, dx$

3. $\displaystyle\int_1^4 (2x^2 - 4x + 1) \, dx$

4. $\displaystyle\int_0^3 (x^3 - 2x^2 + x) \, dx$

5. $\displaystyle\int_0^4 (\sqrt{x} - x) \, dx$

6. $\displaystyle\int_1^8 (\sqrt[3]{x^2} + 2) \, dx$

7. $\displaystyle\int_{-3}^{-1} (2x^2 - x) \, dx$

8. $\displaystyle\int_{-4}^{-2} (x^3 + 7) \, dx$

9. $\displaystyle\int_1^2 \frac{1}{x^2} \, dx$

10. $\displaystyle\int_1^4 \frac{1}{\sqrt{x}} \, dx$

11. $\displaystyle\int_2^3 \frac{x^2 - 2}{x^2} \, dx$

12. $\displaystyle\int_1^4 \frac{x + 1}{\sqrt{x}} \, dx$

13. $\displaystyle\int_1^2 (x^2 - 1)^2 \, dx$

14. $\displaystyle\int_0^1 (1 - 2x^3)^2 x \, dx$

15. $\displaystyle\int_0^1 (2 + x^2)^2 x \, dx$

16. $\displaystyle\int_{-1}^2 (x^3 + 2)^2 x^2 \, dx$

17. $\displaystyle\int_0^3 \sqrt{x^2 + 4} \, x \, dx$

18. $\displaystyle\int_0^2 (x^3 - 1)^{2/3} x^2 \, dx$

19. $\displaystyle\int_1^2 \sqrt{1 + 2x} \, dx$

20. $\displaystyle\int_{-1}^6 (2 + x)^{1/3} \, dx$

21. $\displaystyle\int_{-3}^{-1} \frac{x+1}{(x^2+2x+3)^2}\,dx$

22. $\displaystyle\int_{2}^{4} \frac{x}{(x^2-1)^3}\,dx$

23. $\displaystyle\int_{1}^{2} \frac{4x+1}{\sqrt{4x^2+2x}}\,dx$

24. $\displaystyle\int_{-1}^{1} \frac{2x+1}{\sqrt[3]{1+3x+3x^2}}\,dx$

STATUS CHECK

Now that you are at the end of Chapter 6, you should be able to:

section **6-1**   Use the power rule to find the indefinite integral of an algebraic function.
Find *y* as a function of *x*, given *dy/dx* and a point on the curve.

section **6-2**   Approximate the area under a curve by computing A_n for a given value of *n*.
Write the definite integral expression for the area under a curve $y = f(x)$ from $x = a$ to $x = b$.

section **6-3**   Evaluate a definite integral using the Fundamental Theorem of Calculus and the power rule.

6-4 More Exercises for Chapter **6**

Exercises 1–10: Integrate:

1. $\displaystyle\int (2x^3 - 7x)\,dx$ 2. $\displaystyle\int (4 - x^2 + x^4)\,dx$

3. $\displaystyle\int x^{3/4}\,dx$ 4. $\displaystyle\int \sqrt[5]{x^2}\,dx$

5. $\displaystyle\int \sqrt{1+3x}\,dx$ 6. $\displaystyle\int \frac{1}{(2x-1)^{1/3}}\,dx$

7. $\displaystyle\int (2 - 4x^2)^3 x\,dx$

8. $\displaystyle\int \sqrt{2x^3 - 7} x^2\,dx$

9. $\displaystyle\int \frac{x-2}{(x^2 - 4x + 2)^3}\,dx$

10. $\displaystyle\int \frac{x}{\sqrt[3]{7x^2 + 2}}\,dx$

Exercises 11–20: Evaluate:

11. $\displaystyle\int_{1}^{2} \left(x - \frac{2}{x^2}\right)dx$ 12. $\displaystyle\int_{1}^{4} \left(\frac{1}{\sqrt{x}} - x^3\right)dx$

13. $\displaystyle\int_{1}^{8} (x^{2/3} - 1)\,dx$ 14. $\displaystyle\int_{1}^{9} (4 - \sqrt{x})\,dx$

15. $\displaystyle\int_{0}^{1} (x^3 - 1)^2 x\,dx$ 16. $\displaystyle\int_{-1}^{0} (2x^2 + 4)^2 x^2\,dx$

17. $\displaystyle\int_{1}^{5} \sqrt{1+3x}\,dx$ 18. $\displaystyle\int_{0}^{1} \sqrt[3]{1+2x^2} x\,dx$

19. $\displaystyle\int_{0}^{3} \frac{4+2x}{\sqrt{4x+x^2}}\,dx$

20. $\displaystyle\int_{-1}^{1} \frac{x+2}{(x^2+4x+7)^2}\,dx$

Exercises 21–22: Write the definite integral expression for each quantity.

21. The area under the curve $y = \sqrt{1+x^2}$ from $x = 1$ to $x = 5$.

22. The area under the curve $y = 1/x^2$ from $x = -4$ to $x = -2$.

23. Find the equation of a curve that has slope $4\sqrt{2x-1}$ and passes through the point $(1, \frac{1}{3})$.

24. Find the equation of a curve that has slope $(2x+1)/(x^2+x-2)^2$ and passes through the point $(0, 2)$.

chapter 7

Using the Integral

Integration can be used to find the force exerted by water against a submerged surface.

7-1 Indefinite Integral

In this chapter we pursue the use of integration as a problem-solving tool. First, we consider some applications of the indefinite integral.

Recall that the indefinite integral is an antiderivative, and

$$\int f(x) \, dx = F(x) + C$$

where $F(x)$ is any function such that $dF(x)/dx = f(x)$. Therefore, if we have an expression for the derivative dy/dx of any quantity y with respect to a variable x, we can find the general form of the function y by taking the indefinite integral. This introduces a constant of integration, and some additional facts about the relationship between y and x must be known in order to evaluate this constant.

A classic example of the use of the indefinite integral occurs in the case of a moving body, where acceleration as a function of time is known. Because acceleration a is the time rate of change of velocity v, or

$$a = \frac{dv}{dt}$$

it follows that

$$\int a \, dt = v$$

Also, because velocity is the time rate of change of position s, or

$$v = \frac{ds}{dt}$$

it follows that

$$\int v \, dt = s$$

Each of these integrations introduces a constant.

EXAMPLE 1. The acceleration of an object is given by $a = 4t^2$. Find an expression for velocity as a function of time if $v = 5$ when $t = 1$.

To find velocity, we integrate acceleration.

$$v = \int a \, dt = \int 4t^2 \, dt = \frac{4t^3}{3} + C$$

Substituting $v = 5$ and $t = 1$,

$$5 = \frac{4(1)}{3} + C, \quad \text{so} \quad C = \frac{11}{3}$$

Therefore,

$$v = \frac{4t^3}{3} + \frac{11}{3} \quad \blacksquare$$

EXAMPLE 2. An object starting from rest has acceleration $a = 3.4t$. Find an expression for s as a function of t if $s = 21.6$ when $t = 3$.

To find s requires two integrations. We use C_1 and C_2 to distinguish the two constants of integration.

$$v = \int a \, dt = \int 3.4t \, dt = 1.7t^2 + C_1$$

The fact that the object started from rest means that $v = 0$ at $t = 0$. Substituting these values, we compute C_1.

$$0 = 1.7(0) + C_1, \quad \text{so} \quad C_1 = 0$$

and

$$v = 1.7t^2$$

Now

$$s = \int v \, dt = \int 1.7t^2 \, dt = \frac{1.7t^3}{3} + C_2$$

Using the given relationship between s and t,

$$21.6 = \frac{1.7}{3}(3)^3 + C_2, \quad \text{so} \quad C_2 = 6.3$$

and

$$s = \frac{1.7t^3}{3} + 6.3 \quad \blacksquare$$

Practice 1. The velocity of a moving body is given by $v = 4t + 2$. Find s as a function of t if $s = 3$ when $t = 1$.

A special case of the moving-body problem is that of a falling object. In this case, the only acceleration acting upon the object is due to the force of gravity; it has a constant value of -9.8 m/s^2 (-32 ft/s^2). The negative sign indicates that the force acts in a downward direction.

EXAMPLE 3. A ball is thrown upward from the ground with an initial velocity of 10 m/s. What is its distance above the ground at $t = 1$ s?

Here $a = -9.8$.

$$v = \int a\ dt = \int -9.8\ dt = -9.8t + C_1$$

The initial velocity (the velocity at $t = 0$) is 10 m/s; this is a positive quantity because the ball was thrown upward.

$$10 = -9.8(0) + C_1, \quad \text{so} \quad C_1 = 10$$
$$v = -9.8t + 10$$
$$s = \int v\ dt = \int (-9.8t + 10)\ dt = -4.9t^2 + 10t + C_2$$

The ball is thrown from ground level, so $s = 0$ when $t = 0$.

$$0 = 0 + 0 + C_2, \quad \text{so} \quad C_2 = 0$$
$$s = -4.9t^2 + 10t$$

This is the general expression for $s(t)$; we want $s(1)$.

$$s(1) = -4.9 + 10 = 5.1$$

The ball is 5.1 m above ground at $t = 1$ s. ∎

EXAMPLE 4. A stone is thrown upward from the top of a 13-m building. At $t = 2$ s, the stone is 3 m above ground. Find the height of the stone at $t = 0.5$ s. Is the stone moving upward or downward at $t = 0.5$ s?

Here again, the only acceleration is due to gravity.

$$v = \int a\ dt = \int -9.8\ dt = -9.8t + C_1$$

We do not have enough information to find C_1 yet, so we press on.

$$s = \int v\ dt = \int (-9.8t + C_1)\ dt = -4.9t^2 + C_1t + C_2$$

We know that $s(0) = 13$ and $s(2) = 3$. Putting these values in the expression for s,

$$13 = 0 + 0 + C_2, \quad \text{so} \quad C_2 = 13$$

and

$$3 = -4.9(2)^2 + C_1(2) + C_2$$

or

$$3 = -19.6 + 2C_1 + 13, \quad \text{so} \quad C_1 = 4.8$$

Therefore,

$$s = -4.9t^2 + 4.8t + 13$$

To find the height of the stone at $t = 0.5$ s, we compute $s(0.5)$.

$$s(0.5) = -4.9(0.5)^2 + 4.8(0.5) + 13 = 14.175$$

The general expression for $v(t)$ is

$$v = -9.8t + C_1 = -9.8t + 4.8$$

and $v(0.5)$ is negative. This indicates that s is decreasing at $t = 0.5$ s, so the stone has already reached its maximum height and is on the way down. ■

Practice 2.   An object thrown upward from the ground has a velocity of 6 m/s at $t = 1$ s. What is the height of the object at $t = 1$ s?

An electric circuit containing an inductor produces a set of equations similar to the acceleration-velocity-position equations of a moving body. The induced voltage V in the circuit is proportional to the time rate of change of current i, so

$$V = L \frac{di}{dt}$$

from which

$$i = \int \frac{V}{L} \, dt = \frac{1}{L} \int V \, dt$$

The constant L is the inductance. Also, the current i in an electric circuit is the time rate of change of the total charge q that has passed a given point in the circuit. Therefore,

$$i = \frac{dq}{dt}$$

or

$$q = \int i \, dt$$

Each of these integrations produces a constant.

EXAMPLE 5.   The current i in a particular circuit as a function of time t is given by the equation $i = t + 3$. Find the total charge q (in coulombs) passing a given point in the circuit in the first 3 s.

To find charge, we integrate current.

$$q = \int i \, dt = \int (t + 3) \, dt = \frac{t^2}{2} + 3t + C$$

At $t = 0$ s, zero total charge q has passed a given point. This fact enables us to evaluate the constant of integration.

$$0 = \frac{(0)^2}{2} + 3(0) + C, \quad \text{so} \quad C = 0$$

and

$$q = \frac{t^2}{2} + 3t$$

We want the value of q when $t = 3$ s.

$$q = \frac{(3)^2}{2} + 3(3) = \frac{27}{2}$$

The total charge passing a given point in the circuit in the first 3 s is $\frac{27}{2}$ C.   ■

EXAMPLE 6.   The voltage V induced in a 0.5-H inductor is 0.8 V. If 1.2 C of charge passes a given point in 2 s, find the total charge which passes the point in 5 s.

We first integrate voltage to get current.

$$i = \frac{1}{L} \int V \, dt = \frac{1}{0.5} \int 0.8 \, dt = \frac{0.8}{0.5} t + C_1 = 1.6t + C_1$$

We do not have the information needed to compute C_1 at this time, but we can do the next integration.

$$q = \int i \, dt = \int (1.6t + C_1) \, dt = 0.8t^2 + C_1 t + C_2$$

From $q = 0$ at $t = 0$ and $q = 1.2$ at $t = 2$, we get the equations

$$0 = 0 + 0 + C_2$$
$$1.2 = 0.8(2)^2 + C_1(2) + C_2$$

The solution to this system is $C_2 = 0$ and $C_1 = -1$. The equation for q as a function of t is

$$q = 0.8t^2 - t$$

For $t = 5$, $q = 0.8(5)^2 - 5 = 15$. A total of 15 C of charge passes the point in 5 s.   ■

Practice 3. The voltage V induced in a 2.0-H inductor as a function of time t is given by the equation $V = 4t$. At $t = 1$ s, the current in the circuit is 4 A. Find the current at $t = 4$ s.

Remember that any case in which we know the rate of change of a function and want to know the original function calls for integration.

In economics, if the cost c of producing x units of a product is a function $c(x)$, then the derivative function dc/dx is called the **marginal cost function**.

EXAMPLE 7. The Super Solder Shop finds that the marginal cost of producing x soldering irons is given by

$$c'(x) = \frac{2}{\sqrt{x}} + 0.024x$$

and the fixed cost (overhead expenses which must be paid even if no items are produced) is $2000. Find the total cost of producing 750 soldering irons.

To find $c(x)$, we integrate $c'(x)$.

$$c(x) = \int c'(x)\, dx = \int \left(\frac{2}{\sqrt{x}} + 0.024x \right) dx = 4\sqrt{x} + 0.012x^2 + C$$

Because $c = 2000$ when $x = 0$, we can evaluate the constant of integration C.

$$2000 = 0 + 0 + C, \quad \text{so} \quad C = 2000$$

Therefore,

$$c(x) = 4\sqrt{x} + 0.012x^2 + 2000$$

and

$$c(750) = 4\sqrt{750} + 0.012(750)^2 + 2000 = 8859.54$$

The total cost for 750 units is $8859.54, or about $11.81 per unit.   ■

Exercises / Section **7-1**

1. The acceleration of an object is given by $a = 2\sqrt{t}$. Find an expression for velocity as a function of time if $v = 20$ when $t = 4$.

2. The velocity of an object is given by $v = 3t^2 + 2t$. Find an expression for s as a function of t if $s = 10$ when $t = 1$.

3. An object moves in a straight line with acceleration $a = 35$ m/s^2. How far from the starting point is the object after 5 s if its velocity is 14 m/s when $t = 2$ s?

4. An object has an initial velocity of 120 cm/s and travels with an acceleration $a = 1/\sqrt{t+1}$

cm/s^2. What is s when $t = 8$ s if $s = 200$ cm when $t = 3$ s?

5. The acceleration of an object is given by $a = 2t + 4$. Find an expression for s as a function of t given that $s = 0$ when $t = 0$ and $s = 5$ when $t = 1$.

6. The acceleration of an object is given by $a = 3t^2 + 1$. Find an expression for s as a function of t given that $s = 19$ when $t = 2$ and $s = 91$ when $t = 4$.

7. What constant acceleration should be applied to a car at rest in order for it to travel 100 m in 5 s?

8. A car passes a certain point going at a speed of 20 m/s (roughly 40 mi/h). What constant deceleration will bring the car to a stop in 40 m? How long will this take?

9. A stone is dropped from a height of 10 m. How long will it take to hit the ground?

10. A ball dropped from a cliff hits the ground in 3 s. How high is the cliff?

11. An object is thrown upward from the ground with an initial velocity of 6 m/s. Find the time it takes to hit the ground and the maximum height the object reaches.

12. A stone thrown upward from the top of a tower with a velocity of 35 m/s hits the ground with a velocity of -50 m/s. Find the height of the tower to the nearest meter.

13. What upward initial velocity is needed to make an object rise exactly 30 m before it falls?

14. A ball is thrown upward from a point 32 m above the ground. It is the same height above the ground after 3 s as it was after 2 s. Find the time it takes to hit the ground.

15. The current in a certain circuit is 3.2 A. How many coulombs of charge pass a given point in the circuit in the first 5 s?

16. The current i in one circuit as a function of time t is given by the equation $i = t\sqrt{t^2 + 4}$. Find the total charge q passing a given point in the circuit in the first 2 s.

17. The voltage induced in a 0.8-H inductor varies as a function of time according to the equation $V = 2t^2 + t$. After 2 s, the current in the circuit is 3 A. Find the total charge which passes a given point in the circuit in 3 s.

18. A voltage V given by $V = 0.8t$ is induced in a 0.2-H inductor. If 3 C of charge pass a given point in the circuit in 1 s, find the total charge that passes the point in 2 s.

19. A voltage of 1.5 V is induced in a 1.2-H inductor. At $t = 2$ s, the current in the circuit is 2 A. Find a general expression for the current as a function of time.

20. A voltage V given by $V = \sqrt{t^3}$ is induced in a 2-H inductor. If the current in the circuit is 4 A when $t = 1$ s, what is the current when $t = 3$ s?

21. The voltage V induced in a certain inductor varies as a function of time according to the equation $V = 3t + 2$. After 1 s, the current in the circuit is 1 A, and after 3 s, the current in the circuit is 5 A. What is the inductance?

22. The voltage V induced in a certain inductor is given by $V = \sqrt{t}$. After 1 s, the current in the circuit is 1 A, and after 4 s, the current is 3 A. Find the current at $t = 6$ s.

23. The marginal cost for one company to produce x items is given by $c'(x) = 0.04x^2 - x$; the fixed cost is $500. Find the total cost to produce 100 items.

24. The marginal cost to manufacture x mechanical pencils is given by $c'(x) = 4/\sqrt{0.2x + 3}$; the fixed cost is $200. Find (to the nearest penny) the cost per item to produce 500 pencils.

25. The voltage V across a capacitor of capacitance C is given by

$$V = \frac{1}{C} \int i \, dt$$

where i is the current in the circuit. The voltage across a particular 0.002-F capacitor is zero when a current of 0.3 A is applied to the circuit. What is the voltage after 1.5 s?

26. Power in an electrical system is defined to be the time rate of change of energy, $P = dW/dt$, where W is in joules and P is in watts. A given system, which initially has zero energy, draws power P given by $P = 2t + 1$ watts. What is the energy in the system after 5 s?

27. Angular velocity ω for a rotating object is the time rate of change of angular displacement θ, $\omega = d\theta/dt$. A flywheel rotates at a rate given by $\omega = 2t^2 - t + 5$ revolutions per minute. Find the number of revolutions in the first 30 s.

28. Angular acceleration α for a rotating object is the time rate of change of angular velocity, $\alpha = d\omega/dt$ (see Exercise 27 above). For $\alpha = t^{2/3}$, find an expression for angular displacement θ as a function of t, given that $\theta = 0$ and $\omega = 0$ when $t = 0$.

29. The inner and outer surfaces of the wall of a building are at different temperatures. The temperature T within the wall is a function of the distance x away from the outer surface, and the rate of change dT/dx is given by

$$\frac{dT}{dx} = x - 0.18x^2$$

If the outer surface has a temperature of 35°C and the wall is 10 cm thick, find the temperature of the inner surface.

30. An offshore oil well after t days of drilling produced oil at the rate of $200 + 3t - 0.2t^2$ barrels per day. Find the total output of the well for the first 10 days of drilling.

31. A certain chemical is dumped into a river as part of industrial waste. The concentration (in grams per standard sample) changes as a function of the distance x (in meters) downstream of the dump site, and the rate of change is given by the equation

$$\frac{dC}{dx} = 400 + 15x - 0.45x^2$$

If the concentration at the dump site is 50,000 g per sample, find the concentration 100 m downstream.

32. At time $t = 0$, a patient is given 25 cc of medication. The medication is neutralized at a variable rate; at any time t, the rate (in cubic centimeters per hour) of medication being neutralized is given by

$$\frac{dm}{dt} = \frac{2}{\sqrt{t + 1}}$$

After how many hours will all of the medication be neutralized?

7-2 Area

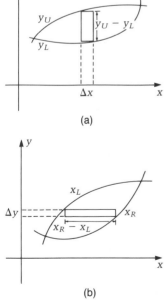

(a)

(b)

Figure 7-1

The definite integral of a function $f(x)$ from a to b is defined by the equation

$$\int_a^b f(x)\ dx = \lim_{n \to \infty} \sum_{i=1}^n f(x_i)\Delta x \qquad (1)$$

It is important to remember that we can *evaluate* the definite integral by the equation

$$\int_a^b f(x)\ dx = F(b) - F(a) \qquad (2)$$

where $F'(x) = f(x)$, but that the *definition* of the definite integral is given by Equation (1). In a given problem, if we can express the quantity we wish to compute in terms of the limit of a sum of that quantity for typical elements, then Equation (1) says we can write a definite integral to solve the problem. We then use Equation (2) to evaluate that definite integral.

The area between two curves can be expressed as a definite integral because we can find the area of a typical rectangular element, sum up such areas, and let the number of elements become infinite. A typical element can be vertical (Figure 7-1(a)), in which case it has width Δx and its height is given by taking the y-value of the upper curve (which we call y_U) minus the y-value of the lower curve (y_L). A typical element could also be horizontal (Figure

7-1(b)), in which case it has thickness Δy and its length is given by taking the x-value of the right-hand curve (x_R) minus the x-value of the left-hand curve (x_L).

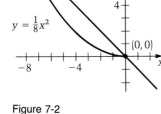

Figure 7-2

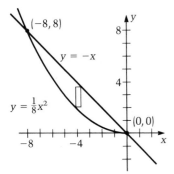

Figure 7-3

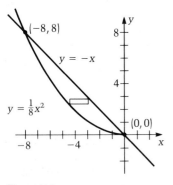

Figure 7-4

EXAMPLE 1. Find the area bounded by the curves $y = \frac{1}{8}x^2$ and $y = -x$.

Figure 7-2 shows the area we want to find. The points of intersection are found by solving the simultaneous system of equations

$$y = \tfrac{1}{8}x^2$$
$$y = -x$$

First we will use vertical elements (Figure 7-3). The typical vertical rectangle has area given by

$$(\text{height})(\text{width}) = (y_U - y_L)\Delta x$$

The limits of integration show the left and right endpoints of the interval from which the typical element can come. Thus

$$A = \int_{-8}^{0} (y_U - y_L)\, dx = \int_{-8}^{0} \left(-x - \tfrac{1}{8}x^2\right) dx$$

Notice that we almost have to draw a graph in order to write this expression correctly. Evaluating the integral,

$$\int_{-8}^{0} \left(-x - \tfrac{1}{8}x^2\right) dx = -\frac{x^2}{2} - \frac{x^3}{24}\Big|_{-8}^{0} = 0 - \left(-32 + \frac{64}{3}\right) = \frac{32}{3}$$

Now let us solve the same problem using horizontal elements (Figure 7-4). The typical horizontal element has area given by

$$(\text{length})(\text{thickness}) = (x_R - x_L)\Delta y$$

Here we must solve each equation for x in terms of y. For the equation $y = \frac{1}{8}x^2$, we take the negative square root when solving for x because we are in the second quadrant, where x is negative. The limits of integration show the lower and upper endpoints of the interval from which the typical element can come. Thus

$$A = \int_{0}^{8} (x_R - x_L)\, dy = \int_{0}^{8} \left[-y - (-\sqrt{8y})\right] dy$$

$$= \int_{0}^{8} \left(-y + (8y)^{1/2}\right) dy$$

$$= \frac{-y^2}{2} + \tfrac{1}{8}(8y)^{3/2} \cdot \frac{2}{3}\Big|_{0}^{8} = \left[-32 + \tfrac{1}{8}(64)^{3/2} \cdot \frac{2}{3}\right] - 0 = -32 + \frac{128}{3}$$

$$= \frac{32}{3}$$

Of course we get the same answer as before. ■

We use the following general procedure to find the area between curves:

1. Sketch a graph of the area, including the coordinates of all points of intersection of the curves bounding the area.
2. Sketch a typical rectangular element, either horizontal or vertical.
3. Write the definite integral. For a vertical element, the integral will take the form

$$\int_{\text{left end}}^{\text{right end}} (y_U - y_L) \, dx$$

and for a horizontal element, the integral will take the form

$$\int_{\text{low end}}^{\text{high end}} (x_R - x_L) \, dy$$

4. Evaluate the definite integral.

Word of Advice

Vertical distances between curves are always measured by $y_U - y_L$ (y-upper minus y-lower); horizontal distances between curves are always measured by $x_R - x_L$ (x-right minus x-left).

Practice 1.   Find the area bounded by the curves $y = 1 - x^2$ and $y = -2x + 1$ by using vertical elements.

Practice 2.   Find the area of Practice 1 by using horizontal elements.

What determines whether we should use vertical or horizontal elements? Sometimes (as in the case of Example 1 and Practices 1 and 2) it does not make any difference which we use. Sometimes it will not seem to make any difference when we begin the problem, but when we try to evaluate the definite integral we have written, we may not be able to find an antiderivative for the integrand. Sometimes it is clear from the graph that one approach is better than the other.

EXAMPLE 2. Find the area bounded by the curves $y = 2 - x^2$ and $y = x$.

Figure 7-5 shows the area we want. Using vertical elements,

$$A = \int_{-2}^{1} (y_U - y_L)\, dx = \int_{-2}^{1} [(2 - x^2) - x]\, dx$$

$$= 2x - \frac{x^3}{3} - \frac{x^2}{2}\Big|_{-2}^{1}$$

$$= 2 - \frac{1}{3} - \frac{1}{2} - \left(-4 + \frac{8}{3} - 2\right) = \frac{9}{2}$$

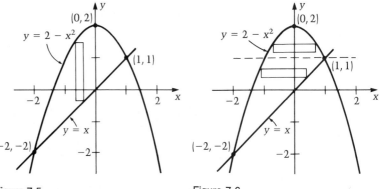

Figure 7-5   Figure 7-6

If we were to attempt horizontal elements, we would have to attack the problem in two parts. The right-hand curve is not the same above $y = 1$ as it is below $y = 1$ (see Figure 7-6). The lower section of area could be found by

$$\int_{-2}^{1} (x_R - x_L)\, dy = \int_{-2}^{1} [y - (-\sqrt{2 - y})]\, dy$$

The upper section of area is bounded by that branch of the curve $y = 2 - x^2$ where x is positive and that branch of the curve $y = 2 - x^2$ where x is negative. Thus we could find this area by

$$\int_{1}^{2} (x_R - x_L)\, dy = \int_{1}^{2} [\sqrt{2 - y} - (-\sqrt{2 - y})]\, dy$$

$$= \int_{1}^{2} 2\sqrt{2 - y}\, dy$$

We can indeed do both of these integrations; the sum of the two results is again $\frac{9}{2}$. However, it is far easier to use vertical elements. ■

Practice 3. Find the area bounded by the curves $y = 1 - x$, $2y = 2 - x$, and $y = 0$.

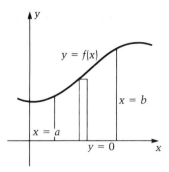

$y = f(x)$

$x = b$

$x = a$

$y = 0$

Figure 7-7

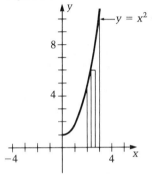

$y = x^2 + 1$

Figure 7-8

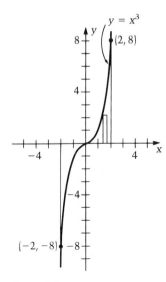

$y = x^3$

$(2, 8)$

$(-2, -8)$

Figure 7-9

In Chapter 6 we talked about finding the area under a curve $y = f(x)$ between $x = a$ and $x = b$. This is simply a special case of the general problem of finding an area bounded by curves; here one of the boundaries has the equation $y = 0$ (Figure 7-7). Using vertical elements, we have

$$A = \int_a^b (y_U - y_L) \, dx = \int_a^b [f(x) - 0] \, dx = \int_a^b f(x) \, dx$$

EXAMPLE 3.   Find the area under the curve $y = x^2 + 1$ from $x = 2$ to $x = 3$.

The area we want is shown in Figure 7-8. Again noting that the lower boundary has the equation $y = 0$, we compute the area by

$$A = \int_2^3 (y_U - y_L) \, dx = \int_2^3 [x^2 + 1 - 0] \, dx$$

$$= \frac{x^3}{3} + x \Big|_2^3 = 9 + 3 - \frac{8}{3} - 2 = \frac{22}{3} \quad \blacksquare$$

If the boundaries of the area we wish to compute change, then we must compute the area in sections, as we discussed in conjunction with Figure 7-6. Otherwise, we may be seriously misled.

EXAMPLE 4.   Find the area bounded by the curves $y = x^3$, $y = 0$, $x = -2$, and $x = 2$.

The area is shown in Figure 7-9, together with a typical element. Suppose we proceed as follows:

$$A = \int_{-2}^2 (y_U - y_L) \, dx = \int_{-2}^2 (x^3 - 0) \, dx = \frac{x^4}{4} \Big|_{-2}^2 = \frac{16}{4} - \frac{16}{4} = 0$$

Now we know the area is not zero, so we reconsider the problem. The difficulty arises because $y = x^3$ is negative between -2 and 0. If $f(x)$ takes on negative values anywhere between $x = a$ and $x = b$, it contributes negative values to the summation represented by $\int_a^b f(x) \, dx$. In this case, because of the symmetry of the curve, the negative contribution from -2 to 0 equals the positive contribution from 0 to 2.

To do the problem correctly, we note that $y = x^3$ is the upper curve only for the right section of area. In the left section of area, $y = 0$ is the upper curve. We compute the area in sections.

$$A = \int_{-2}^0 (y_U - y_L) \, dx + \int_0^2 (y_U - y_L) \, dx$$

$$= \int_{-2}^0 (0 - x^3) \, dx + \int_0^2 (x^3 - 0) \, dx = \frac{-x^4}{4} \Big|_{-2}^0 + \frac{x^4}{4} \Big|_0^2$$

$$= 0 - (-4) + 4 - 0 = 8$$

Because the area is symmetric around the *y*-axis, we could also compute the area between 0 and 2 and double it. $\blacksquare$

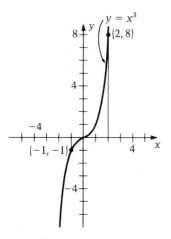

Figure 7-10

If, in Example 4, we had been asked to find the area bounded by $y = x^3$, $y = 0$, $x = -1$, and $x = 2$ (Figure 7-10), we could not have used symmetry; instead we would have to compute the area of each section separately. Also, a careless step—such as trying to do a single integration from -1 to 2—would have produced a wrong answer, but since the answer would not have been zero, we might not have realized we had an error. This illustrates again that we need to draw a careful graph and then pay attention to it.

Practice 4. Find the area bounded by $y = x^2 - 4$, $y = 0$, $x = -1$, and $x = 3$.

Exercises / Section 7-2

Exercises 1–6: Find the indicated area.

1. The area under the curve $y = 3x + 2$ from $x = 0$ to $x = 3$.

2. The area under the curve $y = 4 - x$ from $x = -1$ to $x = 3$.

3. The area under the curve $y = 2x^2$ from $x = 1$ to $x = 2$.

4. The area under the curve $y = \sqrt{x}$ from $x = 0$ to $x = 4$.

5. The area under the curve $y = \dfrac{1}{x^2}$ from $x = -3$ to $x = -1$.

6. The area under the curve $y = \sqrt{2x + 1}$ from $x = 0$ to $x = 4$.

Exercises 7–8: Find the area bounded by the given curves. Do the problem two ways, once with vertical elements and once with horizontal elements.

7. $y = x^2$, $y = 1$, $x = 3$

8. $y = 4 - x^2$, $x = 0$, $y = 3$ (first quadrant)

Exercises 9–28: Find the area bounded by the given curves.

9. $y = 3x$, $y = 2$, $x = 2$, $x = 3$

10. $y = -x$, $y = 5$, $x = -3$, $x = -2$

11. $y = x^2 + 1$, $y = 0$, $x = 0$, $x = 2$

12. $y = \sqrt{x}$, $x = 0$, $y = 4$

13. $y = 2x$, $x = 0$, $y = 1$, $y = 2$

14. $y = 2x$, $y = -3x$, $y = 4$

15. $y = x^2$, $y = x^3$

16. $y = \sqrt{x}$, $y = x^2$

17. $y = 1 + x^2$, $y = 3 - x$

18. $y = 2 - x^2$, $y = x^3$, $x = 0$

19. $y = \sqrt{x}$, $y = 2 - x$, $y = 0$

20. $y = x^2$, $y = 2 - x^2$, $y = 0$ (first quadrant)

21. $y = 2x$, $y = -x^2$, $y = -4$

22. $y = -2x^2$, $y = 2x^2 - 2$

23. $y = x^2 + \sqrt{2}$, $y = \sqrt{x + 2}$, $x = -2$

24. $y = x^3$, $y = -3x$, $y = -3$

25. $y = 2x$, $y = 0$, $x = -1$, $x = 1$

26. $y = x^3$, $y = x$

27. $y = x^3 - x$, $y = 0$

28. $y = x$, $y = 3x$, $x = -1$, $x = 2$

29. A stamping machine is computer-controlled. If it is set to cut a piece of sheet metal bounded by the curves $y = 4 - 2x^2$ and $y = 0$, find (to two decimal places) the area (in square meters) of the piece stamped out.

30. A company finds that the marginal cost of producing x units of its product is given by the equation

$$c'(x) = x^2 - 14x + 105$$

The cost function $c(x)$ is therefore given by the indefinite integral $c(x) = \int c'(x)\,dx$, and $c(b) - c(a)$ is the definite integral $\int_a^b c'(x)\,dx$, which is the area under the marginal cost curve. Find the total cost of increasing production from 20 to 30 units.

31. Chemical waste from a dump site spreads through ground water at the rate of $v = \sqrt{t} + 2$ m/day. Find the total distance it spreads between 8 days and 12 days after it is dumped by finding the area under a curve (see Exercise 30).

32. For a fluid of density ρ rotating at an angular velocity ω about the vertical axis of a cylindrical container, the pressure variation at a given depth as a function of the distance r from the axis is given by

$$\frac{dp}{dr} = \rho \omega^2 r$$

Suppose a fluid has density $\rho = -0.9 \times 10^3$ kg/m^3 and is rotating at an angular velocity of 34 rad/s^2. Find the total pressure change (in newtons per square meter, or pascals) between a distance of 0.03 m from the axis and a distance of 0.05 m from the axis by finding ω where $\omega^2 =$ area under a curve (see Exercise 30).

7-3 Volume

A solid of revolution is obtained whenever an area in the xy-plane is rotated about a fixed horizontal or vertical line in the plane. The fixed line is an axis of symmetry of the resulting solid (see Figure 7-11).

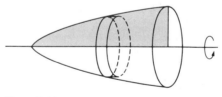

Figure 7-11

Calculus can be used to find the volume of such solids. A thin cylindrical slice, or disk, perpendicular to the axis of symmetry can be used as a typical element; such an element is shown in Figure 7-11. Cutting the shape into n such slices, adding the volumes of these slices, and then letting the number of slices become infinite gives us the volume we want. This is exactly the type of situation that calls for a definite integral.

The face of the typical slice is a circle, so the volume of the slice is the area of the circular face multiplied by the thickness of the slice. The area of the circular face is π times the square of the radius of the circle. Thus the volume of the typical element is

$$\pi(\text{radius})^2(\text{thickness})$$

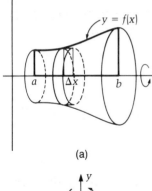

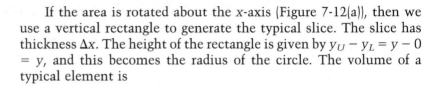

If the area is rotated about the x-axis (Figure 7-12(a)), then we use a vertical rectangle to generate the typical slice. The slice has thickness Δx. The height of the rectangle is given by $y_U - y_L = y - 0 = y$, and this becomes the radius of the circle. The volume of a typical element is

$$\pi y^2 \Delta x$$

and the volume of the solid is given by the expression

$$V = \int_a^b \pi y^2 \, dx$$

If the area is rotated about the y-axis (Figure 7-12(b)), then we use a horizontal rectangle to generate a typical slice. The slice has thickness Δy. The length of the rectangle is given by $x_R - x_L = x - 0 = x$, and this becomes the radius of the circle. The volume of the typical element is

$$\pi x^2 \Delta y$$

and the volume of the solid is given by the expression

$$V = \int_c^d \pi x^2 \, dy$$

In either case, a typical rectangle is chosen perpendicular to the axis of rotation.

Figure 7-12

EXAMPLE 1. Find the volume generated by rotating the area in the first quadrant bounded by $y = 9 - x^2$, $x = 0$, and $y = 0$ about the x-axis.

We use a vertical rectangle of height $y_U - y_L = y - 0 = y$ and thickness Δx (Figure 7-13). The volume of a slice is

$$\pi r^2 (\text{thickness}) = \pi y^2 \Delta x$$

and the definite integral is

$$V = \int_0^3 \pi y^2 \, dx = \pi \int_0^3 (9 - x^2)^2 \, dx$$

$$= \pi \int_0^3 (81 - 18x^2 + x^4) \, dx$$

$$= \pi \left(81x - 6x^3 + \frac{x^5}{5}\right)\Big|_0^3 = \frac{648}{5}\pi \quad \blacksquare$$

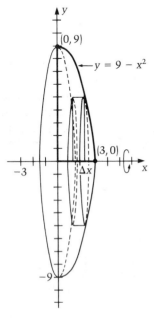

Figure 7-13

Practice 1. Find the volume generated by rotating the area bounded by $y = -2x + 2$, $x = 0$, and $y = 0$ about the x-axis.

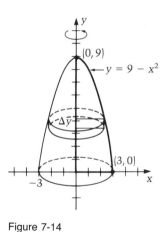

Figure 7-14

EXAMPLE 2. Find the volume generated by rotating the area of Example 1 about the y-axis.

We use a horizontal rectangle of length $x_R - x_L = x - 0 = x$ and thickness Δy (Figure 7-14). The volume of a slice is

$$\pi r^2(\text{thickness}) = \pi x^2 \Delta y$$

and the definite integral is

$$V = \int_0^9 \pi x^2 \, dy = \pi \int_0^9 (9 - y) \, dy = \pi\left(9y - \frac{y^2}{2}\right)\Big|_0^9 = \frac{81}{2}\pi \quad \blacksquare$$

Practice 2. Find the volume generated by rotating the area of Practice 1 about the y-axis.

The axis of rotation need not be the x-axis or y-axis. If it is not, this will simply change the expression for the horizontal or vertical distance which is the length of the typical rectangle (and the radius of the circular slice). If we follow the rule that vertical distance is always $y_U - y_L$ and that horizontal distance is always $x_R - x_L$, we shall have no trouble.

Figure 7-15

EXAMPLE 3. Find the volume generated by rotating the area bounded by $y = x^3$, $y = 0$, and $x = 1$ about the line $x = 1$.

From Figure 7-15, we see that a circular slice will be produced by choosing a typical rectangle to be horizontal. The length of the rectangle, following the $x_R - x_L$ rule, is $1 - x$, and this is also the radius of the circular slice; the thickness of the slice is Δy. The volume of a slice is

$$\pi r^2(\text{thickness}) = \pi(1 - x)^2 \Delta y$$

and the definite integral is

$$V = \pi \int_0^1 (1 - x)^2 \, dy = \pi \int_0^1 (1 - y^{1/3})^2 \, dy$$

$$= \pi \int_0^1 (1 - 2y^{1/3} + y^{2/3}) \, dy = \pi\left(y - \frac{3}{2}y^{4/3} + \frac{3}{5}y^{5/3}\right)\Big|_0^1$$

$$= \frac{1}{10}\pi \quad \blacksquare$$

In order to use circular slices to find the volume of a solid of revolution, we need to keep three facts in mind.

1. The typical rectangle is perpendicular to the axis of rotation.
2. The length of the rectangle is given by $y_U - y_L$ for a vertical rectangle and $x_R - x_L$ for a horizontal rectangle. In either case, the length of the rectangle is the radius of the circular slice.
3. The volume of a circular slice is

$$\pi r^2(\text{thickness})$$

Practice 3.   Find the volume generated by rotating the area in the first quadrant bounded by $y = 2x^2$, $x = 0$, and $y = 8$ about the line $y = 8$.

In all the problems we have done so far, the axis of rotation has been one of the boundaries of the area. If this is not the case, then the volume generated will have a "hole" in it. The volume can still be found by computing the volume of the larger shape and then subtracting the volume of the hole.*

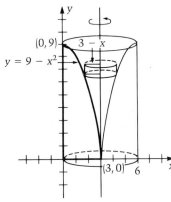

$y = 9 - x^2$

$(0, 9)$

$3 - x$

$(3, 0)$ 6

Figure 7-16

EXAMPLE 4.   Find the volume generated by rotating the area of Example 1 about the line $x = 3$.

From Figure 7-16, we see that the larger volume is a circular cylinder of radius 3 and height 9; its volume is $\pi(3)^2 9 = 81\pi$. The volume of the hole can be found using horizontal slices, where the radius is $3 - x$.

$$V_{\text{hole}} = \int_0^9 \pi(3 - x)^2 \, dy$$

$$= \pi \int_0^9 (3 - \sqrt{9 - y})^2 \, dy$$

$$= \pi \int_0^9 [9 - 6(9 - y)^{1/2} + (9 - y)] \, dy$$

$$= \pi \int_0^9 [18 - y - 6(9 - y)^{1/2}] \, dy$$

$$= \pi \left[18y - \frac{y^2}{2} + 6 \cdot \frac{2}{3}(9 - y)^{3/2}\right]\Bigg|_0^9 = \frac{27}{2}\pi$$

The volume generated is thus

$$81\pi - \frac{27}{2}\pi = \frac{135}{2}\pi \quad \blacksquare$$

*Another approach is to use a flat, circular washer as a typical element. Its volume will be $\pi[(\text{outer radius})^2 - (\text{inner radius})^2](\text{thickness})$.

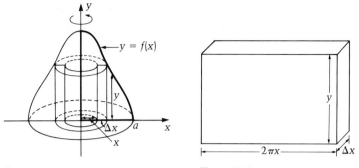

Figure 7-17

Figure 7-18

There is another approach to finding the volume of a solid of revolution. Instead of approximating the volume by a succession of circular cross sections, we can approximate it by a nested set of thin-walled cylinders, called *cylindrical shells*. In Figure 7-17, a typical cylindrical shell is generated by a vertical rectangle of height $y - 0 = y$ and thickness Δx. To find the volume of this cylindrical shell, we can imagine cutting a seam down the side, unrolling the shell, and laying it out flat (Figure 7-18). The shape is then a rectangular solid of height y, thickness Δx, and width the circumference of the circle, $2\pi r$—or in this case, $2\pi x$. Therefore, the volume of the shell is

$$(2\pi r)(\text{height})(\text{thickness}) = (2\pi x)(y)(\Delta x)$$

The volume of the solid shown in Figure 7-17 is given by the expression

$$V = \int_0^a 2\pi x y \, dx$$

If the area is to be rotated about the *x*-axis (Figure 7-19), then we use a horizontal rectangle to generate the typical shell. The cylindrical shell has radius y, length $x - 0 = x$, and thickness Δy. The volume of the solid shown is given by the expression

$$V = \int_0^c 2\pi y x \, dy.$$

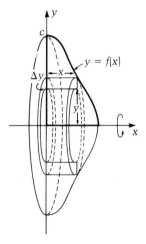

Figure 7-19

EXAMPLE 5. Find the volume of Example 2 by using shells.

We generate a cylindrical shell by using a vertical rectangle of height y and thickness Δx (Figure 7-20). Note that this rectangle is parallel to the axis of rotation. The volume of a shell is

$$(2\pi r)(\text{height})(\text{thickness}) = 2\pi x y \Delta x$$

and the definite integral is

$$V = \int_0^3 2\pi x y \, dx = 2\pi \int_0^3 x(9 - x^2) \, dx = 2\pi \left(\frac{9x^2}{2} - \frac{x^4}{4} \right) \Big|_0^3 = \frac{81}{2}\pi$$

(Of course, this is the same answer we got for Example 2.) ■

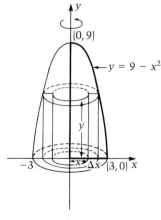

Figure 7-20

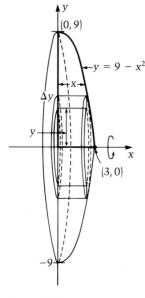

Figure 7-21

Practice 4. Find the volume of Practice 2 by using shells.

EXAMPLE 6. Find the volume of Example 1 by using shells.

A cylindrical shell is generated by a horizontal rectangle of length x and thickness Δy (Figure 7-21). Notice that this rectangle is parallel to the axis of rotation. The volume of a shell is

$$(2\pi r)(\text{length})(\text{thickness}) = 2\pi yx\Delta y$$

and the definite integral is

$$V = \int_0^9 2\pi yx \, dy = 2\pi \int_0^9 (9 - y)^{1/2}y \, dy$$

Now we are stuck. We do not know how to evaluate this integral because we do not know an antiderivative for the integrand. Our choice of integration patterns is quite limited at the moment (limited, in fact, to the power rule, which does not apply here). However, the function can be integrated; the value of the definite integral is, of course, the same as the answer for Example 1. ■

As in the case of finding plane areas by integration, the choice of the best typical element to use—circular slices or cylindrical shells—may be influenced by one of several factors. We may be able to solve the problem equally well with either approach (as in Examples 2 and 5). There may be geometric considerations that make one choice superior, or it may be—as in Example 6—that we do not know how to evaluate the definite integral resulting from one approach, so we must try the other approach.

Practice 5. Find the volume of Practice 1 by using shells.

If the axis of rotation is not the x-axis or the y-axis, this simply affects the expression for the height or length of the rectangle. The general rules $y_U - y_L$ or $x_R - x_L$ apply.

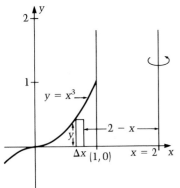

Figure 7-22

EXAMPLE 7. Use shells to find the volume generated by rotating the area bounded by $y = x^3$, $y = 0$, and $x = 1$ about the line $x = 2$.

From Figure 7-22, we see that this volume will have a hole in it. Unlike Example 4, however, where we had to subtract the hole, we can do a single calculation here by using shells. The typical rectangle is vertical with height y and thickness Δx. The radius of the cylinder, using the $x_R - x_L$ rule, is $2 - x$. The volume of a shell is

$$(2\pi r)(\text{height})(\text{thickness}) = 2\pi(2 - x)(y)\Delta x$$

and the definite integral is

$$V = \int_0^1 2\pi(2 - x)y\ dx = 2\pi \int_0^1 (2 - x)x^3\ dx$$

$$= 2\pi \int_0^1 (2x^3 - x^4)\ dx$$

$$= 2\pi \left(\frac{x^4}{2} - \frac{x^5}{5}\right)\Big|_0^1 = \frac{3}{5}\pi$$

Notice that the upper limit of integration here is 1; this is the farthest right that a typical element can be, even though the radius of the shell is measured from $x = 2$. ■

   In order to use cylindrical shells to find the volume of a solid of revolution, we need to keep three facts in mind.

   1. The typical rectangle is parallel to the axis of rotation.
   2. The height of a vertical rectangle is given by $y_U - y_L$ and the length of a horizontal rectangle by $x_R - x_L$.
   3. The volume of a cylindrical shell is

$$(2\pi r)(\text{height or length})(\text{thickness})$$

   If the rectangle is vertical, the radius is a horizontal distance, and if the rectangle is horizontal, the radius is a vertical distance.

Practice 6.   Use shells to find the volume of Practice 3.

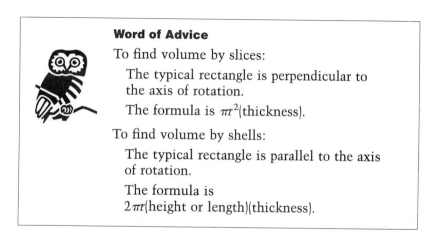

Word of Advice

To find volume by slices:

   The typical rectangle is perpendicular to the axis of rotation.

   The formula is $\pi r^2(\text{thickness})$.

To find volume by shells:

   The typical rectangle is parallel to the axis of rotation.

   The formula is $2\pi r(\text{height or length})(\text{thickness})$.

Exercises / Section 7-3

Exercises 1–8: Find the volume generated by rotating the area bounded by the given curves about the axis specified. Use the method shown.

1. $4y = 8 - x$, $x = 0$, $y = 0$; rotated about the x-axis (slices).

2. $4y = 8 - x$, $x = 0$, $y = 0$; rotated about the y-axis (slices).

3. $4y = 8 - x$, $x = 0$, $y = 0$; rotated about the y-axis (shells).

4. $4y = 8 - x$, $x = 0$, $y = 0$; rotated about the x-axis (shells).

5. $y = \sqrt{4 - x}$, $x = 0$, $y = 0$; rotated about the x-axis (slices).

6. $y = \sqrt{4 - x}$, $x = 0$, $y = 0$; rotated about the x-axis (shells).

7. $y = 1 - x^3$, $x = 0$, $y = 0$; rotated about the y-axis (shells).

8. $y = 1 - x^3$, $x = 0$, $y = 0$; rotated about the y-axis (slices).

Exercises 9–22: Find the volume generated by rotating the area bounded by the given curves about the line specified. Use whichever method (slices or shells) seems easier.

9. $y = \sqrt{4 - x}$, $x = 0$, $y = 0$; rotated about the y-axis.

10. $y = x^2 + 1$, $x = 0$, $y = 0$, $x = 3$; rotated about the x-axis.

11. $y = \sqrt{x - 1}$, $y = 0$, $x = 0$, $y = 2$; rotated about the x-axis.

12. $y = x^3 + 1$, $x = 0$, $y = 0$, $x = 2$; rotated about the y-axis.

13. $y = 1 - x$, $y = 1$, $x = 1$; rotated about the x-axis.

14. $y = 1 - x$, $y = 1$, $x = 1$; rotated about the y-axis.

15. $y = 1 - x$, $y = 1$, $x = 1$; rotated about the line $x = 1$.

16. $y = 1 - x$, $y = 1$, $x = 1$; rotated about the line $x = 2$.

17. $y = x^2$, $y = x^3$; rotated about the line $x = 1$.

18. $y = x^2$, $y = x^3$; rotated about the line $y = 1$.

19. $y = 2x^2$, $y = 0$, $x = 1$, $x = 2$; rotated about the x-axis.

20. $y = 2x^2$, $y = 0$, $x = 1$, $x = 2$; rotated about the line $x = 2$.

21. $y = x^3$, $x = 1$, $x = 2$, $y = 1$; rotated about the line $y = 1$.

22. $y = x^3$, $x = 1$, $x = 2$, $y = 1$; rotated about the line $x = 2$.

23. Derive the formula for the volume of a sphere of radius a.

24. Derive the formula for the volume of a right circular cone of height h and base radius a.

25. A spherical oil storage tank has a radius of 5 m. When the oil depth at the deepest part measures 3 m, what is the volume of oil contained in the tank?

26. A mixing bowl with circular horizontal cross-sections has a parabolic outline; its radius at the top is 10 cm and its depth at the deepest part is 18 cm. What is the volume of the bowl? (See Figure 7-23.)

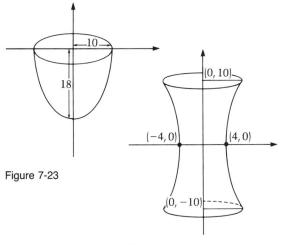

Figure 7-23

Figure 7-24

27. Cooling towers at a nuclear reactor plant are hyperbolic in outline with circular, horizontal cross sections. Find the volume of the tower shown in Figure 7-24.

28. During its manufacture, soap is stored for 18 h in vats that are inverted right circular cones. The radius at the top is 1.3 m and the depth is 1.8 m. Find (to two decimal places) the volume of such a vat.

7-4 Moments

In Figure 7-25, an object of mass m_1 is located at a point with coordinates (x_1, y_1). (A mass located at a point is called a *point mass*. Since every object occupies some space, we can only approximate point masses in real life.) Let us imagine the dotted line to represent a rigid, weightless rod (these do not really exist either!) free to rotate about the *x*-axis. If we then place the coordinate system in a horizontal plane, we know that mass m_1 would rotate about the *x*-axis as it dropped. This tendency to rotate is measured by the *first moment* of the mass. The **first moment** of a point mass with respect to an axis is the product of the mass and the directed distance of the mass from the axis. (The **directed distance** of a point from the *x*-axis is the *y*-coordinate of the point; the **directed distance** of a point from the *y*-axis is the *x*-coordinate of the point.) Thus the first moment of mass m_1 with respect to the *x*-axis is $m_1 y_1$. The first moment of m_1 with respect to the *y*-axis is $m_1 x_1$. That the tendency of a mass to rotate about an axis should involve the distance from the axis is illustrated by opening a heavy book; it is easier to open the book if you stick your thumb in near the outside edge than if you stick your thumb in near the binding.

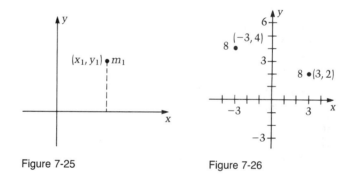

Figure 7-25   Figure 7-26

Given more than one mass, the first moment of the *system* of point masses with respect to an axis is the sum of the separate first moments. Directed distance, which can be either positive or negative, is used in computing first moments because the direction of rotation changes. Thus in the system shown in Figure 7-26, the first moment of the system with respect to the *y*-axis is given by

$$8(-3) + 8(3) = 0$$

There is no tendency for the system to rotate about the *y*-axis; it is "balanced" with respect to the *y*-axis. The first moment with respect to the *x*-axis is

$$8(4) + 8(2) = 48$$

Suppose we have computed the first moments of a system with respect to the *x*- and *y*-axes. Then suppose we find a point such that if the mass of the entire system were concentrated at that point and both moments recomputed, their values would be the same as before. Such a point is called the **center of mass**, or **centroid**, of the system. For the system of Figure 7-26, the center of mass is located at (0, 3). To see this, suppose we put the mass of the entire system (8 + 8 = 16 units) at (0, 3). Then the first moment with respect to the y-axis is $16 \cdot 0 = 0$, and the first moment with respect to the x-axis is $16 \cdot 3 = 48$. These are the same quantities as before.

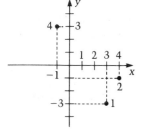

Figure 7-27

EXAMPLE 1. Find the first moment of the system shown in Figure 7-27 with respect to each axis, and find the coordinates of the center of mass of the system.

The first moment of the system with respect to the y-axis, M_y, is

$$4(-1) + 2(4) + 1(3) = 7$$

The first moment of the system with respect to the x-axis, M_x, is

$$4(3) + 2(-1) + 1(-3) = 7$$

The entire mass of the system is $4 + 2 + 1 = 7$ units. If 7 units were placed at $(\bar{x}, \bar{y})$ in such a way that M_y and M_x for this single-point mass system would have the same value as before, then

$$7 \cdot \bar{x} = M_y = 7$$

and

$$7 \cdot \bar{y} = M_x = 7$$

Solving for $(\bar{x}, \bar{y})$, we find the centroid is located at (1, 1). ∎

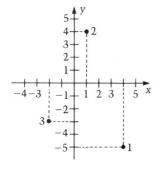

Figure 7-28

Practice 1. Find the coordinates of the center of mass of the system shown in Figure 7-28.

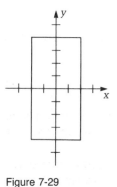

Figure 7-29

Now consider an area that has constant density k throughout. The mass of the area is then kA, where A is the value of the area. We do an easy case first, the rectangle of Figure 7-29. Each point of mass in the figure contributes to the first moment of the area with respect to the y-axis, but by symmetry, its contribution is canceled by a corresponding point mass across the y-axis. Therefore, the total first moment with respect to the y-axis. M_y, is zero. Similarly, $M_x = 0$. The centroid of the area must therefore be located at $(\bar{x}, \bar{y})$, where

$$kA \cdot \bar{x} = 0$$
$$kA \cdot \bar{y} = 0$$

or $(\bar{x}, \bar{y}) = (0, 0)$. The point (0, 0) is the geometric center of the rectangle. It is also the point upon which the rectangle could be

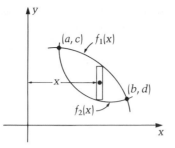

Figure 7-30

balanced. The centroid of any area of constant density is its geometric center, or "balancing point," but if the area is not symmetric, then we must use integration to find this point.

If we consider any area of constant density k, we find M_y, its first moment with respect to the y-axis, as follows. We use vertical rectangles for typical elements (see Figure 7-30). The mass of the element is

$$k(y_U - y_L)\Delta x$$

and its centroid is at its center, a distance x from the y-axis. Therefore, the first moment of the element with respect to the y-axis is

$$xk(y_U - y_L)\Delta x$$

Summing the moments of all such elements and taking the limit of the sum, we get the first moment of the area:

$$M_y = \int_a^b xk(y_U - y_L)\ dx = k \int_a^b x(y_U - y_L)\ dx$$

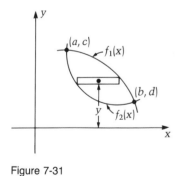

Figure 7-31

To compute M_x, the first moment with respect to the x-axis, we use a similar argument with horizontal rectangles (Figure 7-31) and get

$$M_x = \int_d^c yk(x_R - x_L)\ dy = k \int_d^c y(x_R - x_L)\ dy$$

The total mass of the area is

$$M = \int_a^b k(y_U - y_L)\ dx = k \int_a^b (y_U - y_L)\ dx$$

We then find the coordinates $(\bar{x}, \bar{y})$ of the centroid by solving the equations

$$M\bar{x} = M_y$$
$$M\bar{y} = M_x$$
(1)

The constant k is a factor on both sides of each equation, so it can be divided out. This means we never need to know k and can ignore it from now on.

In the system of equations (1), there are three distinct quantities to be computed, M, M_x, and M_y. Each requires it own integral.

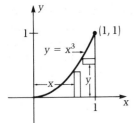

Figure 7-32

EXAMPLE 2. Find the coordinates of the centroid of the area bounded by the curves $y = x^3$, $y = 0$, and $x = 1$.

The area, together with typical rectangles, is shown in Figure 7-32. Performing the integration required for the three quantities M, M_y, and M_x, we get

$$M = \int_0^1 y \, dx = \int_0^1 x^3 \, dx = \frac{x^4}{4}\Big|_0^1 = \frac{1}{4}$$

$$M_y = \int_0^1 xy \, dx = \int_0^1 x \cdot x^3 \, dx = \int_0^1 x^4 \, dx = \frac{x^5}{5}\Big|_0^1 = \frac{1}{5}$$

$$M_x = \int_0^1 y(1 - y^{1/3}) \, dy = \int_0^1 (y - y^{4/3}) \, dy = \frac{y^2}{2} - \frac{3y^{7/3}}{7}\Big|_0^1$$

$$= \frac{1}{2} - \frac{3}{7} = \frac{1}{14}$$

To find the coordinates $(\bar{x}, \bar{y})$ of the centroid, we solve the following equations.

$$M\bar{x} = M_y \qquad M\bar{y} = M_x$$
$$\tfrac{1}{4}\bar{x} = \tfrac{1}{5} \qquad \tfrac{1}{4}\bar{y} = \tfrac{1}{14}$$
$$\bar{x} = \tfrac{4}{5} \qquad \bar{y} = \tfrac{4}{14} = \tfrac{2}{7}$$

The centroid is located at $(\tfrac{4}{5}, \tfrac{2}{7})$. ∎

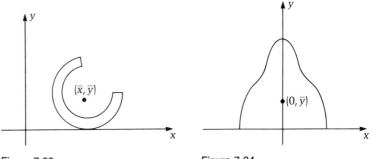

Figure 7-33   Figure 7-34

Notice that the answer in Example 2 seems to make sense; namely, the centroid is fairly close to the x-axis and to the line $x = 1$, where most of the mass is concentrated. Sometimes there can be surprising results. For example, the center of mass of an area like the one shown in Figure 7-33 is not even inside the area being considered.

We can sometimes use geometric ideas to reduce the amount of work required. For the area of Figure 7-34, we know from the symmetry involved that the centroid lies on the y-axis, with $\bar{x} = 0$. We need to find only $\bar{y}$.

Practice 2. Find the coordinates of the centroid of the area bounded by $y = 3 - 3x$, $y = 3$, and $x = 1$.

We can also locate centroids for solids of revolution. In solids of constant density, we know from the symmetry of the figure that the centroid lies along the axis of revolution. (We can also ignore the density, just as we did for area.) To compute $\bar{x}$ in the solid of Figure 7-35, we need to know the first moment M_{yz} of the solid with respect to the yz-plane. We choose a slice parallel to the yz-plane as a typical element of volume; by symmetry of the slice, its centroid

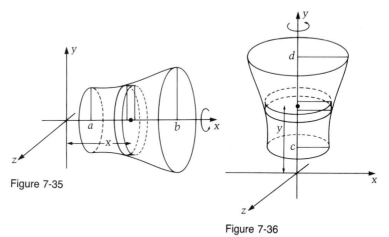

Figure 7-35

Figure 7-36

is located at $(x, 0)$, a distance x from the yz-plane. Therefore, the first moment of the slice with respect to the yz-plane is

$$(\text{distance})(\pi r^2)(\text{thickness})$$

or

$$x\pi y^2 \Delta x$$

and the first moment of the volume with respect to the yz-plane is

$$M_{yz} = \int_a^b x\pi y^2 \, dx$$

The mass M is the volume

$$M = \int_a^b \pi y^2 \, dx$$

and $\bar{x}$ is the solution to the equation

$$M\bar{x} = M_{yz}$$

For a solid of revolution about the y-axis (Figure 7-36), we find $\bar{y}$ by solving

$$M\bar{y} = M_{xz}$$

where

$$M = \int_c^d \pi x^2 \, dy$$

and

$$M_{xz} = \int_c^d y\pi x^2 \, dy$$

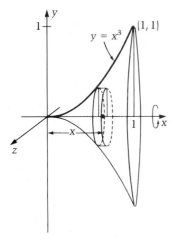

Figure 7-37

EXAMPLE 3.   Find the coordinates of the centroid of the volume generated by rotating the area of Example 2 about the x-axis.

From symmetry (see Figure 7-37), we know $\bar{y} = 0$. To find $\bar{x}$, we need the mass (volume), which is

$$M = \int_0^1 \pi y^2 \, dx = \pi \int_0^1 (x^3)^2 \, dx = \pi \int_0^1 x^6 \, dx$$

$$= \frac{\pi x^7}{7}\Big|_0^1 = \frac{\pi}{7}$$

We also need M_{yz}, which is

$$M_{yz} = \int_0^1 x \pi y^2 \, dx = \pi \int_0^1 x(x^3)^2 \, dx = \pi \int_0^1 x^7 \, dx$$

$$= \frac{\pi x^8}{8}\Big|_0^1 = \frac{\pi}{8}$$

Then we solve

$$M\bar{x} = M_{yz}$$

$$\frac{\pi}{7}\bar{x} = \frac{\pi}{8}$$

$$\bar{x} = \frac{7}{8}$$

The center of mass is at $(\frac{7}{8}, 0)$.   ■

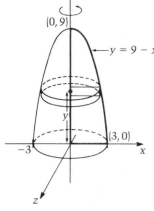

Figure 7-38

EXAMPLE 4.   Find the coordinates of the centroid of the volume generated by rotating the area bounded by $y = 9 - x^2$, $y = 0$, and $x = 0$ about the y-axis.

The volume is shown in Figure 7-38. We know that $\bar{x} = 0$. The quantities we need in order to find $\bar{y}$ are

$$M = \int_0^9 \pi x^2 \, dy = \pi \int_0^9 (9 - y) \, dy = \pi\left(9y - \frac{y^2}{2}\right)\Big|_0^9 = \frac{81}{2}\pi$$

$$M_{xz} = \int_0^9 y\pi x^2 \, dy = \pi \int_0^9 y(9 - y) \, dy = \pi \int_0^9 (9y - y^2) \, dy$$

$$= \pi\left(\frac{9y^2}{2} - \frac{y^3}{3}\right)\Big|_0^9 = \frac{243}{2}\pi$$

Finally,

$$M\bar{y} = M_{xz}$$

$$\frac{81}{2}\pi\bar{y} = \frac{243}{2}\pi$$

$$\bar{y} = 3$$

The coordinates of the centroid are $(0, 3)$.   ■

Practice 3. Find the coordinates of the centroid of the volume generated by rotating the area bounded by $y = -x + 1$, $y = 0$, and $x = 0$ about the x-axis.

We noted in the beginning of this section that the first moment of a point mass with respect to an axis is the product of the mass and the directed distance of the mass from the axis. The **second moment**, or **moment of inertia**, of a point mass with respect to an axis is the product of the mass and the square of its distance from the axis. The second moment of a system is the sum of the individual second moments. The moment of inertia measures the tendency of the system to resist a change in rotational motion about the axis.

The difference between first and second moments can be illustrated by thinking of a rotating ice skater. When the skater pulls in his or her arms, the second moment of the system is reduced and the skater's rotational speed increases. (Remember that the second moments of a point on the left and a point symmetric to it on the right are added together, because the distances are squared.) The centroid of the figure has not changed. (The first moments of two symmetric points have a sum of zero.)

We compute the second moment of an area with respect to an axis much the same way we computed the first moment, except that the factor representing the distance of the element from the axis is squared.

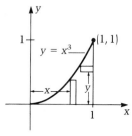

$y = x^3$

$(1, 1)$

Figure 7-39

EXAMPLE 5. Find the moment of inertia of the area of Example 2 with respect to the y-axis and with respect to the x-axis, denoted by I_y and I_x, respectively.

From Figure 7-39, using vertical elements,

$$I_y = \int_0^1 x^2 ky \, dx = k \int_0^1 x^2 (x^3) \, dx = k \int_0^1 x^5 \, dx$$

$$= \frac{kx^6}{6} \Big|_0^1 = \frac{k}{6}$$

For I_x, we use horizontal elements and get

$$I_x = \int_0^1 y^2 k(1 - x) \, dy = k \int_0^1 y^2 (1 - y^{1/3}) \, dy$$

$$= k \int_0^1 (y^2 - y^{7/3}) \, dy = k \left(\frac{y^3}{3} - \frac{3y^{10/3}}{10} \right) \Big|_0^1 = \frac{k}{30} \quad \blacksquare$$

Practice 4. Find I_y and I_x for the area of Practice 2.

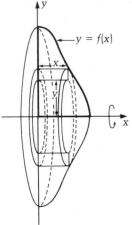

Figure 7-40

Moments of inertia can also be computed for solids of revolution. Because we do not have the "canceling-out" effect of symmetry that worked for us in computing first moments, we need to choose as a typical volume element one for which every point is equidistant from the axis. This suggests using shells; then we can find the second moment of a solid of revolution with respect to its axis of rotation as follows. In Figure 7-40, every particle in the cylindrical shell is a distance y from the x-axis. The second moment of the shell with respect to the x-axis is, therefore,

$$y^2k2\pi yx\Delta y$$

and, for the entire solid,

$$I_x = \int_0^c y^2k(2\pi y)x \; dy$$

Similarly, in Figure 7-41, the second moment with respect to the y-axis is

$$I_y = \int_0^a x^2k(2\pi x)y \; dx$$

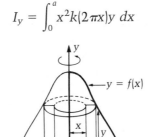

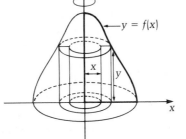

Figure 7-41

EXAMPLE 6. Find the moment of inertia with respect to the y-axis of the volume generated by revolving the area bounded by $y = x^2$, $x = 0$, and $y = 4$ about the y-axis.

In order to generate a shell, we use a vertical rectangle at a distance x from the y-axis (Figure 7-42). The moment of inertia of the shell is

$$(\text{distance})^2(\text{density})(2\pi r)(\text{height})(\text{thickness})$$

or

$$(x^2)k(2\pi x)(4 - y)\Delta x$$

and for the entire solid,

$$I_y = \int_0^2 (x^2)k(2\pi x)(4 - y) \; dx = 2\pi k \int_0^2 x^3(4 - x^2) \; dx$$

$$= 2\pi k \int_0^2 (4x^3 - x^5) \; dx = 2\pi k\left(x^4 - \frac{x^6}{6}\right)\bigg|_0^2 = \frac{32}{3}\pi k \quad \blacksquare$$

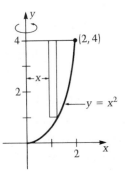

Figure 7-42

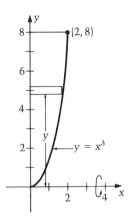

Figure 7-43

EXAMPLE 7. Find I_x for the volume generated by rotating the area bounded by $y = x^3$, $x = 0$, and $y = 8$ about the x-axis.

From Figure 7-43, we see that

$$I_x = \int_0^8 y^2 k(2\pi y)x \, dy = 2\pi k \int_0^8 y^3(y^{1/3}) \, dy$$

$$= 2\pi k \int_0^8 y^{10/3} \, dy = 2\pi k \left. \frac{3y^{13/3}}{13} \right|_0^8 = \frac{3}{13}\pi k2^{14} \quad \blacksquare$$

Practice 5. Find I_x for the volume of Practice 3.

Exercises / Section **7-4**

Exercises 1–4: Find the coordinates of the center of mass of the system of point masses described.

1. 2 units at $(2, 1)$, 3 units at $(0, 5)$, 4 units at $(-1, -3)$.

2. 3 units at $(1, -3)$, 1 unit at $(4, 2)$, 4 units at $(-3, 2)$, 1 unit at $(2, -1)$.

3. 2 units at $(2, 5)$, 3 units at $(-6, 1)$, 5 units at $(-2, -1)$, 3 units at $(-1, 3)$.

4. 3 units at $(2, 0)$, 4 units at $(1, 5)$, 8 units at $(-1, 1)$.

Exercises 5–10: Find the coordinates of the centroid of the area bounded by the given curves.

5. $y = 2x$, $y = 0$, $x = 2$

6. $y = \sqrt{x}$, $x = 0$, $y = 2$

7. $y = x^3$, $x = 0$, $y = -8$

8. $y = x^2$, $y = 5$

9. $y = \sqrt{x}$, $y = x^2$

10. $y = x^2$, $y = x^3$

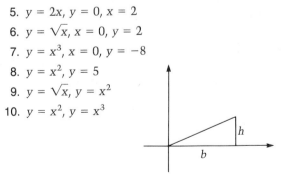

Figure 7-44

11. Find the coordinates of the centroid of a right triangle with two legs b and h, as in Figure 7-44.

12. Locate the centroid of a semicircle of radius a.

13. Find the coordinates of the center of mass of the volume generated by rotating the area bounded by $y = 3x$, $x = 0$, and $y = 6$ about the y-axis.

14. Find the coordinates of the center of mass of the volume generated by rotating the area bounded by $y = 2\sqrt{x}$, $y = 0$, and $x = 4$ about the x-axis.

15. Find the coordinates of the centroid of the volume generated by rotating the area bounded by $y = x^3$, $y = 0$, $x = 1$, and $x = 2$ about the x-axis.

16. Locate the centroid of the volume generated by rotating the area bounded by $y = 1/x^3$, $x = 0$, $y = 1$, and $y = 3$ about the y-axis.

17. Locate the centroid of a right circular cone of base radius a and height h.

18. Find the coordinates of the centroid of a hemisphere of radius a.

Exercises 19–36: Assume a constant density k.

19. Find I_y and I_x for the area of Exercise 5.

20. Find I_y and I_x for the area of Exercise 6.

21. Find I_y and I_x for the area of Exercise 7.

22. Find I_y and I_x for the area of Exercise 8.

23. Find I_y for the volume of Exercise 13.

24. Find I_x for the volume of Exercise 14.

25. Find the moment of inertia with respect to the x-axis of the volume generated by revolving the area bounded by $y = x$ and $y = x^2$ about the x-axis.

26. Find the moment of inertia with respect to the y-axis of the volume generated by revolving the area bounded by $y = x$ and $y = x^2$ about the y-axis.

Exercises 27–36: The **radius of gyration** of an area or volume with respect to an axis is a distance R such that if the entire mass were placed at distance R from the axis, the moment of inertia with respect to that axis would stay the same. To find the radius of gyration of a mass M with respect to the y-axis, R_y, we solve the equation

$$MR_y^2 = I_y$$

To find the radius of gyration of a mass M with respect to the x-axis, R_x, we solve the equation

$$MR_x^2 = I_x$$

27. Find R_y and R_x for the area of Exercise 5.

28. Find R_y and R_x for the area of Exercise 6.

29. Find R_y and R_x for the area of Exercise 7.

30. Find R_y and R_x for the area of Exercise 8.

31. Find the radius of gyration with respect to the y-axis of the area bounded by $y = \sqrt{x}$ and $y = x^2$.

32. Find the radius of gyration with respect to the x-axis of the area bounded by $y = x^2$ and $y = x^3$.

33. Find R_y for the volume of Exercise 13.

34. Find R_x for the volume of Exercise 14.

35. Find the radius of gyration with respect to the x-axis of the volume generated by revolving the area bounded by $y = x$ and $y = x^2$ about the x-axis.

36. Find the radius of gyration with respect to the y-axis of the volume generated by revolving the area bounded by $y = x$ and $y = x^2$ about the y-axis.

7-5 Fluid Pressure and Force

Suppose that a plate is suspended below the surface of a liquid, as shown in Figure 7-45. The liquid exerts a force upon the plate, and the force increases as the depth below the surface increases. If we choose a typical rectangular element parallel to the surface, then we may consider its depth to be everywhere the same. The element is at location y and its depth is the vertical distance from the surface of the liquid to the element. Using $y_U - y_L$, the depth is $0 - y = -y$. (Because the plate is below the y-axis, y is negative and $-y$ is positive.)

The force on the element is the product of the weight w per unit volume of the liquid, the depth of the element, and the area of the element. The force on the element in Figure 7-45 is, therefore,

$$w(\text{depth})(\text{area}) = w(-y)(x_R - x_L)\Delta y$$

Summing the forces on all such elements and taking the limit of the sum, we get the force F on the entire plate:

$$F = \int_c^d w(-y)(x_R - x_L)\,dy$$

Notice that the limits of integration run from bottom to top, as is usual when we integrate with respect to y. Also, F represents the force on one side of the submerged plate, which is balanced by an equal force on the other side.

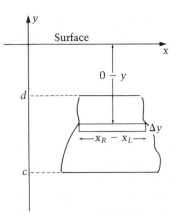

Figure 7-45

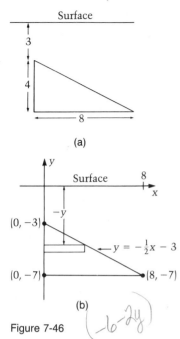

Surface

(a)

Figure 7-46

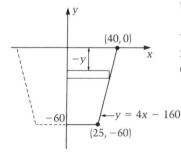

Surface

$(0, -3)$

$-y$

$y = -\frac{1}{2}x - 3$

$(0, -7)$

$(8, -7)$

(b)

The SI unit of weight or force is the newton. As a comparison with the British system of units, 1 lb is about 4.448 N. Although it is popular to refer to the "weight" of an object in kilograms, the kilogram is actually a unit of mass.

EXAMPLE 1. Find the force on a plate submerged in water, as shown in Figure 7-46(a). For water the weight per unit volume, w, is 9800 N/m^3. We use a coordinate system as shown in Figure 7-46(b). This allows us to determine the equation of the line because we know two points that the line goes through. Then the force is

$$F = \int_{-7}^{-3} w(-y)x \, dy = w \int_{-7}^{-3} -y(-6 - 2y) \, dy$$

$$= w \int_{-7}^{-3} (6y + 2y^2) \, dy$$

$$= w\left(3y^2 + \frac{2y^3}{3}\right)\Big|_{-7}^{-3} = \frac{272}{3}w = \frac{272}{3}(9800) = \frac{2,665,600}{3} \, N \ \blacksquare$$

EXAMPLE 2. The end of a settling tank in a city sanitation system is the shape of a trapezoid with upper edge 80 m long, lower edge 50 m long, and depth 60 m. Find the force on the end of the tank if the tank is full of water.

We set up a coordinate system as in Figure 7-47. Again we can write the equation of the straight line through two points. We will make use of the symmetry of the figure by finding the force on half of it and doubling the result. The force is

$$F = 2\int_{-60}^{0} w(-y)x \, dy = 2w \int_{-60}^{0}(-y)\left(\frac{y}{4} + 40\right) dy$$

$$= 2w \int_{-60}^{0}\left(\frac{-y^2}{4} - 40y\right) dy = 2w\left(\frac{-y^3}{12} - 20y^2\right)\Big|_{-60}^{0}$$

$$= 180,000w = 180,000(9800) = 1.764 \times 10^9 \, N \ \blacksquare$$

y

$(40, 0)$

x

$-y$

-60

$y = 4x - 160$

$(25, -60)$

Figure 7-47

Practice 1. Find the force on a rectangular plate of width $\frac{1}{2}$ m and height 1 m if it is submerged vertically in water with its upper edge 2 m below the surface.

Exercises / Section **7-5**

1. A square metal plate 3 m on an edge is submerged in water with one edge in the surface. Find the force on the plate.

2. Find the force on the plate of Exercise 1 if it is submerged with its upper edge 2 m below the surface of the water.

3. A rectangular plate 2 m high and 4 m wide is submerged in water with one edge in the surface. Find the force on the plate.

4. Find the force on the plate of Exercise 3 if it is submerged with its upper edge 1.5 m below the surface of the water.

5. The end of a trough is shaped like an isosceles triangle pointing downward; its horizontal upper edge is 1 m and its depth is 1 m. Find the force on the end of the trough when the trough is full of water.

6. Find the force on the end of the trough of Exercise 5 when the water in the trough only measures $\frac{1}{2}$ m deep at the deepest point.

7. The end of an oil tank car is a circle of radius 2 m. Find the force on the end of the car if it is half filled with oil of weight density 9066 N/m^3.

8. What is the force on the end of the car of Exercise 7 if the oil only measures 1 m deep at the deepest point?

9. A cylindrical tank is lying on its side; the radius of the circular cross section is 3 m. Find the force on one end of the tank if the tank contains water 1 m deep at the deepest point covered with a layer of oil (weight density: 9066 N/m^3) 0.5 m deep.

10. The end of a boat has a parabolic shape opening upward. The horizontal upper edge has a width of 2 m, and the depth at the deepest part is 4 m. What is the force on the end of the boat if it is submerged in seawater (weight density: 10,045 N/m^3) with its upper edge in the surface? (*Hint:* Set up a coordinate system with the vertex of the parabola at the origin.)

7-6   Other Applications

In this section we consider four more applications of the definite integral. Each of the four, as have all previous applications of the definite integral, depends upon being able to express the quantity we wish to compute as the limit of a sum of that quantity for typical elements. The four applications are work done by a variable force, the length of a curve whose equation is known, the surface area of a solid of revolution, and the area within a curve whose equation is given in polar coordinates.

If a constant force F applied to an object moves the object through a distance d, the **work** W done in moving the object is defined to be

$$W = Fd$$

Thus, for example, lifting an object that weighs 1 N through a vertical distance of 1 m results in 1 N · m, or 1 J, of work being done. According to the definition, one could hold a heavy object all day without doing any work!

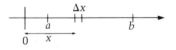

Suppose an object is being moved along the x-axis, as in Figure 7-48, and that the force F required to move the object is not constant, but varies with x. Thus $F = F(x)$. If we take as a typical element a short distance Δx, we may assume the force $F(x)$ to have a constant value over that interval. The work done to move the object through the distance Δx is, therefore, $F(x)\Delta x$. Adding all such bits of work and taking the limit, the work done to move the object from a to b is

Figure 7-48

$$W = \int_a^b F(x) \, dx$$

EXAMPLE 1. According to Hooke's law, the force required to stretch a spring beyond its natural length is proportional to the distance it is stretched. Thus $F(x) = kx$, where the constant k depends on the particular spring and is called the **spring constant**. If a spring of natural length 0.1 m requires a force of 2 N to stretch it 0.02 m beyond its natural length, find the work done in stretching it 0.05 m beyond its natural length.

Using the given information in the equation $F(x) = kx$, we get

$$2 = k(0.02)$$

or

$$k = 100 \text{ N/m}$$

The work done is

$$W = \int_0^{0.05} F(x) \, dx = \int_0^{0.05} 100x \, dx = 50x^2 \Big|_0^{0.05} = 0.125 \text{ J} \quad \blacksquare$$

Practice 1. How much work is done on the spring of Example 1 in stretching it from a length of 0.1 m to a length of 0.18 m?

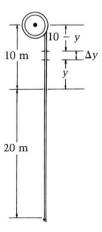

Figure 7-49

EXAMPLE 2. A 30-m rope weighing 8 N/m is hung from a winch. Find the work done in winding up 10 m of the rope.

From Figure 7-49, we see that the bottom 20 m of rope all moves a distance of 10 m, so we can treat this part of the problem separately. The work to move the bottom 20 m is

$$(\text{weight})(\text{distance}) = 8(20)(10) = 1600 \text{ J}$$

Within the upper 10 m of rope, a small section of rope of length Δy (and weight $8\Delta y$) moves a distance $10 - y$. The work done to lift this section is therefore $(10 - y)(8\Delta y)$. Writing the definite integral, the work for the upper part of the rope is

$$\int_0^{10} (10 - y)8 \, dy = 80y - 4y^2 \Big|_0^{10} = 400 \text{ J}$$

The total amount of work done is 1600 J + 400 J = 2000 J. $\quad \blacksquare$

Practice 2. For the rope of Example 2, what is the work required to wind up 15 m of the rope?

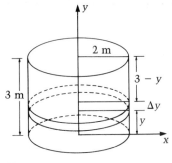

Figure 7-50

EXAMPLE 3. A cylindrical tank of radius 2 m and height 3 m is filled with water (which weighs 9800 N/m³). Find the work required to pump all the water out of the top of the tank.

In the case of the rope problem, we moved a section of length; here we move a "section" of volume. The volume of a typical horizontal slice (Figure 7-50) is $\pi(2)^2\Delta y$, and the work required to move it out the top of the tank is

$$(\text{weight})(\text{distance}) = 9800[\pi(2)^2\Delta y](3 - y).$$

The total work required is

$$W = \int_0^3 9800(4\pi)(3 - y) \, dy = 19,600\pi \text{ J} \quad \blacksquare$$

We can also use integration to find the length of a section of a curve. If $y = f(x)$ is the equation of the curve, then the length s of the curve from the point where $x = a$ to the point where $x = b$, called the *length of arc*, is given by

$$s = \int_a^b \sqrt{1 + (y')^2} \, dx \tag{1}$$

While we shall not show how this equation is derived, its basis—as in all uses of the definite integral—lies in breaking up the desired quantity into a sum of approximating quantities.

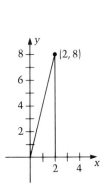

Figure 7-51

EXAMPLE 4. Find the length of arc of the curve $y = 4x$ from $x = 0$ to $x = 2$.

From Figure 7-51, we see that the length of arc requested is the hypotenuse of the right triangle. Its length, by the Pythagorean theorem, is

$$\sqrt{2^2 + 8^2} = \sqrt{68} = \sqrt{4 \cdot 17} = 2\sqrt{17}$$

Now using Equation (1) with $y = 4x$ and $y' = 4$,

$$s = \int_0^2 \sqrt{1 + 4^2} \, dx = \int_0^2 \sqrt{17} \, dx$$

$$= \sqrt{17}x \Big|_0^2 = 2\sqrt{17} \quad \blacksquare$$

EXAMPLE 5. Find the length of arc of the curve $y = \frac{2}{3}x^{2/3}$ from $x = 0$ to $x = 1$.

Using Equation (1) with $y' = \frac{4}{9}x^{-1/3}$, we get

$$s = \int_0^1 \sqrt{1 + \left(\frac{4}{9}x^{-1/3}\right)^2} \, dx = \int_0^1 \sqrt{1 + \frac{16}{81}x^{-2/3}} \, dx$$

$$= \int_0^1 \sqrt{1 + \frac{16}{81x^{2/3}}} \, dx$$

$$= \int_0^1 \sqrt{\frac{81x^{2/3} + 16}{81x^{2/3}}} \, dx = \int_0^1 \frac{1}{9}x^{-1/3}\sqrt{81x^{2/3} + 16} \, dx$$

We integrate this using the power rule with $u = 81x^{2/3} + 16$ and $du = 54x^{-1/3}$.

$$s = \tfrac{1}{9} \cdot \tfrac{1}{54} \int_0^1 54x^{-1/3}(81x^{2/3} + 16)^{1/2}\, dx$$

$$= \tfrac{1}{9} \cdot \tfrac{1}{54} \cdot \tfrac{2}{3}(81x^{2/3} + 16)^{3/2} \Big|_0^1$$

$$= \tfrac{1}{729}[(97)^{3/2} - (16)^{3/2}] = 1.223 \quad \blacksquare$$

Practice 3. Find the length of arc of the curve $y = x^{3/2}$ from $x = 0$ to $x = 12$.

Now let $y = f(x)$ be the equation of a curve; if we take the area beneath the curve from $x = a$ to $x = b$ and rotate it about the x-axis, we generate a solid of revolution. The surface area A of this solid is given by the equation

$$A = 2\pi \int_a^b y\sqrt{1 + (y')^2}\, dx \qquad (2)$$

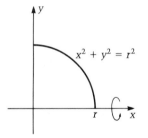

Figure 7-52

EXAMPLE 6. Verify that the surface area of a sphere of radius r is $A = 4\pi r^2$.

The solid of revolution generated by the area of Figure 7-52 is a hemisphere of radius r; we need only show that its surface area is $2\pi r^2$. Using Equation (2) with $y = \sqrt{r^2 - x^2}$ and $y' = \tfrac{1}{2}(r^2 - x^2)^{-1/2}(-2x)$, we get

$$A = 2\pi \int_0^r (\sqrt{r^2 - x^2})\sqrt{1 + x^2(r^2 - x^2)^{-1}}\, dx$$

$$= 2\pi \int_0^r \sqrt{r^2 - x^2}\,\sqrt{1 + \frac{x^2}{r^2 - x^2}}\, dx$$

$$= 2\pi \int_0^r \sqrt{r^2 - x^2}\,\sqrt{\frac{r^2 - x^2 + x^2}{r^2 - x^2}}\, dx = 2\pi \int_0^r r\, dx$$

$$= 2\pi r x \Big|_0^r = 2\pi r^2 \quad \blacksquare$$

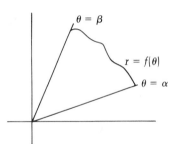

Figure 7-53

Practice 4. Find the surface area of the solid of revolution generated by rotating the area bounded by $y = x$ from $x = 0$ to $x = 1$ about the x-axis.

Our final application of the definite integral concerns finding the area enclosed within a section of curve whose equation is given in polar coordinates. In Figure 7-53, the area is bounded by the curves $r = f(\theta)$, $\theta = \alpha$, and $\theta = \beta$. When we found the area beneath a

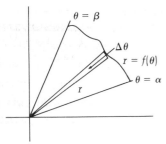

Figure 7-54

curve given in rectangular coordinates, we used rectangles as typical elements. Here an appropriate typical element is a small circular sector of radius r and central angle $\Delta\theta$ (see Figure 7-54). The area of such a sector is $\frac{1}{2}r^2\Delta\theta$ (see Exercise 39, Section 3-2). The total area we want is then obtained by adding the areas of the typical elements and letting the number of elements become infinite. The total area A is, therefore,

$$A = \int_\alpha^\beta \tfrac{1}{2}r^2 \, d\theta \tag{3}$$

EXAMPLE 7. Find the area bounded by the curves $r = 3$, $\theta = 0$, and $\theta = \pi/2$.

From Figure 7-55, we see that the area is a quarter circle, of which the radius is 3, and the answer is, therefore, $A = \frac{1}{4}\pi r^2 = \frac{9}{4}\pi$. Now, using Equation (3),

$$A = \int_0^{\pi/2} \frac{1}{2} \cdot 9 \, d\theta = \frac{9}{2}\theta \Big|_0^{\pi/2} = \frac{9\pi}{4} \quad \blacksquare$$

Figure 7-55

Exercises / Section 7-6

1. A spring of natural length 0.4 m stretches an additional 0.2 m when a weight of 10 N is hung on it. What is the work required to stretch the spring 0.3 m beyond its natural length?

2. Find the work done in stretching the spring of Exercise 1 from a length of 0.5 m to a length of 0.6 m.

3. A certain spring stretches 0.5 m under a weight of 80 N. Find the work done by attaching a weight of 20 N to the spring.

4. It requires 30 J of work to stretch a particular spring from a length of 0.5 m to a length of 0.6 m, and another 50 J of work to stretch it from a length of 0.6 m to a length of 0.7 m. Find the spring constant and the natural length of the spring.

5. Find the work done in winding up a 20 m rope weighing 12 N/m.

6. What is the work done in winding up 5 m of the rope of Exercise 5?

7. A cable 10 m long and weighing 15 N/m is supporting a load of plywood weighing 2000 N. Find the work done to raise the load 4 m by winding up the cable.

8. A uniform 15 m cable weighing 300 N is hung from a winch. The cable supports a load of 800 N. Find the work required to raise the load 8 m.

9. A cylindrical cooling tank has radius 4 m and height 6 m. If the tank is full of water, find the work required to pump all the water out the top.

10. Find the work required to empty the tank of Exercise 9 if it is initially only half full.

11. A vat in a chemical plant is in the shape of a circular cone with its vertex at the bottom; the radius of the top of the vat is 1 m and the depth is 1.4 m (see Figure 7-56). At the end of a certain processing operation, the vat is filled to a depth of 0.8 m with a sludge weighing 12,820 N/m³. Find the work required to empty the vat by pumping the sludge out the top.

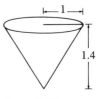

Figure 7-56

12. A spherical oil tank of radius 2 m is filled to a depth of 1 m at the deepest part with oil of weight density 9066 N/m^3. What is the work required to pump the oil out the top of the tank?

13. According to Newton's law of gravitation, the force F of attraction between two objects of mass m_1 and m_2 that are s units apart is given by $F = cm_1m_2/s^2$, where c is a constant. Given two objects of mass 3 kg and 7 kg that are initially 10 m apart, find the work required to separate them until they are 40 m apart. (Give your answer in terms of c.)

14. A rocket of mass 8000 kg is fired upward from the surface of a satellite space station of mass 2×10^8 kg. Assume that the mass of the satellite is concentrated at its center and that its radius is 300 m. What work is required to lift the rocket to a height of 70 m above the surface? (See Exercise 13; give your answer in terms of c.)

15. The force between two electrical charges q_1 and q_2 that are s units apart is given by $F = -pq_1q_2/s^2$, where p is a constant; if the charges are of the same sign, the force is one of repulsion, but if the charges are of opposite sign, the force is one of attraction. Find the work required to move two charges of 1.4×10^{-12} C and 1.2×10^{-12} C from a distance of 1 m apart to a distance of 0.3 m apart. (Give your answer in terms of p.)

16. What is the work required to separate a charge of 2×10^{-18} C and a charge of -3.1×10^{-16} C from a distance of 4.2×10^{-6} m to a distance of 1.4×10^{-4} m? (See Exercise 15; give your answer in terms of p.)

17. Find the length of arc of the curve $y = \frac{1}{2}x$ from $x = 2$ to $x = 8$.

18. Find the length of arc of the curve $y = \frac{3}{4}x^{3/2} + 1$ from $x = 0$ to $x = 2$.

19. Find the length of arc of the curve $y = 2(x + 1)^{3/2}$ from $x = 3$ to $x = 5$.

20. Find the length of arc of the curve $y = x^3/2 + 1/(6x)$ from $x = 1$ to $x = 2$.

21. Find the surface area of the solid of revolution generated by rotating the area bounded by $y = 3x$ from $x = 0$ to $x = 4$ about the x-axis.

22. Find the surface area of the solid of revolution generated by rotating the area bounded by $y = \frac{1}{3}x^3$ from $x = 2$ to $x = 4$ about the x-axis.

23. Find the surface area of the solid of revolution generated by rotating the area bounded by $y = \sqrt{x}$ from $x = 0$ to $x = 2$ about the x-axis.

24. Find the surface area of the solid of revolution generated by rotating the area bounded by $y = \sqrt{2 + x}$ from $x = -1$ to $x = 2$ about the x-axis.

25. Find the area bounded by the curves $r = 2$, $\theta = \pi/6$, and $\theta = \pi/3$.

26. Find an integral expression for the area bounded by one loop of the curve $r^2 = \cos 2\theta$.

7·7 **Double Integrals**

The area under the curve $y = f(x)$ from $x = a$ to $x = b$ can be found by computing $\int_a^b f(x)\,dx$. This integral represents the limit of the sum of the areas of typical rectangular elements. Now suppose that we have a three-dimensional surface represented by the function of two variables $z = f(x, y)$, and that we want the volume beneath this surface bounded by the curves $y = g_1(x)$, $y = g_2(x)$, $x = a$, and $x = b$ (see Figure 7-57). Notice that this is not a solid of revolution, so we cannot find the volume by using slices or shells. We can approximate the volume by using as a typical element a rectangular solid of volume $f(x, y)\Delta x\Delta y$, as in Figure 7-58. If we temporarily hold x fixed and sum up in the y-direction from $y = g_1(x)$ to $y = g_2(x)$ the volumes of all the rectangular solids for this fixed value of x, we would approximate the volume of the slice of Figure 7-59. Indeed, integrating in the y-direction while holding x fixed would give us exactly the volume of the slice. Finally, an integration in the x-direction from $x = a$ to $x = b$ would take the limit of the sum of all these slices and give the volume of the entire solid. In summary, we want to integrate $f(x, y)$ with respect to y while holding x fixed and then integrate the results with respect to x.

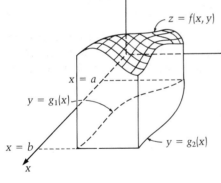

Figure 7-57

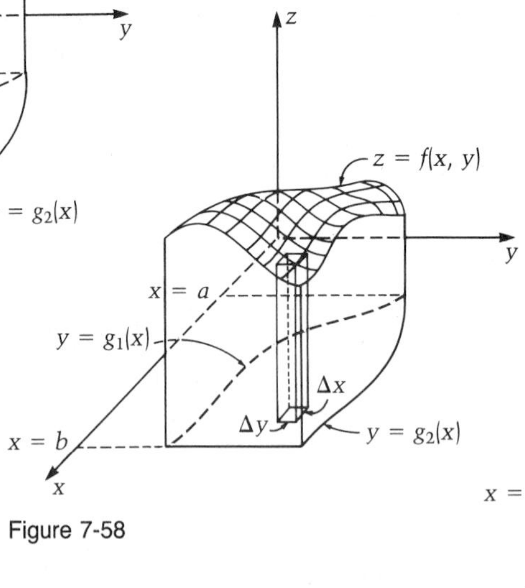

Figure 7-58

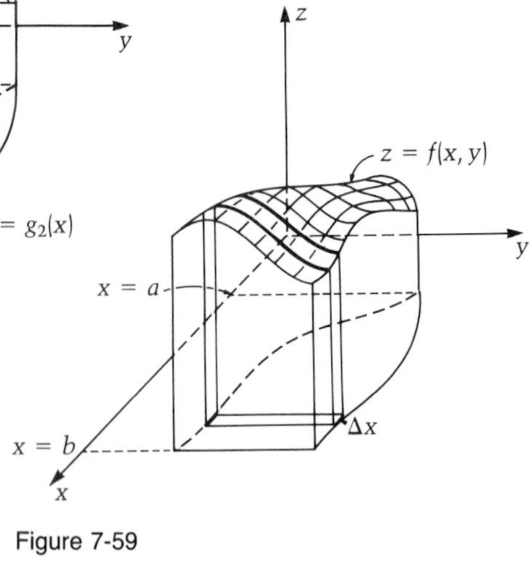

Figure 7-59

The **double integral** of $f(x, y)$, denoted by

$$\int_a^b \int_{g_1(x)}^{g_2(x)} f(x, y)\, dy\, dx$$

represents the process we have just described. The inner integral sign is paired with the symbol dy and indicates integration with respect to y holding x fixed. The limits on the inner integration—which will be substituted for y in evaluating the inner integral—are functions of x. Therefore, the result of the inner integration is some function of x. The outer integration is then done in the usual way.

EXAMPLE 1. Evaluate $\int_2^3 \int_x^{x^2} (2x + y)\, dy\, dx$.

We do the inner integral first, integrating $2x + y$ with respect to y while treating x as a constant.

$$\int_2^3 \int_x^{x^2} (2x + y)\, dy\, dx = \int_2^3 \left(2xy + \frac{y^2}{2} \Big|_x^{x^2} \right) dx$$

$$= \int_2^3 \left(2x^3 + \frac{x^4}{2} - 2x^2 - \frac{x^2}{2} \right) dx$$

$$= \int_2^3 \left(2x^3 + \frac{x^4}{2} - \frac{5}{2}x^2 \right) dx \qquad \begin{array}{l}\text{At this point it}\\ \text{is an ordinary}\\ \text{integral.}\end{array}$$

$$= \frac{2x^4}{4} + \frac{x^5}{10} - \frac{5x^3}{6} \Big|_2^3 = 37.77$$

The inner integration is performed first on a double integral, whatever the name of the variable. In particular, we might integrate first with respect to x, holding y constant, and then with respect to y. ∎

EXAMPLE 2. Evaluate $\int_0^1 \int_2^{3y} xy^2\, dx\, dy$.

Here we integrate first with respect to x, treating y as a constant.

$$\int_0^1 \int_2^{3y} xy^2\, dx\, dy = \int_0^1 \left(\frac{x^2 y^2}{2} \Big|_2^{3y} \right) dy = \int_0^1 \left(\frac{9y^4}{2} - 2y^2 \right) dy$$

$$= \frac{9y^5}{10} - \frac{2y^3}{3} \Big|_0^1 = \frac{7}{30} \quad ∎$$

Practice 1. Evaluate $\int_1^2 \int_{2x}^{4x} (x^2 - y)\, dy\, dx$.

Now we use the double integral to find the volume under a surface.

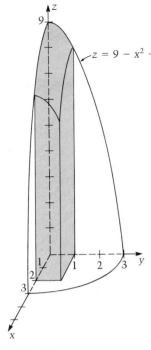

Figure 7-60

EXAMPLE 3. Find the volume in the first octant under the surface $z = 9 - x^2 - y^2$ and between the planes $y = 1$ and $x = 2$.

Figure 7-60 shows the required volume. Integrating first with respect to y, the volume is given by the double integral

$$V = \int_0^2 \int_0^1 (9 - x^2 - y^2)\, dy\, dx$$

$$= \int_0^2 \left[(9 - x^2)y - \frac{y^3}{3} \Big|_0^1 \right] dx$$

$$= \int_0^2 \left(9 - x^2 - \frac{1}{3} \right) dx$$

$$= 9x - \frac{x^3}{3} - \frac{1}{3}x \Big|_0^2$$

$$= \frac{44}{3}$$

In this problem, it is equally easy to integrate first with respect to x. Then

$$V = \int_0^1 \int_0^2 (9 - x^2 - y^2)\, dx\, dy$$

$$= \int_0^1 \left(9x - \frac{x^3}{3} - y^2 x \Big|_0^2 \right) dy$$

$$= \int_0^1 \left(18 - \frac{8}{3} - 2y^2 \right) dy$$

$$= \frac{46}{3}y - \frac{2y^3}{3} \Big|_0^1 = \frac{44}{3} \quad \blacksquare$$

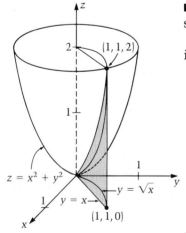

Figure 7-61

EXAMPLE 4. Find the volume in the first octant under the surface $z = x^2 + y^2$ and bounded by the curves $y = x$ and $y = \sqrt{x}$.

Figure 7-61 shows the desired volume. We write the double integral:

$$V = \int_0^1 \int_x^{\sqrt{x}} (x^2 + y^2)\, dy\, dx$$

$$= \int_0^1 \left(x^2 y + \frac{y^3}{3} \Big|_x^{x^{1/2}} \right) dx$$

$$= \int_0^1 \left(x^{5/2} + \frac{x^{3/2}}{3} - x^3 - \frac{x^3}{3} \right) dx$$

$$= \frac{2}{7}x^{7/2} + \frac{2}{15}x^{5/2} - \frac{x^4}{4} - \frac{x^4}{12} \Big|_0^1 = \frac{3}{35} \quad \blacksquare$$

Practice 2.   Find the volume in the first octant under the plane $z = -\dfrac{x}{2} - y + 2$.

Exercises / Section 7-7

Exercises 1–10: Evaluate the given integral.

1. $\int_0^2 \int_y^{y^2} 2x \, dx \, dy$

2. $\int_0^2 \int_x^{x^2} 2x \, dy \, dx$

3. $\int_0^2 \int_{x^2}^{2x^2} 3xy \, dy \, dx$

4. $\int_1^4 \int_y^{3y} (x - y) \, dx \, dy$

5. $\int_1^5 \int_1^x \sqrt{x - y} \, dy \, dx$

6. $\int_3^5 \int_{y^2}^{y^4} \frac{1}{\sqrt{x - y^2}} \, dx \, dy$

7. $\int_2^5 \int_0^{y^{3/2}} \frac{x}{y^2} \, dx \, dy$

8. $\int_0^1 \int_{\sqrt{x}}^{x^2} 3x^2\sqrt{y} \, dy \, dx$

9. $\int_1^4 \int_x^{x^2} \sqrt{xy} \, dy \, dx$

10. $\int_0^1 \int_0^y \sqrt{1 - y^2} \, dx \, dy$

11. Find the volume in the first octant under the plane $z = -\dfrac{x}{3} - \dfrac{y}{2} + 1$.

12. Find the volume in the first octant under the plane $z = y$ and bounded by the planes $y = x$, $y = 2x$, and $y = 4$. (Integrate first with respect to x.)

13. Find the volume in the first octant bounded by the surface $z = x^2$ and the planes $x = 2$ and $y = 3$.

14. Find the volume in the first octant bounded by the planes $z = -y + 1$ and $y = x$.

15. Find the volume in the first octant bounded by the cylinder $x^2 + z^2 = 3$ and the plane $y = x$.

16. Find the volume in the first octant bounded by the surface $z = x^2 + y^2$ and the plane $y = -x + 1$.

STATUS CHECK

Now that you are at the end of Chapter 7, you should be able to:

section **7-1**   Given an algebraic equation for the acceleration of a moving object as a function of time and two conditions relating time and the motion of the object, find an expression for the distance of the object from the starting point as a function of time; evaluate this expression for a specific value of time.

Given two conditions relating time and the motion of a falling object, find an expression for the distance of the object from the starting point as a function of time; evaluate this expression for a specific value of time.

Given an algebraic equation for induced voltage in an electric circuit as a function of time and conditions about the circuit at specific times, find expressions for current in the circuit and total charge passing a fixed point in the circuit as functions of time.

From an algebraic equation for the rate of change of one quantity with respect to another and a condition relating the two quantities, find an expression for the first quantity as a function of the second.

section **7-2**   Find the area bounded by algebraic curves.

section **7-3**   Find the volume of a solid of revolution using either the method of circular slices or the method of cylindrical shells.

section **7-4**   Find the coordinates of the center of mass of a system of point masses.

Find the coordinates of the centroid of an area bounded by algebraic curves.

Find the coordinates of the centroid of a solid of revolution.

Find the moment of inertia with respect to each axis of an area bounded by algebraic curves.

Find the moment of inertia with respect to the axis of rotation of a solid of revolution.

section **7-5**   Find the fluid force on a vertical submerged area bounded by algebraic curves.

section **7-6**   Find the work done in moving an object subject to a variable force through a given distance (examples are work done in stretching a spring, winding up a rope, and pumping liquid out the top of a tank).

Find the length of arc of an algebraic curve between two given points.

Find the surface area of a solid of revolution.

Write an integral expression for the area bounded by a curve in polar coordinates.

section **7-7**   Evaluate double integrals.

Using double integrals, find the volume under a surface given by an algebraic equation.

7-8 More Exercises for Chapter 7

1. The acceleration of a moving object is given by $a = 2t^2$. What is s when $t = 6$ if the initial velocity is 3 and $s = 4$ when $t = 2$?

2. An object moves according to an acceleration $a = 3t + 2$. Find an expression for s as a function of t, given that $s = 25.5$ when $t = 1$ and $s = 42$ when $t = 2$.

3. What initial velocity will cause an object thrown into the air to rise exactly 12 m before it falls?

4. What constant acceleration is required to move a car at rest a distance of 35 m in 3 s?

5. The current i in a certain circuit as a function of time t is given by the equation $i = 3t^2 + t$. What is the total charge that passes a given point in the circuit in the first 2 s?

6. The voltage V induced in a 0.3-H inductor varies as a function of time according to the equation $V = 3.2t$. At $t = 1$ s, the current in the circuit is 0.8 A. What is the current when $t = 3$ s?

7. A voltage V given by $V = \sqrt{t} + 1$ is induced in a 0.5-H inductor. If 2 C of charge passes a given point in the circuit in 1 s, what is the total charge which passes the point in 3 s?

8. A voltage of 3 V is induced in a certain inductor. At $t = 1$ s, the current in the circuit is 0.8 A, and at $t = 2$ s, the current is 1.4 A. What is the inductance?

9. The marginal cost to produce x units of a given item is $c'(x) = 0.01x^2 - x/3$; the fixed cost is $1000. Find the total cost to produce x items.

10. Angular velocity ω for a rotating object is the time rate of change of angular displacement θ; $\omega = d\theta/dt$. The angular velocity of a particular object varies with time according to the equation $\omega = t^2 - t + 2$. If $\theta = 0$ when $t = 0$, find an expression for the angular displacement as a function of time.

11. Find the area bounded by $y = 4x - x^2$ and $y = 0$.

12. Find the area bounded by $y = 1/x^2$, $x = \frac{1}{2}$, and $y = x$.

13. Find the smaller area bounded by $y^2 = -x - 2$, $y = 2x + 7$, and $y = 0$.

14. Find the total area bounded by $y = 2x^2$, $y = x + 3$, $x = -2$, and $x = 0$.

15. Find the volume generated by rotating the area in the first quadrant bounded by $y = 3x^2$, $x = 0$, and $y = 12$ about the y-axis.

16. Find the volume generated by rotating the area of Exercise 11 about the x-axis.

17. Find the volume generated by rotating the area bounded by $y = x^4$, $y = 0$, and $x = 1$ about the line $x = 1$.

18. Find the volume generated by rotating the area bounded by $y = x$, $y = x^2$, and $x = \frac{1}{2}$ about the y-axis.

19. Find the coordinates of the centroid of the area bounded by $y = x^2$, $y = 0$, and $x = 3$.

20. Find the coordinates of the centroid of the area bounded by $y = x^{4/3}$ and $y = 2x$.

21. Find the coordinates of the centroid of the volume generated by revolving the area of Exercise 19 about the x-axis.

22. Find the coordinates of the centroid of the volume generated by revolving the area bounded by $y = x^2 + 1$, $x = 0$, and $y = 2$ about the y-axis.

23. Find the moment of inertia with respect to the y-axis and with respect to the x-axis for the area of Exercise 19.

24. Find the moment of inertia with respect to the y-axis for the area in the first quadrant bounded by $y = 4 - x^2$, $y = 0$, and $x = 0$.

25. Find the moment of inertia with respect to the x-axis for the volume generated by revolving the area of Exercise 19 about the x-axis.

26. Find the moment of inertia with respect to the y-axis for the volume generated by revolving the area of Exercise 24 about the y-axis.

27. A plate in the shape of a right triangle is suspended in a liquid of weight density ω newtons per cubic meter. One leg of the triangle is 3 m long and is parallel to the liquid surface at a depth of 2 m. The other leg of the triangle extends vertically downward and is 1 m long. Find the force on the plate.

28. A section of a floodgate is in the shape of a trapezoid. The upper edge is 8 m long and lies in the surface of the water. The lower edge is 4 m long and the section is 2 m deep. Find the force on the floodgate (for water, $\omega = 9800$ N/m^3).

29. A spring of natural length 2 m stretches 0.3 m under a weight of 850 N. What is the work done in stretching the spring from a length of 2.5 m to a length of 2.8 m?

30. What work is required to wind up 12 m of an 18 m rope that weighs 5 N/m?

31. What is the amount of work done in pumping all the water out of the top of a cylindrical tank of radius 0.5 m and height 2 m if the tank initially contains water to a depth of 1 m?

32. The force of attraction due to gravity between two objects of mass m_1 and m_2 that are s units apart is given by $F = cm_1m_2/s^2$, where c is a constant. Find the work required to separate two objects of mass 50 kg and 80 kg from 20 m apart to 100 m apart. (Give your answer in terms of c.)

33. What is the length of arc of the curve $y = \frac{3}{2}x^{2/3} - 4$ from $x = 1$ to $x = 8$?

34. Find the surface area of the solid of revolution generated by rotating the area bounded by $y = x^3$ from $x = 0$ to $x = 2$ about the x-axis.

35. Evaluate $\int_1^2 \int_{2y}^{y^2} (x + y^2) \, dx \, dy$.

36. Evaluate $\int_1^2 \int_0^x \sqrt{1 + 2x^2} \, dy \, dx$.

37. Find the volume in the first octant under the plane $z = -\frac{4}{3}x - 2y + 4$.

38. Find the volume in the first octant bounded by the surface $z = 9 - x^2$ and the plane $x + y = 3$.

chapter 8

Derivatives of Transcendental Functions

Wavy curves like the one shown on the oscilloscope are graphs of transcendental functions.

8-1 Quick Trigonometry Review

In all our differentiation and integration work so far, we have dealt only with algebraic functions. The trigonometric and logarithmic functions are examples of transcendental (nonalgebraic) functions. In this chapter, we learn how to differentiate these functions. First, we do a final brushup on trigonometry.

Recall that angles in calculus are measured in terms of radians and that no units are written with radian measure. A scientific calculator computes trigonometric functions of angles; if the angles are in radian measure, then the calculator must be set in radian mode.

If we make a table of values of the sine function for many different angles in radian measure, we can use these values to sketch the graph of the sine function $y = \sin \theta$. The graph is shown in Figure 8-1. The sine function is **periodic** in nature with period

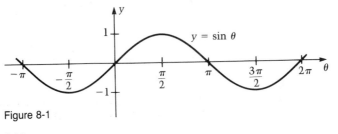

Figure 8-1

246

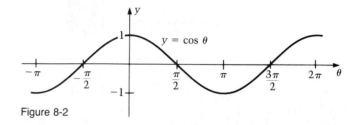

Figure 8-2

2π—that is, the pattern of values repeats itself every 2π units.*

The graph of the function $y = \cos \theta$ is shown in Figure 8-2.

The graph of the sine function is symmetric about the origin because

$$\sin(-\theta) = -\sin \theta \tag{1}$$

The graph of the cosine function is symmetric about the y-axis because

$$\cos(-\theta) = \cos \theta \tag{2}$$

Equations (1) and (2) are **identities**, equations that are true for any value of θ. Many other trigonometric identities can be derived.

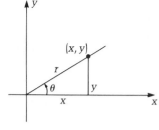

Figure 8-3

EXAMPLE 1. Prove the trigonometric identities

$$\sin \theta = \frac{1}{\csc \theta}$$

$$\tan \theta = \frac{\sin \theta}{\cos \theta}$$

$$\sin^2\theta + \cos^2\theta = 1$$

For any angle θ let (x, y) be a point on its terminal side with radius vector r (see Figure 8-3). Then $x^2 + y^2 = r^2$. From the definitions of the trigonometric functions,

$$\sin \theta = \frac{y}{r} = \frac{1}{r/y} = \frac{1}{\csc \theta}$$

Also,

$$\tan \theta = \frac{y}{x} = \frac{y/r}{x/r} = \frac{\sin \theta}{\cos \theta}$$

Finally,

$$\sin^2\theta + \cos^2\theta = \frac{y^2}{r^2} + \frac{x^2}{r^2} = \frac{y^2 + x^2}{r^2} = 1 \quad \blacksquare$$

---

*The **period** of a function $f(x)$ is the smallest positive number k, if such a number exists, such that $f(x + k) = f(x)$ for all x.

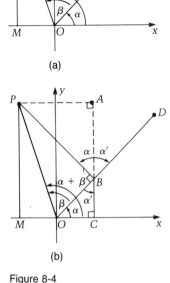

(a)

(b)

Figure 8-4

EXAMPLE 2. Prove the identity

$$\sin(\alpha + \beta) = \sin \alpha \cos \beta + \cos \alpha \sin \beta \tag{3}$$

Figure 8-4(a) shows $\alpha + \beta$ in standard position; $\sin(\alpha + \beta) = \overline{PM}/\overline{OP}$. In order to express distances in terms of functions of α and β, we add the lines shown in Figure 8-4(b). Note in Figure 8-4(b) that angle OBC is the complement of α, so angle ABD is the complement of α and thus angle ABP is α. We then have

$$\sin(\alpha + \beta) = \frac{\overline{PM}}{\overline{OP}} = \frac{\overline{BC}}{\overline{OP}} + \frac{\overline{AB}}{\overline{OP}}$$

But $\overline{BC}/\overline{OB} = \sin \alpha$ and $\overline{AB}/\overline{PB} = \cos \alpha$. Therefore, $\overline{BC} = \overline{OB} \sin \alpha$, $\overline{AB} = \overline{PB} \cos \alpha$, and

$$\sin(\alpha + \beta) = \frac{\overline{OB} \sin \alpha}{\overline{OP}} + \frac{\overline{PB} \cos \alpha}{\overline{OP}}$$

Finally, $\overline{OB}/\overline{OP} = \cos \beta$ and $\overline{PB}/\overline{OP} = \sin \beta$. Thus

$$\sin(\alpha + b) = \cos \beta \sin \alpha + \sin \beta \cos \alpha$$
$$= \sin \alpha \cos \beta + \cos \alpha \sin \beta$$

In our picture, $\alpha + \beta$ is a second-quadrant angle, but a similar argument holds for any α and β. ∎

Once a trigonometric identity has been proved, it may be used to prove other identities.

EXAMPLE 3. Using Equations (2) and (3), prove the identity

$$\sin\left(\frac{\pi}{2} - \beta\right) = \cos \beta \tag{4}$$

We use Equation (3), replacing α by $\pi/2$ and β by $-\beta$. We get

$$\sin\left(\frac{\pi}{2} - \beta\right) = \sin \frac{\pi}{2} \cos(-\beta) + \cos \frac{\pi}{2} \sin(-\beta)$$
$$= 1 \cdot \cos(-\beta) + 0 \cdot \sin(-\beta)$$
$$= \cos \beta \qquad \text{By Equation (2).} \quad ∎$$

Practice 1. Use Equation (4) (replacing β by $\pi/2 - \beta$) to prove the identity

$$\cos\left(\frac{\pi}{2} - \beta\right) = \sin \beta$$

There are many trigonometric identities. Below is a list of some of the most frequently used identities, including those we have already derived.

$$\sin(-\theta) = -\sin\theta \tag{5}$$

$$\cos(-\theta) = \cos\theta \tag{6}$$

$$\sin\theta = \frac{1}{\csc\theta} \tag{7}$$

$$\cos\theta = \frac{1}{\sec\theta} \tag{8}$$

$$\tan\theta = \frac{1}{\cot\theta} \tag{9}$$

$$\tan\theta = \frac{\sin\theta}{\cos\theta} \tag{10}$$

$$\cot\theta = \frac{\cos\theta}{\sin\theta} \tag{11}$$

$$\sin^2\theta + \cos^2\theta = 1 \tag{12}$$

$$1 + \tan^2\theta = \sec^2\theta \tag{13}$$

$$\cot^2\theta + 1 = \csc^2\theta \tag{14}$$

$$\sin(\alpha + \beta) = \sin\alpha\cos\beta + \cos\alpha\sin\beta \tag{15}$$

$$\sin(\alpha - \beta) = \sin\alpha\cos\beta - \cos\alpha\sin\beta \tag{16}$$

$$\sin\left(\frac{\pi}{2} - \beta\right) = \cos\beta \tag{17}$$

$$\cos\left(\frac{\pi}{2} - \beta\right) = \sin\beta \tag{18}$$

$$\cos(\alpha + \beta) = \cos\alpha\cos\beta - \sin\alpha\sin\beta \tag{19}$$

$$\cos(\alpha - \beta) = \cos\alpha\cos\beta + \sin\alpha\sin\beta \tag{20}$$

$$\sin 2\alpha = 2\sin\alpha\cos\alpha \tag{21}$$

$$\cos 2\alpha = \cos^2\alpha - \sin^2\alpha \tag{22}$$

$$\cos 2\alpha = 2\cos^2\alpha - 1 \tag{23}$$

$$\cos 2\alpha = 1 - 2\sin^2\alpha \tag{24}$$

$$\sin\alpha - \sin\beta = 2\cos\frac{\alpha + \beta}{2}\sin\frac{\alpha - \beta}{2} \tag{25}$$

$$\sin\frac{\alpha}{2} = \pm\sqrt{\frac{1 - \cos\alpha}{2}} \tag{26}$$

$$\cos\frac{\alpha}{2} = \pm\sqrt{\frac{1 + \cos\alpha}{2}} \tag{27}$$

$$\tan\frac{\alpha}{2} = \pm\sqrt{\frac{1 - \cos\alpha}{1 + \cos\alpha}} \tag{28}$$

Believe it or not, many of these identities will be needed in conjunction with differentiation and integration of trigonometric functions because they allow us to rewrite trigonometric expressions in alternate forms that are easier to work with.

Exercises / Section 8-1

Exercises 1–16: Prove the identities listed in this section.

1. Equation (8).

2. Equation (9).

3. Equation (11).

4. Equation (13).

5. Equation (14).

6. Equation (16).   (*Hint:* Replace β with $-\beta$ in Equation 15.)

7. Equation (19).   (*Hint:* Replace β with $\alpha + \beta$ in Equation 17.)

8. Equation (20).

9. Equation (21).

10. Equation (22).

11. Equation (23).

12. Equation (24).

13. Equation (25).   (*Hint:* The right side of Equation (25) can be found by replacing α with $\alpha/2$ and β with $\beta/2$ in Equations (19) and (16), multiplying out the results, and multiplying by 2.)

14. Equation (26).   (*Hint:* Replace α with $\alpha/2$ in Equation (24).)

15. Equation (27).   (*Hint:* Replace α with $\alpha/2$ in Equation (23).)

16. Equation (28).

Exercises 17–28: Prove the given trigonometric identity.

17. $\cos \theta \tan \theta = \sin \theta$

18. $(1 - \cos^2\theta)(1 + \cot^2\theta) = 1$

19. $\tan^2\theta - \sin^2\theta = \tan^2\theta \sin^2\theta$

20. $\dfrac{1}{1 + \sin \theta} + \dfrac{1}{1 - \sin \theta} = 2 \sec^2\theta$

21. $\dfrac{\cos \theta}{\tan \theta + \sec \theta} - \dfrac{\cos \theta}{\tan \theta - \sec \theta} = 2$

22. $4 \sin \theta + \tan \theta = \dfrac{4 + \sec \theta}{\csc \theta}$

23. $\sin(\alpha + \beta) \sin(\alpha - \beta) = \sin^2\alpha - \sin^2\beta$

24. $\cos(\alpha + \beta) \cos(\alpha - \beta) = \cos^2\alpha - \sin^2\beta$

25. $(\sin \theta + \cos \theta)^2 = 1 + \sin 2\theta$

26. $\cos^4\theta - \sin^4\theta = \cos 2\theta$

27. $\dfrac{2}{1 - \cos 2\theta} = \csc^2\theta$

28. $\dfrac{1}{\tan \theta + \cot 2\theta} = \sin 2\theta$

29. The expression for the luminance from an extended light source of area A in the direction θ from a perpendicular to A is

$$B = \frac{I}{A \cos \theta}$$

Show that this expression can be written

$$B = \frac{I}{A} \sec \theta$$

30. Computing flux density on the axis of a particular solenoid requires the expression

$$B = k(\cos \theta_1 - \cos \theta_2)$$

Show that this expression can be written

$$B = \frac{k}{2}\left[\frac{\sin 2\theta_1 + \sin 2\theta_2 - 2 \sin(\theta_1 + \theta_2)}{\sin \theta_1 - \sin \theta_2}\right]$$

8-2   Derivative of the Sine Function

In order to find the derivative of the sine function, we go back to the definition of the derivative and use the four-step delta process. Once we have the expression for the derivative of the sine function, we use trigonometric identities to help us find the derivatives of the remaining trigonometric functions without having to resort to the definition.

An expression of the form

$$\lim_{\theta \to 0} \frac{\sin \theta}{\theta}$$

will occur as we proceed, so we evaluate this limit first. Both numerator and denominator approach 0 as $\theta \to 0$; therefore, we cannot simply substitute to evaluate the limit. We can, however, substitute progressively smaller values of θ to enable us to estimate the limiting value.

Practice 1.   Estimate $\lim_{\theta \to 0} (\sin \theta)/\theta$ by computing the values in the following table. Here θ is measured in radians.

θ	0.5	0.1	0.01	-0.5	-0.1	-0.01
$\sin \theta$	.479	.099	.009	$-.479$	$-.099$	$-.009$
$\dfrac{\sin \theta}{\theta}$	.959	.998	.999	.958	.998	.999

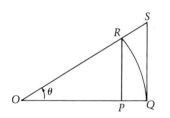

Figure 8-5

From Practice 1, the value of $(\sin \theta)/\theta$ appears to be approaching 1 as $\theta \to 0$. We can use a geometric argument to prove that this limit is correct. Figure 8-5 shows a sector of a circle of radius 1 with inscribed angle θ (in radians), together with two right triangles. From Exercise 39 of Section 3-2, the area of the circular sector is $\frac{1}{2}\theta r^2 = \frac{1}{2}\theta$. Distances in Figure 8-5 are as follows:

$$\overline{OR} = \overline{OQ} = r = 1$$
$$\overline{RP} = \overline{OR} \sin \theta = \sin \theta$$
$$\overline{SQ} = \overline{OQ} \tan \theta = \tan \theta$$
$$\overline{OP} = \overline{OR} \cos \theta = \cos \theta$$

It is easy to see that

area of triangle $OPR \leq$ area of circular sector $\leq$ area of triangle OQS

or

$$\tfrac{1}{2}(\overline{RP})(\overline{OP}) \leq \tfrac{1}{2}\theta \leq \tfrac{1}{2}(\overline{SQ})(\overline{OQ})$$
$$\tfrac{1}{2} \sin \theta \cos \theta \leq \tfrac{1}{2}\theta \leq \tfrac{1}{2} \tan \theta$$

Dividing through by $\frac{1}{2} \sin \theta$ (a positive value), we get

$$\cos \theta \leq \frac{\theta}{\sin \theta} \leq \frac{1}{\cos \theta}$$

Now we take the limit of all three expressions as $\theta \to 0$:

$$\lim_{\theta \to 0} \cos \theta \leq \lim_{\theta \to 0} \frac{\theta}{\sin \theta} \leq \lim_{\theta \to 0} \frac{1}{\cos \theta}$$

Because $\lim_{\theta \to 0} \cos \theta = 1$, we get

$$1 \le \lim_{\theta \to 0} \frac{\theta}{\sin \theta} \le 1$$

and, therefore, $\lim_{\theta \to 0} \theta/(\sin \theta) = 1$. Finally,

$$\lim_{\theta \to 0} \frac{\sin \theta}{\theta} = \lim_{\theta \to 0} \frac{1}{\theta/(\sin \theta)} = \frac{1}{1} = 1$$

This geometric argument shows that $\lim_{\theta \to 0} (\sin \theta)/\theta = 1$ for any positive angle θ. If θ is negative, $-\theta$ is positive; also, $\theta \to 0$ exactly when $-\theta \to 0$. Therefore, we can write

$$\lim_{\theta \to 0} \frac{\sin \theta}{\theta} = \lim_{\theta \to 0} \frac{-\sin \theta}{-\theta} = \lim_{\theta \to 0} \frac{\sin (-\theta)}{-\theta} = \lim_{-\theta \to 0} \frac{\sin (-\theta)}{-\theta} = 1$$

We now have the result that

$$\boxed{\lim_{\theta \to 0} \frac{\sin \theta}{\theta} = 1}$$

where θ is any angle in radian measure. (The simplicity of this limit is one reason radian measure for angles is so convenient.)

We are finally ready to find the derivative of the sine function. We let $y = f(x) = \sin x$ and we compute

$$\frac{dy}{dx} = \lim_{\Delta x \to 0} \frac{f(x + \Delta x) - f(x)}{\Delta x}$$

1. $f(x + \Delta x) = \sin(x + \Delta x)$

2. $f(x + \Delta x) - f(x) = \sin(x + \Delta x) - \sin x$

$$= 2 \cos\left(\frac{x + \Delta x + x}{2}\right) \sin\left(\frac{x + \Delta x - x}{2}\right)$$

By Equation (25) of Section 8-1.

$$= 2 \cos\left(\frac{2x + \Delta x}{2}\right) \sin \frac{\Delta x}{2}$$

$$= 2 \cos\left(x + \frac{\Delta x}{2}\right) \sin \frac{\Delta x}{2}$$

3. $\dfrac{f(x + \Delta x) - f(x)}{\Delta x} = \dfrac{2 \cos(x + \Delta x/2) \sin(\Delta x/2)}{\Delta x}$

$$= \frac{\cos(x + \Delta x/2) \sin(\Delta x/2)}{\Delta x/2}$$

4. $\lim_{\Delta x \to 0} \dfrac{f(x + \Delta x) - f(x)}{\Delta x} = \lim_{\Delta x \to 0} \dfrac{\cos(x + \Delta x/2) \sin(\Delta x/2)}{\Delta x/2}$

Because $\Delta x \to 0$ exactly when $\Delta x/2 \to 0$, we may replace $\lim_{\Delta x \to 0}$ with $\lim_{\Delta x/2 \to 0}$ in the above expression. This results in

$$\frac{dy}{dx} = \lim_{\Delta x/2 \to 0} \cos\left(x + \frac{\Delta x}{2}\right) \frac{\sin(\Delta x/2)}{\Delta x/2}$$

Clearly the first factor has a limit of cos x. Applying the limit to the second factor results in the form $\lim_{\theta \to 0} (\sin \theta)/\theta$, which has the value 1. Therefore,

$$\frac{d(\sin x)}{dx} = \cos x \tag{1}$$

where x is an angle in radian measure.

Now suppose that $y = \sin u$, where u is a function of x. The chain rule applies to this situation; it tells us that to find dy/dx, we first find dy/du and then du/dx. The equation is

$$\frac{dy}{dx} = \frac{dy}{du} \cdot \frac{du}{dx}$$

If $y = \sin u$, then by Equation (1), $dy/du = \cos u$. Therefore,

$$\boxed{\frac{d(\sin u)}{dx} = \cos u \frac{du}{dx}}$$

This is the "sine rule" for differentiation.

EXAMPLE 1. Differentiate $y = \sin x^2$.
This is of the form $y = \sin u$, where $u = x^2$. The derivative is

$$\frac{dy}{dx} = \cos x^2 \left[\frac{d(x^2)}{dx} \right] = (\cos x^2)2x = 2x \cos x^2 \quad \blacksquare$$

Word of Advice

To find the derivative with respect to x of $y = \sin u$ don't forget the du/dx after cos u.

EXAMPLE 2. Differentiate $y = \sin(3x + 1)$.
Here $y = \sin u$ with $u = 3x + 1$ and

$$\frac{dy}{dx} = \cos(3x + 1) \frac{d(3x + 1)}{dx} = 3 \cos(3x + 1) \quad \blacksquare$$

Practice 2. Differentiate $y = \sin(2 - x)$.

The pattern for differentiating the sine function can also be used in combination with all of our previous differentiation patterns.

EXAMPLE 3.   Find the derivative of $y = \sin^2(x^3)$.

The basic pattern is a function to a power (remember that $\sin^2(x^3)$ means $[\sin(x^3)]^2$), and we therefore begin with the power rule.

$$y' = 2 \sin(x^3) \frac{d(\sin x^3)}{dx} \qquad \text{Power rule.}$$

$$= 2 \sin x^3 \cos x^3 \frac{d(x^3)}{dx} \qquad \text{Sine rule.}$$

$$= 2 \sin x^3 \cos x^3 (3x^2) \qquad \text{Polynomial.}$$

$$= 6x^2 \sin x^3 \cos x^3 \quad \blacksquare$$

EXAMPLE 4.   Differentiate $y = 4x^2 \sin 5x$.

$$y' = 4x^2 \frac{d(\sin 5x)}{dx} + \sin 5x \frac{d(4x^2)}{dx} \qquad \text{Product rule.}$$

$$= 4x^2 \cos 5x(5) + \sin 5x(8x)$$

$$= 20x^2 \cos 5x + 8x \sin 5x \quad \blacksquare$$

With practice, you will probably be able to skip writing the first line in our solution to Example 4 and go directly to the second line. However, the first line represents the patterns you should be applying mentally, if not actually in writing.

Practice 3.   Find the derivative of $y = 2 \sin^3 2x$.

Exercises / Section 8-2

Exercises 1–20: Find the derivative.

1. $y = \sin 2x$

2. $y = \sin(4x - 1)$

3. $y = \sin(x^2 - 2x)$

4. $y = \sin(x^3 + x)$

5. $y = 3x^2 - \sin x^2$

6. $y = \sin(2x - 1) + \sqrt{x}$

7. $y = 3x \sin(2x + 1)$

8. $y = x^2 \sin x^2$

9. $y = \sqrt{x} \sin x$

10. $y = (2x^2 + 4x) \sin(4 - x)$

11. $y = \dfrac{\sin x}{x^2}$

12. $y = \dfrac{2x^2 - 1}{\sin 2x}$

13. $y = \sin^2 3x$

14. $y = \sin^3(2x + 1)$

15. $y = \sqrt{\sin(1 - 2x)}$

16. $y = \sqrt{\sin^3(x^2)}$

17. $y = \sin^3 2x - \sin^2 3x$

18. $y = \sin \sqrt{x} - \sin^2 x$

19. $y = \sin(\sin 2x)$

20. $y = \sin(\sin x^2)$

21. If $y = \sin 2x$, show that $y' = 2 - 2y \tan x$.

22. If $y = \sin^2 2x$, show that $y' = 4\sqrt{y}\sqrt{1 - y}$. (Assume that $2x$ is in the first quadrant so that all trigonometric functions are positive.)

23. Find the slope of a tangent line to the curve $y = \sin 3x$ at $x = \pi/3$.

24. Find the slope of a tangent line to the curve $y = \sin^2(x + 1)$ at $x = 1$.

25. Distance (in meters) as a function of time (in seconds) for a particular object is given by the equation $s = t^3 - \sin 2t$. Find the velocity at $t = 3$ s.

26. Distance (in kilometers) as a function of time (in minutes) for a particular object is given by the equation $s = \sqrt{t} + \sin^2\sqrt{t}$. Find the velocity at $t = 2$ min.

27. Current i (in amperes) as a function of time t (in seconds) in a certain circuit is given by the equation $i = 3\sin(2\pi t + \pi)$. Find the rate of change of current with respect to time at $t = 2$ s.

28. Energy W (in joules) supplied to an electrical circuit as a function of time t (in seconds) is given by $W = 31\sin^2(2t)$. Find an expression for the power (dW/dt) required by the circuit.

8-3 Derivatives of Other Trigonometric Functions

The trigonometric identities will enable us to find formulas for the derivatives of the remaining five trigonometric functions, now that we know how to differentiate $y = \sin u$. Throughout this section, u represents a function of x.

To find the derivative of $y = \cos u$, we write $\cos u$ as $\sin(\pi/2 - u)$ (Equation (17) of Section 8-1). Then

$$\frac{dy}{dx} = \frac{d\sin(\pi/2 - u)}{dx} = \cos\left(\frac{\pi}{2} - u\right)\frac{d(\pi/2 - u)}{dx}$$

$$= \cos\left(\frac{\pi}{2} - u\right)\left(-\frac{du}{dx}\right)$$

$$= -\sin u \frac{du}{dx}$$

Equation (18) of Section 8-1.

Therefore,

$$\boxed{\frac{d(\cos u)}{dx} = -\sin u \frac{du}{dx}}$$

EXAMPLE 1. Differentiate $y = \cos 3x^2$.
This is of the form $y = \cos u$ with $u = 3x^2$.

$$\frac{dy}{dx} = -\sin 3x^2 \frac{d(3x^2)}{dx} = -6x \sin 3x^2 \quad \blacksquare$$

EXAMPLE 2. Differentiate $y = \sin^2 x \cos 2x$.

$$\frac{dy}{dx} = \sin^2 x \frac{d(\cos 2x)}{dx} + \cos 2x \frac{d(\sin^2 x)}{dx}$$

Product rule.

$$= \sin^2 x \, (-\sin 2x) \frac{d(2x)}{dx} + \cos 2x (2) \sin x \frac{d(\sin x)}{dx}$$

Cosine rule and power rule.

$$= \sin^2 x \, (-\sin 2x)(2) + \cos 2x (2) \sin x \cos x$$

Polynomial and sine rule.

$$= -2 \sin^2 x \sin 2x + \cos 2x \sin 2x \quad \blacksquare$$

To find the derivative of $y = \tan u$, we write $\tan u$ as $\sin u/\cos u$ $\sin(\pi/2 - u)$ (Equation (17) of Section 8-1). Then

$$\frac{dy}{dx} = \frac{d}{dx}\left(\frac{\sin u}{\cos u}\right) = \frac{\cos u \cos u \, du/dx - \sin u(-\sin u) \, du/dx}{\cos^2 u}$$

$$= \frac{(\cos^2 u + \sin^2 u)}{\cos^2 u} \frac{du}{dx}$$

$$= \frac{1}{\cos^2 u} \frac{du}{dx}$$

$$= \sec^2 u \frac{du}{dx}$$

Practice 1. Prove that $d(\cot u)/dx = -\csc^2 u \, du/dx$ by writing $\cot u$ as $\cos u/\sin u$ and using the quotient rule.

The two new rules we have developed are

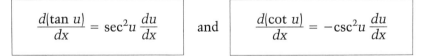

$$\boxed{\frac{d(\tan u)}{dx} = \sec^2 u \frac{du}{dx}} \quad \text{and} \quad \boxed{\frac{d(\cot u)}{dx} = -\csc^2 u \frac{du}{dx}}$$

Of course, we can use these new rules in conjunction with the old ones.

EXAMPLE 3. Find the derivative of $y = \tan^2 3x$.

$$\frac{dy}{dx} = 2 \tan 3x \frac{d(\tan 3x)}{dx} \qquad \text{Power rule.}$$

$$= 2 \tan 3x \sec^2 3x \frac{d(3x)}{dx} \qquad \text{Tangent rule.}$$

$$= 2 \tan 3x \sec^2 3x(3) \qquad \text{Polynomial.}$$

$$= 6 \tan 3x \sec^2 3x \quad \blacksquare$$

EXAMPLE 4. Find the derivative of $y = x \cot x^2$.

$$\frac{dy}{dx} = x \frac{d(\cot x^2)}{dx} + \cot x^2 \frac{d(x)}{dx} \qquad \text{Product rule.}$$

$$= x(-\csc^2 x^2) \frac{d(x^2)}{dx} + \cot x^2 \qquad \text{Cotangent rule.}$$

$$= x(-\csc^2 x^2)(2x) + \cot x^2 \qquad \text{Polynomial.}$$

$$= -2x^2 \csc^2 x^2 + \cot x^2 \quad \blacksquare$$

Practice 2. Differentiate $y = \cos^2 x + \tan 2x$.

To find the derivative of $y = \sec u$, we write $\sec u$ as $1/\cos u = (\cos u)^{-1}$ and use the power rule.

$$\frac{dy}{dx} = \frac{d(\cos u)^{-1}}{dx} = (-1)(\cos u)^{-2}(-\sin u)\frac{du}{dx}$$

$$= \frac{\sin u}{\cos^2 u}\frac{du}{dx}$$

$$= \frac{1}{\cos u} \cdot \frac{\sin u}{\cos u}\frac{du}{dx}$$

$$= \sec u \tan u \frac{du}{dx}$$

Practice 3. Prove that $d(\csc u)/dx = -\csc u \cot u \, du/dx$ by writing $\csc u$ as $1/\sin u = (\sin u)^{-1}$ and using the power rule.

The last two trigonometric rules are, therefore,

$$\boxed{\frac{d(\sec u)}{dx} = \sec u \tan u \frac{du}{dx}}$$

and

$$\boxed{\frac{d(\csc u)}{dx} = -\csc u \cot u \frac{du}{dx}}$$

EXAMPLE 5. Differentiate $y = \sec^3(2x - 1)$.

$$y' = 3 \sec^2(2x - 1) \sec(2x - 1) \tan(2x - 1)(2)$$

$$\text{Power rule, secant rule, polynomial.}$$

$$= 6 \sec^3(2x - 1) \tan(2x - 1) \quad \blacksquare$$

EXAMPLE 6. Find the derivative of $y = \dfrac{\csc(1 - x)}{2x}$.

$$\frac{dy}{dx} = \frac{2x[-\csc(1 - x)\cot(1 - x)(-1)] - \csc(1 - x)2}{4x^2}$$

Quotient rule, cosecant rule, polynomial.

$$= \frac{2x\csc(1 - x)\cot(1 - x) - 2\csc(1 - x)}{4x^2} \quad \blacksquare$$

Practice 4. Differentiate $y = (x - \csc 2x)^2$.

Word of Advice

Be careful of signs when differentiating the trigonometric functions. The cosine rule, cotangent rule, and cosecant rule all introduce minus signs.

Exercises / Section **8-3**

Exercises 1–38: Find the derivative.

1. $y = \cos(3x + 2)$
2. $y = \cos(4x^2 + 2x)$
3. $y = 3\tan 4x$
4. $y = 4\tan(2x^2 + 1)$
5. $y = \cot x^2$
6. $y = \cot(1 - 2x)^2$
7. $y = \sec 3x$
8. $y = 2\sec x^2$
9. $y = 2\csc(1 - x)$
10. $y = \csc\sqrt{x}$
11. $y = \cos^2(2x + 4)$
12. $y = \sqrt{\cos x}$
13. $y = \tan^2(4x - 2)$
14. $y = \tan^3 2x^2$
15. $y = \cot^2(1 - x)$
16. $y = 3\cot^4 2x$
17. $y = \sec^4 2x$
18. $y = \sqrt{\sec(x + 1)}$
19. $y = \csc^2(3x + 1)$
20. $y = (\csc x^2)^{-1}$
21. $y = x\csc x^2$
22. $y = x^2\tan x$
23. $y = x^3\sec^2 x$
24. $y = x^2\cos^2(x - 1)$
25. $y = \cot(x + 1)\sec(x - 1)$
26. $y = \tan^2 2x\cos x^2$
27. $y = \dfrac{\cot x}{x}$
28. $y = \dfrac{\sec^2 x}{2x}$
29. $y = \dfrac{\sin^2 2x}{\cos x}$
30. $y = \dfrac{1 - \cot^2 x}{\tan 2x}$

31. $y = \tan^3 3x - \sin^2 2x$
32. $y = \cos^2(2x + 1) - \cot^2 2x$
33. $y = (\sec^2 x + 1)^2$
34. $y = (1 - \csc x)^3$
35. $y = [\sec^2 2x - \tan(x + 1)]^3$
36. $y = [\csc^3(x - 1) + \cos^2 x - \sin 2x]^2$
37. $y = (1 + \sin^2 x)^2\cos 2x$
38. $y = (\sec^2 x - \cos^2 x)\sin^2(x + 1)$

39. If $y = \sin x$, show that $d^4y/dx^4 = y$.
40. If $y = \cos x\sin x$, show that $y' = 1 - 2(\tan x)y$.
41. Find the slope of a tangent line to the curve $y = 3\cos(2x - 1)$ at $x = \pi/2$.
42. Find the slope of a tangent line to the curve $y = \cot^2 x^2$ at $x = 0.5$.
43. Distance s as a function of time t for a particular object is given by the equation $s = \tan^2 2t + t + 2$. Find the velocity at $t = 1$.
44. Distance s as a function of time t for a particular object is given by the equation $s = t\sqrt{\sec 2t}$. Find the velocity at $t = 0.2$.

8-4 Inverse Trigonometric Functions and Their Derivatives

When we write the function $y = \sin x$, x is the independent variable and y is the dependent variable. Given any value for x (an angle) we find the corresponding value for y (the sine of that angle). Suppose we reverse the role of dependent and independent variable; if we are given the sine of an angle, can we find the corresponding angle? If so, we would have a function of the form $y = f(x)$, where x is the sine of an angle and y is the angle. But a moment's thought should convince us that this is not going to work. For example, if $x = 0$, what would the corresponding y-value be? We know that y must be an angle such that $\sin y = 0$, but there are many such angles. We could say $y = 0$, $y = \pi$, $y = 2\pi$, $y = -\pi$, and so on. We do not satisfy the definition of a *function*, because for a given value of x, the independent variable, there is more than one corresponding value for y, the dependent variable.

We can see the dilemma graphically. If we consider points of the form (x, y) where x is the sine of an angle and y is the angle, then the coordinates of these points satisfy the equation $\sin y = x$. We can graph $\sin y = x$ by running the sine curve up and down the y-axis, as in Figure 8-6. The graph fails the vertical-line test for a function.

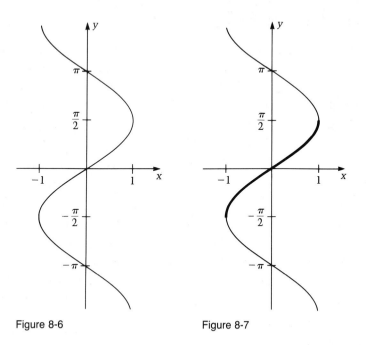

Figure 8-6   Figure 8-7

We solve the problem by somewhat arbitrarily selecting a single section of the graph that does satisfy the vertical-line test for a function. The section is shown in Figure 8-7; it means we are now considering only points whose coordinates satisfy the two equations

Figure 8-8

sin $y = x$ and $-\pi/2 \leq y \leq \pi/2$. With this restriction, we do have a function, denoted by $y = $ Arcsin x.*

What has happened so far? We have defined a new function, $y = $ Arcsin x, which starts with a number and gives us an angle. It is the inverse of the sine function. Keep in mind, however, that there is really nothing mysterious about this new function. In fact, just remember that

$$y = \text{Arcsin } x \quad \text{means} \quad \sin y = x \quad \text{and} \quad -\frac{\pi}{2} \leq y \leq \frac{\pi}{2}$$

The restriction on the y-values can also be remembered by noting that y represents angles that, in standard position on a rectangular coordinate system, have terminal sides in quadrants I or IV (see Figure 8-8).

EXAMPLE 1. Find Arcsin($\frac{1}{2}$).

In the Arcsine function, we start with a number and end up with an angle. Here we want the angle between $-\pi/2$ and $\pi/2$ whose sine is $\frac{1}{2}$. That angle is $\pi/6$. ■

You can use your calculator to compute the value of the Arcsine function, either by using a SIN^{-1} key if one is available, or by using first the INV or ARC key and then the SIN key. Be sure that you are operating in radian mode if you want the answer to be an angle in radian measure.

Practice 1. Use your calculator to find Arcsin(0.3), Arcsin (0.73), and Arcsin(−0.8).

Practice 2. Try to find Arcsin(2) on your calculator. What happens? Why?

We can define inverse trigonometric functions for the remaining five trigonometric functions. In each case, restrictions are imposed on the values of the angle to satisfy the definition of function, just as in the case of the Arcsine function.

$$y = \text{Arccos } x \quad \text{means} \quad \cos y = x \quad \text{and} \quad 0 \leq y \leq \pi$$

Quadrants I or II.

$$y = \text{Arctan } x \quad \text{means} \quad \tan y = x \quad \text{and} \quad -\frac{\pi}{2} < y < \frac{\pi}{2}$$

Quadrants I or IV.

*Another common notation for $y = $ Arcsin x is $y = $ Sin^{-1}x, which is read y is the inverse sine of x; this does *not* mean that Sin x is being raised to the negative one power.

$$y = \text{Arccot } x \quad \text{means} \quad \cot y = x \quad \text{and} \quad 0 < y < \pi$$

Quadrants I or II.

$$y = \text{Arcsec } x \quad \text{means} \quad \sec y = x \quad \text{and} \quad 0 \le y \le \pi$$

Quadrants I or II.

$$y = \text{Arccsc } x \quad \text{means} \quad \csc y = x \quad \text{and} \quad -\frac{\pi}{2} \le y \le \frac{\pi}{2}$$

Quadrants I or IV.

The Arccotangent, Arcsecant, and Arccosecant functions are seldom used, and we concentrate on Arcsine, Arccosine, and Arctangent.

EXAMPLE 2.   Find cos(Arctan 2).

We note first that the problem makes sense. Beginning with the number 2, we apply the Arctangent function; this will result in an angle, and we may then take the cosine of that angle. A calculator, once again, computes these functions easily.

$$\cos(\text{Arctan } 2) = \cos(1.107) = 0.4472 \quad \blacksquare$$

Practice 3.   Find sin(Arccos 0.6) and Arctan(sin 1.2). (Use your calculator, of course.)

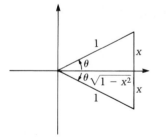

Figure 8-9

EXAMPLE 3.   Find tan(Arcsin x).

A calculator certainly will not help us here, but drawing a picture will. Arcsin x is an angle θ in quadrant I or IV whose sine is $x = x/1$. (The angle will be positive if x is positive and negative if x is negative.) Figure 8-9 illustrates both possibilities for θ. In either case, the third side of the triangle is $\sqrt{1 - x^2}$, by the Pythagorean theorem. Thus, in either case, $\tan \theta = x/\sqrt{1 - x^2}$ and tan(Arcsin x) $= x/\sqrt{1 - x^2}$. $\blacksquare$

Practice 4.   Find sin(Arccos x). (Remember that Arccos x will be an angle in quadrant I or II.)

Next we turn our attention to differentiation. In order to find dy/dx for the function $y = \text{Arcsin } u$, $u = u(x)$, we write $\sin y = u$ and use implicit differentiation.

$$\sin y = u$$

$$\cos y \frac{dy}{dx} = \frac{du}{dx}$$

Solving for dy/dx,

$$\frac{dy}{dx} = \frac{1}{\cos y}\frac{du}{dx} = \frac{1}{\sqrt{1 - \sin^2 y}}\frac{du}{dx} = \frac{1}{\sqrt{1 - u^2}}\frac{du}{dx}$$

Here we choose the positive square root for cos y because y, according to the restrictions of the Arcsine function, comes from either quadrant I or quadrant IV, and in either case cos y is nonnegative. Thus

$$\frac{d(\text{Arcsin } u)}{dx} = \frac{1}{\sqrt{1 - u^2}}\frac{du}{dx}$$

The same procedure works to find the derivative of $y = \text{Arccos } u$. Letting cos $y = u$, we differentiate and solve for dy/dx:

$$\cos y = u$$

$$-\sin y\,\frac{dy}{dx} = \frac{du}{dx}$$

$$\frac{dy}{dx} = \frac{-1}{\sin y}\frac{du}{dx} = \frac{-1}{\sqrt{1 - \cos^2 y}}\frac{du}{dx} = \frac{-1}{\sqrt{1 - u^2}}\frac{du}{dx}$$

We choose the positive root for sin y because y comes from quadrant I or quadrant II, and in either case sin y is nonnegative. The result is

$$\frac{d(\text{Arccos } u)}{dx} = \frac{-1}{\sqrt{1 - u^2}}\frac{du}{dx}$$

Finally, we find dy/dx for $y = \text{Arctan } u$.

$$\tan y = u$$

$$\sec^2 y\,\frac{dy}{dx} = \frac{du}{dx}$$

$$\frac{dy}{dx} = \frac{1}{\sec^2 y}\frac{du}{dx} = \frac{1}{1 + \tan^2 y}\frac{du}{dx} = \frac{1}{1 + u^2}\frac{du}{dx}$$

Therefore,

$$\frac{d(\text{Arctan } u)}{dx} = \frac{1}{1 + u^2}\frac{du}{dx}$$

EXAMPLE 4.   Differentiate $y = \text{Arcsin}(2x^2)$.

This is of the form $y = \text{Arcsin } u$ with $u = 2x^2$. Following the pattern,

$$\frac{dy}{dx} = \frac{1}{\sqrt{1 - (2x^2)^2}} \cdot \frac{d(2x^2)}{dx} = \frac{4x}{\sqrt{1 - 4x^4}} \quad \blacksquare$$

EXAMPLE 5. Differentiate $y = \text{Arctan}(3x + 1)$.

$$\frac{dy}{dx} = \frac{1}{1 + (3x + 1)^2} \cdot 3 = \frac{3}{2 + 6x + 9x^2} \quad \blacksquare$$

EXAMPLE 6. Find the derivative of $y = \text{Arccos}^3 3x$.

$$\frac{dy}{dx} = 3 \text{ Arccos}^2 3x \left(\frac{-1}{\sqrt{1 - 9x^2}} \right) \cdot 3$$

$$= \frac{-9 \text{ Arccos}^2 3x}{\sqrt{1 - 9x^2}} \quad \blacksquare$$

Practice 5. Differentiate $y = 3 \text{ Arcsin}(1 - x)$.

Exercises / Section 8-4

Exercises 1–8: Find the indicated value. Do not use a calculator.

1. Arcsin 1
2. $\text{Arcsin}(-1)$
3. Arccos 0
4. $\text{Arccos}(-1)$
5. Arctan 1
6. $\text{Arctan}(-1)$
7. $\sin(\text{Arcsin } 0.42)$
8. $\text{Arccos}(\cos 1)$

Exercises 9–22: Find the indicated value, using a calculator.

9. $\text{Arcsin}(-0.6)$
10. $\text{Arcsin}(0.9)$
11. $\text{Arccos}(0.12)$
12. $\text{Arccos}(-0.4)$
13. $\text{Arctan}(\sqrt{7})$
14. $\text{Arctan}(-14.1)$
15. $\cos(\text{Arctan } 1.5)$
16. $\sin(\text{Arccos } 0.8)$
17. $\tan(\text{Arcsin } 0.62)$
18. $\cos(\text{Arcsin } 0.13)$
19. $\text{Arcsin}(\tan 0.3)$
20. $\text{Arctan}(\cos 3.1)$
21. $\text{Arccos}(\sin 2)$
22. $\text{Arcsin}(\cos 4.3)$

Exercises 23–28: Find an expression in terms of x.

23. $\sin(\text{Arctan } x)$
24. $\cos(\text{Arcsin } x)$
25. $\tan(\text{Arccos } x)$
26. $\cos(\text{Arctan } x)$
27. $\sin(2 \text{ Arccos } x)$ (*Hint:* Use a trigonometric identity.)

28. $\cos(2 \text{ Arcsin } x)$ (*Hint:* Use a trigonometric identity.)

Exercises 29–48: Find the derivative.

29. $y = \text{Arcsin}(1 - 2x)$
30. $y = 2 \text{ Arcsin}(x^2 - 1)$
31. $y = 4 \text{ Arcsin}(4x^4)$
32. $y = \text{Arcsin } \sqrt{x}$
33. $y = \text{Arccos } 3x$
34. $y = 2 \text{ Arccos}\left(\frac{x + 1}{2}\right)$
35. $y = \frac{1}{2} \text{ Arccos } \sqrt{1 - x}$
36. $y = \text{Arccos}(x^2 - 2x)$
37. $y = \text{Arctan}(1 - x)$
38. $y = 2 \text{ Arctan}\left(\frac{1}{x}\right)$
39. $y = \text{Arcsin}^2(x^2)$
40. $y = \text{Arctan}^3(2x)$
41. $y = \sqrt{\text{Arccos}(1 - x)}$
42. $y = \frac{1}{\text{Arcsin } 3x}$
43. $y = x \text{ Arcsin } x^2$

44. $y = x^2 \text{ Arccos } x$

45. $y = \text{Arctan}\left(2x + \dfrac{1}{1 + 4x^2}\right)$

46. $y = \text{Arcsin } x - \sqrt{1 - x^2}$

47. $y = \dfrac{\text{Arccos } x}{x}$

48. $y = \dfrac{\sin x}{\text{Arcsin } x}$

49. Prove that $\dfrac{d(\text{Arccot } u)}{dx} = \dfrac{-1}{1 + u^2} \dfrac{du}{dx}$.

50. Prove that $\dfrac{d(\text{Arcsec } u)}{dx} = \dfrac{1}{\sqrt{u^2(u^2 - 1)}} \dfrac{du}{dx}$.

51. Prove that $\dfrac{d(\text{Arccsc } u)}{dx} = \dfrac{-1}{\sqrt{u^2(u^2 - 1)}} \dfrac{du}{dx}$.

52. If $y = \text{Arctan } 2x$, show that $(1 + \tan^2 y)\dfrac{dy}{dx} = 2$.

53. Find the slope of a tangent line to the curve $y = \text{Arcsin}^2 2x$ at $x = 0.3$.

54. Distance s (in meters) as a function of time t (in seconds) for a particular object is given by the equation $s = \sin(\text{Arctan } 2t)$. Find the velocity at $t = 3$ s.

8-5 Applications

In Chapter 5 we studied a number of problem types that could be solved by use of the derivative. These types included writing equations of tangent lines and normal lines to curves, sketching graphs of functions, maximizing or minimizing quantities, finding velocity and acceleration vectors, finding the time rate of change of a quantity, and using differentials to estimate changes in quantities. Now that we know how to differentiate trigonometric and inverse trigonometric functions, we can do any of these problem types that involve such functions.

EXAMPLE 1.   Find the equation of the tangent line to the curve $y = \sin^2(1 - 2x)$ at $x = 2$.

The point on the curve has x-coordinate $x = 2$; the corresponding y-value is $\sin^2(-3) = 0.0199$. To find the slope of the tangent line, we differentiate.

$$y = \sin^2(1 - 2x)$$

$$\frac{dy}{dx} = 2 \sin(1 - 2x) \cos(1 - 2x)(-2)$$

and

$$\left.\frac{dy}{dx}\right|_{x=2} = -0.5588$$

Using the point-slope form for the equation of a line, we get

$$y - 0.0199 = -0.5588(x - 2)$$

Simplifying,

$$y = -0.5588x + 1.1375 \quad \blacksquare$$

EXAMPLE 2. Sketch the graph of the function $y = \sin x - \cos x$ for $0 \leq x \leq 2\pi$. Indicate any maximum points, minimum points, and inflection points.

First we find the derivatives.

$$\frac{dy}{dx} = \cos x + \sin x \qquad \frac{d^2y}{dx^2} = -\sin x + \cos x$$

To find critical points, we solve

$$y' = \cos x + \sin x = 0$$
$$\sin x = -\cos x$$
$$\frac{\sin x}{\cos x} = \tan x = -1$$
$$x = \frac{3\pi}{4}, \frac{7\pi}{4}$$

Using the second derivative test,

$$y''\Big|_{x=3\pi/4} = \frac{-2}{\sqrt{2}} < 0 \qquad \text{maximum at } \left(\frac{3\pi}{4}, \frac{2}{\sqrt{2}}\right) = \left(\frac{3\pi}{4}, 1.414\right)$$

$$y''\Big|_{x=7\pi/4} = \frac{2}{\sqrt{2}} > 0 \qquad \text{minimum at } \left(\frac{7\pi}{4}, \frac{-2}{\sqrt{2}}\right) = \left(\frac{7\pi}{4}, -1.414\right)$$

Inflection points are found by setting d^2y/dx^2 equal to zero.

$$-\sin x + \cos x = 0$$
$$\sin x = \cos x$$
$$\tan x = 1$$
$$x = \frac{\pi}{4}, \frac{5\pi}{4}$$

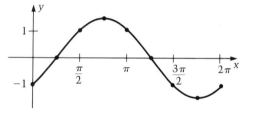

Figure 8-10

Inflection points occur at $(\pi/4, 0)$ and $(5\pi/4, 0)$. The points in the table are other points on the graph that are easily found.

x	0	$\pi/2$	π	$3\pi/2$	2π
y	-1	1	1	-1	-1

The graph is shown in Figure 8-10. ∎

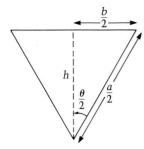

Figure 8-11

EXAMPLE 3. A gutter with a triangular cross section is to be made by creasing a piece of metal of width a lengthwise and bending up the sides. Find the angle between the sides that makes the cross sectional area a maximum.

Figure 8-11 illustrates the situation. We want the value of θ that maximizes the area. Our first task is, therefore, to express area A as a function of θ: $A = \frac{1}{2}bh$, where

$$\frac{b/2}{a/2} = \sin\frac{\theta}{2} \quad \text{and} \quad \frac{h}{a/2} = \cos\frac{\theta}{2}$$

so that

$$b = a\sin\frac{\theta}{2} \quad \text{and} \quad h = \frac{a}{2}\cos\frac{\theta}{2}$$

Therefore,

$$A = \frac{1}{2}\left(a\sin\frac{\theta}{2}\right)\left(\frac{a}{2}\cos\frac{\theta}{2}\right) = \frac{1}{8}a^2\left(2\sin\frac{\theta}{2}\cos\frac{\theta}{2}\right) = \frac{1}{8}a^2\sin\theta$$

Next, we find $dA/d\theta$ and set it equal to zero, remembering that a is constant.

$$\frac{dA}{d\theta} = \frac{1}{8}a^2\cos\theta$$

$$\frac{1}{8}a^2\cos\theta = 0$$

$$\theta = \frac{\pi}{2}$$

Note that $d^2A/d\theta^2 = -\frac{1}{8}a^2\sin\theta$, which is negative when $\theta = \pi/2$, so that A is indeed a maximum when $\theta = \pi/2$. ∎

EXAMPLE 4. The parametric equations of a moving object are $x = \cos^2 t$, $y = \sin 2t$. Find the velocity and acceleration vectors for $t = \pi/4$.

We differentiate x and y to find the magnitudes v_x and v_y of the velocity components, $\mathbf{v}_x$ and $\mathbf{v}_y$.

$$v_x = \frac{dx}{dt} = -2\cos t\sin t$$

$$v_x\bigg|_{t=\pi/4} = -1$$

$$v_y = \frac{dy}{dt} = 2\cos 2t$$

$$v_y\bigg|_{t=\pi/4} = 0$$

$$v = \sqrt{v_x{}^2 + v_y{}^2} = \sqrt{1 + 0} = 1$$

$$\tan\alpha = \frac{v_y}{v_x} = \frac{0}{-1}$$

$$\alpha = \pi$$

α equals π rather than 0 because v_x is negative.

To find acceleration, we differentiate again.

$$a_x = \frac{dv_x}{dt} = -2(\cos^2 t - \sin^2 t) = -2 \cos 2t$$

$$a_x \bigg|_{t=\pi/4} = 0$$

$$a_y = \frac{dv_y}{dt} = -4 \sin 2t$$

$$a_y \bigg|_{t=\pi/4} = -4$$

$$a = \sqrt{a_x^2 + a_y^2} = \sqrt{0 + 16} = 4$$

$$\tan \beta = \frac{a_y}{a_x} = \frac{-4}{0}$$

$$\beta = -\pi/2$$

The velocity vector has magnitude 1 and direction π, and the acceleration vector has magnitude 4 and direction $-\pi/2$. ∎

EXAMPLE 5. A particle is moving along a curve whose polar equation is $r = 4 \sin 2\theta$ (r is measured in centimeters). The angular velocity $d\theta/dt$ has a constant value of 3 rad/s. Find the magnitude of the velocity vector (in centimeters per second) when $\theta = \pi/6$.

We recall that the magnitudes of the perpendicular components of velocity in polar coordinates are given by the equations

$$v_r = \frac{dr}{dt} \qquad v_\theta = r \frac{d\theta}{dt}$$

Differentiating r with respect to t, we get

$$v_r = \frac{dr}{dt} = 8 \cos 2\theta \frac{d\theta}{dt}$$

$$v_\theta = r \frac{d\theta}{dt} = 4 \sin 2\theta \frac{d\theta}{dt}$$

We evaluate these equations for $\theta = \pi/6$ and $d\theta/dt = 3$.

$$v_r = 8 \cdot \frac{1}{2} \cdot 3 = 12$$

$$v_\theta = 4 \cdot \frac{\sqrt{3}}{2} \cdot 3 = 6\sqrt{3}$$

Finally,

$$v = \sqrt{v_r^2 + v_\theta^2} = 15.87 \text{ cm/s} \quad ∎$$

EXAMPLE 6. One member of a surveying team, walking at the rate of 10 m/min, travels east for 8 min and then turns south. A second surveyor stands at the original point. How fast is the line connecting the two people rotating 12 min after the start?

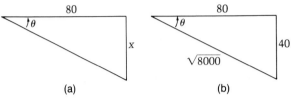

(a)   (b)

Figure 8-12

In Figure 8-12(a), 80 m is constant (10 m/min times 8 min), but x and θ are variables. We are given $dx/dt = 10$ and want to find $d\theta/dt$. We need a relationship between x and θ. From the figure,

$$\frac{x}{80} = \tan \theta$$

$$x = 80 \tan \theta$$

Differentiating with respect to t,

$$\frac{dx}{dt} = 80 \sec^2\theta \frac{d\theta}{dt}$$

At $t = 12$, $x = 40$; from Figure 8-12(b), $\sec \theta = \sqrt{8000}/80 = \sqrt{5}/2$. Substituting into the equation and solving for $d\theta/dt$,

$$10 = 80\left(\frac{\sqrt{5}}{2}\right)^2 \frac{d\theta}{dt}$$

$$\frac{d\theta}{dt} = 0.1$$

The line rotates at the rate of 0.1 rad/min.

Another way to express the relationship between x and θ is by the equation

$$\theta = \text{Arctan } \frac{x}{80}$$

Differentiating with respect to t,

$$\frac{d\theta}{dt} = \frac{1}{1 + (x/80)^2} \cdot \frac{1}{80} \cdot \frac{dx}{dt}$$

Substituting the known values,

$$\frac{d\theta}{dt} = \frac{1}{1 + (40/80)^2} \cdot \frac{1}{80} \cdot 10 = 0.1$$

Here is a case where use of the inverse trigonometric function simplifies the work involved. ∎

EXAMPLE 7. The position of a certain object moving in simple harmonic motion is given by

$$s = 2 \sin 3t$$

Estimate the change in s as t changes from 5.00 to 5.05.

We use the differential ds to approximate the change Δs.

$$\begin{aligned} ds &= 6 \cos 3t \; dt \\ &= 6 \cos(15)(0.05) \\ &= -0.228 \quad \blacksquare \end{aligned}$$

EXAMPLE 8. An angle θ is measured by computing $\tan \theta = y/x$. If y is measured to be 3 cm $\pm$ 0.02 cm and x is measured to be 5 cm $\pm$ 0.02 cm, estimate the maximum error in computing θ.

We know that $\theta = \text{Arctan } y/x$. The maximum error is estimated by using the total differential.

$$d\theta = \frac{\partial \theta}{\partial x} \, dx + \frac{\partial \theta}{\partial y} \, dy$$

$$= \frac{1}{1 + (y/x)^2} \left(\frac{-y}{x^2} \right) dx + \frac{1}{1 + (y/x)^2} \left(\frac{1}{x} \right) dy$$

Evaluating this expression with $y = 3$, $dy = 0.02$, $x = 5$, and $dx = 0.02$, we get

$$d\theta = 0.0012$$

The maximum error in computing θ is 0.0012 rad. $\quad \blacksquare$

Exercises / Section 8-5

1. Find the equation of the tangent line to the curve $y = \sin 2x \cos 3x$ at $x = 1.2$.

2. Find the equation of the normal line to the curve $y = \sin 2x \cos 3x$ at $x = 1.2$.

3. Find the equation of the tangent line to the curve $y = \text{Arctan}(2x + 1) + \tan x$ at $x = 1$.

4. Find the equation of the normal line to the curve $y = \text{Arctan}(2x + 1) + \tan x$ at $x = 1$.

Exercises 5–8: Sketch the graph of the function for $0 \le x \le 2\pi$. Indicate any maximum points, minimum points, and inflection points.

5. $y = \sin^2\left(\frac{x}{2}\right)$

6. $y = x + \sin x$

7. $y = \sin^2 x + \cos x$

8. $y = x - 2 \cos x$

9. Current i in a certain electrical circuit t seconds after the start of an experiment is given by the expression $i = 40 \sin 3t$. Find the first time after the start of the experiment when the current peaks at a maximum.

10. A certain individual's blood pressure at time t is given by the expression $P = 90 + 25 \sin 6t$. Find the values of the maximum and minimum pressure. When do these values occur?

11. Use trigonometric functions to find the area of the largest rectangle that can be inscribed in a circle of radius a.

12. Use trigonometric functions to find the shape of the rectangle with maximum perimeter that can be inscribed in a circle of radius a.

13. A mural 4 m high is painted with the bottom edge 1 m above an observer's eye level. The

most pleasing view is obtained when the observer stands so that the vertical angle in his or her line of sight subtended by the mural is a maximum. Use inverse trigonometric functions to find how far away from the wall the observer should stand.

14. A ray of light travels in air with velocity v_1 from P_1 to a point S on the surface of the water and then in water with velocity v_2 to point P_2 (see Figure 8-13). The ray of light will follow the path that takes the least time to travel.

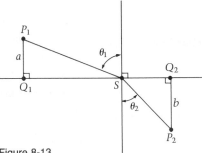

Figure 8-13

Show that the minimum travel time occurs when

$$\frac{\sin \theta_1}{\sin \theta_2} = \frac{v_1}{v_2}$$

(*Hint:* Find an equation for time as a function of θ_1 and θ_2; find a second equation relating θ_1 and θ_2 to the distance $\overline{Q_1Q_2}$. Differentiate both expressions with respect to θ_1, using the fact that distance $\overline{Q_1Q_2}$ is a constant.)

15. The parametric equations of a moving object are $x = \sin 2t$, $y = t^2$. Find the velocity vector for $t = \pi/2$.

16. Find the acceleration vector at $t = \pi/2$ for the object in Exercise 15.

17. A particle is moving along the curve $r = 2 \sec \theta$ at a constant angular velocity of 2 rad/s (r is measured in meters). Find the magnitude of the velocity vector when $\theta = \pi/4$.

18. An object moving with constant angular velocity of 0.3 rad/s traces the path $r = -\cos 3\theta$ (r is measured in centimeters). Find the magnitude of the velocity vector when $\theta = 1.4$.

19. A hot-air balloon is rising vertically from a point 100 m away from an observer. The angle

of elevation is increasing at the rate of 0.8 rad/min. Find the rate of the ascension of the balloon when the angle of elevation is 0.5 rad.

20. A forest ranger at the top of a 20-m observation tower tracks the progress of a fire. When the angle of depression is 0.3 rad, it is increasing at the rate of 0.02 rad/min. How fast is the fire approaching the bottom of the tower?

21. A person fishing hooks a fish from the bow of a boat that is 2 m above the water; the fish moves away from the boat along the surface of the water. The angle of depression of the line decreases at the rate of 0.1 rad/s. How fast is the fish traveling when the angle of depression is 0.4 rad?

22. A lighthouse is on a rock 1000 m out from a straight shoreline. If the light revolves at the rate of 2 rpm, how fast does the beam of light travel along the shore at a point 4000 m away from the point directly opposite the lighthouse?

23. An airplane at an altitude of 8 km travels east at a speed of 800 km/h. Twelve minutes after the plane passes directly over an observer, what is the rate at which the angle of elevation from the observer to the plane is changing?

24. One car travels west from point P at 50 km/h, while a second car, beginning at the same time from a point 80 km south of P, travels north at 60 km/h. What is the rate of rotation of a line drawn between the two cars at $t = 1$ h?

25. Current in a certain electric circuit is given by the equation $i = 25 \cos^2 t$, where i is measured in amperes and t in seconds. Use differentials to estimate the change in i as t changes from 2.00 s to 2.03 s.

26. The height of a tree is established by measuring the angle of elevation of the top of the tree from a point 40 m from the base of the tree. If the angle is measured as 0.5 rad ± 0.02 rad, use differentials to estimate the maximum error in computing the height of the tree.

27. The index of refraction n of a substance is defined by $n = \sin \theta_1/\sin \theta_2$, where θ_1 is the angle of incidence of light upon the surface and θ_2 is the angle of refraction. Estimate the maximum error in computing n if θ_1 is measured as 0.5 ± 0.02 and θ_2 is measured as 0.4 ± 0.01.

8-6 The Exponential and Logarithmic Functions

In addition to the trigonometric functions, other transcendental functions with many applications in technology and science are the exponential and logarithmic functions. For the rest of this chapter, we study the properties of these functions and how to differentiate them.

The **exponential function** is a function of the form

$$y = b^x$$

where b is a constant (the **base**) and x, the independent variable, is the power or **exponent**. For example, if $b = 2$, then selected values of the independent variable x and corresponding values of the dependent variable y are

x	-2	-1	0	1	2	3
y	$\frac{1}{4}$	$\frac{1}{2}$	1	2	4	8

We would like the exponential function to be defined and have real number values for all real number values of x, but this puts certain restrictions on the allowable values for the base. If b is negative—for instance, $b = -3$—then $(-3)^{1/2}$ would represent $\sqrt{-3}$, which is not a real number. Therefore, we rule out negative values for the base. Also, the base cannot be zero, because 0^0 is undefined. Finally, we require that $b \neq 1$ simply because $1^x = 1$ for any value of x, and this is a boring function. Therefore, the base in the exponential function must be a positive number different from 1, and we assume from now on that b has such a value.

Suppose we again choose a base of 2 and consider the exponential function $y = 2^x$. We have already computed the value of this function for certain integer values of x; these points on the graph of the function are shown in Figure 8-14. It is easy to compute 2^3 and not too difficult to at least approximate $2^{1/2} = \sqrt{2}$, but what is the meaning of $2^{\sqrt{2}}$ or 2^{π}, for example? We assign values to such expressions so that the function $y = 2^x$ is continuous. That is, we "fill in" the graph of Figure 8-14 to get Figure 8-15, and then this graph defines the value of $y = 2^x$ for any real number x. The values of $y = b^x$ for any allowable base b are defined by the same procedure we have used here for $b = 2$.

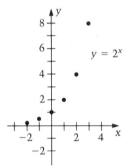

Figure 8-14

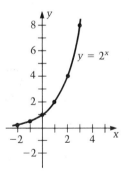

Figure 8-15

Practice 1. Sketch a graph of the function $y = 3^x$.

Whatever base b is used in the exponential function, $b^0 = 1$ and $b^x > 0$ for any real number x.

Rules of exponents, with which we are familiar for integer or rational exponents, still hold for any real number exponent. Thus

$$b^v b^w = b^{v+w}$$

$$\frac{b^v}{b^w} = b^{v-w}$$

$$(b^v)^w = b^{vw}$$

In the exponential function $y = b^x$, given any value for x, the independent variable (which is the exponent), we find the corresponding value for y, the dependent variable (which is the number b raised to that power). Suppose we reverse the roles of the dependent and independent variable; if we are given the number that is b raised to some power, can we find the corresponding power? This gives us a function of the form $y = f(x)$, where x is the value of b raised to a power and y is that power. This function is called the **logarithmic function to the base b**, denoted by $y = \log_b x$. Therefore,

$$y = \log_b x \quad \text{means} \quad b^y = x$$

EXAMPLE 1. The exponential equation $4^2 = 16$ becomes $2 = \log_4 16$ in logarithmic form. The logarithmic equation $\frac{1}{3} = \log_8 2$ becomes $8^{1/3} = 2$ in exponential form. ■

Practice 2. Write $16^{1/2} = 4$ in logarithmic form. Write $3 = \log_2 8$ in exponential form.

EXAMPLE 2. Solve the equation $2 = \log_3 x$.

By putting the logarithmic equation in exponential form, we get $3^2 = x$; the solution is $x = 9$. ■

Practice 3. Solve the equation $y = \log_2 32$.

Word of Advice

If you remember that

$$y = \log_b x \quad \text{and} \quad b^y = x$$

mean the same thing, then logarithms will not seem difficult.

Because the logarithmic equation $y = \log_b x$ is the exponential equation $b^y = x$, we can deduce certain facts about the logarithmic function. The base b must be a positive number different from 1. Whatever the base, $\log_b 1 = 0$. Also, because $b^y > 0$ for all y, when we write $b^y = x$ or $y = \log_b x$, we see that the logarithmic function is defined only for positive values of the independent variable.

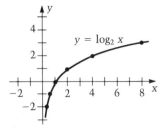

Figure 8-16

EXAMPLE 3. Graph the function $y = \log_2 x$.

Writing the logarithmic equation in exponential form, we get $2^y = x$. Next, we get a table of values for (x, y) by choosing values for y and computing corresponding values for x.

x	$\frac{1}{4}$	$\frac{1}{2}$	1	2	4	8
y	-2	-1	0	1	2	3

Connecting these points in a continuous curve, we obtain the graph shown in Figure 8-16. Note that the function is defined only for positive values of x. ∎

We know that $y = 2^x$ and $y = \log_2 x$ represent two functions whose dependent and independent variables have been reversed. A comparison of the two graphs (Figure 8-17) shows this. Each graph is a *reflection* of the other through the line $y = x$.

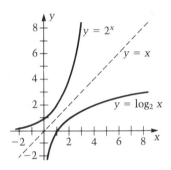

Figure 8-17

Practice 4. Sketch a graph of $y = \log_3 x$ by making use of your answer to Practice 1.

There are three rules of logarithms that are direct consequences of the definition of the logarithmic function and the three rules of exponents mentioned earlier in this section. These rules, which we prove shortly, are

$$\log_b rs = \log_b r + \log_b s \tag{1}$$

$$\log_b\left(\frac{r}{s}\right) = \log_b r - \log_b s \tag{2}$$

$$\log_b r^n = n \log_b r \tag{3}$$

To prove these rules, let

$$\log_b r = v \quad \text{and} \quad \log_b s = w \tag{4}$$

Then

$$b^v = r \quad \text{and} \quad b^w = s$$

so that

$$rs = b^v b^w = b^{v+w} \qquad \frac{r}{s} = \frac{b^v}{b^w} = b^{v-w} \qquad r^n = (b^v)^n = b^{vn}$$

We now have three exponential equations, which can be rewritten in logarithmic form. Thus

$$rs = b^{v+w} \quad \text{means} \quad \log_b rs = v + w$$

but from Equations (4), $v + w = \log_b r + \log_b s$. Therefore,

$$\log_b rs = \log_b r + \log_b s$$

Similarly, $\dfrac{r}{s} = b^{v-w}$ means $\log_b\left(\dfrac{r}{s}\right) = v - w = \log_b r - \log_b s$

and $r^n = b^{vn}$ means $\log_b r^n = vn = n \log_b r$

This proves the three rules of logarithms. Note that Equations (1), (2), and (3) all involve logarithms to the same base. These rules of logarithms can be helpful because they allow us to expand or to condense expressions involving logarithms.

EXAMPLE 4.   Write as a sum of logarithms: $\log_b xy^2$.
Using Equations (1) and (3),

$$\log_b xy^2 = \log_b x + \log_b y^2 = \log_b x + 2 \log_b y \quad \blacksquare$$

EXAMPLE 5.   Write as a single logarithm: $\log_b 3 - \frac{1}{2} \log_b x$.
Using Equations (3) and (2),

$$\log_b 3 - \frac{1}{2} \log_b x = \log_b 3 - \log_b \sqrt{x} = \log_b \frac{3}{\sqrt{x}} \quad \blacksquare$$

Practice 5.   Write as a single logarithm: $\log_b 2 + \log_b 5 - \log_b w$.

Equations (1)–(3) can be remembered by chanting the following mystic phrases:

The log of a product is the sum of the logs.

The log of a quotient is the difference of the logs.

The log of something to a power is the power times the log of the something.

But one should not get too carried away with this "poetry."

Word of Advice
Note that there is no rule for the logarithm of a sum, nor for the product of logarithms, nor for the quotient of logarithms.

While the base of the logarithm can be any positive number other than 1, only two bases are widely used. Logarithms to the base 10 are called **common logarithms** and are denoted by log x; if no base is indicated, it is assumed to be 10. Common logarithms are extremely useful as computational aids, a role significantly reduced with the advent of electronic calculators. (However, it should be noted that some computations within a calculator are performed by using logarithms.)

The other base is an irrational number denoted by e. The value of e is about 2.7183. This number arises in calculus much the same way that π arises in geometry; it just turns out to be convenient to have it around. More formally, e is defined to be a limit—

$$e = \lim_{z \to 0} (1 + z)^{1/z}$$

(Here we cannot evaluate the limit by direct substitution because the exponent would be undefined.)

Practice 6. Complete the following table to convince yourself that the number e exists and has a value close to 2.7183.

z	1	0.1	0.01	0.001	0.0001	-0.1	-0.01	-0.001	-0.0001
$(1 + z)^{1/z}$	2	2.5937							

Logarithms to the base e are called **natural logarithms** and are denoted by ln x.

Although we shall deal only with natural logarithms, other bases are occasionally encountered; it is useful to be able to change base, such as from base b to base a. Suppose we want to find $\log_a x$ and we have available logarithms to the base b. Letting $p = \log_a x$ gives the equation $a^p = x$. We then take the logarithm to the base b of both sides of this equation.

$$p \log_b a = \log_b x \qquad \text{or} \qquad p = \frac{\log_b x}{\log_b a}$$

$$\text{Therefore,} \qquad \log_a x = \frac{\log_b x}{\log_b a}$$

Your calculator, if equipped with e^x and ln x functions, allows you to easily evaluate these functions for a given value of x. You may have to evaluate e^x by using first the INV key and then the LN key.

Exercises / Section **8-6**

Exercises 1–8: Change to logarithmic form.

1. $5^2 = 25$
2. $4^3 = 64$
3. $6^0 = 1$
4. $4^{1/2} = 2$
5. $27^{1/3} = 3$
6. $8^{2/3} = 4$
7. $4^{-2} = \frac{1}{16}$
8. $100^{-1/2} = \frac{1}{10}$

Exercises 9–16: Change to exponential form.

9. $\log_3 81 = 4$

10. $\log_2 32 = 5$

11. $\log_4 \frac{1}{4} = -1$

12. $\log_2 \frac{1}{4} = -2$

13. $\log_{36} 6 = \frac{1}{2}$

14. $\log_{27} 9 = \frac{2}{3}$

15. $\log_9 \frac{1}{3} = -\frac{1}{2}$

16. $\log_{16} \frac{1}{2} = -\frac{1}{4}$

Exercises 17–26: Solve the equation.

17. $4 = \log_2 x$

18. $\frac{2}{3} = \log_{27} x$

19. $y = \log_2 \frac{1}{8}$

20. $y = \log_8 16$

21. $\frac{1}{2} = \log_b 7$

22. $-2 = \log_b 4$

23. $\log_3 6 - \log_3 18 = x$

24. $\log_4 2 + 3 \log_4 2 = x$

25. $\log_2 x + \log_2 2x = 3$

26. $\log_3(x^2 - 1) - \log_3(x + 1) = 2$

Exercises 27–30: Write as a sum or difference of logarithms.

27. $\log_b r^2 s$

28. $\log_b(rs)^2$

29. $\log_b \frac{r^2}{s}$

30. $\log_b \frac{r}{s^2}$

Exercises 31–36: Write as a single logarithm.

31. $\log_b 3 - \log_b 7$

32. $\log_b 6 + \log_b x$

33. $3 \log_b x + 2 \log_b y$

34. $\frac{1}{2} \log_b x - \log_b y$

35. $2 \log_b x - 4 \log_b x^2 + \log_b y$

36. $\log_b x + 2 \log_b y + \frac{1}{2} \log_b x$

Exercises 37–40: Use your calculator.

37. Under certain conditions, wind velocity v (in centimeters per second) at a height of x centi-

meters above the ground is given by $v = 200 \ln(x/0.8)$. What is the velocity when $x = 40$ cm?

38. A manufacturing company estimates that its profit P (in dollars) on producing x units of an item is given by $P = 150 \ln(x + 1)$. What is the total profit on production of 200 units?

39. The voltage V (in volts) across a certain capacitor at any time t (in seconds) is given by $V = e^{-0.004t}$. Find V at $t = 7$ s.

40. When a heated object is immersed in a cooling solution, its temperature T (in degrees Celsius) after t seconds is given by $T = 18 + 35e^{-0.5t}$. Find T at $t = 10$ s.

Exercises 41–44: Functions known as **hyperbolic functions** have a number of applications. The **hyperbolic sine** of x, sinh x, is defined to be

$$\sinh x = \frac{e^x - e^{-x}}{2}$$

and the **hyperbolic cosine** of x, cosh x, is defined to be

$$\cosh x = \frac{e^x + e^{-x}}{2}$$

Hyperbolic functions behave much like trigonometric functions. Prove the following hyperbolic identities.

41. $\cosh^2 x - \sinh^2 x = 1$

42. $\sinh 2x = 2 \sinh x \cosh x$

43. $\cosh 2x = 2 \cosh^2 x - 1$

44. $\cosh 2x = \cosh^2 x + \sinh^2 x$

8-7   Derivatives of Logarithmic and Exponential Functions

We find the derivative of the logarithmic function by returning to the definition of the derivative. We let $y = f(x) = \log_b x$ (this implies $x > 0$), and use the four-step delta process to compute

$$\frac{dy}{dx} = \lim_{\Delta x \to 0} \frac{f(x + \Delta x) - f(x)}{\Delta x}$$

1. $f(x + \Delta x) = \log_b(x + \Delta x)$

2. $f(x + \Delta x) - f(x) = \log_b(x + \Delta x) - \log_b x = \log_b \frac{x + \Delta x}{x}$

$$= \log_b\left(1 + \frac{\Delta x}{x}\right)$$

3. $\dfrac{f(x + \Delta x) - f(x)}{\Delta x} = \dfrac{1}{\Delta x} \log_b\left(1 + \dfrac{\Delta x}{x}\right)$

Now here comes a nice maneuver. We multiply and divide by x and then use one of the rules of logarithms:

$$\frac{1}{\Delta x} \log_b\left(1 + \frac{\Delta x}{x}\right) = \frac{1}{x} \cdot \frac{x}{\Delta x} \log_b\left(1 + \frac{\Delta x}{x}\right) = \frac{1}{x} \log_b\left(1 + \frac{\Delta x}{x}\right)^{x/\Delta x}$$

4. $\displaystyle\lim_{\Delta x \to 0} \dfrac{f(x + \Delta x) - f(x)}{\Delta x} = \lim_{\Delta x \to 0} \dfrac{1}{x} \log_b\left(1 + \dfrac{\Delta x}{x}\right)^{x/\Delta x}$

Because for $x > 0$, $\Delta x \to 0$ exactly when $\Delta x/x \to 0$, we may replace $\lim_{\Delta x \to 0}$ with $\lim_{\Delta x/x \to 0}$ in the above expression. This results in

$$\frac{dy}{dx} = \lim_{\Delta x/x \to 0} \frac{1}{x} \log_b\left(1 + \frac{\Delta x}{x}\right)^{x/\Delta x} = \frac{1}{x} \log_b \lim_{\Delta x/x \to 0} \left(1 + \frac{\Delta x}{x}\right)^{x/\Delta x}$$

We now have a limit of the form

$$\lim_{z \to 0} (1 + z)^{1/z}$$

From Section 8-6, this limit exists and is denoted by e. Thus

$$\frac{d}{dx}(\log_b x) = \frac{1}{x} \log_b e$$

If we choose the base b of the logarithm function to be e, then $\log_e e = 1$ and

$$\frac{d}{dx}(\ln x) = \frac{1}{x}$$

This simplification is the reason why e is the "natural" logarithm base.

Finally, if $y = \ln u$, where u is a function of x, then by the chain rule,

$$\boxed{\frac{d}{dx}(\ln u) = \frac{1}{u}\frac{du}{dx}}$$

EXAMPLE 1.   Differentiate $y = \ln \sin x$.
This is of the form $y = \ln u$, where $u = \sin x$. The derivative is

$$\frac{dy}{dx} = \frac{1}{\sin x} \frac{d(\sin x)}{dx} = \frac{1}{\sin x} \cdot \cos x = \cot x \quad \blacksquare$$

EXAMPLE 2.   Find the derivative of $y = \ln\sqrt{1 - 2x}$.
Rather than treat this as an instance of $y = \ln u$ with $u = \sqrt{1 - 2x}$ (which would work but takes longer), let us use a rule of logarithms.

$$y = \ln(1 - 2x)^{1/2} = \tfrac{1}{2}\ln(1 - 2x)$$

Differentiating,

$$\frac{dy}{dx} = \frac{1}{2} \cdot \frac{1}{1-2x}(-2) = \frac{-1}{1-2x} \quad \blacksquare$$

EXAMPLE 3.   Find the derivative of $y = \ln^2(1 + 3x)$.

This expression does not denote the logarithm of something to a power; rather, it says to take the logarithm of something and then raise the result to a power. Unlike Example 2, no rules of logarithms apply, and we differentiate by using the power rule.

$$\frac{dy}{dx} = 2 \ln(1 + 3x) \cdot \frac{1}{1 + 3x} \cdot 3 = \frac{6 \ln(1 + 3x)}{1 + 3x} \quad \blacksquare$$

Practice 1.   Differentiate $y = \ln(5x + 2)$.

Practice 2.   Differentiate $y = \ln x^3$ and $y = \ln^3 x$.

In order to differentiate the exponential function $y = b^x$, we take the natural logarithm of both sides of the equation.

$$\ln y = \ln b^x = x \ln b$$

We then differentiate the resulting equation with respect to x, using implicit differentiation.

$$\ln y = x \ln b \qquad \frac{1}{y}\frac{dy}{dx} = \ln b$$

Solving for dy/dx gives

$$\frac{dy}{dx} = y \ln b = b^x \ln b$$

If we choose the base b of the exponential function to be e, then $\ln e = 1$ and

$$\frac{d(e^x)}{dx} = e^x$$

Finally, if $y = e^u$, where u is a function of x, then by the chain rule,

$$\boxed{\frac{d(e^u)}{dx} = e^u \frac{du}{dx}}$$

EXAMPLE 4. Find the derivative of $y = e^{3x}$.

This is of the form $y = e^u$ where $u = 3x$. The derivative is

$$\frac{dy}{dx} = e^{3x}\frac{d(3x)}{dx} = 3e^{3x} \quad \blacksquare$$

EXAMPLE 5. Differentiate $y = e^{\sin 2x}$.

$$\frac{dy}{dx} = e^{\sin 2x}\frac{d(\sin 2x)}{dx} = e^{\sin 2x}(\cos 2x)(2) = 2(\cos 2x)e^{\sin 2x} \quad \blacksquare$$

EXAMPLE 6. Find the derivative of $y = \ln \tan e^{x^2}$.

$$\frac{dy}{dx} = \frac{1}{\tan e^{x^2}} \cdot \sec^2 e^{x^2} \cdot e^{x^2} \cdot 2x = 2xe^{x^2} \cdot \frac{1}{\cos^2 e^{x^2}} \cdot \frac{\cos e^{x^2}}{\sin e^{x^2}}$$

$$= \frac{2xe^{x^2}}{\cos e^{x^2} \sin e^{x^2}} \quad \blacksquare$$

Practice 3. Differentiate $y = e^{4x^2}$.

Practice 4. Find the derivative of $y = xe^{2x}$.

Exercises / Section **8-7**

Exercises 1–30: Find the derivative.

1. $y = \ln(3x - 4)$

2. $y = \ln 2x^2$

3. $y = \ln \sqrt{x}$

4. $y = \ln \sqrt{4x + 7}$

5. $y = \ln \tan x$

6. $y = \ln \sec x$

7. $y = x^2\ln x^2$

8. $y = x \ln x^3$

9. $y = \dfrac{\ln(2x + 1)}{\ln(2x - 1)}$

10. $y = \dfrac{\ln x}{x}$

11. $y = \ln \sin^2 x$

12. $y = \ln(\ln x)$

13. $y = \ln \dfrac{x^2 - 1}{2x + 4}$

14. $y = \ln \dfrac{x^2}{\sin 2x}$

15. $y = \ln^2 \tan x$

16. $y = \sqrt{\ln(3x - 4)}$

17. $y = e^{2x}$

18. $y = 3e^{x^2}$

19. $y = e^{\sqrt{x}}$

20. $y = \sqrt{e^x}$

21. $y = xe^{\sin x}$

22. $y = (\cos x)e^{x^2}$

23. $y = \dfrac{e^x}{2x}$

24. $y = \dfrac{e^x}{e^{2x} - e^{-2x}}$

25. $y = \sin(\tan e^{x^2})$

26. $y = \ln \cos e^{3x}$

27. $y = \ln \dfrac{e^{3x}}{x + e^x}$

28. $y = e^{2x} \ln x$

29. $y = \text{Arcsin } e^{3x}$

30. $y = \ln \text{Arctan } 2x$

31. If $y = x - x \ln x$, show that $xy' + x - y = 0$.

32. If $y = \ln^2 x$, show that $(y')^2 = 4y/x^2$.

33. If $y = e^x + e^{2x}$, show that $y'' - 3y' + 2y = 0$.

34. If $y = e^x \cos x$, show that $y'' - 2y' + 2y = 0$.

35. Distance s as a function of time t for a particular object is given by the equation $s = e^{-t^2}$. Find the velocity at $t = 1$.

36. Distance s as a function of time t for a particular object is given by the equation $s = e^{-0.1t} \cos 0.4t$. Find the velocity at $t = 0.8$.

37. Distance s as a function of time t for a particular object is given by the equation $s = \ln(t^2 + 7)$. Find the velocity at $t = 4$.

38. Distance s as a function of time t for a particular object is given by the equation $s = \sin \ln(t^2 + 2t)$. Find the velocity at $t = 1.3$.

39. Find the slope of a tangent line to the curve $y = e^{x^2}$ at $x = 1$.

40. Find the slope of a tangent line to the curve $y = \ln(1 + \sin^2 x)$ at $x = \pi/4$.

Exercises 41–42: Read instructions for Exercises 41–44 of Section 8-6 for the definitions of the sinh and cosh functions.

41. Show that $\dfrac{d(\sinh x)}{dx} = \cosh x$.

42. Show that $\dfrac{d(\cosh x)}{dx} = \sinh x$.

Exercises 43–46: Apply the natural logarithm function to both sides of the equation and then use implicit differentiation to find dy/dx.

43. $y = x^x$ **44.** $y = x^{e^x}$

45. $y = (\ln x)^x$ **46.** $y = x^{\ln x}$

8·8 Applications

All of the usual applications of the derivative apply to problems involving logarithmic and exponential functions.

EXAMPLE 1. The population density of a certain animal species in a particular geographic area is estimated to vary as a function of time t (in centuries) from a fixed starting time by means of the equation $P(t) = \ln(10 + 9t - t^2)$, $0 \le t \le 6$. At what time is the population density greatest?

To find the critical points, we set dP/dt equal to zero.

$$\frac{dP}{dt} = \frac{1}{10 + 9t - t^2}(9 - 2t) = 0$$

$$9 - 2t = 0$$

$$t = \frac{9}{2}$$

The only critical point occurs at $t = \frac{9}{2}$. As a check that this point does produce a maximum value for P, we can compute P'' and evaluate it at $t = \frac{9}{2}$. The value of $P''(\frac{9}{2})$ is negative, and the population density is greatest at $t = \frac{9}{2}$. ∎

EXAMPLE 2. Sketch a graph of the function $y = \dfrac{\ln x}{x}$.

The logarithmic function is defined only for values of x greater than zero. An intercept occurs when

$$y = \frac{\ln x}{x} = 0$$

$$\ln x = 0$$

$$x = 1$$

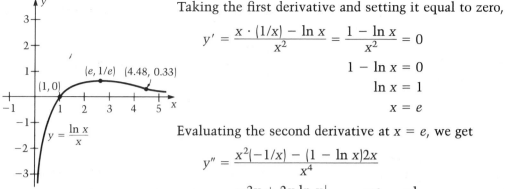

Figure 8-18

Taking the first derivative and setting it equal to zero,

$$y' = \frac{x \cdot (1/x) - \ln x}{x^2} = \frac{1 - \ln x}{x^2} = 0$$

$$1 - \ln x = 0$$

$$\ln x = 1$$

$$x = e$$

Evaluating the second derivative at $x = e$, we get

$$y'' = \frac{x^2(-1/x) - (1 - \ln x)2x}{x^4}$$

$$= \frac{-3x + 2x \ln x}{x^4}\Big|_{x=e} = \frac{-e}{e^4} = \frac{-1}{e^3} < 0$$

A maximum occurs at $(e, 1/e)$. Points of inflection occur where $y'' = 0$, or where $\ln x = \frac{3}{2}$ ($x = 4.48$). The graph is shown in Figure 8-18. ∎

EXAMPLE 3. Current as a function of voltage at a *p-n* junction of a germanium semiconductor at room temperature is given by the equation $I = I_0(e^{39V} - 1)$, where I_0 is a constant. Find the rate of change of current with respect to voltage.

$$\frac{dI}{dV} = I_0(39e^{39V}) \quad ∎$$

Exercises / Section **8-8**

1. Find the equation of the tangent line to the curve $y = e^{1/x}$ at $x = 2$.

2. Find the equation of the normal line to the curve $y = e^{1/x}$ at $x = 2$.

3. Find the equation of a line tangent to $y = x \ln x$ and parallel to the line $4x - 2y + 5 = 0$.

4. Find the equation of a line tangent to $y = \ln x^2$ and perpendicular to the line $2x + 8y + 9 = 0$.

Exercises 5–8: Sketch the graph of the function.

5. $y = e^x - x$ 6. $y = x^2 e^x$

7. $y = \ln^2 x$ 8. $y = x^2 \ln x$

9. An equation of damped vibratory motion is $y = e^{-0.3t} \sin 2\pi t$. Find the smallest value of t for which y is a maximum.

10. A certain utility company estimates customer demand D for electric power each day (in kilowatts) as a function of t, the number of hours past midnight, $0 \leq t \leq 12$. The equation is $D = 94 \ln(2 + 14t - t^2)$. When does maximum demand occur?

11. The parametric equations of a moving object are $x = \ln(t^2 + e^t)$, $y = e^{-0.02t}$. Find the velocity vector for $t = 0.4$.

12. The position of a moving object at any time t is given by the equations $x = e^{t^2 + \tan t}$, $y = \sin e^t$. Find the velocity vector at $t = 0$.

13. Voltage (in volts) in a particular electric circuit as a function of time t (in seconds) is given by $V = 120(1 - e^{-0.3t})$. Find an expression for the time rate of change of V.

14. The population P of a bacteria colony at time t (in minutes) is given by $P = 35,000e^{0.7t}$. Find an expression for the rate of growth.

15. If a rocket ejects gases at a constant velocity u, $v(t)$ is the velocity of the rocket at time t after

lift-off and $m(t)$ is the mass of the rocket at time t, then

$$v(t) = v(0) + u \ln \frac{m(0)}{m(t)}$$

Show that the acceleration $a(t)$ of the rocket satisfies the equation

$$a(t)m(t) = -um'(t)$$

16. During an epidemic, the number I of infected individuals as a function of weeks t since the beginning of the epidemic is given by

$$I = \frac{M}{1 + Be^{-kt}}$$

where M, B, and k are constants. Show that

$$\frac{dI}{dt} = \frac{k}{M}I(M - I)$$

17. The equation of free oscillations has the form

$$y'' + 2by' + a^2y = 0$$

where a and b are constants and y is the displacement of the object. Show that the function

$$y(t) = e^{-bt}\frac{\sinh(ct)}{c}$$

satisfies this equation, where $c = (b^2 - a^2)^{1/2}$. (See the exercises in Sections 8-6 and 8-7 for the definition and properties of the sinh function.)

18. The basic equation for the theory of thermal breakdown of a dielectric is of the form

$$\frac{d^2v(x)}{dx^2} + ke^{v(x)} = 0$$

for k a constant. Show that the temperature function

$$v(x) = v(0) - 2 \ln \cosh\left(x\sqrt{\frac{k}{2}e^{v(0)}}\right)$$

satisfies this equation. (See the exercises in Sections 8-6 and 8-7 for the definition and properties of the cosh function.)

19. The velocity (in centimeters per second) of an object moving through a certain medium in which it encounters resistance is given by

$$v = 64(1 - e^{-0.02t})$$

Use differentials to estimate the change in velocity as t changes from 1 s to 1.03 s.

20. A company estimates its profit P per item on a certain item as a function of the number x of items produced. The equation is

$$P = 4 \ln(2 + x^2 + 0.01x^3)$$

for $10 \le x \le 10,000$. Use differentials to estimate the change in P as x changes from 2000 to 2100.

21. Find the total differential for $z = e^{(x + y^2)}\ln(x^2 + 1)$.

22. The current i in a certain electric circuit containing a variable resistor R is given by $i = 8(1 - e^{-Rt/12})$. Use the total differential to estimate the change in i (in amperes) if R changes from 1 Ω to 1.02 Ω and t changes from 15 s to 16 s.

STATUS CHECK

Now that you are at the end of Chapter 8, you should be able to:

section **8-1**   Prove trigonometric identities.

section **8-2**   Differentiate functions that involve the sine function.
Find the slope of the tangent line at a given point to a function that involves the sine function.
Given an expression for distance as a function of time that involves the sine function, find the velocity at a specific time.

section **8-3**   Differentiate functions that involve the trigonometric functions.
Find the slope of the tangent line at a given point to a function that involves the trigonometric functions.

Given an expression for distance as a function of time that involves the trigonometric functions, find the velocity at a specific time.

section **8-4**   Evaluate an inverse trigonometric function at a given number.

Find expressions in terms of x for trigonometric functions of inverse trigonometric functions of x.

Differentiate functions that involve the inverse trigonometric functions.

section **8-5**   Using functions that involve trigonometric functions or inverse trigonometric functions:
   • Find the equations of the tangent line and normal line to a curve at a given point.
   • Sketch the graph of a function, indicating maximum points, minimum points, and inflection points.
   • Find the maximum or minimum value of a quantity.
   • Find the velocity and acceleration vectors at a specific time from parametric equations for a moving object.
   • Find the magnitude of the velocity vector at a specific time from polar parametric equations for a moving object.
   • Find the time rate of change of a quantity given the time rate of change of a related quantity.
   • Estimate changes in quantities by means of differentials.

section **8-6**   Change an exponential equation to its corresponding logarithmic form.

Change a logarithmic equation to its corresponding exponential form.

Solve equations involving logarithms.

Using the rules of logarithms, expand or condense an expression involving logarithms.

section **8-7**   Differentiate functions that involve logarithmic or exponential functions.

Find the slope of the tangent line at a given point to a function involving logarithmic or exponential functions.

Given an expression for distance as a function of time that involves logarithmic or exponential functions, find the velocity at a specific time.

section **8-8**   Using functions that involve logarithmic or exponential functions:
   • Find the equations of the tangent line and normal line to a curve at a given point.
   • Sketch the graph of a function, indicating maximum points, minimum points, and inflection points.
   • Find the maximum or minimum value of a quantity.
   • Find the velocity vector at a specific time from parametric equations for a moving object.
   • Find the rate of change of one quantity with respect to a related quantity.
   • Estimate changes in quantities by means of differentials.

8-9 More Exercises for Chapter 8

Exercises 1–6: Prove the given trigonometric identity.

1. $\sec \theta (1 - \sin^2 \theta) = \cos \theta$ 2. $\tan \theta (\tan \theta + \cot \theta) = \sec^2 \theta$ 3. $\tan^2 \theta \cos^2 \theta + \cot^2 \theta \sin^2 \theta = 1$

4. $\dfrac{1 + \cos \theta}{\sin \theta} = \dfrac{\sin \theta}{1 - \cos \theta}$ 5. $2 \sin \theta + \sin 2\theta = \dfrac{2 \sin^3 \theta}{1 - \cos \theta}$ 6. $\sin 3\theta = 3 \sin \theta - 4 \sin^3 \theta$

7. Write $3^{-2} = \frac{1}{9}$ in logarithmic form.

8. Write $16^{3/4} = 8$ in logarithmic form.

9. Write $\log_4 \frac{1}{2} = -\frac{1}{2}$ in exponential form.

10. Write $\log_{25} 125 = \frac{3}{2}$ in exponential form.

11. Solve for b: $\frac{1}{3} = \log_b 2$.

12. Solve for x: $\log_6 4 + 2 \log_6 3 = x$.

Exercises 13–52: Differentiate.

13. $y = \sin x^3$

14. $y = \tan(2 - x^2)$

15. $y = \cos x \tan x^2$

16. $y = x^3 \sec x^2$

17. $y = \dfrac{\sec x}{x}$

18. $y = \sin^3 2x$

19. $y = \text{Arcsin}(1 - 3x^2)$

20. $y = \text{Arccos}(3x^4 + 2x)$

21. $y = \text{Arctan} \sqrt{x + 1}$ 22. $y = (\text{Arctan } 2x)^2$

23. $y = x^2 \text{ Arcsin } x$

24. $y = \cos x \text{ Arccos } 2x$

25. $y = e^{x^3}$ 26. $y = e^{-1/x}$

27. $y = x^2 e^x$ 28. $y = xe^{2x-1}$

29. $y = \sqrt{e^{2x} + 4}$ 30. $y = \dfrac{e^{2x}}{e^x - 1}$

31. $y = \ln(2x - 7)$

32. $y = \ln(x^2 + 7x - 2)$ 33. $y = \ln\sqrt{x^4 - 2}$

34. $y = \ln^2(x^3 + 2x)$ 35. $y = \ln(x^3 \ln x)$

36. $y = \ln \dfrac{3x^2 + 2}{\sqrt{1 + x}}$ 37. $y = e^{\sin 2x}$

38. $y = e^x \sin x$ 39. $y = \text{Arctan } e^{x^2}$

40. $y = \text{Arcsin} \ln 2x$ 41. $y = \ln \dfrac{\sin x}{x^2}$

42. $y = \ln^2(x^2 + e^x)$ 43. $y = e^{x^2} \ln x$

44. $y = \sin x \ln 3x$ 45. $y = \ln(x^2 + \sin x)$

46. $y = \ln \tan e^x$ 47. $y = \dfrac{\ln x}{e^{\sin x}}$

48. $y = \dfrac{e^{x^3}}{\sin \ln x}$ 49. $e^x \ln y = x^2 y$

50. $\ln xy = e^{x+y}$ 51. $ye^{\sin^2 x} = y^3 \cos x$

52. $e^{y^2} = \sin(x^2 + y^2)$

53. Find the equation of the tangent line and normal line to the curve $y = \text{Arcsin}(1 - x^2)$ at $x = 0.4$.

54. Find the equation of the tangent line and normal line to the curve $y = x \cos^3(x^2 - 2)$ at $x = 1$.

55. Find the equation of the tangent line to the curve $y = \sin xe^{2x}$ at $x = \pi/2$.

56. Find the equation of the normal line to the curve $y = \ln(x^2 + 4x)$ at $x = 1$.

57. Sketch a graph of the function $y = 2 \sin 2x$ for $0 \le x \le 2\pi$.

58. Sketch a graph of the function $y = 2 \cos x + \sin 2x$ for $0 \le x \le 2\pi$.

59. Sketch a graph of the function $y = e^{-1/x}$.

60. Sketch a graph of the function $y = \ln \sin x$, $0 \le x \le 2\pi$.

61. Routes 23 and 84 cross at right angles. Trigtown is 10 km north of Route 23 on a road parallel to but 8 km west of Route 84 (see Figure 8-19). Find the shortest straight road passing through Trigtown that can connect Route 23 with Route 84.

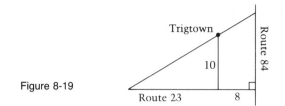

Figure 8-19

62. Conical paint filters are made by removing a circular sector from a circular piece of filter paper of radius a and gluing the two straight edges together. Find the maximum volume of a filter obtained in this way. What is the size of the angle of the circular sector to be removed from the original paper to obtain the maximum volume?

63. Find the maximum area of a rectangle in the first quadrant with two sides along the axes which can be fitted under the curve $y = -\ln x$.

64. A piece of sheet metal is approximated by the area in the first quadrant under the curve $y = e^{-x^2}$. Find the area of the largest rectangle that can be cut from this piece, as shown in Figure 8-20.

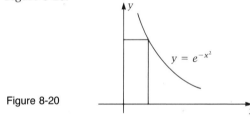

Figure 8-20

65. A particle moves according to the parametric equations $x = t + \cos t$, $y = \cos^2 t$. Find the velocity vector for $t = 1$.

66. Find the acceleration vector at $t = 1$ for the particle of Exercise 65.

67. An object moving with constant angular velocity of 3 rad/min traces the path $r = \sin \theta - 1$ (r is measured in meters). Find the magnitude of the velocity when $\theta = \pi$.

68. A particle moves along the curve $r = \tan \theta$ at a constant angular velocity of 0.5 rad/s (r is measured in centimeters). Find the magnitude of the velocity vector when $\theta = 1$.

69. A tree that is 15 m high casts a shadow on the ground. The angle of elevation of the sun is decreasing at the rate of 0.3 rad/h. How fast is the shadow lengthening when the angle of elevation is 0.6?

70. A person standing atop a 20-m cliff watches a boat approaching the base of the cliff at 3 m/min. Find the rate of change of the angle of depression when the boat is 40 m out from the base of the cliff.

71. At a construction site, two pulleys are located 7 m apart on poles 5 m high. Ropes over these pulleys are pulled to hoist an object midway between the pulleys (see Figure 8-21). The angle θ is decreasing at the rate of 0.6 rad/min. How fast is the object rising when it is 1 m off the ground?

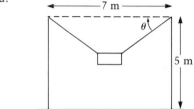

Figure 8-21

72. A tracking camera is located on the ground 135 m from a launch pad. If a rocket is launched vertically with a velocity of 250 m/s, what is the angular rate of rotation of the tracking camera 2 s after the launch?

73. The concentration c in the blood of a certain drug administered by muscle injection is given by $c = 2.8(e^{-0.02t} - e^{-0.6t})$ at time t after the injection. Find an expression for the rate of change of concentration.

74. The rate i of emission of electrons from the surface of metal heated to temperature T is given by

$$\ln i + \frac{b}{T} = A + 2 \ln T$$

where A and b are constants. Use implicit differentiation to find an expression for di/dT.

75. The force on a certain straight conductor at an angle θ to a uniform magnetic field is given by $F = 0.002 \sin \theta$. Use differentials to estimate the change in F as θ changes from 0.7 to 0.72 radians.

76. An angle θ is measured by computing $\sin \theta = y/r$. If y is measured to be 14 m $\pm$ 0.1 m and r is measured to be 17 m $\pm$ 0.2 m, estimate the maximum error in computing θ. (Take the absolute value of your answer.)

77. The amount A of a certain nutrient in soil samples from a particular locale as a function of time t is given by $A = 280(1 + e^{-0.008t})$. Use differentials to estimate the change in A as t changes from 10 to 11.

78. Find the total differential for $z = \sin e^x \cos y$.

chapter 9

Patterns for Integration

With practice, finding the right integration pattern for your problem becomes child's play!

9-1 The Power Rule

We return in this chapter to the integration problem: We wish to compute a definite integral of the form

$$\int_a^b f(x)\, dx$$

By the Fundamental Theorem of Calculus

$$\int_a^b f(x)\, dx = F(b) - F(a)$$

where $F(x)$ is an antiderivative for $f(x)$. Much of this chapter is devoted to developing and using integration rules to help us find antiderivatives of various functions. Transcendental functions, such as logarithmic or trigonometric functions, may be involved. We shall eventually have a whole list of integration rules. When faced with finding an antiderivative $\int f(x)\, dx$, we can then compare our problem against the entries in our list in an attempt to find a matching pattern.

So far our list has only one entry, the power rule (from Chapter 6).

$$\int u^n \, du = \frac{u^{n+1}}{n+1} + C \qquad n \neq -1$$

We repeat here the advice given in Chapter 6 for applying the power rule.

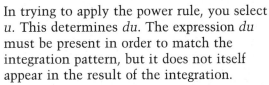

Word of Advice

In trying to apply the power rule, you select u. This determines du. The expression du must be present in order to match the integration pattern, but it does not itself appear in the result of the integration.

EXAMPLE 1.   Integrate: $\int \tan^3 x \sec^2 x \, dx$.

In matching this integral with the power rule, we observe that there are two functions raised to a power. We could choose $u = \tan x$ and $n = 3$, or we could choose $u = \sec x$ and $n = 2$. Suppose we choose

$$u = \tan x \qquad n = 3$$

Then

$$du = \sec^2 x \, dx$$

which is exactly the rest of our integrand. We have matched the pattern $\int u^n \, du$ in the form $\int (\tan x)^3 (\sec^2 x \, dx)$. The result is

$$\frac{\tan^4 x}{4} + C$$

Suppose, however, that we had chosen

$$u = \sec x \qquad n = 2$$

Then

$$du = \sec x \tan x \, dx$$

This does not match the remainder of our integrand, which is $\tan^3 x$. This attempt to use the power rule fails.   ■

EXAMPLE 2.   Integrate: $\int \frac{\ln^2 x \, dx}{x}$.

Let

$$u = \ln x \qquad n = 2$$

Then

$$du = \frac{1}{x} dx$$

This is exactly the rest of our integrand, so we have matched the power rule pattern.

$$\int \frac{\ln^2 x}{x} dx = \frac{\ln^3 x}{3} + C \quad \blacksquare$$

EXAMPLE 3. Integrate: $\int \sqrt{1 + e^{4x^3}}\, e^{4x^3}\, x^2\, dx$.

The obvious choice here, in trying to match the power rule, is to let

$$u = 1 + e^{4x^3} \qquad n = \tfrac{1}{2}$$

Then

$$du = e^{4x^3} \cdot 12x^2\, dx$$

The rest of our integrand contains all of du except for the constant factor 12. We take care of this problem in the usual way.

$$\int \sqrt{1 + e^{4x^3}}\, e^{4x^3} x^2\, dx = \frac{1}{12} \int (1 + e^{4x^3})^{1/2}\, e^{4x^3}(12x^2)\, dx$$

$$= \frac{1}{12} \cdot \frac{2}{3} (1 + e^{4x^3})^{3/2} + C$$

$$= \frac{1}{18} (1 + e^{4x^3})^{3/2} + C \quad \blacksquare$$

Practice 1. Integrate: $\int \sqrt{\cos 2x + 1}\, \sin 2x\, dx$.

EXAMPLE 4. Evaluate: $\int_1^e \frac{(\cos \ln x)(\sin \ln x)}{x} dx$.

Let

$$u = \cos \ln x \qquad n = 1$$

Then

$$du = -\sin \ln x \cdot \frac{1}{x} dx$$

Adjusting for the missing factor -1, we have

$$\int_1^e \frac{(\cos \ln x)(\sin \ln x)}{x} \, dx = -1 \int_1^e \frac{(\cos \ln x)(-\sin \ln x)}{x} \, dx$$

$$= -\frac{\cos^2(\ln x)}{2} \Big|_1^e$$

$$= -\frac{\cos^2(\ln e)}{2} - \left(-\frac{\cos^2(\ln 1)}{2} \right)$$

$$= \frac{-\cos^2 1 + \cos^2 0}{2} = \frac{-\cos^2 1 + 1}{2}$$

But, you may be asking, why can't we do the following? Let

$$u = \sin \ln x \qquad n = 1$$

Then

$$du = \cos \ln x \cdot \frac{1}{x} \, dx$$

and

$$\int_1^e \frac{(\cos \ln x)(\sin \ln x)}{x} \, dx = \frac{\sin^2(\ln x)}{2} \Big|_1^e = \frac{\sin^2(\ln e)}{2} - \frac{\sin^2(\ln 1)}{2}$$

$$= \frac{\sin^2 1}{2} - \frac{\sin^2 0}{2} = \frac{\sin^2 1}{2}$$

This is also correct. From the identity $\sin^2 x + \cos^2 x = 1$, we get

$$\frac{\sin^2 1}{2} = \frac{1 - \cos^2 1}{2}$$

and the two answers are the same. ∎

Exercises / Section **9-1**

Exercises 1–26: Integrate:

1. $\displaystyle \int x^2 \sqrt{1 + 2x^3} \, dx$

2. $\displaystyle \int \frac{2x^2}{(1 + x^3)^2} \, dx$

3. $\displaystyle \int \sin^2 x \cos x \, dx$

4. $\displaystyle \int \cos^4 x \sin x \, dx$

5. $\displaystyle \int \sqrt{\sin x} \cos x \, dx$

6. $\displaystyle \int \frac{\cos x}{\sin^2 x} \, dx$

7. $\displaystyle \int \sqrt{\tan x} \sec^2 x \, dx$

8. $\displaystyle \int \sec^5 x \tan x \, dx$ (*Hint:* Write $\sec^5 x \tan x$ as $\sec^4 x \sec x \tan x$.)

9. $\displaystyle \int_0^{\pi/6} \sin^3 2x \cos 2x \, dx$

10. $\displaystyle \int_0^{\pi/4} \frac{\cos 2x}{(1 + \sin 2x)^3} \, dx$

11. $\displaystyle \int \frac{(\text{Arcsin } x)^2}{\sqrt{1 - x^2}} \, dx$

12. $\displaystyle \int \frac{\text{Arctan } 3x}{1 + 9x^2} \, dx$

13. $\displaystyle \int \frac{3 \ln^4 x}{x} \, dx$

14. $\displaystyle \int \frac{\ln 2x}{x} \, dx$

15. $\displaystyle \int \frac{(1 - \ln x)^2}{x} \, dx$

16. $\displaystyle \int \frac{(\ln 3x + 1)^3}{3x} \, dx$

17. $\displaystyle \int_0^1 (e^{x^2} - 1)^2 x e^{x^2} \, dx$

18. $\displaystyle \int_0^2 \frac{e^x}{\sqrt{1 + 5e^x}} \, dx$

19. $\displaystyle \int e^x \sin^2 e^x \cos e^x \, dx$

20. $\displaystyle\int (e^{\sin x})^2 e^{\sin x} \cos x \, dx$

21. $\displaystyle\int \tan x \ln \sec x \, dx$

22. $\displaystyle\int \frac{\tan x}{(1 + \ln \cos x)^{3/2}} \, dx$

23. $\displaystyle\int (e^{2x} - e^{-2x})\sqrt{e^{2x} + e^{-2x}} \, dx$

24. $\displaystyle\int \frac{(2 - 4e^{-x})^2}{e^x} \, dx$

25. $\displaystyle\int x \cot x^2 \ln \sin x^2 \, dx$

26. $\displaystyle\int \frac{\sqrt{1 - \ln^2 x} \, \ln x}{x} \, dx$

27. Find the area of the region bounded by the curves $y = (\ln x)/x$, $y = 0$, and $x = e$.

28. The velocity of an object at time t is given by $v = \sin t \cos^2 t$. Find an expression for s as a function of t if $s = 0$ when $t = 0$.

29. The electric potential V of a point at distance d along the axis of a uniformly charged circular disk of radius a is given by $[V = k \int_0^a (y^2 + d^2)^{-1/2} y \, dy]$, where k is a constant. Find the value of V.

30. The current i in a particular electrical circuit as a function of time t is given by $i = e^{3t}$. Find an expression for the total charge q that has passed a given point in the circuit at time t, where $q = \int i \, dt$. (*Hint:* Write e^{3t} as $(e^t)^2 e^t$.)

9-2   The Logarithmic Form

In Chapter 8 we learned some new differentiation rules. Every differentiation rule also produces an integration rule, because if we know that

$$\frac{dF(x)}{dx} = f(x)$$

then we also know that

$$\int f(x) \, dx = F(x) + C$$

The differentiation rule for the logarithmic function, which we developed in Chapter 6, is

$$\frac{d \ln u}{dx} = \frac{1}{u} \frac{du}{dx} \tag{1}$$

where $u = u(x)$. Because the logarithmic function is defined only for positive values, Equation (1) is valid only for values of x for which $u(x) > 0$. We can generalize Equation (1) by considering $|u|$. For $u \neq 0$, $\ln|u|$ is defined. If $u > 0$, then $|u| = u$ and

$$\frac{d \ln|u|}{dx} = \frac{d \ln u}{dx} = \frac{1}{u} \frac{du}{dx}$$

by Equation (1). If $u < 0$, then $|u| = -u$, $-u > 0$, and

$$\frac{d \ln|u|}{dx} = \frac{d \ln(-u)}{dx} = \frac{1}{(-u)} \frac{d(-u)}{dx} = \frac{1}{(-u)} (-1) \frac{du}{dx} = \frac{1}{u} \frac{du}{dx}$$

The generalization of Equation (1) is, therefore,

$$\frac{d \ln|u|}{dx} = \frac{1}{u} \frac{du}{dx} \tag{2}$$

Equation (2) leads us to the integration rule

$$\int \frac{1}{u}\, du = \ln|u| + C \tag{3}$$

Notice that $\int (1/u)\, du$ can also be written as $\int u^{-1}\, du$. Equation (3) therefore takes care of integrating $\int u^n\, du$ for $n = -1$, the one case excluded by the power rule.

EXAMPLE 1. Integrate: $\int \dfrac{x}{x^2 + 1}\, dx.$

If we attempt to match the pattern $(1/u)\, du$, we choose

$$u = x^2 + 1$$

and then

$$du = 2x\, dx$$

Except for the constant factor of 2, this is exactly what we have.

$$\int \frac{x}{x^2 + 1}\, dx = \frac{1}{2} \int \frac{2x}{x^2 + 1}\, dx = \frac{1}{2} \ln|x^2 + 1| + C$$

In this particular case, we can remove the absolute value signs because $x^2 + 1$ is always positive. The final answer is, therefore, $\frac{1}{2} \ln(x^2 + 1) + C.$ ∎

EXAMPLE 2. Integrate: $\int \dfrac{\sin x}{\cos x}\, dx.$

Trying to match the logarithmic pattern, we let

$$u = \cos x$$

and then

$$du = -\sin x\, dx$$

The integrand matches the pattern (except for a factor of -1) and

$$\int \frac{\sin x}{\cos x}\, dx = -\int -\frac{\sin x}{\cos x}\, dx = -\ln|\cos x| + C \quad ∎$$

Word of Advice

All the integration patterns are expressed in terms of u and du, where u is some function of x. In comparing your integral with a pattern, *write down* your choice for u and then determine du. Trying to do this mentally leads to errors.

Practice 1. Integrate: $\int \dfrac{dx}{1-x}$.

EXAMPLE 3. Integrate: $\int \dfrac{x}{(x^2+1)^2}\, dx$.

This looks like the problem in Example 1, right? Wrong! If we try to force this problem into the pattern $(1/u)\, du$, we must take the entire denominator for u—that is,

$$u = (x^2+1)^2$$

Then

$$du = 2(x^2+1)2x$$

by the power rule for differentiation. We certainly do not have all of this in our integrand, but we can match this integral to the power form $\int u^n\, du$, where

$$u = x^2 + 1 \qquad n = -2$$
$$du = 2x\, dx$$

Thus

$$\int \frac{x}{(x^2+1)^2}\, dx = \frac{1}{2} \int (x^2+1)^{-2}\, 2x\, dx = \frac{1}{2}\,\frac{(x^2+1)^{-1}}{-1} + C$$

by the power rule for integration. ■

EXAMPLE 4. Evaluate: $\displaystyle\int_0^1 \frac{e^{2x}}{e^{2x}-3}\, dx$.

We let

$$u = e^{2x} - 3$$

and then

$$du = 2e^{2x}\, dx$$

Therefore,

$$\int_0^1 \frac{e^{2x}}{e^{2x}-3}\, dx = \frac{1}{2} \int_0^1 \frac{2e^{2x}}{e^{2x}-3}\, dx = \frac{1}{2} \ln|e^{2x}-3|\Big|_0^1$$

$$= \frac{1}{2}(\ln|e^2-3| - \ln|1-3|) = \frac{1}{2}[\ln(e^2-3) - \ln 2]$$

$$= \frac{1}{2} \ln \frac{e^2-3}{2} \qquad ■$$

EXAMPLE 5. Integrate: $\int \dfrac{1 - \sec^2 x}{x - \tan x}\, dx$.

This matches the form $\int (1/u)\, du$ directly, so we write

$$\int \frac{1 - \sec^2 x}{x - \tan x}\, dx = \ln|x - \tan x| + C \quad \blacksquare$$

Exercises / Section 9-2

Exercises 1–26: Integrate:

1. $\displaystyle\int \frac{dx}{3x + 1}$

2. $\displaystyle\int \frac{dx}{1 - 4x}$

3. $\displaystyle\int \frac{x}{x^2 + 1}\, dx$

4. $\displaystyle\int \frac{x^3}{1 + x^4}\, dx$

5. $\displaystyle\int_0^2 \frac{x + 1}{x^2 + 2x + 3}\, dx$

6. $\displaystyle\int_1^2 \frac{x^2 - 2}{x^3 - 6x + 1}\, dx$

7. $\displaystyle\int_0^{\pi/6} \frac{\cos x}{1 - \sin x}\, dx$

8. $\displaystyle\int_0^{\pi/4} \frac{\sec^2 x}{\tan x - 2}\, dx$

9. $\displaystyle\int \frac{\cos x}{\sin^2 x}\, dx$

10. $\displaystyle\int \frac{\sec x \tan x}{\sqrt{\sec x + 1}}\, dx$

11. $\displaystyle\int \tan x\, dx$

12. $\displaystyle\int \cot x\, dx$

13. $\displaystyle\int \frac{1}{x \ln x}\, dx$

14. $\displaystyle\int \frac{1}{x(\ln x + 2)}\, dx$

15. $\displaystyle\int \frac{e^x}{1 - e^x}\, dx$

16. $\displaystyle\int \frac{e^{2x}}{e^{2x} + 4}\, dx$

17. $\displaystyle\int \frac{1 - e^{2x}}{2x - e^{2x}}\, dx$

18. $\displaystyle\int \frac{e^x - e^{-x}}{e^x + e^{-x}}\, dx$

19. $\displaystyle\int \frac{e^{\sin x} \cos x + 1}{e^{\sin x} + x}\, dx$

20. $\displaystyle\int \frac{xe^{x^2}}{e^{x^2} + 1}\, dx$

21. $\displaystyle\int \frac{dx}{\sqrt{x}}$

22. $\displaystyle\int \frac{dx}{x \ln^2 x}$

23. $\displaystyle\int \frac{dx}{\sqrt{x}(1 + \sqrt{x})}$

24. $\displaystyle\int \frac{dx}{\sqrt{x}(\sqrt{3x} + 1)}$

25. $\displaystyle\int_{-1}^{1} \int_1^{e^x} \frac{1}{xy}\, dy\, dx$

26. $\displaystyle\int_0^1 \int_0^y \frac{1}{(y + 1)(x + 1)}\, dx\, dy$

27. Find the area bounded by the curves $xy = 1$, $x = 1$, $x = e$, and $y = 0$.

28. Find the area bounded by the curves $y = e^x/(e^x + 1)$, $y = 0$, $x = 0$, and $x = 2$.

29. Find the volume generated by revolving the area bounded by $y = 1/(x^2 + 4)$, $y = 0$, $x = 0$, and $x = 1$ about the y-axis; use cylindrical shells.

30. Find the volume generated by revolving the area bounded by $y = 1/(x^2 \ln x)$, $y = 1/x$, $x = 3$, and $x = 4$ about the y-axis; use cylindrical shells.

31. When a metal bar is deformed by elongation, the natural strain ϵ is defined as

$$\epsilon = \int_{L_0}^{L_1} \frac{dL}{L}$$

where L_0 is the original length and L_1 is the new length. Find the value for ϵ if a bar is stretched 2% of its original length.

32. A capacitor consists of two concentric right circular cylinders of radii a and b. If a charge q is put on the capacitor, then its capacitance C is given by $C = q/V$, where

$$V = \int_a^b \frac{2q}{kx}\, dx$$

and k is a constant. If $a = 1$, $b = 2e$, $q = 3$, and $k = 4$, what is C?

33. If a heated object is cooled by placing it in a solution at a temperature of 14°C, the rate of change of the temperature T of the object with respect to time t satisfies the equation

$$\frac{dT}{dt} = k(T - 14)$$

or

$$\frac{dT}{T - 14} = k\, dt$$

where k is a constant. Integrate both sides of the equation and find an expression for T as a function of t.

34. The velocity v of a sphere dropped through a viscous medium at time t after the start of the fall satisfies the equation

$$\int_0^v \frac{1}{(m'g/k) - x} \, dx = \int_0^t \frac{k}{m} \, dt$$

where m is the mass of the sphere, m' is its mass adjusted for buoyancy, g is the gravitational constant, and k is a constant depending upon the viscosity of the medium. Integrate both sides of the equation and find an expression for v as a function of t.

9-3   The Exponential Form

The differentiation rule for the exponential function, which we learned in Chapter 6, is

$$\frac{d(e^u)}{dx} = e^u \frac{du}{dx}$$

The corresponding integration rule is

$$\int e^u \, du = e^u + C$$

The exponential form is one of the easiest integration patterns to recognize.

EXAMPLE 1.   Integrate: $\int xe^{x^2} \, dx$.

Matching this to the exponential pattern, we let $u = x^2$; then $du = 2x \, dx$. We are missing only the constant factor of 2.

$$\int xe^{x^2} \, dx = \tfrac{1}{2} \int 2xe^{x^2} \, dx = \tfrac{1}{2}e^{x^2} + C \quad \blacksquare$$

In Example 1, $2x$ is required in the integrand in order to match the pattern $\int e^u \, du$. We were able to supply the missing 2 because it is a constant factor, and its reciprocal could be moved through the integral sign. If the problem had been $\int 2e^{x^2} \, dx$, we would not have been able to supply the missing x, because it is not a constant factor. We could not have done the integration.

Word of Advice

Only *constant factors* can be moved through integral signs!

Practice 1. Integrate: $\int e^{2x}\,dx$.

EXAMPLE 2. Integrate: $\int \dfrac{1}{e^x}\,dx$.

At first we might be tempted to try the logarithmic pattern $\int (1/u)\,du$. If so, we let $u = e^x$. But then $du = e^x\,dx$, and we do not have (nor can we get) this. Rewriting the integrand in the form $\int e^{-x}\,dx$, we recognize the $\int e^u\,du$ pattern with $u = -x$, and $du = -dx$.

$$\int e^{-x}\,dx = -\int -e^{-x}\,dx = -e^{-x} + C \quad \blacksquare$$

The integrand of Example 2 required some rewriting in order to help us determine how to do the integration. We are already familiar with such rewriting procedures as replacing $\sqrt{1 + x}$ in an integrand with $(1 + x)^{1/2}$. As our list of integration patterns grows and we are able to handle more complex integrands, rewriting tactics will play a larger role in our integration-solving arsenal.

Practice 2. Integrate: $\int (e^{2x})^3\,dx$.

EXAMPLE 3. Integrate: $\int \sin x\, e^{\cos x}\,dx$.

Following the exponential pattern $\int e^u\,du$ with $u = \cos x$, we write

$$\int \sin x\, e^{\cos x}\,dx = -\int (-\sin x)e^{\cos x}\,dx = -e^{\cos x} + C. \quad \blacksquare$$

EXAMPLE 4. Integrate: $\int e^x(e^x + 1)^3\,dx$.

This does not really fit the exponential form at all, even though e^x is in it. Instead, it suggests a power form, with $u = e^x + 1$ and $du = e^x\,dx$. Thus

$$\int e^x(e^x + 1)^3\,dx = \frac{(e^x + 1)^4}{4} + C \quad \blacksquare$$

EXAMPLE 5. Integrate: $\int \dfrac{e^x}{1 + e^x}\, dx$.

The logarithmic form $\int (1/u)\, du$ can be applied, with $u = 1 + e^x$ and $du = e^x\, dx$.

$$\int \frac{e^x}{1 + e^x}\, dx = \ln(1 + e^x) + C$$

(Absolute value signs are not needed here because $1 + e^x$ is always positive.) ∎

Exercises / Section 9-3

Exercises 1–22: Integrate:

1. $\displaystyle\int e^{3x}\, dx$

2. $\displaystyle\int e^{-2x}\, dx$

3. $\displaystyle\int_0^1 x^2 e^{x^3}\, dx$

4. $\displaystyle\int_0^2 xe^{2x^2}\, dx$

5. $\displaystyle\int e^{2x+3}\, dx$

6. $\displaystyle\int (2x - 1)e^{x^2 - x}\, dx$

7. $\displaystyle\int_0^{\pi/4} (\sec^2 x)e^{\tan x}\, dx$

8. $\displaystyle\int_0^{\pi/3} (\sec x \tan x)e^{\sec x}\, dx$

9. $\displaystyle\int e^x e^{e^x}\, dx$

10. $\displaystyle\int e^x e^{(1 - e^x)}\, dx$

11. $\displaystyle\int \frac{e^{\sqrt{x}}}{\sqrt{x}}\, dx$

12. $\displaystyle\int \frac{e^{1/x}}{x^2}\, dx$

13. $\displaystyle\int (e^{2x} - e^{-2x})\, dx$

14. $\displaystyle\int (e^x - e^{-x})^2\, dx$

15. $\displaystyle\int (e^{2x} - x)^2(2e^{2x} - 1)\, dx$

16. $\displaystyle\int e^x \sqrt{e^x + 1}\, dx$

17. $\displaystyle\int e^{4x} e^x\, dx$

18. $\displaystyle\int (e^2 + e^{2x})e^x\, dx$

19. $\displaystyle\int \frac{e^{2x}}{2 - e^{2x}}\, dx$

20. $\displaystyle\int \frac{1 + e^{3x}}{3x + e^{3x}}\, dx$

21. $\displaystyle\int_0^1 \int_0^y e^{x+y}\, dx\, dy$

22. $\displaystyle\int_1^2 \int_0^{x^2} xe^y\, dy\, dx$

23. Find the area of the region bounded by the curves $y = e^{-2x}$, $y = -1$, $x = 0$, and $x = 2$.

24. Find the area of the region bounded by the curves $y = e^{3x}$, $y = -e^{2x}$, $x = -1$, and $x = 1$.

25. Find the volume generated by rotating the area bounded by $y = e^{x/2}$, $y = 0$, $x = 0$, and $x = 2$ about the x-axis.

26. Find the volume generated by rotating the area bounded by $y = e^{2x^2}$, $y = 0$, $x = 0$, and $x = 1$ about the y-axis.

27. The velocity of a moving object satisfies the equation

$$v = \frac{\sin t\, e^{\tan^2 t}}{\cos^3 t}$$

$0 \le t < \pi/2$. Find the expression for s as a function of t if $s = 5$ when $t = 0$.

28. Find the first moment with respect to the y-axis of the area of constant density k bounded by $y = 1/e^{x^2}$, $y = 0$, $x = 0$, and $x = 1$.

29. If the world rate of oil consumption (in billions of barrels per year) since 1970 is given by the equation $do/dt = 16e^{0.05t}$, where 1970 corresponds to $t = 0$, find the total oil that will be consumed between 1970 and 2000 by computing $\int_0^{30} 16e^{0.05t}\, dt$.

30. A piece of electronic equipment has a probability p of failing after a months and before b months of use, where $p = k \int_a^b e^{-kt}\, dt$. If there is a 40% chance of failure within 1 year, what is the value of k?

9-4 Basic Trigonometric Forms

By reversing the differentiation rules for the six trigonometric functions, we obtain six integration patterns.

$$\frac{d(\sin u)}{dx} = \cos u \, \frac{du}{dx}$$

$$\int \cos u \, du = \sin u + C$$

$$\frac{d(\cos u)}{dx} = -\sin u \, \frac{du}{dx}$$

$$\int \sin u \, du = -\cos u + C$$

$$\frac{d(\tan u)}{dx} = \sec^2 u \, \frac{du}{dx}$$

$$\int \sec^2 u \, du = \tan u + C$$

$$\frac{d(\cot u)}{dx} = -\csc^2 u \, \frac{du}{dx}$$

$$\int \csc^2 u \, du = -\cot u + C$$

$$\frac{d(\sec u)}{dx} = \sec u \tan u \, \frac{du}{dx}$$

$$\int \sec u \tan u \, du = \sec u + C$$

$$\frac{d(\csc u)}{dx} = -\csc u \cot u \, \frac{du}{dx}$$

$$\int \csc u \cot u \, du = -\csc u + C$$

Word of Advice

The sign change for the integration of the sine and cosine functions works in the opposite way as in the differentiation of these functions. The derivative of the sine function is the cosine function, but the integral of the sine function is the negative of the cosine function. The derivative of the cosine function is the negative of the sine function, but the integral of the cosine function is the sine function.

EXAMPLE 1. Integrate: $\int \cos 2x \, dx$.

This clearly fits the pattern $\int \cos u \, du$, where $u = 2x$ and $du = 2 \, dx$. Thus

$$\int \cos 2x \, dx = \tfrac{1}{2} \int \cos 2x(2) \, dx = \tfrac{1}{2} \sin 2x + C \quad \blacksquare$$

EXAMPLE 2. Integrate: $\int x^2 \sec x^3 \tan x^3 \, dx$.

Here we match the pattern $\int \sec u \tan u \, du$, with $u = x^3$ and $du = 3x^2$.

$$\int x^2 \sec x^3 \tan x^3 \, dx = \tfrac{1}{3} \int 3x^2 \sec x^3 \tan x^3 \, dx = \tfrac{1}{3} \sec x^3 + C \quad \blacksquare$$

Practice 1. Integrate: $\int \sec^2(3x + 1) \, dx$.

EXAMPLE 3. Integrate: $\int \sec^2 x \tan x \, dx$.

This does not match the form $\int \sec^2 u \, du$, nor the form $\int \sec u \tan u \, du$. Even though the integrand contains trigonometric functions, it is integrable by the power rule. Letting $u = \tan x$, $n = 1$, we see that $du = \sec^2 x \, dx$, and

$$\int \sec^2 x \tan x \, dx = \frac{\tan^2 x}{2} + C$$

We can also use the power rule on the secant function if we rewrite the integrand.

$$\int \sec^2 x \tan x \, dx = \int \sec x(\sec x \tan x) \, dx = \frac{\sec^2 x}{2} + C \quad \blacksquare$$

Two antiderivatives of a given function should differ only by a constant (see Practice 2 below).

Practice 2. Show that the two antiderivatives of $\sec^2 x \tan x$ found in Example 3 differ only by a constant.

In developing a list of integration rules for integrands that contain trigonometric functions, it seems we should at least be able to handle the six trigonometric functions themselves. Thus we would like rules to handle the integration of the sine, cosine, tangent, cotangent, secant, and cosecant functions. So far, we have formulas to integrate $\int \sin u \, du$ and $\int \cos u \, du$, but not the others. We shall work on the rest now.

In order to integrate $\int \tan u \, du$, we use the identity $\tan u = \sin u / \cos u$. In this fractional form, we recognize the logarithmic pattern $\int (1/u) \, du$. Thus

$$\int \tan u \, du = \int \frac{\sin u}{\cos u} \, du = -\int \frac{-\sin u \, du}{\cos u} = -\ln|\cos u| + C$$

The same sort of rewriting method also reduces $\int \cot u \, du$ to the logarithmic form.

$$\int \cot u \, du = \int \frac{\cos u}{\sin u} \, du = \ln|\sin u| + C$$

A fancier rewriting approach is required in order to integrate the secant function. It involves multiplying by a special form of 1, namely, $(\tan u + \sec u)/(\sec u + \tan u)$.

$$\int \sec u \, du = \int \frac{\sec u(\tan u + \sec u)}{\sec u + \tan u} \, du$$

$$= \int \frac{\sec u \tan u + \sec^2 u}{\sec u + \tan u} \, du$$

This now fits the logarithmic pattern $\int (1/u) \, du$. The result is

$$\int \sec u \, du = \ln|\sec u + \tan u| + C$$

A similar idea works for the cosecant function.

$$\int \csc u \, du = \int \frac{\csc u(\cot u + \csc u)}{\csc u + \cot u} \, du$$

$$= -\int \frac{-\csc u \cot u - \csc^2 u}{\csc u + \cot u} \, du$$

$$= -\ln|\csc u + \cot u| + C$$

The new integration rules are

$$\int \tan u \, du = -\ln|\cos u| + C$$

$$\int \cot u \, du = \ln|\sin u| + C$$

$$\int \sec u \, du = \ln|\sec u + \tan u| + C$$

$$\int \csc u \, du = -\ln|\csc u + \cot u| + C$$

EXAMPLE 4. Integrate: $\int \cot 3x \, dx$.

This is easy; we use $\int \cot u \, du$ with $u = 3x$.

$$\int \cot 3x \, dx = \tfrac{1}{3} \int \cot 3x(3) \, dx = \tfrac{1}{3} \ln|\sin 3x| + C \quad \blacksquare$$

EXAMPLE 5. Integrate: $\int_1^{4/\pi} \dfrac{\sec(1/x)}{x^2} \, dx$.

Letting $u = 1/x$, we get $du = -1/x^2$ and

$$\int_1^{4/\pi} \frac{\sec(1/x)}{x^2} \, dx = -\int_1^{4/\pi} -\frac{\sec(1/x)}{x^2} \, dx = -\ln\left|\sec \frac{1}{x} + \tan \frac{1}{x}\right|\Big|_1^{4/\pi}$$

$$= -\ln\left|\sec \frac{\pi}{4} + \tan \frac{\pi}{4}\right| + \ln|\sec 1 + \tan 1|$$

$$= -\ln(2.414) + \ln(3.408)$$

$$= 0.345 \quad \blacksquare$$

Practice 3. Integrate: $\int x \tan x^2 \, dx$.

When faced with an integration problem, we attempt to match the problem with one of the integration patterns in our basic list. We run down the list, making obvious eliminations; for example, if the integrand does not contain the sine function, we rule out the pattern $\int \sin u \, du$. However, as Example 3 shows, an integrand can contain trigonometric functions and still not be integrable by one of the trigonometric forms.

Sometimes a little rewriting or algebraic manipulation on the integrand is required in order for us to match the basic integration patterns.

EXAMPLE 6. Integrate: $\int \dfrac{dx}{1 + \cos 2x}$.

This does not match any of our integration patterns. (If $1 + \cos 2x$ appeared in the numerator instead of the denominator, we would be in good shape—but it does not.) We employ a rewriting strategy.

$$\int \frac{1}{1 + \cos 2x} \, dx = \int \frac{1}{1 + \cos 2x} \cdot \frac{1 - \cos 2x}{1 - \cos 2x} \, dx$$

$$= \int \frac{1 - \cos 2x}{1 - \cos^2 2x} \, dx$$

By the trigonometric identity $\sin^2 x + \cos^2 x = 1$, we get

$$\int \frac{1 - \cos 2x}{1 - \cos^2 2x}\, dx = \int \frac{1 - \cos 2x}{\sin^2 2x}\, dx$$

$$= \int \left(\frac{1}{\sin^2 2x} - \frac{\cos 2x}{\sin^2 2x} \right) dx$$

$$= \int [\csc^2 2x - \cos 2x\, (\sin 2x)^{-2}]\, dx$$

Now at last we can integrate each term and get

$$-\tfrac{1}{2} \cot 2x + \tfrac{1}{2}(\sin 2x)^{-1} + C \quad \blacksquare$$

The equations of curves in polar coordinates often involve trigonometric functions. We recall from Section 7-6 that the area A bounded by the curves $r = r(\theta)$, $\theta = \alpha$, and $\theta = \beta$ is

$$A = \int_\alpha^\beta \tfrac{1}{2} r^2\, d\theta$$

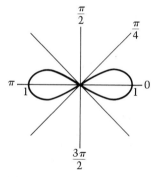

EXAMPLE 7. The area bounded by one loop of the curve $r^2 = \cos 2\theta$ (see Figure 9-1) is expressed by

$$A = 2 \int_0^{\pi/4} \tfrac{1}{2} \cos 2\theta\, d\theta$$

where we have made use of symmetry. Thus

$$A = \int_0^{\pi/4} \cos 2\theta\, d\theta$$

$$= \frac{1}{2} \sin 2\theta \Big|_0^{\pi/4}$$

$$= \frac{1}{2} \left(\sin \frac{\pi}{2} - \sin 0 \right)$$

$$= \frac{1}{2} \quad \blacksquare$$

Figure 9-1

Exercises / Section **9-4**

Exercises 1–30: Integrate.

1. $\displaystyle\int \sin(1 - 2x)\, dx$

2. $\displaystyle\int x \cos x^2\, dx$

3. $\displaystyle\int \sec^2 5x\, dx$

4. $\displaystyle\int \csc^2(\tfrac{1}{2}x)\, dx$

5. $\displaystyle\int e^x \sec(e^x) \tan(e^x)\, dx$

6. $\displaystyle\int \csc(1 + 3x) \cot(1 + 3x)\, dx$

7. $\displaystyle\int e^x \cot e^x\, dx$

8. $\displaystyle\int \sec 3x\, dx$

9. $\displaystyle\int \csc \frac{x}{2}\, dx$

10. $\displaystyle\int x^2 \tan x^3\, dx$

11. $\displaystyle\int_0^1 x \sin(1 - x^2)\, dx$

12. $\displaystyle\int_1^e \frac{\sec^2(\ln x)}{x}\, dx$

13. $\displaystyle\int \frac{\cos(e^{-x})}{e^x}\, dx$

14. $\displaystyle\int x \sec(\pi - x^2) \tan(\pi - x^2)\, dx$

15. $\displaystyle\int \frac{x}{\sin^2 x^2}\, dx$

16. $\displaystyle\int \tan^2 x\, dx$

17. $\displaystyle\int \frac{1 + \cot x}{\cos x}\, dx$

18. $\displaystyle\int \frac{\sin 2x + \cos 2x}{\sin 2x}\, dx$

19. $\displaystyle\int (\csc x - 1)^2\, dx$

20. $\displaystyle\int (\sin^4 x - \cos^4 x)\, dx$

21. $\displaystyle\int \frac{\cos^2 x}{1 - \sin x}\, dx$

22. $\displaystyle\int \frac{\cos^3 x}{1 - \sin x}\, dx$

23. $\displaystyle\int e^{\sin x} \cos x\, dx$

24. $\displaystyle\int \sec^2 2x \tan^2 2x\, dx$

25. $\displaystyle\int \frac{\cos^2 x}{\sin^4 x}\, dx$

26. $\displaystyle\int \frac{\cos x}{\sin^4 x}\, dx$

27. $\displaystyle\int \frac{1 - \sin x}{1 + \cos x}\, dx$

28. $\displaystyle\int \frac{1 - \sec x \tan x}{x - \sec x}\, dx$

29. $\displaystyle\int_0^\pi \int_{\pi/2}^y \cos x\, dx\, dy$

30. $\displaystyle\int_0^{\pi/3} \int_0^x \sec x \sec^2 y\, dy\, dx$

31. Find the area bounded by the curves $y = \sin x$, $y = 0$, and $x = \pi/2$.

32. Find the area bounded by the curves $y = \sec^2 x$, $y = \frac{1}{2}$, $x = 0$, and $x = \pi/4$.

33. Find the total area bounded by the curve whose equation in polar coordinates is $r^2 = 2 \sin \theta$.

34. Find the total area bounded by the curve whose equation in polar coordinates is $r^2 = \sin \theta + \cos \theta$.

35. Find the volume generated by rotating the area bounded by $y = \sec x$, $y = 0$, $x = -\pi/4$, and $x = \pi/4$ about the x-axis.

36. Find the volume generated by rotating the area bounded by $y = \sin x^2$, $y = 0$, and $x = 1$ about the y-axis.

37. Find the first moment with respect to the y-axis of the area bounded by $y = \tan x^2$, $y = 0$, and $x = 1$; assume constant density k.

38. The acceleration of a moving particle is given by the equation $a = \sin(1 + t)$. What is the velocity at $t = 2$ if $v(0) = 0$?

39. The surface tension T on a globule of mercury on a glass plate satisfies an equation of the form

$$gk \int_0^a y\, dy = T \int_0^a \sin x\, dx$$

where g, k, and a are constants. Evaluate the integrals to find an expression for T.

40. The average pressure on an eardrum between time $t = 0$ and $t = 1$ is $\int_0^1 k \cos(2\pi a t)$, where k and a are constants. Find the average pressure.

9-5 Tricks for Trigonometric Integrals

We already have some patterns for integrating expressions which contain trigonometric functions. Certain integrals which contain trigonometric functions but which do not match these patterns can be rewritten, by using trigonometric identities, into forms that do match existing patterns.
    Integrals of the form

$$\int \sin^n u \cos^m u\, du$$

can be handled under one of two cases.

Case 1: At least one power n or m is odd.

In this case we take the function that is raised to the odd power, factor out a first power of that function, and use the identity

$$\sin^2 u + \cos^2 u = 1$$

on the remaining (even) power of that function. This allows us to express the integrand as the sum of terms which look like $\sin u \cos^p u$ or $\cos u \sin^p u$, and these can be integrated by the power rule. An example will make this clear.

EXAMPLE 1. Integrate: $\int \cos^3 x \sin^3 x \, dx$.

Here one of the powers (in fact both powers) is odd. We could factor out $\cos x$ or $\sin x$. Suppose we factor out $\cos x$. For the remaining $\cos^2 x$ factor, we substitute $1 - \sin^2 x$. Then

$$\int \cos^3 x \sin^3 x \, dx = \int \cos x \cos^2 x \sin^3 x \, dx$$

$$= \int \cos x \, (1 - \sin^2 x) \sin^3 x \, dx$$

$$= \int (\cos x \sin^3 x - \cos x \sin^5 x) \, dx$$

$$= \frac{\sin^4 x}{4} - \frac{\sin^6 x}{6} + C \quad \blacksquare$$

An integral of the form $\int \cos^4 x \sin x \, dx$ falls into the case we are considering, but it is already in a form in which the power rule on the cosine function applies, so we would not do any rewriting before we integrated. Also, an integral of the form $\int \sin^3 x \, dx$ falls under this case.

Practice 1. Integrate: $\int \sin^3 x \, dx$.

Case 2: Both powers n and m are even.

To deal with this case, we use the identities

$$\sin^2 u = \frac{1 - \cos 2u}{2}$$

$$\cos^2 u = \frac{1 + \cos 2u}{2}$$

to reduce the powers. We may have to use these identities more than once.

EXAMPLE 2. Integrate: $\int \sin^4 2x \, dx$.

Note that $\int \sin^4 2x \, dx = \int \sin^4 2x \cos^0 2x \, dx$, so the integral does fit the case we are considering (recall that zero is an even number).

$$\int \sin^4 2x \, dx = \int (\sin^2 2x)^2 \, dx = \int \left(\frac{1 - \cos 4x}{2}\right)^2 dx$$

(Here $u = 2x$, and so $2u$ becomes $4x$). Continuing,

$$\int \left(\frac{1 - \cos 4x}{2}\right)^2 dx = \int \frac{1 - 2\cos 4x + \cos^2 4x}{4} \, dx$$

$$= \frac{1}{4} \int (1 - 2\cos 4x + \cos^2 4x) \, dx$$

We use one of the two identities again to get rid of the $\cos^2 4x$ term.

$$\frac{1}{4} \int (1 - 2\cos 4x + \cos^2 4x) \, dx = \frac{1}{4} \int \left(1 - 2\cos 4x + \frac{1 + \cos 8x}{2}\right) dx$$

$$= \frac{1}{4} \left[x - \frac{2}{4} \sin 4x + \frac{1}{2}x + \frac{1}{2} \cdot \frac{1}{8} \sin 8x\right] + C$$

$$= \frac{1}{4}x - \frac{1}{8} \sin 4x + \frac{1}{8}x + \frac{1}{64} \sin 8x + C$$

$$= \frac{3}{8}x - \frac{1}{8} \sin 4x + \frac{1}{64} \sin 8x + C \quad \blacksquare$$

Obviously one would not like to have to integrate $\int \sin^{42} 2x \, dx$ by this procedure—but it would work!

Practice 2. Integrate: $\int \cos^2 3x \, dx$.

For integrals that involve the secant function and the tangent function together, we attempt to arrange certain favorable combinations. Powers of the tangent function together with $\sec^2 u$ can be integrated by the power rule on the tangent function. Powers of the secant function together with $\sec u \tan u$ can be integrated by the power rule on the secant function. The same ideas work for integrals that are combinations of the cosecant function and the cotangent function. The identities

$$1 + \tan^2 u = \sec^2 u$$
$$1 + \cot^2 u = \csc^2 u$$

may help us produce these favorable combinations.

EXAMPLE 3. Integrate: $\int \sec^4 x \, dx$.

Using an identity,

$$\int \sec^4 x \, dx = \int \sec^2 x \sec^2 x \, dx = \int \sec^2 x \, (1 + \tan^2 x) \, dx$$

$$= \int \sec^2 x + \int \sec^2 x \tan^2 x \, dx$$

$$= \tan x + \frac{\tan^3 x}{x} + C \quad \blacksquare$$

Practice 3. Integrate: $\int \cot^3 x \, dx$.

If all else fails, it is sometimes helpful to express all the trigonometric functions in terms of the sine and cosine functions. However, there are still trigonometric integrals that we will not be able to reduce to basic forms no matter how clever we are with trigonometric identities.

EXAMPLE 4. Integrate: $\int \dfrac{\tan^2 x}{\sec^3 x} \, dx$.

Writing the integral in terms of sines and cosines, we get

$$\int \frac{\tan^2 x}{\sec^3 x} \, dx = \int \frac{\sin^2 x/\cos^2 x}{1/\cos^3 x} \, dx = \int \frac{\sin^2 x}{\cos^2 x} \cdot \cos^3 x \, dx$$

$$= \int \sin^2 x \cos x \, dx = \frac{\sin^3 x}{3} + C \quad \blacksquare$$

Exercises / Section **9-5**

Exercises 1–30: Integrate:

1. $\int \cos^3 2x \, dx$

2. $\int \sin^5 x \, dx$

3. $\int \sin^2 x \cos^3 x \, dx$

4. $\int \sin^3 x \cos^2 x \, dx$

5. $\int_0^{\pi/4} \cos^2 2x \, dx$

6. $\int_0^{0.1} \sin^2 5x \, dx$

7. $\int \sin^2 2x \cos^2 2x \, dx$

8. $\int \cos^2 x \sin^4 x \, dx$

9. $\int \sqrt{\cos x} \sin^3 x \, dx$

10. $\int \cos^3 x \sin^7 x \, dx$

11. $\int \cos^3 x \, dx$

12. $\int \sin^5 2x \, dx$

13. $\int \tan x \sec^2 x \, dx$

14. $\int \cot^2 x \csc^4 x \, dx$

15. $\int \tan^4 2x \, dx$

16. $\int \cot^4 x \, dx$

17. $\int \tan^5 x \, dx$

18. $\int \csc^4 x \, dx$

19. $\int_0^{\pi/6} \tan^3 x \sec^3 x \, dx$

20. $\int_0^{\pi/4} \tan^3 x \sec^4 x \, dx$

21. $\displaystyle\int \cot x \, \csc^4 x \, dx$ 22. $\displaystyle\int \cot^3 x \, \csc^4 x \, dx$

23. $\displaystyle\int \tan x \, \sqrt{\sec x} \, dx$ 24. $\displaystyle\int \frac{\tan^2 x - 1}{\sec^2 x} \, dx$

25. $\displaystyle\int \frac{\cos^3 x}{\sin^4 x} \, dx$

26. $\displaystyle\int (\sin 2x + \cos 2x)^2 \, dx$

27. $\displaystyle\int \frac{dx}{1 + \sin x}$ 28. $\displaystyle\int \frac{\cos x}{2 - \sin x} \, dx$

29. $\displaystyle\int \frac{\sec^2 x}{(1 + \tan x)^2} \, dx$ 30. $\displaystyle\int \cot^2 x \, \sec x \, dx$

31. Find the area under the curve $y = \sin^2 x$ from $x = 0$ to $x = \pi$.

32. Find the area bounded by the curve whose equation in polar coordinates is $r = 2 - \sin \theta$.

33. Find the length of the curve $y = \ln \sec x$ from $x = 0$ to $x = \pi/6$.

34. What is the volume generated by rotating the area bounded by $y = \cos x$, $x = 0$, and $y = 0$ about the x-axis?

9-6   Inverse Trigonometric Forms

In the last chapter we learned how to differentiate inverse trigonometric functions. Applying these differentiation rules where $u = u(x)$ and a is constant, we get

$$\frac{d(\text{Arcsin } u/a)}{dx} = \frac{1}{\sqrt{1 - (u/a)^2}} \frac{d(u/a)}{dx} = \frac{1}{\sqrt{(a^2 - u^2)/a^2}} \frac{1}{a} \frac{du}{dx}$$

$$= \frac{a}{\sqrt{a^2 - u^2}} \cdot \frac{1}{a} \frac{du}{dx} = \frac{1}{\sqrt{a^2 - u^2}} \frac{du}{dx}$$

and

$$\frac{d(\text{Arctan } u/a)}{dx} = \frac{1}{1 + (u/a)^2} \frac{d(u/a)}{dx} = \frac{1}{(a^2 + u^2)/a^2} \cdot \frac{1}{a} \frac{du}{dx}$$

$$= \frac{a^2}{a^2 + u^2} \cdot \frac{1}{a} \frac{du}{dx} = \frac{a}{a^2 + u^2} \frac{du}{dx}$$

By reversing these differentiation formulas, we get two new integration rules.

$$\int \frac{1}{\sqrt{a^2 - u^2}} \, du = \text{Arcsin } \frac{u}{a} + C$$

$$\int \frac{1}{a^2 + u^2} \, du = \frac{1}{a} \text{Arctan } \frac{u}{a} + C$$

EXAMPLE 1. Integrate: $\displaystyle\int \frac{1}{\sqrt{4 - x^2}} \, dx$.

This matches the pattern $\int 1/\sqrt{a^2 - u^2} \, du$, where $a = 2$ and $u = x$. Therefore,

$$\int \frac{1}{\sqrt{4 - x^2}} \, dx = \text{Arcsin } \frac{x}{2} + C \quad \blacksquare$$

EXAMPLE 2. Integrate: $\dfrac{3}{\sqrt{9 - 4x^2}}\, dx$.

The denominator matches the form $\sqrt{a^2 - u^2}$ if we set $a = 3$ and $u = 2x$. We take the constant 3 in the numerator out of the integral, and then we need a constant 2 in order to have du.

$$\int \frac{3}{\sqrt{9 - 4x^2}}\, dx = \frac{3}{2} \int \frac{2}{\sqrt{(3)^2 - (2x)^2}}\, dx = \frac{3}{2}\, \text{Arcsin}\, \frac{2x}{3} + C \quad \blacksquare$$

EXAMPLE 3. Integrate: $\dfrac{x}{\sqrt{4 - x^2}}\, dx$.

Although this integral resembles that of Example 1, the x in the numerator prevents it from matching the Arcsine pattern. However, the presence of the x does allow the integral to be done by the power rule. Thus

$$\int \frac{x}{\sqrt{4 - x^2}}\, dx = \int x(4 - x^2)^{-1/2}\, dx = -\frac{1}{2} \int (-2x)(4 - x^2)^{-1/2}\, dx$$

$$= -\frac{1}{2} \frac{(4 - x^2)^{1/2}}{\frac{1}{2}} + C = -(4 - x^2)^{1/2} + C \quad \blacksquare$$

Word of Advice

Your integral must match an integration pattern *exactly* in order for that pattern to apply.

Practice 1. Integrate: $\displaystyle\int \frac{1}{\sqrt{4 - 9x^2}}\, dx$.

EXAMPLE 4. Integrate: $\dfrac{1}{36 + 9x^2}\, dx$.

This matches the Arctangent pattern, with $a = 6$, $u = 3x$, and $du = 3\, dx$.

$$\int \frac{1}{36 + 9x^2}\, dx = \int \frac{1}{(6)^2 + (3x)^2}\, dx = \frac{1}{3} \int \frac{3}{(6)^2 + (3x)^2}\, dx$$

$$= \frac{1}{3} \cdot \frac{1}{6}\, \text{Arctan}\, \frac{3x}{6} + C = \frac{1}{18}\, \text{Arctan}\, \frac{x}{2} + C \quad \blacksquare$$

Practice 2. Integrate: $\displaystyle\int \frac{1}{4 + x^2}\, dx$.

In order to use the Arctangent integration rule, we need an expression of the form $u^2 + a^2$ in the denominator. Quadratic expressions in x that are not of this form can be rewritten in this form by completing the square.

EXAMPLE 5.   Rewrite $x^2 + 4x + 13$ in the form $u^2 + a^2$, where $u = u(x)$ and a is a constant.

We can think of $x^2 + 4x$ as the first two terms in the expression $(x + k)^2$, where k is a constant equal to half of the coefficient of the x-term. Thus, $(x + 2)^2 = x^2 + 4x + 4$. We write

$$x^2 + 4x + 13 = x^2 + 4x + 4 + 9$$
$$= (x + 2)^2 + 9$$
$$= (x + 2)^2 + (3)^2 \quad \blacksquare$$

EXAMPLE 6.   Integrate: $\displaystyle\int \frac{1}{x^2 + 4x + 13}\, dx$.

Using the result of Example 5 and the Arctangent integration pattern,

$$\int \frac{1}{x^2 + 4x + 13}\, dx = \int \frac{1}{(x + 2)^2 + (3)^2}\, dx$$
$$= \frac{1}{3} \operatorname{Arctan} \frac{x + 2}{3} + C \quad \blacksquare$$

Practice 3.   Integrate: $\displaystyle\int \frac{1}{x^2 + 2x + 5}\, dx$.

Exercises / Section 9-6

Exercises 1–22: Integrate.

1. $\displaystyle\int \frac{1}{\sqrt{9 - x^2}}\, dx$

2. $\displaystyle\int \frac{1}{\sqrt{36 - x^2}}\, dx$

3. $\displaystyle\int \frac{4}{\sqrt{4 - 4x^2}}\, dx$

4. $\displaystyle\int \frac{7}{\sqrt{25 - 9x^2}}\, dx$

5. $\displaystyle\int \frac{2x}{\sqrt{16 - 7x^4}}\, dx$

6. $\displaystyle\int \frac{3x}{\sqrt{25 - 10x^4}}\, dx$

7. $\displaystyle\int \frac{e^x}{\sqrt{9 - e^{2x}}}\, dx$

8. $\displaystyle\int \frac{\sec^2 x}{\sqrt{16 - \tan^2 x}}\, dx$

9. $\displaystyle\int \frac{e^{2x}}{\sqrt{1 - e^{2x}}}\, dx$

10. $\displaystyle\int \frac{\sin 2x}{\sqrt{1 - \sin^2 x}}\, dx$

11. $\displaystyle\int \frac{1}{16 + 25x^2}\, dx$

12. $\displaystyle\int \frac{1}{25 + 4x^2}\, dx$

13. $\displaystyle\int \frac{x}{9 + x^2}\, dx$

14. $\displaystyle\int \frac{3x}{10 + 16x^2}\, dx$

15. $\int_0^1 \dfrac{2x}{4 + x^4}\, dx$

16. $\int_0^2 \dfrac{x^2}{9 + x^6}\, dx$

17. $\int \dfrac{\cos x}{9 + \sin^2 x}\, dx$

18. $\int \dfrac{e^x}{25 + 4e^{2x}}\, dx$

19. $\int \dfrac{1}{x^2 + 6x + 13}\, dx$

20. $\int \dfrac{2}{x^2 + 8x + 25}\, dx$

21. $\int \dfrac{4}{x^2 - 6x + 10}\, dx$

22. $\int \dfrac{1}{x^2 - 10x + 29}\, dx$

23. Find the area of the region bounded by $y = 1/\sqrt{9 - x^2}$, $x = -1$, $x = 1$, and $y = 0$.

24. Find the area of the region bounded by $y = x/(1 + x^4)$, $y = 0$, $x = 0$, and $x = 2$.

25. Find the moment with respect to the y-axis of the area bounded by $y = e^{x^2}/(1 + e^{2x^2})$, $y = 0$, $x = 0$, and $x = 2$. (Assume a constant density k.)

26. Find the moment with respect to the y-axis of the area bounded by $y = 1/\sqrt{16 - x^4}$, $y = 0$, $x = 0$, and $x = 1.3$. (Assume a constant density k.)

27. The velocity v of a moving object is given by $v = 1/\sqrt{4 - t^2}$. Find the position of the object at time $t = 2$ if $s = 10$ when $t = 0$.

28. Find the volume of the solid of revolution obtained by rotating the area bounded by $y = 1/\sqrt{x^2 + 4}$, $x = 0$, $y = 0$, and $x = 1$ about the x-axis.

29. The current i in a certain electric circuit is given by
$$i = \frac{\cos t}{1 + \sin^2 t}$$
What is the total charge that has passed a given point in the circuit in the first second?

30. Find the equation of the curve for which
$$\frac{dy}{dx} = \frac{x^2}{\sqrt{100 - x^6}}$$
and that passes through the point $(0, 2)$.

9-7 Integration by Parts

We have already seen that it is sometimes necessary to rewrite an integrand (using trigonometric identities, for example) in order to express the integrand in a form that matches one of the integration patterns. In this section and the next two, we learn some much more subtle approaches to rewriting. Given a little time, we would probably all think of using a trigonometric identity in an integrand with trigonometric functions, but the rewriting ideas coming up would most likely never occur to us. Nonetheless, they do work, and as long as someone else has been clever enough to discover them, we might as well make use of them!

The formula for integration by parts comes from the rule for differentiating a product. If u and v are both functions of x, then

$$\frac{d(uv)}{dx} = u\frac{dv}{dx} + v\frac{du}{dx}$$

Let us integrate both sides of this equation with respect to x. The left side of this equation is written as the derivative of a function with respect to x, so that integration gives back that function. On the right side, we symbolically represent the integration term by term.

$$\int \frac{d(uv)}{dx}\, dx = \int u\, \frac{dv}{dx}\, dx + \int v\, \frac{du}{dx}\, dx$$

$$uv = \int u\, dv + \int v\, du \tag{1}$$

As a matter of fact, the left side of Equation (1) should be $uv + C$, where C is the arbitrary constant of integration. The C is omitted, however, because constants of integration will be introduced by the integrals on the right side of Equation (1), and these can all be lumped together as one arbitrary constant at the end.

Rewriting Equation (1), we get

$$\boxed{\int u\, dv = uv - \int v\, du}$$

This is the formula for integration by parts. Example 1 shows how to apply it.

EXAMPLE 1.   Integrate: $\int xe^x\, dx$.

We first note that this is not integrable by any of our basic forms. It fails to match the form $\int e^u\, du$ because of the factor of x. We attempt to match the integral with the $\int u\, dv$ part of the formula for integration by parts. We let

$$u = x \qquad dv = e^x\, dx$$

The remainder of the formula requires us to know du and v, so we compute these.

$$u = x \qquad dv = e^x\, dx$$
$$du = dx \qquad v = e^x$$

(Again, the constant that goes along with integrating dv will be included in the final constant.) Applying the integration-by-parts formula, we have

$$\int xe^x\, dx = xe^x - \int e^x\, dx$$

The integral on the right is one we can do (note that the troublesome x is now gone). Finally,

$$\int xe^x\, dx = xe^x - e^x + C \quad \blacksquare$$

In order to use integration by parts, we must rewrite our integrand in the form $u\,dv$. This means we must choose which part is to be u, and the remaining part is then dv. Two factors govern this choice.

1. We must be able to integrate dv to find v. (We must also be able to differentiate u to find du, but this is not a problem; we can always differentiate, but we cannot always integrate.)
2. Assuming that we can integrate dv, the new integral $\int v\,du$ should not be any "worse" than the original one. Let us reconsider the integral of Example 1.

EXAMPLE 2. Integrate: $\displaystyle\int xe^x\,dx$.

Suppose our choice for u and dv had been as follows:

$$u = e^x \qquad\qquad dv = x\,dx$$

Then

$$du = e^x\,dx \qquad\qquad v = \frac{x^2}{2}$$

and

$$\int xe^x\,dx = e^x\frac{x^2}{2} - \int \frac{x^2}{2}\,e^x\,dx$$

Our new integral,

$$\int \frac{x^2}{2}\,e^x\,dx$$

is even further from being integrable than the original one; originally we wished we did not have the factor of x, and now we have x^2! We made the wrong choice for u and dv. ■

EXAMPLE 3. Integrate: $\displaystyle\int x^2 \ln x\,dx$.

We choose

$$u = x^2 \qquad\qquad dv = \ln x\,dx$$

Then

$$du = 2x^2\,dx \qquad\qquad v = ?$$

Wrong choice again! We cannot integrate $\ln x$. Try

$$u = \ln x \qquad\qquad dv = x^2\,dx$$

Then

$$du = \frac{1}{x}\,dx \qquad\qquad v = \frac{x^3}{3}$$

Applying the formula,

$$\int x^2 \ln x = \frac{x^3}{3} \ln x - \int \frac{x^3}{3} \cdot \frac{1}{x} \, dx$$

$$= \frac{x^3}{3} \ln x - \frac{1}{3} \int x^2 \, dx$$

$$= \frac{x^3}{3} \ln x - \frac{1}{3} \frac{x^3}{3} + C$$

$$= \frac{x^3}{3} \ln x - \frac{x^3}{9} + C$$

Success!   ■

By this time you may be wondering how you will ever get the hang of choosing u and dv. Practice helps, but there are also some general guidelines. The choice for u should, if possible, be a function that becomes less complex by differentiation (for example, x^7, whose power is reduced by differentiation, or $\ln x$, which changes from a transcendental to an algebraic function when differentiated). The choice for v should, if possible, be a function that gets no more complex when integrated (for example, e^x, $\sin x$, or $\cos x$).

Word of Advice

In using integration by parts, write u and dv, then compute du and v; do not try to do this in your head.

EXAMPLE 4.   Integrate: $\int x \sin x \, dx$.

We choose

$$u = x \qquad dv = \sin x \, dx$$
$$du = dx \qquad v = -\cos x$$

Then

$$\int x \sin x \, dx = -x \cos x - \int -\cos x \, dx$$

$$= -x \cos x + \int \cos x \, dx$$

$$= -x \cos x + \sin x + C \quad ■$$

Practice 1.   Integrate: $\int x \cos x \, dx$.

Sometimes integration by parts must be done several times.

EXAMPLE 5. Integrate: $\int_0^2 x^2 e^{-x}\, dx$.

We write

$$u = x^2 \qquad\qquad dv = e^{-x}\, dx$$
$$du = 2x\, dx \qquad\qquad v = -e^{-x}$$

$$\int_0^2 x^2 e^{-x}\, dx = -x^2 e^{-x}\Big|_0^2 - \int_0^2 -2x e^{-x}\, dx$$

$$= -x^2 e^{-x}\Big|_0^2 + 2\int_0^2 x e^{-x}\, dx$$

Another integration by parts is needed on the new integral.

$$u = x \qquad\qquad dv = e^{-x}\, dx$$
$$du = dx \qquad\qquad v = -e^{-x}$$

Thus

$$\int_0^2 x^2 e^{-x}\, dx = -x^2 e^{-x}\Big|_0^2 + 2\left[-x e^{-x}\Big|_0^2 - \int_0^2 -e^{-x}\, dx\right]$$

$$= -x^2 e^{-x}\Big|_0^2 - 2x e^{-x}\Big|_0^2 + 2\int_0^2 e^{-x}\, dx$$

$$= -x^2 e^{-x}\Big|_0^2 - 2x e^{-x}\Big|_0^2 - 2e^{-x}\Big|_0^2$$

$$= -0.541 - 0.541 - 0.271 + 2$$

$$= 0.647 \quad \blacksquare$$

EXAMPLE 6. Integrate: $\int \sin 2x \cos x\, dx$.

What to choose here for u and dv is not clear. We try

$$u = \sin 2x \qquad\qquad dv = \cos x\, dx$$
$$du = 2\cos 2x\, dx \qquad\qquad v = \sin x$$

Then

$$\int \sin 2x \cos x\, dx = \sin 2x \sin x - 2\int \sin x \cos 2x\, dx$$

Here the new integral does not look any better than the original one, but it does not look any worse either. We try integration by parts again on the new integral.

$$u = \cos 2x \qquad\qquad dv = \sin x\, dx$$
$$du = -2\sin 2x\, dx \qquad\qquad v = -\cos x$$

We get

$$\int \sin 2x \cos x \, dx = \sin 2x \sin x$$
$$- 2\left[-\cos 2x \cos x - \int 2 \sin 2x \cos x \, dx\right]$$

or

$$\int \sin 2x \cos x \, dx = \sin 2x \sin x$$
$$+ 2 \cos 2x \cos x + 4 \int \sin 2x \cos x \, dx$$

Our original integral has reappeared. Subtracting this integral from both sides of the above equation and rearranging terms gives

$$-\sin 2x \sin x - 2 \cos 2x \cos x = 3 \int \sin 2x \cos x \, dx$$

Solving for $\int \sin 2x \cos x \, dx$, we get

$$\int \sin 2x \cos x \, dx = -\tfrac{1}{3}(\sin 2x \sin x + 2 \cos 2x \cos x) \quad \blacksquare$$

What happened in Example 6 seems quite boggling. Perhaps you should read it again. This tactic (integration by parts twice with the original integral reappearing so that it can be solved for algebraically) can be used when both u and dv retain the same level of complexity when integrated or differentiated. Such combinations as $\sin ax \cos bx$, $\sin ax \sin bx$, $\cos ax \cos bx$, $e^{ax} \sin bx$, and $e^{ax} \cos bx$ fall into this category.

Word of Advice

Occasionally, when you try integration by parts twice and carry out the resulting algebra, you end up with the true but unhelpful equation $0 = 0$. This indicates that your choices for u and dv in the second integration by parts were ineffective; go back and reverse them.

Practice 2. Integrate: $\int \cos 3x \sin x \, dx$.

Exercises / Section **9-7**

Exercises 1–18: Integrate:

1. $\int x e^{2x} \, dx$

2. $\int 3x e^{-x} \, dx$

3. $\int_{\pi/2}^{\pi} x \sin 2x \, dx$

4. $\int_{0}^{\pi/2} x \cos 3x \, dx$

5. $\int x \ln x \, dx$

6. $\int \text{Arctan } x \, dx$

7. $\int x e^{x^2} \, dx$

8. $\int x \sin x^2 \, dx$

9. $\int \frac{x^3}{\sqrt{x^2 + 1}} \, dx$

10. $\int \frac{x^5}{\sqrt{1 - x^3}} \, dx$

11. $\int x \sec^2 x \, dx$

12. $\int x \sec x \tan x \, dx$

13. $\int x^2 \cos 2x \, dx$

14. $\int x^2 e^{3x} \, dx$

15. $\int e^x \cos x \, dx$

16. $\int e^x \sin x \, dx$

17. $\int \sin x \sin 2x \, dx$

18. $\int \cos x \cos 2x \, dx$

19. Find the area of the region bounded by $x = 1$, $y = 0$, and $y = \text{Arcsin } x$.

20. Find the volume generated by rotating the area bounded by $y = \ln x$, $y = 0$, and $x = e$ about the x-axis.

21. The velocity of a moving object is given by the equation $v = t\sqrt[3]{1 + t}$. If $s = 0$ when $t = 0$, what is s when $t = 1$?

22. Find the equation of the curve with derivative $dy/dx = x \csc^2 x$ that passes through the point $(\pi/4, 2)$.

23. Find the first moment with respect to the y-axis of the area bounded by $y = \sin x$, $y = 0$, and $x = \pi/2$; assume constant density k.

24. Find the moment of inertia with respect to the y-axis of the area bounded by $y = x/\sqrt{1 - x^2}$, $y = 0$, and $x = 0.5$; assume constant density k.

25. The total population (in thousands of people) within a radius of 3 mi of the center of a certain urban area is estimated by the integral

$$\int_{0}^{3} 300 \pi r e^{-0.12r} \, dr$$

Find the value of this integral.

26. One end of a holding tank in a marine biology research laboratory is shaped like the area bounded by the two curves $y = \ln|x|$, and by the x-axis and the line $y = -1$. If the tank is filled with seawater (weight density: 10045 N/m³), what is the force (in newtons) on the end of the tank?

9-8 Integration by Trigonometric Substitution

Integrals that contain certain expressions with radical signs can sometimes be solved by the strategy of using a temporary new variable. The new variable represents an angle, and the integrand is rewritten in terms of trigonometric functions of this new variable. We substitute a new variable for the original one, and turn an algebraic integral into a trigonometric one—hence the name *trigonometric substitution* for this technique. Of course, once an antiderivative has been found in terms of the new variable, we must reverse the substitution in order to find the antiderivative in terms of the original variable.

There are three types of radical expressions for which trigonometric substitution may be helpful. The three expressions, and the substitution to use for each case, are given below; a is a constant.

$$
\begin{array}{ll}
\sqrt{a^2 - x^2}; & \text{let } x = a \sin \theta. \\
\sqrt{a^2 + x^2}; & \text{let } x = a \tan \theta. \\
\sqrt{x^2 - a^2}; & \text{let } x = a \sec \theta.
\end{array}
$$

In each case, the suggested substitution replaces x with an expression involving a new variable θ. The substitution must be carried out consistently throughout the integrand; thus if x becomes $a \sin \theta$, this dictates a new expression for any other place in the integrand where x appears, including dx.

Why might these strange-looking substitutions be helpful? The answer is that each leads to a simplification by making use of a trigonometric identity, as the following examples will show.

EXAMPLE 1. Integrate: $\displaystyle\int \frac{\sqrt{9 - x^2}}{x^2}\, dx$.

This integral, which is not integrable by any of our previous techniques, does contain a radical of the form $\sqrt{a^2 - x^2}$, where $a = 3$. The suggested substitution is

$$x = 3 \sin \theta$$

Then

$$x^2 = 9 \sin^2\theta$$
$$dx = 3 \cos \theta\, d\theta$$

Making all these substitutions in the original integral results in the new integral

$$\int \frac{\sqrt{9 - 9 \sin^2\theta}}{9 \sin^2\theta} \cdot 3 \cos \theta\, d\theta$$

The variable of integration is now θ, and the integrand involves trigonometric functions. The simplification that uses a trigonometric identity is

$$\sqrt{9 - 9 \sin^2\theta} = \sqrt{9(1 - \sin^2\theta)} = \sqrt{9 \cos^2\theta} = 3 \cos \theta$$

The new integral is, therefore,

$$\int \frac{3 \cos \theta}{9 \sin^2\theta} \cdot 3 \cos \theta\, d\theta = \int \frac{\cos^2\theta}{\sin^2\theta}\, d\theta = \int \cot^2\theta\, d\theta$$
$$= \int (\csc^2\theta - 1)\, d\theta = -\cot \theta - \theta + C$$

We now must reverse the substitution to get the answer in terms of x. The original substitution

$$x = 3 \sin \theta$$

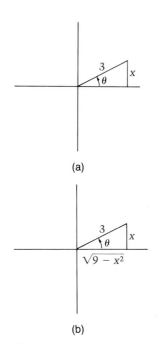

(a)

(b)

Figure 9-2

may be written

$$\frac{x}{3} = \sin \theta$$

or

$$\theta = \text{Arcsin} \frac{x}{3}$$

If we draw a picture, as in Figure 9-2(a), we can complete the triangle using the Pythagorean theorem (Figure 9-2(b)) and determine $-\cot \theta$ in terms of x. The integral finally becomes

$$\int \frac{\sqrt{9 - x^2}}{x^2} \, dx = -\cot \theta - \theta + C = -\frac{\sqrt{9 - x^2}}{x} - \text{Arcsin} \frac{x}{3} + C$$

Notice that the original radical $\sqrt{9 - x^2}$ is also a part of the answer. ■

Practice 1. Integrate: $\displaystyle\int \frac{dx}{x^2 \sqrt{4 - x^2}}$.

EXAMPLE 2. Integrate: $\displaystyle\int \frac{dx}{x^2 \sqrt{4 + x^2}}$.

This integral contains the radical $\sqrt{4 + x^2}$. The appropriate substitution is to let $x = 2 \tan \theta$. Then $x^2 = 4 \tan^2\theta$ and $dx = 2 \sec^2\theta \, d\theta$. Substituting,

$$\int \frac{dx}{x^2 \sqrt{4 + x^2}} = \int \frac{2 \sec^2\theta \, d\theta}{4 \tan^2\theta \sqrt{4 + 4 \tan^2\theta}}$$

For the simplification step,

$$\sqrt{4 + 4 \tan^2\theta} = \sqrt{4(1 + \tan^2\theta)} = \sqrt{4 \sec^2\theta} = 2 \sec \theta$$

We therefore have

$$\int \frac{2 \sec^2\theta \, d\theta}{4 \tan^2\theta \sqrt{4 + \tan^2\theta}} = \int \frac{2 \sec^2\theta \, d\theta}{4 \tan^2\theta \cdot 2 \sec \theta}$$

$$= \int \frac{\sec \theta \, d\theta}{4 \tan^2\theta}$$

$$= \frac{1}{4} \int \frac{1/\cos \theta}{\sin^2\theta/\cos^2\theta} \, d\theta = \frac{1}{4} \int \frac{\cos \theta}{\sin^2\theta} \, d\theta$$

$$= \frac{1}{4} \int \sin^{-2}\theta \cos \theta \, d\theta$$

Using the power rule for integration,

$$\frac{1}{4} \int \sin^{-2}\theta \cos \theta \, d\theta = -\frac{1}{4} \sin^{-1}\theta + C = -\frac{1}{4 \sin \theta} + C$$

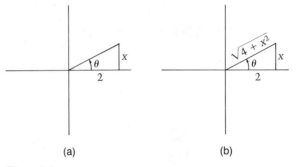

(a)   (b)

Figure 9-3

To reverse the substitution, we draw the triangle that represents $x = 2 \tan \theta$, or $\theta = \text{Arctan}(x/2)$ (Figure 9-3(a)) and solve for $\sin \theta$ from Figure 9-3(b). The integral is

$$\int \frac{dx}{x^2\sqrt{4 + x^2}} = -\frac{1}{4 \sin \theta} + C = -\frac{1}{4x/\sqrt{4 + x^2}} + C$$

$$= -\frac{\sqrt{4 + x^2}}{4x} + C$$

Again, the original radical has reappeared. ■

Practice 2. Integrate: $\int \frac{dx}{\sqrt{1 + x^2}}$.

EXAMPLE 3. Integrate: $\int \frac{x^3}{\sqrt{x^2 - 1}}\, dx$.

Let $x = \sec \theta$; then $x^2 = \sec^2\theta$, $x^3 = \sec^3\theta$, and $dx = \sec \theta \tan \theta\, d\theta$.

$$\int \frac{x^3}{\sqrt{x^2 - 1}}\, dx = \int \frac{\sec^3\theta}{\sqrt{\sec^2\theta - 1}} \cdot \sec \theta \tan \theta\, d\theta$$

$$= \int \frac{\sec^3\theta}{\sqrt{\tan^2\theta}} \cdot \sec \theta \tan \theta\, d\theta = \int \sec^4\theta\, d\theta$$

$$= \int \sec^2\theta \sec^2\theta\, d\theta = \int (1 + \tan^2\theta)\sec^2\theta\, d\theta$$

$$= \int \sec^2\theta + \int \tan^2\theta \sec^2\theta\, d\theta = \tan \theta + \frac{\tan^3\theta}{3} + C$$

$$= \sqrt{x^2 - 1} + \frac{(x^2 - 1)^{3/2}}{3} + C$$

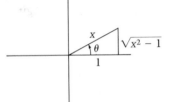

Figure 9-4

The reverse substitution in the last step is made with the help of Figure 9-4. ■

Practice 3. Integrate: $\int \dfrac{\sqrt{x^2 - 16}}{x}\, dx$.

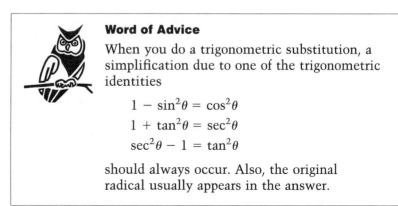

Word of Advice

When you do a trigonometric substitution, a simplification due to one of the trigonometric identities

$$1 - \sin^2\theta = \cos^2\theta$$
$$1 + \tan^2\theta = \sec^2\theta$$
$$\sec^2\theta - 1 = \tan^2\theta$$

should always occur. Also, the original radical usually appears in the answer.

Of course, not every integral involving a radical requires the use of trigonometric substitution. Many can be done using previous techniques.

EXAMPLE 4. Integrate: $\int \dfrac{x}{\sqrt{x^2 - 1}}\, dx$.

While this problem seems similar to Example 3, a simple rewriting shows it to be integrable by the power rule.

$$\int \frac{x}{\sqrt{x^2 - 1}}\, dx = \int (x^2 - 1)^{-1/2}x\, dx = \frac{1}{2}\int (x^2 - 1)^{-1/2} 2x\, dx$$
$$= \frac{1}{2}\frac{(x^2 - 1)^{1/2}}{\frac{1}{2}} + C = \sqrt{x^2 - 1} + C \quad \blacksquare$$

If we are evaluating a definite integral by the method of trigonometric substitution, we must reverse the substitution *before* evaluating at the limits, because the limits apply to the original variable.

EXAMPLE 5. Integrate: $\displaystyle\int_2^3 \frac{dx}{(x^2 - 1)^{3/2}}$.

A radical sign does not explicitly appear in the integrand, but the fractional exponent $\frac{3}{2}$ indicates the presence of the radical. We let

$x = \sec\theta$. Then $x^2 = \sec^2\theta$ and $dx = \sec\theta\tan\theta\,d\theta$. Making these substitutions,

$$\int \frac{dx}{(x^2 - 1)^{3/2}} = \int \frac{\sec\theta\tan\theta\,d\theta}{(\sec^2\theta - 1)^{3/2}}$$

$$= \int \frac{\sec\theta\tan\theta\,d\theta}{(\tan^2\theta)^{3/2}}$$

$$= \int \frac{\sec\theta\tan\theta\,d\theta}{\tan^3\theta}$$

$$= \int \frac{\sec\theta}{\tan^2\theta}\,d\theta$$

$$= \int \sin^{-2}\theta\,\cos\theta\,d\theta$$

$$= -\frac{1}{\sin\theta}$$

$$= -\frac{x}{\sqrt{x^2 - 1}} \qquad \text{By Figure 9-5.}$$

Figure 9-5

Now that we are back to the original variable, we consider the limits of integration.

$$\left. \frac{-x}{\sqrt{x^2 - 1}} \right|_2^3 = -\frac{3}{\sqrt{8}} + \frac{2}{\sqrt{3}} = 0.094 \quad \blacksquare$$

Exercises / Section **9-8**

Exercises 1–20: Integrate:

1. $\displaystyle\int \frac{dx}{\sqrt{9 + x^2}}$

2. $\displaystyle\int \frac{dx}{x\sqrt{1 - x^2}}$

3. $\displaystyle\int \frac{dx}{\sqrt{x^2 - 9}}$

4. $\displaystyle\int \frac{x^3}{\sqrt{4 - x^2}}\,dx$

5. $\displaystyle\int \frac{dx}{(1 + x^2)^{3/2}}$

6. $\displaystyle\int \frac{x^2}{(4 - x^2)^{3/2}}\,dx$

7. $\displaystyle\int x^3(9 - x^2)^{3/2}\,dx$

8. $\displaystyle\int \frac{x^3}{(1 + x^2)^{3/2}}\,dx$

9. $\displaystyle\int \frac{dx}{x\sqrt{x^2 - 4}}$

10. $\displaystyle\int \sqrt{25 - x^2}\,dx$

11. $\displaystyle\int \frac{x}{\sqrt{16 + x^2}}\,dx$

12. $\displaystyle\int x(x^2 - 4)^{3/2}\,dx$

13. $\displaystyle\int_1^2 \frac{dx}{x\sqrt{9 - x^2}}$

14. $\displaystyle\int_5^6 \frac{dx}{x^2(x^2 - 16)^{3/2}}$

15. $\displaystyle\int \frac{x^2\,dx}{(25x^2 + 16)^{3/2}}$

16. $\displaystyle\int \frac{dx}{\sqrt{1 - \sqrt{x}}}$

17. $\displaystyle\int \frac{\cos x\,dx}{\sqrt{9 + \sin^2 x}}$

18. $\displaystyle\int \frac{\cos x\,dx}{(1 + \sin^2 x)^{3/2}}$

19. $\displaystyle\int \frac{dx}{\sqrt{x^2 + 2x + 2}}$

20. $\displaystyle\int \frac{dx}{\sqrt{21 - x^2 - 4x}}$

21. Use integration by trigonometric substitution to find the area of a circle of radius 5.

22. A clock mechanism uses a stainless steel cam in the shape of an ellipse. If the equation of the edge of the ellipse is given by $9x^2 + 16y^2 = 144$, find its area.

23. Find the volume generated when the area bounded by $y = 1/(x^2 + 4)$, $x = 0$, $x = 2$, and $y = 0$ is rotated about the x-axis.

24. Find the volume generated when the area bounded by $y = x^2/(9 - x^2)^{3/2}$, $y = 0$, $x = 1$, and $x = 2$ is rotated about the y-axis.

25. Find the first moment with respect to the y-axis of the area bounded by $y = 1/x^4(x^2 + 1)$, $y = 0$, $x = 1$, and $x = 4$; assume a constant density k.

26. Find the length of arc of the curve $y = x^2$ from $x = 0$ to $x = 2$. (*Hint:* Use integration by parts after the trigonometric substitution.)

9-9   Integration by Partial Fractions

Our last integration method applies when the integrand is a quotient of two polynomials. Suppose, for example, that our problem is

$$\int \frac{5x + 1}{x^2 + x - 2} \, dx$$

We can factor the denominator and write

$$\frac{5x + 1}{x^2 + x - 2} = \frac{5x + 1}{(x + 2)(x - 1)}$$

This fraction can be obtained by adding two fractions, one with denominator $x + 2$ and the other with denominator $x - 1$. We can check that

$$\frac{5x + 1}{(x + 2)(x - 1)} = \frac{3}{x + 2} + \frac{2}{x - 1}$$

Here we have expressed the fraction $(5x + 1)/(x + 2)(x - 1)$ as a sum of **partial fractions**. Then

$$\int \frac{5x + 1}{(x + 2)(x - 1)} \, dx = \int \frac{3}{x + 2} \, dx + \int \frac{2}{x - 1} \, dx$$
$$= 3 \ln|x + 2| + 2 \ln|x - 1| + C$$

and we are done. The question is, given a quotient of polynomials, how do we find its **partial fraction decomposition**? For the case of $(5x + 1)/(x + 2)(x - 1)$, we can guess that the partial fraction decomposition has the form

$$\frac{5x + 1}{(x + 2)(x - 1)} = \frac{A}{x + 2} + \frac{B}{x - 1}$$

where A and B are unknown constants. Then, working backwards, we add:

$$\frac{A}{x + 2} + \frac{B}{x - 1} = \frac{A(x - 1) + B(x + 2)}{(x + 2)(x - 1)}$$
$$= \frac{Ax - A + Bx + 2B}{(x + 2)(x - 1)}$$
$$= \frac{(A + B)x + 2B - A}{(x + 2)(x - 1)}$$

Now setting this expression equal to our original expression,

$$\frac{5x + 1}{(x + 2)(x - 1)} = \frac{(A + B)x + (2B - A)}{(x + 2)(x - 1)}$$

we get two equal fractions with the same denominators. The numerators must therefore be equal, and we write

$$5x + 1 = (A + B)x + (2B - A) \tag{1}$$

The coefficients of x on each side of Equation (1) and the constant terms on each side of Equation (1) must be equal. Thus

$$5 = A + B$$
$$1 = 2B - A$$

Solving this system of equations gives us

$$A = 3$$
$$B = 2$$

and

$$\frac{5x + 1}{(x + 2)(x - 1)} = \frac{3}{x + 2} + \frac{2}{x - 1}$$

What we have done here suggests a general procedure. The steps in this procedure are given below.

Remember that this method of partial fraction decomposition is to be used only on integrands of the form $\frac{p(x)}{q(x)}$, where $p(x)$ and $q(x)$ are polynomials (and only when these integrals do not fit some easier integration pattern, such as the power rule).

We proceed as follows; the steps will be clearer after you study the examples.

1. Use long division, if necessary, to rewrite the problem as the sum of a polynomial plus a fraction with the degree of the numerator less than the degree of the denominator. Concentrate on this resulting fraction.

2. Factor the denominator into linear factors (those of the form $ax + b$) or irreducible quadratic factors (those of the form $ax^2 + bx + c$ that cannot be factored further).

3. Write the fraction as a sum of partial fractions with unknown constants.

   a. For every denominator factor of the form $(ax + b)^n$, include a sum of the form

   $$\frac{A_1}{ax + b} = \frac{A_2}{(ax + b)^2} + \cdots + \frac{A_n}{(ax + b)^n}$$

   where $A_1, A_2, \ldots, A_n$ are unknown constants.

   b. For every denominator factor of the form $(ax^2 + bx + c)^n$, include a sum of the form

   $$\frac{A_1 x + B_1}{ax^2 + bx + c} + \frac{A_2 x + B_2}{(ax^2 + bx + c)^2} + \cdots + \frac{A_n x + B_n}{(ax^2 + bx + c)^n}$$

   where $A_1, A_2, \ldots, A_n$ and $B_1, B_2, \ldots, B_n$ are unknown constants.

4. Add the fractions from Step 3, set the result equal to the original fraction, and then set the two numerators equal. Use this equation in Step 5.

5. Solve for the unknown constants by equating the co-efficients of equal powers of x on each side of the equation and solving the resulting system of equations.

6. Replace the unknown constants in the sum of Step 3 with their known values from Step 5 and integrate the result.

Actually, the whole process is not nearly as bad as it sounds!

EXAMPLE 1. Integrate: $\displaystyle\int \frac{x^3 - 5x^2 + 7x - 4}{x^2 - 6x + 9}\,dx$.

The integrand is a fraction, where the degree of the numerator is not less than the degree of the denominator. We can do long division (Step 1):

$$
\begin{array}{r}
x + 1 \\
x^2 - 6x + 9 \,\overline{\big)\, x^3 - 5x^2 + 7x - 4} \\
\underline{x^3 - 6x^2 + 9x} \\
x^2 - 2x - 4 \\
\underline{x^2 - 6x + 9} \\
4x - 13
\end{array}
$$

Therefore,

$$\frac{x^3 - 5x^2 + 7x - 4}{x^2 - 6x + 9} = x + 1 + \frac{4x - 13}{x^2 - 6x + 9}$$

We include $x + 1$ in the final step, but for now we factor the denominator of the fractional part of this expression (Step 2).

$$\frac{4x - 13}{x^2 - 6x + 9} = \frac{4x - 13}{(x - 3)^2}$$

The denominator has one linear factor to the second power. (Note that the denominator is *not* an irreducible quadratic factor.) From Step 3(a), the form of the partial fraction decomposition is

$$\frac{4x - 13}{(x - 3)^2} = \frac{A_1}{x - 3} + \frac{A_2}{(x - 3)^2}$$

Adding the fractions on the right side of this equation, we get

$$\frac{4x - 13}{(x - 3)^2} = \frac{A_1(x - 3) + A_2}{(x - 3)^2}$$

Setting the numerators equal (Step 4),

$$4x - 13 = A_1(x - 3) + A_2$$

or

$$4x - 13 = A_1 x + (A_2 - 3A_1)$$

From Step 5, we equate the coefficients of x and the constant terms from each side of this equation. This results in the system of equations

$$4 = A_1$$
$$-13 = A_2 - 3A_1$$

The solution to this system is

$$A_1 = 4$$
$$A_2 = -1$$

The integral then becomes (Step 6)

$$\int \frac{x^3 - 5x^2 + 7x + 4}{x^2 - 6x + 9} = \int \left(x + 1 + \frac{4}{x - 3} - \frac{1}{(x - 3)^2} \right) dx$$

$$= \frac{x^2}{2} + x + 4 \ln|x - 3| - \frac{(x - 3)^{-1}}{-1} + C$$

$$= \frac{x^2}{2} + x + 4 \ln|x - 3| + \frac{1}{x - 3} + C \quad \blacksquare$$

EXAMPLE 2. Integrate: $\displaystyle\int \frac{5x^2 - 2x + 1}{(x^2 + 1)(x - 1)} \, dx.$

Here the degree of the numerator is two and the degree of the denominator (if multiplied out) is three. Therefore, we do not need to divide. Also, the denominator is already factored into one irreducible quadratic factor and one linear factor. Following the rules outlined in Step 3(a) and (b), we write

$$\frac{5x^2 - 2x + 1}{(x^2 + 1)(x - 1)} = \frac{Ax + B}{x^2 + 1} + \frac{C}{x - 1}$$

Adding the two fractions on the right side of this equation, we get

$$\frac{5x^2 - 2x + 1}{(x^2 + 1)(x - 1)} = \frac{(Ax + B)(x - 1) + C(x^2 + 1)}{(x^2 + 1)(x - 1)}$$

The numerators must be equal; therefore, we can write

$$5x^2 - 2x + 1 = (Ax + B)(x - 1) + C(x^2 + 1)$$
$$= Ax^2 + Bx - Ax - B + Cx^2 + C$$
$$= (A + C)x^2 + (B - A)x + (C - B)$$

The coefficients of x^2 and of x and the constant term on each side of the equal sign must be the same. Thus

$$5 = A + C$$
$$-2 = B - A$$
$$1 = C - B$$

The solution to this system of equations is

$$A = 3$$
$$B = 1$$
$$C = 2$$

We now put these values back into the original partial fraction decomposition.

$$\int \frac{5x^2 - 2x + 1}{(x^2 + 1)(x - 1)} \, dx = \int \left(\frac{3x + 1}{x^2 + 1} + \frac{2}{x - 1} \right) dx$$

$$= \int \left(\frac{3x}{x^2 + 1} + \frac{1}{x^2 + 1} + \frac{2}{x - 1} \right) dx$$

$$= \int \frac{3x}{x^2 + 1} \, dx + \int \frac{1}{x^2 + 1} \, dx + \int \frac{2}{x - 1} \, dx$$

The first and third integrals are natural logarithm forms and the second is an Arctangent form. The result is

$$\int \frac{5x^2 - 2x + 1}{(x^2 + 1)(x - 1)} = \frac{3}{2} \ln|x^2 + 1| + \text{Arctan } x$$
$$+ 2 \ln|x - 1| + C \quad \blacksquare$$

Practice 1. In the integral $\int \dfrac{x^4 + x^3 - 3x - 2}{x^3 + x^2 - 2x}$, use long division to rewrite the integrand as the sum of a polynomial plus a fraction with the degree of the numerator less than the degree of the denominator.

Practice 2. Factor the denominator in the fraction obtained as part of the answer to Practice 1.

Practice 3. Write the fraction from Practice 2 as the sum of partial fractions with unknown constants.

Practice 4. Solve for the unknown constants in the answer to Practice 3.

Practice 5. Using the results from Practices 1–4, find

$$\int \frac{x^4 + x^3 - 3x - 2}{x^3 + x^2 - 2x} \, dx$$

Once we have found the partial fraction decomposition of an integrand, how do we know we will always be able to do the resulting integrations? Each resulting integral fits one of the patterns with which we have already dealt. There are only four possibilities.

1. Terms of the form $A/(ax + b)$, which come from linear factors. Terms such as this fit the natural logarithm pattern, although we may have to fiddle with the constants. Thus

$$\int \frac{3}{4x + 2} \, dx = 3 \int \frac{1}{4x + 2} \, dx = \frac{3}{4} \int \frac{4}{4x + 2} \, dx$$

$$= \frac{3}{4} \ln|4x + 2| + C$$

2. Terms of the form $A/(ax + b)^n$, $n > 1$, which come from repeated linear factors. Such terms, again with appropriate adjustment of constants, can be integrated by the power rule. For example,

$$\int \frac{5}{(2x + 1)^3} \, dx = 5 \int \frac{1}{(2x + 1)^3} \, dx = \frac{5}{2} \int 2(2x + 1)^{-3} \, dx$$

$$= \frac{5}{2} \frac{(2x + 1)^{-2}}{-2} + C = \frac{-5}{4(2x + 1)^2} + C$$

3. Terms of the form $(Ax + B)/(ax^2 + bx + c)$, which arise from quadratic factors. Here, we ultimately split the integral into two parts, one of the form $(2ax + b)/(ax^2 + bx + c)$, which fits the natural logarithm pattern, and one of the form $m/(ax^2 + bx + c)$, where m is a constant. We can integrate this second term by using the Arctangent integration rule; we may have to first complete the square, as in Examples 5 and 6 of Section 9-6. Again, there is some work to be done to get the constants the way we want them.

$$\int \frac{5x + 10}{2x^2 + 6x + 5} \, dx = 5 \int \frac{x + 2}{2x^2 + 6x + 5} \, dx$$

$$= \frac{5}{4} \int \frac{4(x + 2)}{2x^2 + 6x + 5} \, dx$$

$$= \frac{5}{4} \int \frac{4x + 8}{2x^2 + 6x + 5} \, dx$$

So far we have succeeded in obtaining $4x$ in the numerator, but we want $4x + 6$ in order to have the derivative of the denominator. Continuing,

$$\frac{5}{4} \int \frac{4x + 8}{2x^2 + 6x + 5} \, dx = \frac{5}{4} \int \frac{4x + 6 + 2}{2x^2 + 6x + 5} \, dx$$

$$= \frac{5}{4} \left[\int \frac{4x + 6}{2x^2 + 6x + 5} \, dx \right.$$

$$\left. + \int \frac{2}{2x^2 + 6x + 5} \, dx \right]$$

Integrating the first integral and factoring 2 out of the numerator and denominator of the second integrand, we get

$$\frac{5}{4}\left(\ln|2x^2 + 6x + 5| + \frac{2}{2}\int \frac{1}{x^2 + 3x + \frac{5}{2}}\, dx\right)$$

We now concentrate on completing the square in the integral

$$\int \frac{1}{x^2 + 3x + \frac{5}{2}}\, dx$$

Recall from Section 9-6 that we try to introduce $(x + k)^2$, where k is one-half the coefficient of the x-term. Thus

$$\int \frac{1}{x^2 + 3x + \frac{5}{2}}\, dx = \int \frac{1}{x^2 + 3x + (\frac{3}{2})^2 - (\frac{3}{2})^2 + \frac{5}{2}}\, dx$$

$$= \int \frac{1}{(x + \frac{3}{2})^2 - \frac{9}{4} + \frac{5}{2}}\, dx$$

$$= \int \frac{1}{(x + \frac{3}{2})^2 + \frac{1}{4}}\, dx$$

$$= \int \frac{1}{(x + \frac{3}{2})^2 + (\frac{1}{2})^2}\, dx = \frac{1}{\frac{1}{2}}\, \text{Arctan}\, \frac{x + \frac{3}{2}}{\frac{1}{2}}$$

The final result is

$$\frac{5}{4}\left[\ln|2x^2 + 6x + 5| + 2\, \text{Arctan}(2x + 3)\right] + C$$

While this was a long process, we were able to complete the integration successfully.

4. Terms of the form $(Ax + B)/(ax^2 + bx + c)^n$, $n > 1$, which come from repeated quadratic factors. Here we eventually split the integral into two parts, one of the form $(2ax + b)/(ax^2 + bx + c)^n$, which fits the power rule, and one of the form $m/(ax^2 + bx + c)^n$ where m is a constant. Manipulating the constants to split the integral in this way works just as it did in the example for Case 3. Finally, we integrate the second term by completing the square in the denominator and then using a trigonometric substitution that introduces $\tan \theta$. We shall not do an example here.

Exercises / Section **9-9**

Exercises 1–20: Integrate:

1. $\displaystyle\int \frac{5x - 3}{x^2 - x}\, dx$

2. $\displaystyle\int \frac{x - 17}{x^2 + x - 6}\, dx$

3. $\displaystyle\int \frac{6x + 17}{2x^2 + 7x - 15}\, dx$

4. $\displaystyle\int \frac{-x + 13}{12x^2 + 11x - 5}\, dx$

5. $\displaystyle\int \frac{6x^2 + 5x - 3}{x^3 - x}\, dx$

6. $\displaystyle\int \frac{3x^2 - 16x + 1}{(x - 1)(x + 2)(x - 3)}\, dx$

7. $\displaystyle\int \frac{4x^2 - x - 2}{x^3 - x^2}\, dx$

8. $\int \dfrac{3x^2 + 11x + 18}{x^3 + 6x^2 + 9x}\, dx$

9. $\int \dfrac{-7x^3 + 6x^2 - x - 6}{x^3(x - 2)}\, dx$

10. $\int \dfrac{x^3 - 9x^2 - 40x + 16}{x^2(x + 4)^2}\, dx$

11. $\int_1^4 \dfrac{2x^3 + 9x^2 - 12x + 3}{x^2(2x - 1)^2}\, dx$

12. $\int_2^3 \dfrac{x^3 + 5x + 10}{(x^2 - 1)^2}\, dx$

13. $\int \dfrac{2x^3 + x^2 - 5x - 1}{2x^2 + 5x + 3}\, dx$

14. $\int \dfrac{2x^4 + 2x^3 + 2x + 3}{x^3 + x^2}\, dx$

15. $\int \dfrac{7x^2 - x + 4}{x^3 + x}\, dx$

16. $\int \dfrac{3x^3 + 4x^2 - 3x + 12}{x^2(3 + x^2)}\, dx$

17. $\int \dfrac{8x^2 + 42x + 50}{x(x^2 + 6x + 10)}\, dx$

18. $\int \dfrac{11x^3 + 7x^2 + 11x - 5}{x^2(x^2 + 4x + 5)}\, dx$

19. $\int \dfrac{6x^3 - 9x^2 + 10x - 9}{(x^2 + 1)^2}\, dx$

20. $\int \dfrac{3x^3 + 5x^2 + 13x + 20}{(4 + x^2)^2}\, dx$

21. Find the area bounded by the curves $y = 1/(2x^2 - 3x)$, $y = 0$, $x = 2$, and $x = 4$.

22. The area bounded by the curves $y = \dfrac{4}{(x + 1)(x - 3)}$, $y = 0$, $x = 4$, and $x = 5$ is rotated about the x-axis. What is the resulting volume?

23. The production rate of a particular coal mine in the 1980s is estimated to be $y = \dfrac{10^8}{t(t + 2)^2}$ tons per year, where t is the number of years since 1970. Estimate the total number of tons that the mine will produce from 1980 to 1985 by computing $\int_{10}^{15} y\, dt$.

24. The force F applied to a moving object at a distance s units from a fixed point, $s \geq 5$, is given by $F = \dfrac{12}{s(s^2 + 4)}$. Find the work done in moving the object from 5 units to 10 units from the fixed point.

9-10 Integration Tables

We now have a list of basic integration rules. We also know some techniques for transforming integrals from one form to another, so that an integral that does not match any integration pattern can perhaps be rewritten in an integrable form. There are many more integration rules, often listed in long tables of integrals, which are useful references for scientists, engineers, and technicians. A short table of integrals is provided at the back of this book. This table includes the basic rules we already know, plus new ones.

At this point you may be inclined to say, as a student of mine once did, "You mean all these rules are in a table in the back of this very book? Wow, why did we have to spend all this time learning all this stuff?" The answer, of course, is that having a list of integration rules available does not in itself solve the problem. You must be able to recognize which integration rule applies or what rewriting tactics might be used to transform the integral into a form to which one of the rules applies. Remember that for a given integration rule to apply, your integral must *exactly* match the pattern of that integration rule. Tables of integrals generally group the integration rules into categories, which makes it easier to locate a matching pattern if one exists.

The following examples use the table of integrals at the back of this book.

EXAMPLE 1. $\int \dfrac{1}{x(3 - 4x)}\,dx$.

This matches the pattern of Rule 20, where $u = x$, $a = 3$, and $b = -4$. The result is

$$\int \frac{1}{x(3 - 4x)}\,dx = -\frac{1}{3}\ln\left|\frac{3 - 4x}{x}\right| + C \quad \blacksquare$$

EXAMPLE 2. $\int \dfrac{\sqrt{9x^2 - 16}}{x}\,dx$.

Because of the presence of $\sqrt{9x^2 - 16}$, we look for forms containing $\sqrt{u^2 - a^2}$. This integral almost matches the pattern of Rule 36, with $u = 3x$, $du = 3\,dx$, and $a = 4$. We write

$$\int \frac{\sqrt{9x^2 - 16}}{x}\,dx = \int \frac{\sqrt{(3x)^2 - (4)^2}}{3x} \cdot 3\,dx$$

$$= \sqrt{9x^2 - 16} - 4 \operatorname{Arccos}\frac{4}{3x} + C \quad \blacksquare$$

EXAMPLE 3. $\int x \tan^3 x^2\,dx$.

We look for a matching pattern under trigonometric forms. We almost have what is required to apply Rule 49, where $n = 3$, $u = x^2$, and $du = 2x\,dx$; we are missing a constant factor, which can be supplied in the usual way.

$$\int x \tan^3 x^2\,dx = \frac{1}{2}\int \tan^3 x^2(2x\,dx) = \frac{1}{2}\left[\frac{1}{2}\tan^2 x^2 - \int \tan x^2(2x\,dx)\right]$$

Finally, using Rule 10, we get

$$\int x \tan^3 x^2\,dx = \frac{1}{2}(\frac{1}{2}\tan^2 x^2 + \ln|\cos x^2|) + C \quad \blacksquare$$

Rule 49 is an example of a *reduction rule*, an integration rule that replaces an integral containing a power with another integral in which the power has been reduced. Continued application of such a rule brings the power down to the point where the expression may finally be integrable.

Practice 1. Use the table of integrals to integrate

$$\int \frac{x}{\sin x^2 \cos x^2}\,dx$$

Exercises / Section 9-10

Exercises 1–34: Integrate, using the table of integrals at the back of the book.

1. $\displaystyle\int \frac{\sqrt{9 + x^2}}{x}\, dx$

2. $\displaystyle\int \frac{x^2}{\sqrt{25 - x^2}}\, dx$

3. $\displaystyle\int \frac{1}{\sqrt{x^2 - 9}}\, dx$

4. $\displaystyle\int x\sqrt{2 - 6x}\, dx$

5. $\displaystyle\int x \sin x\, dx$

6. $\displaystyle\int e^{2x} \sin 3x\, dx$

7. $\displaystyle\int_1^4 \ln x\, dx$

8. $\displaystyle\int_1^2 \frac{dx}{x^2\sqrt{16 - x^2}}$

9. $\displaystyle\int x^2 \sin x\, dx$

10. $\displaystyle\int x^2 e^{3x}\, dx$

11. $\displaystyle\int \cos^3 x\, dx$

12. $\displaystyle\int \frac{\sqrt{5 + 2x}}{x}\, dx$

13. $\displaystyle\int \frac{dx}{\sqrt{9 + 4x^2}}$

14. $\displaystyle\int \frac{dx}{\sin 2x \cos 2x}$

15. $\displaystyle\int \frac{dx}{x\sqrt{9x^2 - 1}}$

16. $\displaystyle\int x \cos 5x\, dx$

17. $\displaystyle\int x \cot^2 x^2\, dx$

18. $\displaystyle\int x \, \text{Arcsin } x^2\, dx$

19. $\displaystyle\int x^3 e^{2x^2}\, dx$

20. $\displaystyle\int \frac{\sqrt{4 + x^4}}{x}\, dx$

21. $\displaystyle\int \frac{x^5}{\sqrt{4x^4 - 49}}\, dx$

22. $\displaystyle\int x \sin 3x^2 \sin 6x^2\, dx$

23. $\displaystyle\int x^8 \ln x^3\, dx$

24. $\displaystyle\int \frac{x^8}{\sqrt{4 + x^6}}\, dx$

25. $\displaystyle\int_0^1 \frac{x^5}{(2 + 6x^3)^2}\, dx$

26. $\displaystyle\int_1^2 \frac{dx}{x^5(4 - 5x^4)}$

27. $\displaystyle\int \cos^2 5x\, dx$

28. $\displaystyle\int \text{Arctan } 6x\, dx$

29. $\displaystyle\int \frac{x^5}{\sqrt{1 + 2x^2}}\, dx$

30. $\displaystyle\int \frac{dx}{x(3 + 4x^2)^2}$

31. $\displaystyle\int \frac{dx}{x\sqrt{4 + 9x^2}}$

32. $\displaystyle\int x^2 \cos 4x\, dx$

33. $\displaystyle\int \frac{x^5}{5 + 12x^2}\, dx$

34. $\displaystyle\int xe^{2x^2} \cos 3x^2\, dx$

Exercises 35–38: Use the table of integrals to help evaluate the integral expressions.

35. The time rate of change of angular displacement θ is angular velocity, ω. The angular velocity of a particular rotating object is given by $\omega = \dfrac{t^2}{\sqrt{4 + t^2}}$ revolutions per minute. Find the total number of revolutions in the first minute.

36. The area bounded by the curves $y = \sin x$, $y = 0$, and $x = \pi/2$ is revolved about the x-axis. Find the volume generated.

37. The voltage V induced in a 1.5-H inductor as a function of time t is given by $V = t/\sqrt{1.4 + 3.8t}$. At $t = 1$ s, the current in the circuit is 3 A. Find the current at $t = 5$ s.

38. Find the length of arc of the curve $y = x^2$ from $x = 0$ to $x = 3$.

9-11 Approximate Methods

Suppose we wish to evaluate a definite integral, such as $\int_a^b f(x)\, dx$. The ideal situation occurs when $f(x)$ is an integrable function—that is, we can find an antiderivative $F(x)$ and use the Fundamental Theorem of Calculus,

$$\int_a^b f(x)\, dx = F(b) - F(a)$$

All our work in this chapter so far has been concerned with integrating functions by matching them with integration patterns. But not every function is integrable. If $f(x)$ is not integrable (or if it is integrable but we are not clever enough to locate a matching integration pattern), then we cannot use the fundamental theorem and we cannot obtain an exact value for the definite integral. However, we can approximate the value of the definite integral as closely as we wish. There are several approximation techniques. We consider two in this section and another in Chapter 10.

All approximation techniques involve doing a number of easy, but tedious, arithmetic computations. The better we want our approximation value to be—that is, the closer to the actual value of the definite integral—the more of these computations we must do. Repeating simple arithmetic computations a large number of times is one of the tasks at which computers excel, so that approximation techniques are more widely used and therefore more important today than when all calculations had to be done by hand. Make sure your calculator is well charged when you do the exercises in this section!

We recall from Chapter 6 that the definite integral $\int_a^b f(x)\,dx$ can be interpreted as the area under the curve $y = f(x)$ from $x = a$ to $x = b$, and also that

$$\int_a^b f(x)\,dx = \lim_{n\to\infty} \sum_{i=1}^{n} f(x_i)\Delta x$$

For a fixed value of n, the expression $\sum_{i=1}^{n} f(x_i)\Delta x$ represents the sum of the areas of n rectangular elements. A typical rectangle has width Δx and height $f(x_i)$ (see Figure 9-6). The expression $\sum_{i=1}^{n} f(x_i)\Delta x$ is therefore an approximation to the value of $\int_a^b f(x)\,dx$, and the approximation improves with larger and larger values of n. The two approximation techniques we discuss next also use the sum of the areas of n typical elements. However, instead of using rectangular elements, we use shapes that may come closer to the shape of $f(x)$ and thus reduce the error.

First we approximate the area by using trapezoidal elements, each of width Δx, as in Figure 9-7(a). The area of a trapezoid is one half the sum of the bases times the altitude; for a typical element, the "bases" are $f(x_i)$ and $f(x_{i+1})$, while the "altitude" is Δx (Figure 9-7(b)).

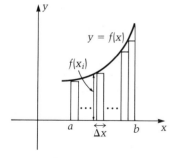

Figure 9-6

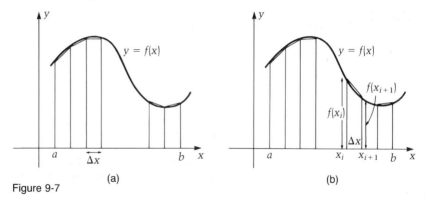

(a)

(b)

Figure 9-7

For n trapezoids, the sum S of the areas is

$$S = \tfrac{1}{2}[f(x_0) + f(x_1)]\Delta x + \tfrac{1}{2}[f(x_1) + f(x_2)]\Delta x$$
$$+ \tfrac{1}{2}[f(x_2) + f(x_3)]\Delta x + \cdots + \tfrac{1}{2}[f(x_{n-2}) + f(x_{n-1})]\Delta x$$
$$+ \tfrac{1}{2}[f(x_{n-1}) + f(x_n)]\Delta x$$

Factoring out Δx and combining like terms, we have

$$S = [\tfrac{1}{2}f(x_0) + f(x_1) + f(x_2) + \cdots + f(x_{n-1}) + \tfrac{1}{2}f(x_n)]\Delta x \qquad (1)$$

Equation (1) is the **trapezoid rule** to approximate the value of the definite integral $\int_a^b f(x)\,dx$. The higher the value of n, the better the approximation.*

EXAMPLE 1. Consider $\displaystyle\int_1^3 \frac{1}{x+2}\,dx$. This does happen to be an integral we can evaluate exactly.

$$\int_1^3 \frac{1}{x+2}\,dx = \ln(x+2)\Big|_1^3 = \ln 5 - \ln 3 = 0.51083$$

This gives us the exact value of the definite integral. Now let us approximate the value of the integral by using the trapezoid rule, with $n = 4$. In order to find Δx, we note that the total length of the interval is $3 - 1 = 2$, and that we are to divide this length into four equal parts, so that $\Delta x = \tfrac{2}{4} = 0.5$. The function $f(x)$ to be used in applying the trapezoid rule is $f(x) = 1/(x + 2)$. From Equation (1),

$$S = \left[\frac{1}{2}f(1) + f(1.5) + f(2) + f(2.5) + \frac{1}{2}f(3)\right](0.5)$$

$$= \left(\frac{1}{2}\cdot\frac{1}{3} + \frac{1}{3.5} + \frac{1}{4} + \frac{1}{4.5} + \frac{1}{2}\cdot\frac{1}{5}\right)(0.5)$$

$$= 0.51230$$

This is a reasonably good approximation to the exact value of 0.51083. We should get a better approximation by increasing the value of n. We use the trapezoid rule again, this time with $n = 8$, and $\Delta x = (3 - 1)/8 = 0.25$.

$$S = \left[\frac{1}{2}f(1) + f(1.25) + f(1.5) + f(1.75) + f(2) + f(2.25) + f(2.5)\right.$$

$$\left. + f(2.75) + \frac{1}{2}f(3)\right](0.25)$$

---

*The following gives an error estimate on the use of the trapezoid rule: If there is a positive number M such that $|f''(x)| \leq M$ everywhere between a and b, then the answer obtained by the trapezoid rule is within $M(b - a)^3/12n^2$ of the value of $\int_a^b f(x)\,dx$.

$$= \left(\frac{1}{2} \cdot \frac{1}{3} + \frac{1}{3.25} + \frac{1}{3.5} + \frac{1}{3.75} + \frac{1}{4} + \frac{1}{4.25} + \frac{1}{4.5} + \frac{1}{4.75}\right.$$
$$\left. + \frac{1}{2} \cdot \frac{1}{5}\right)(0.25)$$
$$= 0.51119$$

This is closer to the exact answer, and we could get closer still by using larger values of n. ∎

EXAMPLE 2. Approximate $\int_2^6 \sqrt{1 + x^3}\, dx$ by using the trapezoid rule with $n = 10$.

The value of Δx is $(6 - 2)/10 = 0.4$, and $f(x) = \sqrt{1 + x^3}$. The values we need to use in the trapezoid rule are:

$$f(2) = \sqrt{1 + (2)^3} = 3 \qquad\qquad f(4) = \sqrt{1 + (4)^3} = 8.062$$
$$f(2.4) = \sqrt{1 + (2.4)^3} = 3.850 \qquad f(4.4) = \sqrt{1 + (4.4)^3} = 9.284$$
$$f(2.8) = \sqrt{1 + (2.8)^3} = 4.791 \qquad f(4.8) = \sqrt{1 + (4.8)^3} = 10.564$$
$$f(3.2) = \sqrt{1 + (3.2)^3} = 5.811 \qquad f(5.2) = \sqrt{1 + (5.2)^3} = 11.900$$
$$f(3.6) = \sqrt{1 + (3.6)^3} = 6.903 \qquad f(5.6) = \sqrt{1 + (5.6)^3} = 13.290$$
$$f(6) = \sqrt{1 + (6)^3} = 14.731$$

$$S = [\tfrac{1}{2}(3) + 3.850 + 4.791 + 5.811 + 6.903 + 8.062 + 9.284$$
$$+ 10.564 + 11.900 + 13.290 + \tfrac{1}{2}(14.731)](0.4)$$
$$= 33.328$$

Note that we cannot do this integral directly. ∎

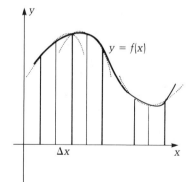

$y = f(x)$

Δx

Figure 9-8

Practice 1. Approximate $\int_0^1 \frac{1}{4 - x^2}\, dx$ by using the trapezoid rule with $n = 5$.

To use the second method to approximate the value of $\int_a^b f(x)\, dx$, *Simpson's rule,* we divide the interval from a to b into an even number of subintervals and then, for each pair of subintervals, assume a "parabolic shape" on top, as in Figure 9-8. This adds some curvature, which may reduce the error accompanying use of the trapezoid rule.

Figure 9-9 shows a pair of subintervals with a parabolic curve $y = rx^2 + sx + t$ going through the three points with coordinates (x_0, y_0), (x_1, y_1), and (x_2, y_2). The area beneath the curve from x_0 to x_2 is

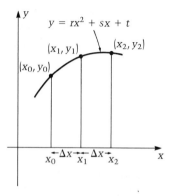

$y = rx^2 + sx + t$

(x_1, y_1) (x_2, y_2)

(x_0, y_0)

$x_0 \;\; x_1 \;\; x_2$

Figure 9-9

$$A = \int_{x_0}^{x_2} y\, dx = \int_{x_0}^{x_2} (rx^2 + sx + t)\, dx = \frac{rx^3}{3} + \frac{sx^2}{2} + tx \Big|_{x_0}^{x_2}$$
$$= \frac{r}{3}x_2^3 + \frac{s}{2}x_2^2 + tx_2 - \left(\frac{r}{3}x_0^3 + \frac{s}{2}x_0^2 + tx_0\right) \qquad (2)$$

We can rewrite this expression for A in terms of x_0, because $x_2 = x_0 + 2\Delta x$. Making this substitution in (2) and rearranging terms (see Exercise 17), we get

$$A = \frac{\Delta x}{3} (6rx_0^2 + 12rx_0\Delta x + 8r(\Delta x)^2 + 6sx_0 + 6s\Delta x + 6t) \tag{3}$$

The three points (x_0, y_0), (x_1, y_1), and (x_2, y_2) lie on the curve $y = rx^2 + sx + t$, and thus we have the three equations

$$y_0 = rx_0^2 + sx_0 + t$$
$$y_1 = rx_1^2 + sx_1 + t$$
$$y_2 = rx_2^2 + sx_2 + t$$

Next (for no apparent reason), we write the expression

$$\frac{\Delta x}{3}(y_0 + 4y_1 + y_2) = \frac{\Delta x}{3}(rx_0^2 + sx_0 + t + 4(rx_1^2 + sx_1 + t)$$
$$+ rx_2^2 + sx_2 + t) \tag{4}$$

Making the substitutions $x_1 = x_0 + \Delta x$ and $x_2 = x_0 + 2\Delta x$ in the above expression and rearranging terms (see Exercise 18), we get

$$\frac{\Delta x}{3}(y_0 + 4y_1 + y_2) = \frac{\Delta x}{3}(6rx_0^2 + 12rx_0\Delta x + 8r(\Delta x)^2 + 6sx_0$$
$$+ 6s\Delta x + 6t \tag{5}$$

Comparing Equations (3) and (5), we see that

$$A = \frac{\Delta x}{3}(y_0 + 4y_1 + y_2)$$

This is the area beneath a parabolic curve from x_0 to x_2. The areas of successive pairs of elements have similar expressions. Thus

area from x_0 to x_2: $\dfrac{\Delta x}{3}(y_0 + 4y_1 + y_2)$

area from x_2 to x_4: $\dfrac{\Delta x}{3}(y_2 + 4y_3 + y_4)$

$$\vdots$$

area from x_{n-2} to x_n: $\dfrac{\Delta x}{3}(y_{n-2} + 4y_{n-1} + y_n)$

To approximate the value of the definite integral, we add these areas. The sum S is

$$S = \frac{\Delta x}{3}(y_0 + 4y_1 + y_2) + \frac{\Delta x}{3}(y_2 + 4y_3 + y_4)$$
$$+ \cdots + \frac{\Delta x}{3}(y_{n-2} + 4y_{n-1} + y_n)$$

or

$$S = \frac{\Delta x}{3}(y_0 + 4y_1 + 2y_2 + 4y_3 + 2y_4 + \cdots + 4y_{n-1} + y_n)$$

which we can write as

$$S = \frac{\Delta x}{3} [f(x_0) + 4f(x_1) + 2f(x_2) + 4f(x_3) + 2f(x_4) + \cdots + 4f(x_{n-1})$$
$$+ f(x_n)] \tag{6}$$

Equation (6) is **Simpson's rule** for approximating the value of the definite integral $\int_b^a f(x)\, dx$. Again, the higher the value of n, the better the approximation; but here n must be an even number.*

EXAMPLE 3. Use Simpson's rule with $n = 4$ to approximate $\int_1^3 \frac{1}{x + 2} \, dx$.

Here $\Delta x = 0.5$ and $f(x) = 1/(x + 2)$. From Equation (6),

$$S = \frac{0.5}{3} [f(1) + 4f(1.5) + 2f(2) + 4f(2.5) + f(3)]$$

$$= \frac{0.5}{3} \left(\frac{1}{3} + 4 \cdot \frac{1}{3.5} + 2 \cdot \frac{1}{4} + 4 \cdot \frac{1}{4.5} + \frac{1}{5} \right)$$

$$= 0.51085$$

From Example 1, we know that the exact value of the definite integral is 0.51083, and the approximation by the trapezoid rule, also with $n = 4$, is 0.51230. Simpson's rule is an improvement. ∎

EXAMPLE 4. Approximate $\int_2^6 \sqrt{1 + x^3} \, dx$ by using Simpson's rule with $n = 10$.

This is the integral of Example 2, and the functional values we computed there for the trapezoid rule are the same values we need now. Using Equation (6) with these values, we get

$$S = \frac{0.4}{3} [3 + 4(3.850) + 2(4.791) + 4(5.811) + 2(6.903) + 4(8.062)$$

$$+ 2(9.284) + 4(10.564) + 2(11.9) + 4(13.29) + 14.731]$$

$$= 33.306 \quad \blacksquare$$

Practice 2. Approximate $\int_1^4 \frac{1}{\sqrt{x} + 1} \, dx$ by using Simpson's rule with $n = 6$.

---

*The following gives an error estimate on the use of Simpson's rule: If there is a positive number M such that $|f^{(4)}(x)| \leq M$ everywhere between a and b, then the answer obtained by Simpson's rule is within $M(b - a)^5/180n^4$ of the value of $\int_a^b f(x)\, dx$.

Use of the trapezoid rule or Simpson's rule requires knowledge of the values of a function at selected points, but it does not require knowing an expression for the function itself. We can, therefore, approximate $\int_a^b f(x)\, dx$ by using experimental data. In practical applications, the actual functional relationship between variables may not be known, and such experimental data are often the only information available.

EXAMPLE 5. An environmental protection group conducts an experiment to determine the amount of wastewater that a local industry is feeding into a lake. The flow rate r, in thousands of gallons per day, is checked at the outlet at various times. Data points are given in the table.

t	1	2	3	4	5
r	8.7	9.1	6.3	7.4	7.8

Use Simpson's rule to approximate the total amount of water dumped during the 4-day observation period.

First we note that a span of 4 days requires 5 observation points. We want to approximate $\int_1^5 r\, dt$. For this problem, Δt (which is Δx in Equation (6)) is 1. We read the functional values directly from the table and use Equation (6).

$$S = \tfrac{1}{3}[8.7 + 4(9.1) + 2(6.3) + 4(7.4) + 7.8]$$
$$= 43.3$$

Approximately 43.3 thousand gallons were dumped over this 4-day period. ■

Exercises / Section 9-11

Exercises 1–6: Use the trapezoid rule and then Simpson's rule, both with $n = 4$, to approximate the value of the given integral. Compare your answers with the exact value found by direct integration.

1. $\displaystyle\int_1^2 (3x^2 - 1)\, dx$

2. $\displaystyle\int_0^4 x^{3/2}\, dx$

3. $\displaystyle\int_0^2 (\sqrt{x} - 1)\, dx$

4. $\displaystyle\int_1^5 x\sqrt{1 + x^2}\, dx$

5. $\displaystyle\int_0^1 e^{2x}\, dx$

6. $\displaystyle\int_0^\pi \sin x\, dx$

Exercises 7–12: Approximate the value of each integral below, using first the trapezoid rule and then Simpson's rule, with the given values of n.

7. $\displaystyle\int_0^1 \sqrt{2 + x^3}\, dx; \quad n = 4$

8. $\displaystyle\int_2^6 x^3\sqrt{x + 1}\, dx; \quad n = 4$

9. $\displaystyle\int_2^3 \frac{1}{x^5 + \sqrt{x}}\, dx; \quad n = 6$

10. $\displaystyle\int_2^5 \frac{\sqrt{5 - x}}{x^3}\, dx; \quad n = 6$

11. $\int_1^3 \ln \sqrt{x}\, dx; \quad n = 10$

12. $\int_0^1 e^{\sqrt{x}}\, dx; \quad n = 10$

Exercises 13–14: Use the trapezoid rule with the given data to approximate the value of the associated definite integral.

13.

x	2	4	6	8	10
$f(x)$	1.8	2.1	2.4	2.5	2.9

14.

x	1.2	1.5	1.8	2.1	2.4	2.7
$f(x)$	31.6	27.4	23.8	24.2	25.5	24.1

Exercises 15–16: Use Simpson's rule with the given data to approximate the value of the associated definite integral.

15.

x	1	2	3	4	5	6	7
$f(x)$	0.03	0.01	0.12	0.24	0.37	0.28	0.31

16.

x	0.6	0.8	1.0	1.2	1.4
$f(x)$	123.4	138.5	152.7	156.1	157.3

17. Show how Equation (3) of this section follows from Equation (2).

18. Show how Equation (5) of this section follows from Equation (4).

STATUS CHECK

Now that you are at the end of Chapter 9, you should be able to:

section **9-1** Integrate using the power rule.
By using the power rule for integration, solve problems such as finding the area between curves, finding the volumes of solids of revolution, and other applications of the definite integral.

section **9-2** Integrate using the logarithm form.
By using the logarithm form for integration, solve problems involving applications of the definite integral.

section **9-3** Integrate using the exponential form.
By using the exponential form for integration, solve problems involving applications of the definite integral.

section **9-4** Integrate using basic trigonometric forms.
By using the basic trigonometric forms for integration, solve problems involving applications of the definite integral.

section **9-5** Use trigonometric identities to convert certain trigonometric integrals into integrable forms and do the resulting integrations.
By using trigonometric identities on definite integrals to convert them to integrable forms, solve problems involving applications of the definite integral.

section **9-6** Integrate using inverse trigonometric forms.
By using the inverse trigonometric forms for integration, solve problems involving applications of the definite integral.

section **9-7** Integrate using the formula for integration by parts (using the formula several times, if necessary).
By using integration by parts, solve problems involving applications of the definite integral.

section **9-8**   Use trigonometric substitution to convert certain algebraic integrals into integrable forms and do the resulting integrations.

By using trigonometric substitution on definite integrals to convert them to integrable forms, solve problems involving applications of the definite integral.

section **9-9**   Use the method of partial fractions to convert certain integrals involving quotients of polynomials into integrable forms and do the resulting integrations.

By using the method of partial fractions on definite integrals to convert them to integrable forms, solve problems involving applications of the definite integral.

section **9-10**   Use tables of integrals to integrate functions.

section **9-11**   Approximate the value of a definite integral by using:
   • The trapezoid rule
   • Simpson's rule
By using trigonometric substitution on definite integrals to convert them to integrable forms, solve problems involving applications of the definite integral.
Given a collection of data points on a function, use the trapezoid rule or Simpson's rule to approximate the value of the definite integral of the function.

9-12 More Exercises for Chapter 9

Exercises 1–38: Integrate (do not use the table of integrals):

1. $\int \tan^2(3x - 1) \sec^2(3x - 1)\, dx$

2. $\int \sec 4x \tan 4x\, dx$

3. $\int \sin^2 x\, dx$

4. $\int x^3 \ln x\, dx$

5. $\int \frac{\sqrt{x^2 - 9}}{x}\, dx$

6. $\int \frac{11x + 2}{3x^2 + 10x - 8}\, dx$

7. $\int x^2 \cos x\, dx$

8. $\int \sin^3 2x \cos^2 2x\, dx$

9. $\int \cos x e^{\sin x}\, dx$

10. $\int \frac{\tan x \sec^2 x}{3 + \tan^2 x}\, dx$

11. $\int \tan 2x\, dx$

12. $\int \frac{2}{16 + 3x^2}\, dx$

13. $\int \frac{\sqrt{x^2 + 1}}{x}\, dx$

14. $\int \frac{x^2}{(x^3 - 1)^2}\, dx$

15. $\int \frac{3x^2 + 21x - 12}{x^2 + 6x}\, dx$

16. $\int \frac{e^{2x}}{e^{2x} - 1}\, dx$

17. $\int \tan x \sec^4 x\, dx$

18. $\int \frac{3}{\sqrt{9 - 25x^2}}\, dx$

19. $\int \sec^3 x\, dx$

20. $\int \frac{1}{x^3 \sqrt{x^2 - 25}}\, dx$

21. $\int \cot x \ln \sin x\, dx$

22. $\int \frac{x - 2}{x^2 - 4x + 1}\, dx$

23. $\int x \cos 4x^2\, dx$

24. $\int e^{4x}\, dx$

25. $\int \frac{\sin x}{\sqrt{1 - \cos^2 x}}\, dx$

26. $\int \frac{10x^2 - 8x + 3}{4x^3 - 4x^2 + x}\, dx$

27. $\int (1 + e^{-2x})^{1/4} e^{-2x}\, dx$

28. $\int \dfrac{e^{\text{Arctan } x}}{x^2 + 1} \, dx$

29. $\int \dfrac{x^{1/3}}{2 + x^{4/3}} \, dx$

30. $\int (1 + \sec x)^2 \, dx$

31. $\int \dfrac{x}{(25 - x^2)^2} \, dx$

32. $\int \dfrac{\sec x \tan x}{1 + \sec^2 x} \, dx$

33. $\int \dfrac{\sin 2x}{1 + \sin^2 x} \, dx$

34. $\int \dfrac{8x^2 + 6x + 20}{(x + 2)(4 + x^2)} \, dx$

35. $\int \dfrac{\sin(\ln x)}{x} \, dx$

36. $\int \tan^3 x \, dx$

37. $\int x^2 e^{-3x} \, dx$

38. $\int \dfrac{\cos^2 x}{\sin 2x} \, dx$

39. Use the trapezoid rule with $n = 6$ to approximate the value of $\int_1^4 \sqrt{x^3 + x} \, dx$.

40. Use Simpson's rule with $n = 8$ to approximate the value of $\int_3^5 \ln(x^2 + 1) \, dx$.

Exercises 41–56: Use integration to solve.

41. Under certain conditions, the total flux ϕ in a coaxial line with inner conductor of radius r_1 and outer conductor of radius r_2 is

$$\phi = \int_{r_1}^{r_2} \dfrac{k}{2\pi r} \, dr$$

where k is a constant. Find the value of ϕ.

42. If the current flowing through a heating device at time x is $C(x)$ and the resistance R is a constant, then according to Joule's law the heat produced at time t is given by

$$H(t) = R \int_0^t [C(x)]^2 \, dx$$

Find an expression for H if

$$C(x) = \sqrt{\dfrac{x^2}{x^3 + 1}}$$

43. A voltage V given by $V = e^{t/2}$ is induced in a 2-H inductor. If the current in the circuit is 5 A when $t = 2$ s, what is the current when $t = 4$ s?

44. Find the equation of a curve whose slope is $(e^{2x+1})^2$ and that passes through the point $(0, e^2)$.

45. Find the volume generated by rotating the region bounded by $y = \sec^2(x^2)$, $y = 0$, $x = 0$, and $x = 1$ about the y-axis.

46. If a rod of length L and uniform cross section is compressed at the ends, the amount $B(x)$ of bending at distance x from an end is given by

$$B(x) = c_1 \cos kx + c_2 \sin kx$$

where c_1, c_2, and k are constants. Find

$$\dfrac{1}{L} \int_0^1 B(x) \, dx$$

which is the average amount of bending of the rod.

47. According to Stokes' theory of the scattering of X-rays, the intensity of scattered radiation in a direction making an angle θ with the primary beam is $I(\theta) = I_m (1 + \cos^2 \theta)$, where I_m is a constant. Find

$$\int_0^{2\pi} I(\theta) \, d\theta$$

which is the total intensity of scattered radiation.

48. Find the area bounded by one loop of the curve whose equation in polar coordinates is $r = 4 \cos 2\theta$.

49. Find the area of the region bounded by the curves $y = 1/(4 + x^2)$, $x = 0$, $y = 0$, and $x = 2$.

50. Find the moment of inertia with respect to the y-axis for the area bounded by $y = 1/\sqrt{1 - x^6}$, $y = 0$, $x = 0.5$, and $x = 0.8$; assume a constant density k.

51. A force of xe^x units is applied to an object at point x on a straight-line path. Find the work done in moving the object from $x = 1$ to $x = 10$.

52. Find the first moment with respect to the y-axis of the area bounded by $y = \sqrt{x + 1}$, $x = 0$, $x = 3$, and $y = 0$; assume a constant density k.

53. Find the equation of the curve for which $dy/dx = 1/(x\sqrt{4 - x^2})$ and that passes through the point $(2, 4)$.

54. Find the length of arc of the curve $y = e^x$ from $x = 0$ to $x = 2$.

55. Find the area of the region bounded by the curves $y = (x + 1)/(x^3 - x^2)$, $y = 0$, $x = 2$, and $x = 4$.

56. The velocity of an object at time t is given by $v = 1/t(t^2 + 1)^2 \, dt$. Find an expression for s as a function of t if $s = \frac{5}{4}$ when $t = 1$.

chapter **10**

Series Expansion of Functions

An infinite series goes on forever, but if it converges, the terms beyond a certain point are so small they do not need to be considered.

10-1 Infinite Series

In this chapter, we will learn a new way of representing functions. This will enable us to approximate the value of definite integrals for those cases where we cannot use the fundamental theorem. This new representation will also allow us to approximate the value of a function of x for a given value of x. For example, the trigonometric functions, such as $\sin x$, have values that can easily be determined for only a few values of x. We know that $\sin \pi/2 = 1$, but what about $\sin 0.8$? We could use a calculator or table of trigonometric functions for a decimal approximation of the value $\sin 0.8$. This approximation, however, is itself obtained by using the methods we develop in this chapter. We begin with some terminology.

An **infinite sequence** of numbers is simply an infinite list of numbers, such as

12, 18, 24, 30, . . .

You are probably familiar with an **arithmetic sequence**, where each item in the sequence (except the first) comes from the preceding one by adding a fixed number. The sequence above is an arithmetic sequence where the fixed number is 6. A **geometric sequence** is a sequence in which each item in the sequence (except the first) comes from the preceding one by multiplying by a fixed number. The sequence

$$1, \frac{1}{3}, \frac{1}{9}, \frac{1}{27}, \ldots$$

is a geometric sequence where the fixed number is $\frac{1}{3}$.

If we add all the items in an infinite sequence, we get an **infinite series**. Thus

$$1 + \frac{1}{3} + \frac{1}{9} + \frac{1}{27} + \cdots \tag{1}$$

is an infinite series. This is not addition in the usual sense of the word, however, since we really cannot actually add an infinite number of things. In order to assign meaning to an expression such as (1), we use a limit.

In general, if the infinite series is

$$a_0 + a_1 + a_2 + a_3 + \cdots \tag{2}$$

we can compute the *partial sums*

$$S_0 = a_0$$
$$S_1 = a_0 + a_1$$
$$S_2 = a_0 + a_1 + a_2$$
$$S_3 = a_0 + a_1 + a_2 + a_3$$
$$\vdots$$
$$S_n = a_0 + a_1 + \cdots + a_n$$

because each of these is a finite sum. We say that the infinite series (2) has a *sum S* (or **converges** to the value S) if

$$\lim_{n \to \infty} S_n = S$$

In order for $\lim_{n \to \infty} S_n$ to have the value S, the values of S_n must get closer and closer to S for larger and larger values of n.

EXAMPLE 1. Estimate the sum of the infinite geometric series

$$1 + \frac{1}{3} + \frac{1}{9} + \frac{1}{27} + \cdots$$

Using a calculator, we compute the partial sums

$$S_0 = 1$$
$$S_1 = 1 + \frac{1}{3} = 1.\overline{3}$$
$$S_2 = 1 + \frac{1}{3} + \frac{1}{9} = 1.\overline{4}$$
$$S_3 = 1 + \frac{1}{3} + \frac{1}{9} + \frac{1}{27} = 1.\overline{481}$$
$$S_4 = 1 + \frac{1}{3} + \frac{1}{9} + \frac{1}{27} + \frac{1}{81} = 1.49383$$
$$S_5 = 1 + \frac{1}{3} + \frac{1}{9} + \frac{1}{27} + \frac{1}{81} + \frac{1}{243} = 1.49795$$
$$S_6 = 1 + \frac{1}{3} + \frac{1}{9} + \frac{1}{27} + \frac{1}{81} + \frac{1}{243} + \frac{1}{729} = 1.49932$$

The values of S_n seem to approach 1.5, so we estimate that the series converges to a sum of 1.5. ∎

Practice 1. Estimate the sum of the infinite series

$$2 + \tfrac{1}{2} + \tfrac{1}{8} + \tfrac{1}{32} + \cdots$$

If $\lim_{n \to \infty} S_n$ does not exist, then the infinite series is said to **diverge**. A divergent series has no sum.

EXAMPLE 2. The series

$$2 + 4 + 6 + 8 + \cdots$$

is a divergent series. The partial sums are

$$S_0 = 2$$
$$S_1 = 2 + 4 = 6$$
$$S_2 = 2 + 4 + 6 = 12$$
$$S_3 = 2 + 4 + 6 + 8 = 20$$
$$S_4 = 2 + 4 + 6 + 8 + 10 = 30$$
$$\vdots$$

These values are not approaching a fixed finite number. ■

EXAMPLE 3. The series

$$1 + (-1) + 1 + (-1) + 1 + \cdots$$

is a divergent series. The partial sums are

$$S_0 = 1$$
$$S_1 = 1 + (-1) = 0$$
$$S_2 = 1 + (-1) + 1 = 1$$
$$S_3 = 1 + (-1) + 1 + (-1) = 0$$

These values do not approach a fixed finite number, they simply oscillate between 0 and 1. ■

In general, it is not an easy task to decide whether a given infinite series converges or diverges, and many tests have been devised for this purpose. In the case of an infinite geometric series, however, there is an easy way to decide whether the series converges or diverges, and if it converges, what the value of the sum is.

An infinite geometric series will have the form

$$a + ar + ar^2 + ar^3 + \cdots \tag{3}$$

where r is the fixed value by which we multiply each term to get the succeeding term. The partial sums are

$$S_0 = a$$
$$S_1 = a + ar$$
$$S_2 = a + ar + ar^2$$
$$\vdots$$
$$S_n = a + ar + ar^2 + \cdots + ar^n$$

Multiplying S_n by r we get

$$rS_n = ar + ar^2 + ar^3 + \cdots + ar^{n+1}$$

Now subtracting rS_n from S_n, we can write

$$S_n - rS_n = a + ar + ar^2 + \cdots + ar^n$$
$$- (ar + ar^2 + ar^3 + \cdots + ar^{n+1})$$

or

$$S_n - rS_n = a - ar^{n+1}$$

Factoring out S_n, we have

$$(1 - r)S_n = a - ar^{n+1} \tag{4}$$

Let us now assume that $r \neq 1$, so that $1 - r \neq 0$, and divide both sides of Equation (4) by $1 - r$.

$$S_n = \frac{a}{1 - r} - \frac{ar^{n+1}}{1 - r}$$

The series would converge if $\lim_{n \to \infty} S_n$ existed. But

$$\lim_{n \to \infty} S_n = \lim_{n \to \infty} \left(\frac{a}{1 - r} - \frac{ar^{n+1}}{1 - r} \right)$$
$$= \frac{a}{1 - r} - \frac{a}{1 - r} \lim_{n \to \infty} r^{n+1}$$

We can do the last step because only part of the expression involves n at all. If r is a fraction—that is, $|r| < 1$—then raising r to higher and higher powers produces smaller and smaller values, so $\lim_{n \to \infty} r^{n+1} = 0$, and

$$\lim_{n \to \infty} S_n = \frac{a}{1 - r}$$

If $|r| > 1$, however, $\lim_{n \to \infty} r^{n+1}$ does not exist. If r is positive, the successive values are rapidly increasing positive values; if r is negative, the values oscillate between positive and negative values, but increase rapidly in absolute value. Because $\lim_{n \to \infty} r^{n+1}$ does not exist, neither does $\lim_{n \to \infty} S_n$.

Therefore, if $|r| < 1$, the geometric series (3) converges to a sum of $a/(1 - r)$; if $|r| > 1$, the series diverges. If $r = 1$, the series becomes

$$a + a + a + a + \cdots$$

which diverges $(\lim_{n \to \infty} S_n = \lim_{n \to \infty} na = \pm\infty)$. If $r = -1$, the series becomes

$$a + (-a) + a + (-a) + \cdots$$

which diverges because the partial sums alternate between a and 0.

To summarize this whole discussion, an infinite geometric series

$$a + ar + ar^2 + \cdots$$

converges to a sum of $a/(1 - r)$ for $|r| < 1$; for $|r| \geq 1$, the series diverges.

EXAMPLE 4.   The series of Example 1,

$$1 + \tfrac{1}{3} + \tfrac{1}{9} + \tfrac{1}{27} + \cdots$$

is an infinite geometric series with $r = \tfrac{1}{3}$. The series converges to

$$\frac{a}{1 - r} = \frac{1}{1 - \tfrac{1}{3}} = \frac{1}{\tfrac{2}{3}} = \frac{3}{2} = 1.5,$$

the answer we had estimated before.   ■

Practice 2.   Find the sum of the infinite geometric series of Practice 1.

A more general form of infinite series is called a **power series in x** and looks like

$$a_0 + a_1 x + a_2 x^2 + a_3 x^3 + \cdots + a_n x^n + \cdots \tag{5}$$

For any given value of x, the power series in x becomes an ordinary infinite series of constants, which may or may not converge. Certainly the power series converges (to a sum of a_0) when x has the value 0. In fact a power series will generally converge for all values of x in some interval around the value $x = 0$. The farther the value of x from 0, however, the less rapid the convergence is apt to be. This simply means that more terms of the series must be added in order to get a close approximation to the sum. The interval of values around $x = 0$ for which the power series converges is called the **interval of convergence**. Some power series may converge for all real-number values of x; in this case the interval of convergence is denoted by $(-\infty, \infty)$.

EXAMPLE 5. Each of the values $x = 0.1$, $x = 0.2$, and $x = 0.5$ lies within the interval of convergence of the power series

$$x - \frac{x^2}{2} + \frac{x^3}{3} - \frac{x^4}{4} + \cdots$$

Estimate the value of the sum in each case.

For $x = 0.1$, the series becomes

$$0.1 - \frac{(0.1)^2}{2} + \frac{(0.1)^3}{3} - \frac{(0.1)^4}{4} + \cdots$$

Using a calculator, we compute the partial sums for this series.

$$S_1 = 0.1$$

$$S_2 = S_1 - \frac{(0.1)^2}{2} = 0.095$$

$$S_3 = S_2 + \frac{(0.1)^3}{3} = 0.095\overline{3}$$

Further partial sums would produce ever-smaller changes in this value. Therefore, by taking only three terms of the series, we can estimate that the sum is close to 0.095.

For $x = 0.2$, the series becomes

$$0.2 - \frac{(0.2)^2}{2} + \frac{(0.2)^3}{3} - \frac{(0.2)^4}{4} + \cdots$$

Computing partial sums, we get

$$S_1 = 0.2$$
$$S_2 = S_1 - \frac{(0.2)^2}{2} = 0.18$$

$$S_3 = S_2 + \frac{(0.2)^3}{3} = 0.182\overline{6}$$

By taking three terms of this series, we are fairly sure that the sum, to the nearest hundredth, is 0.18, but we are not able to guess the next digit with any certainty.

For $x = 0.5$, the series becomes

$$0.5 - \frac{(0.5)^2}{2} + \frac{(0.5)^3}{3} - \frac{(0.5)^4}{4} + \cdots$$

Partial sums are

$$S_0 = 0.5$$

$$S_1 = S_0 - \frac{(0.5)^2}{2} = 0.375$$

$$S_2 = S_1 + \frac{(0.5)^3}{3} = 0.41\overline{6}$$

Three terms of this series does not tell us much at all. We take several more partial sums.

$$S_3 = S_2 - \frac{(0.5)^4}{4} = 0.401$$

$$S_4 = S_3 + \frac{(0.5)^5}{5} = 0.407$$

We estimate the sum to be 0.4 to the nearest tenth. ■

Within the interval of convergence of the power series

$$a_0 + a_1 x + a_2 x^2 + a_3 x^3 + \cdots + a_n x^n + \cdots$$

each value of x produces a unique value for the series sum. The sum is, therefore, a function of x. If we use $f(x)$ to denote this function, we can then write

$$f(x) = a_0 + a_1 x + a_2 x^2 + \cdots + a_n x^n + \cdots$$

Thus in Example 5, if we write

$$f(x) = x - \frac{x^2}{2} + \frac{x^3}{3} - \frac{x^4}{4} + \cdots$$

we have computed that

$$f(0.1) \approx 0.095$$
$$f(0.2) \approx 0.18$$
$$f(0.5) \approx 0.4$$

where $\approx$ denotes *is approximately equal to.*

Practice 3. Within the interval of convergence of the series, let

$$f(x) = 1 + \frac{x}{1} + \frac{x^2}{1 \cdot 2} + \frac{x^3}{1 \cdot 2 \cdot 3} + \frac{x^4}{1 \cdot 2 \cdot 3 \cdot 4} + \cdots$$

Estimate the values of $f(0.1)$ and $f(0.2)$ to the nearest hundredth.

Exercises / Section **10-1**

Exercises 1–12: Decide whether you think the series converges or diverges, and if it converges, estimate the sum.

1. $\frac{2}{3} + \frac{4}{9} + \frac{8}{27} + \cdots$

2. $16 - 6 + \frac{18}{8} - \frac{54}{64} + \cdots$

3. $\frac{1}{1^3} + \frac{1}{2^3} + \frac{1}{3^3} + \frac{1}{4^3} + \cdots$

4. $\frac{1}{3(1) + 1} - \frac{1}{3(2) + 1} + \frac{1}{3(3) + 1}$
   $- \frac{1}{3(4) + 1} + \cdots$

5. $3 - 4 + \frac{16}{3} - \frac{64}{9} + \cdots$

6. $\frac{4}{3} + 2 + \frac{6}{2} + \frac{18}{4} + \cdots$

7. $\frac{1}{5} + \frac{2}{5^2} + \frac{3}{5^3} + \frac{4}{5^4} + \cdots$

8. $1 - \dfrac{1}{10} + \dfrac{2}{10^2} - \dfrac{3}{10^3} + \cdots$

9. $0.99 + (0.99)^2 + (0.99)^3 + \cdots$

10. $5(\frac{10}{11})^3 + 5(\frac{10}{11})^4 + 5(\frac{10}{11})^5 + \cdots$

11. $1 - 2 + 3 - 4 + 5 - 6 + \cdots$

12. $\dfrac{1^2}{1+1} + \dfrac{2^2}{1+2} + \dfrac{3^2}{1+3} + \dfrac{4^2}{1+4} + \cdots$

13. Within the interval of convergence of the series, let

$$f(x) = 1 - x + x^2 - x^3 + \cdots$$

Estimate the value of $f(0.2)$ to the nearest hundredth.

14. Within the interval of convergence of the series, let

$$f(x) = 1 + \frac{x^1}{1^3 + 1} + \frac{x^2}{2^3 + 1} + \frac{x^3}{3^3 + 1}$$
$$+ \frac{x^4}{4^3 + 1} + \cdots$$

Estimate the value of $f(0.3)$ to the nearest hundredth.

10-2 Maclaurin Series

In the previous section, we began with a power series

$$a_0 + a_1 x + a_2 x^2 + a_3 x^3 + a_4 x^4 + \cdots + a_n x^n + \cdots$$

and we used $f(x)$ to represent the sum of this series within its interval of convergence. Thus we wrote

$$f(x) = a_0 + a_1 x + a_2 x^2 + a_3 x^3 + a_4 x^4 + \cdots + a_n x^n + \cdots \quad (1)$$

Suppose now that we begin with a function $f(x)$ and try to find a power series—that is, try to find values for the coefficients a_i such that Equation (1) is valid in some interval around $x = 0$. This simply means that the value of $f(x)$ is given by the sum of the power series for all x-values in this interval. If such a series can be found, it is called the **power series expansion** of $f(x)$, and $f(x)$ is said to be *represented* by its power series expansion.

We assume that a given function $f(x)$ has a power series expansion for some interval around zero, and we do not worry about finding the interval. Based on this assumption, what do the coefficients in the power series expansion have to be?

Our starting assumption is Equation (1),

$$f(x) = a_0 + a_1 x + a_2 x^2 + a_3 x^3 + a_4 x^4 + \cdots$$

This equation is certainly true for $x = 0$, so that

$$f(0) = a_0$$

If we differentiate both sides of Equation (1), we get

$$f'(x) = a_1 + 2a_2 x + 3a_3 x^2 + 4a_4 x^3 + \cdots$$

Evaluating at $x = 0$,

$$f'(0) = a_1$$

We continue differentiating and evaluating for $x = 0$:

$$f''(x) = 2a_2 + 2 \cdot 3a_3 x + 3 \cdot 4a_4 x^2 + \cdots$$
$$f''(0) = 2a_0$$
$$f'''(x) = 2 \cdot 3a_3 + 2 \cdot 3 \cdot 4a_4 x + \cdots$$
$$f'''(0) = 2 \cdot 3a_3$$
$$f^{(4)}(x) = 2 \cdot 3 \cdot 4a_4 + \cdots$$
$$f^{(4)}(0) = 2 \cdot 3 \cdot 4a_4$$
$$\vdots$$

From these results we can solve for the a_i coefficients. It is convenient to use **factorial notation**, where for a positive integer n, $n! = 1 \cdot 2 \cdot 3 \cdots (n-1)n$; $n!$ is read n *factorial*. The coefficients are

$$a_0 = f(0)$$

$$a_1 = f'(0) = \frac{f'(0)}{1!}$$

$$a_2 = \frac{f''(0)}{2} = \frac{f''(0)}{2!}$$

$$a_3 = \frac{f'''(0)}{2 \cdot 3} = \frac{f'''(0)}{3!}$$

$$a_4 = \frac{f^{(4)}(0)}{2 \cdot 3 \cdot 4} = \frac{f^{(4)}(0)}{4!}$$

$$\vdots$$

In general,

$$a_n = \frac{f^{(n)}(0)}{n!}$$

The power series expansion of $f(x)$, Equation (1), can now be written as

$$f(x) = f(0) + \frac{f'(0)}{1!}x + \frac{f''(0)}{2!}x^2 + \frac{f'''(0)}{3!}x^3 + \frac{f^{(4)}(0)}{4!}x^4$$

$$+ \cdots + \frac{f^{(n)}(0)}{n!}x^n + \cdots \tag{2}$$

The right side of Equation (2) is called the **Maclaurin series expansion** of $f(x)$.

What we have shown so far is that if we assume that a function has a power series expansion in some interval around zero, then the power series expansion must be the Maclaurin series. This does not tell us when such an assumption can be made, however. One condition that must certainly be true in order for this assumption to hold (and which we took for granted in finding the a_i's) is that $f(0)$, $f'(0)$, $f''(0)$, . . . , $f^{(n)}(0)$, . . . , must all be defined. All the functions we deal with in this chapter do have Maclaurin series expansions.

EXAMPLE 1. Find the first four terms of the Maclaurin series expansion of $f(x) = e^{2x}$.

In order to use Equation (2), we compute the required derivatives and evaluate them at $x = 0$.

$$f(x) = e^{2x} \qquad f(0) = 1$$
$$f'(x) = 2e^{2x} \qquad f'(0) = 2$$
$$f''(x) = 4e^{2x} \qquad f''(0) = 4$$
$$f'''(x) = 8e^{2x} \qquad f'''(0) = 8$$

From Equation (2),

$$e^{2x} = 1 + \frac{2x}{1!} + \frac{4}{2!}x^2 + \frac{8}{3!}x^3 + \cdots$$
$$= 1 + 2x + 2x^2 + \frac{4}{3}x^3 + \cdots \quad \blacksquare$$

Word of Advice

When computing the coefficients in a Maclaurin series expansion of $f(x)$, organize your work neatly. Compute the derivatives and evaluate the derivatives; then write the series and simplify if necessary. Trying to do it all in one step leads to errors.

Practice 1. Find the first four terms of the Maclaurin series expansion of $f(x) = e^{x/2}$.

EXAMPLE 2. Find the first four terms of the Maclaurin series expansion of $f(x) = 1/(1 + x)$.

For this function,

$$f(x) = \frac{1}{1 + x} \qquad f(0) = 1$$

$$f'(x) = \frac{-1}{(1 + x)^2} \qquad f'(0) = -1$$

$$f''(x) = \frac{2}{(1 + x)^3} \qquad f''(0) = 2$$

$$f'''(x) = \frac{-6}{(1 + x)^4} \qquad f'''(0) = -6$$

Therefore,

$$\frac{1}{1 + x} = 1 - x + \frac{2x^2}{2!} - \frac{6x^3}{3!} + \cdots$$
$$= 1 - x + x^2 - x^3 + \cdots \quad \blacksquare$$

EXAMPLE 3.   Find the first three nonzero terms of the Maclaurin series expansion of $f(x) = \sin 4x$.

$$f(x) = \sin 4x \qquad\qquad f(0) = 0$$
$$f'(x) = 4 \cos 4x \qquad\qquad f'(0) = 4$$
$$f''(x) = -16 \sin 4x \qquad\qquad f''(0) = 0$$
$$f'''(x) = -64 \cos 4x \qquad\qquad f'''(0) = -64$$
$$f^{(4)}(x) = 256 \sin 4x \qquad\qquad f^{(4)}(0) = 0$$
$$f^{(5)}(x) = 1024 \cos 4x \qquad\qquad f^{(5)}(0) = 1024$$

Therefore,

$$\sin 4x = 4x - \frac{64}{3!}x^3 + \frac{1024}{5!}x^5 - \cdots$$

$$= 4x - \frac{32}{3}x^3 + \frac{128}{15}x^5 - \cdots \quad \blacksquare$$

Practice 2.   Find the first three nonzero terms of the Maclaurin series expansion of $f(x) = \cos 2x$.

Exercises / Section **10-2**

Exercises 1–12: Find the first three nonzero terms of the Maclaurin series expansion of the given function.

1. $f(x) = \sin x$

2. $f(x) = \cos x$

3. $f(x) = e^x$

4. $f(x) = \ln(1 + x)$

5. $f(x) = \dfrac{1}{1 - x}$

6. $f(x) = e^{-x}$

7. $f(x) = \sin 2x$

8. $f(x) = \sqrt{1 + x}$

9. $f(x) = \ln(1 - 2x)$

10. $f(x) = \dfrac{1}{\sqrt{1 + x}}$

11. $f(x) = e^{x^2}$

12. $f(x) = e^x \cos x$

Exercises 13–20: Find the first two nonzero terms of the Maclaurin series expansion of the given function.

13. $f(x) = \text{Arctan } x$

14. $f(x) = \sec x$

15. $f(x) = \ln \cos x$

16. $f(x) = \sin^2 x$

17. $f(x) = \sqrt{1 + \sin x}$

18. $f(x) = \tan x$

19. $f(x) = e^{\cos x}$

20. $f(x) = \text{Arcsin } x$

21. Use long division to find a series expansion for $f(x) = 1/(1 + x)$; compare the results with Example 2 of this section.

22. Use long division to find a series expansion for $f(x) = 1/(1 - x)$; compare the results with Exercise 5 above.

23. Show that the Maclaurin series expansion of $f(x) = 2x^2 + 1$ is $f(x)$ itself.

24. Show that the Maclaurin series expansion of $f(x) = 3x^3 - 4x$ is $f(x)$ itself.

Exercises 25–28: Explain why the given function has no Maclaurin series representation.

25. $f(x) = \cot x$

26. $f(x) = \ln x$

27. $f(x) = e^{1/x}$

28. $f(x) = \sqrt{x}$

10-3 Operating with Series

The Maclaurin series expansion of $f(x)$ is a representation of $f(x)$ in a polynomial-like form. (If $f(x)$ is already a polynomial, then its Maclaurin series expansion is $f(x)$ itself; see Exercises 23 and 24 of Section 10-2.) Many operations can be done on power series by treating them essentially as polynomials. Allowable operations are:

substitution	division
addition	differentiation
multiplication	integration

EXAMPLE 1. If

$$f(x) = a_0 + a_1 x + a_2 x^2 + a_3 x^3 + \cdots$$

then

$$f'(x) = a_1 + 2a_2 x + 3a_3 x^2 + \cdots$$

The infinite series has been differentiated term by term, just as if it were a finite polynomial. ■

Use of these operations on Maclaurin series already known to us allows us easily to generate the Maclaurin series expansions for related functions without having to resort to the definition of the Maclaurin series. From the first five exercises of Section 10-2, we already know the following basic Maclaurin series representations:

$$\sin x = x - \frac{x^3}{3!} + \frac{x^5}{5!} - \cdots \tag{1}$$

$$\cos x = 1 - \frac{x^2}{2!} + \frac{x^4}{4!} - \cdots \tag{2}$$

$$e^x = 1 + x + \frac{x^2}{2!} + \frac{x^3}{3!} + \cdots \tag{3}$$

$$\ln(1 + x) = x - \frac{x^2}{2} + \frac{x^3}{3} - \frac{x^4}{4} + \cdots \tag{4}$$

$$\frac{1}{1 - x} = 1 + x + x^2 + x^3 + \cdots \tag{5}$$

With these as a starting point, we can obtain series for other functions.

EXAMPLE 2. Find the Maclaurin series expansion of $f(x) = e^{2x}$. Using the series for e^x, we *substitute* $2x$ for x.

$$e^x = 1 + x + \frac{x^2}{2!} + \frac{x^3}{3!} + \cdots$$

Substituting $2x$ for x,

$$e^{2x} = 1 + (2x) + \frac{(2x)^2}{2!} + \frac{(2x)^3}{3!} + \cdots$$

or

$$e^{2x} = 1 + 2x + 2x^2 + \frac{4}{3}x^3 + \cdots$$

This agrees with the result of Example 1 in Section 10-2, and we avoided having to use the definition of the Maclaurin series. ∎

Word of Advice

When substituting an expression involving x for x in a Maclaurin series expansion, the entire expression, not just x, is raised to the various powers.

EXAMPLE 3.   Find the Maclaurin series expansion for $\tan x$.
    We write $\tan x = \sin x / \cos x$ and *divide* the corresponding series.

$$\tan x = \frac{\sin x}{\cos x} = \frac{x - x^3/3! + x^5/5! - \cdots}{1 - x^2/2! + x^4/4! - \cdots}$$

We use long division.

$$
\begin{array}{r}
x + \dfrac{2x^3}{3!} + \dfrac{36}{5!}x^5 + \cdots \\[2mm]
1 - \dfrac{x^2}{2!} + \dfrac{x^4}{4!} - \cdots \overline{\Big)\; x - \dfrac{x^3}{3!} + \dfrac{x^5}{5!} - \cdots} \\[2mm]
\underline{x - \dfrac{x^3}{2!} + \dfrac{x^5}{4!} - \cdots} \\[2mm]
\dfrac{2x^3}{3!} - \dfrac{4x^5}{5!} \\[2mm]
\underline{\dfrac{2x^3}{3!} - \dfrac{2}{3 \cdot 2!}x^5 + \cdots} \\[2mm]
\dfrac{36}{5!}x^5 + \cdots
\end{array}
$$

$$\left\{ \left(\frac{x^3}{2!} - \frac{x^3}{3!} \right) + \left(\frac{x^5}{5!} - \frac{x^5}{4!} \right) + \cdots \right.$$

$$\left\{ \left(\frac{2}{3 \cdot 2!}x^5 - \frac{4}{5!}x^5 \right) + \cdots \right.$$

The first three nonzero terms are

$$x + \frac{2}{3!}x^3 + \frac{36}{5!}x^5 + \cdots$$

or

$$x + \frac{1}{3}x^3 + \frac{3}{10}x^5 + \cdots \quad ∎$$

Practice 1. Use substitution to find the first three nonzero terms of the Maclaurin series expansion of $f(x) = e^{x^2}$. Compare your result with Exercise 11 of Section 10-2.

Practice 2. Use multiplication to find the first three nonzero terms of the Maclaurin series expansion of $f(x) = e^x \cos x$. Compare your result with Exercise 12 of Section 10-2.

Exercises / Section **10-3**

Exercises 1–18: Find the first three nonzero terms of the Maclaurin series expansion by operating on known series.

1. $f(x) = e^{-x}$ (Exercise 6 of Section 10-2.)

2. $f(x) = \sin 2x$ (Exercise 7 of Section 10-2.)

3. $f(x) = \ln(1 - x^2)$

4. $f(x) = \dfrac{1}{1 + x^3}$

5. $f(x) = \sin \dfrac{x}{2}$

6. $f(x) = \cos x^4$

7. $f(x) = \ln(1 - 5x)$

8. $f(x) = \dfrac{1}{1 - 3x}$

9. $f(x) = e^{-x} \sin x$

10. $f(x) = \sin x \cos x$

11. $f(x) = \sec x$ (*Hint:* $\sec x = \dfrac{1}{\cos x}$.)

12. $f(x) = \ln \dfrac{1 + x}{1 - x}$ (*Hint:* Use a property of logarithms.)

13. $f(x) = \sinh x$, where $\sinh x = \dfrac{e^x - e^{-x}}{2}$.

14. $f(x) = \cosh x$, where $\cosh x = \dfrac{e^x + e^{-x}}{2}$.

15. $f(x) = \sin^2 x$ (*Hint:* $\sin^2 x = (\sin x)(\sin x)$.)

16. $f(x) = \sin^2 x$ (*Hint:* $\sin^2 x = \frac{1}{2}(1 - \cos 2x)$.)

17. $f(x) = \dfrac{1}{1 - \sin x}$

18. $f(x) = e^{\sin x}$

19. Find the Maclaurin series expansion for $\cos x$ by differentiating the series for $\sin x$.

20. Show that differentiating the Maclaurin series expansion for e^x results in the series for e^x.

21. Find the Maclaurin series expansion for $\cosh x$ by differentiating the series for $\sinh x$ (see Exercises 13 and 14 above).

22. Find the Maclaurin series for $f(x) = 1/(1 - x)^2$ by differentiating the series for $1/(1 - x)$.

23. Find the Maclaurin series for $\ln(1 - x)$ by integrating the series for $-1/(1 - x)$ term by term.

24. Find the Maclaurin series for Arctan x by a substitution followed by an integration on the series for $1/(1 - x)$.

25. Find the function whose Maclaurin series expansion is

$$1 \cdot 2 + 2 \cdot 3x + 3 \cdot 4x^2 + 4 \cdot 5x^3 + \cdots$$

(*Hint:* Start with $f(x) = 1/(1 - x)$.)

26. Find the function whose Maclaurin series expansion is

$$1 + 4x + 9x^2 + 16x^3 + 25x^4 + \cdots$$

(*Hint:* See Exercise 22.)

10-4 Computing with Series

The power series representation of a function allows us to approximate the value of the function anywhere in an interval around zero. We simply evaluate a number of terms in the Maclaurin series expansion of the function.

EXAMPLE 1. Approximate the value of $e^{0.2}$.

We are asked to approximate $f(0.2)$ for $f(x) = e^x$. Using the Maclaurin series expansion of e^x,

$$e^x = 1 + x + \frac{x^2}{2!} + \frac{x^3}{3!} + \cdots$$

Substituting 0.2 for x and taking three terms of the series, we get

$$e^{0.2} \approx 1 + (0.2) + \frac{(0.2)^2}{2} = 1.22$$

Using four terms of the series results in

$$e^{0.2} \approx 1 + (0.2) + \frac{(0.2)^2}{2} + \frac{(0.2)^3}{6} = 1.221\overline{3}$$

which does not affect the hundredth place of our answer. Indeed, successive terms of the series are so small that our answer of 1.22 is correct to two decimal places. ∎

Practice 1. Approximate $e^{0.1}$ using three terms of the Maclaurin series for e^x.

EXAMPLE 2. Approximate $\sin(0.1)$.

Using two terms of the Maclaurin series for $\sin x$,

$$\sin x = x - \frac{x^3}{3!} + \cdots$$

with $x = 0.1$, we get

$$\sin 0.1 \approx 0.1 - \frac{(0.1)^3}{3!} = 0.09983$$

Here the second term of the series is very small $(0.0001\overline{6})$ and succeeding terms would be even smaller. Because our value of x is so close to zero, the series converged rapidly to the value of $f(x)$ and only two terms were required to get an answer accurate to three decimal places. ∎

Practice 2. Approximate $\cos(-0.3)$ using two terms of the Maclaurin series for $\cos x$.

Definite integrals of functions for which antiderivatives do not exist may be approximated by integrating the Maclaurin series representations of the functions. Once again, convergence will be more rapid for values (in this case the limits of integration) close to zero.

EXAMPLE 3. Approximate $\int_0^{0.1} e^x \, dx$.

This is a case where we can certainly integrate directly and find the exact value, but we will also use infinite series just to see how it works. Integrating directly,

$$\int_0^{0.1} e^x \, dx = e^x \Big|_0^{0.1} = e^{0.1} - e^0 = e^{0.1} - 1$$

Now using the series approach, we integrate the Maclaurin series for e^x term by term.

$$\int_0^{0.1} e^x \, dx = \int_0^{0.1} \left(1 + x + \frac{x^2}{2!} + \cdots \right) dx = x + \frac{x^2}{2!} + \frac{x^3}{3!} + \cdots \Big|_0^{0.1}$$

$$= (0.1) + \frac{(0.1)^2}{2!} + \frac{(0.1)^3}{3!} + \cdots$$

But the last expression can be written in the form

$$\left[1 + (0.1) + \frac{(0.1)^2}{2!} + \frac{(0.1)^3}{3!} + \cdots \right] - 1$$

which we recognize as the series expansion for $e^{0.1}$ minus 1. This agrees with the answer obtained by using direct integration. ∎

Of course the real value of using the Maclaurin series expansion of $f(x)$ is when $\int_a^b f(x) \, dx$ is not directly integrable.

EXAMPLE 4. Approximate $\int_0^{0.1} e^{x^2} \, dx$.

Here is a function we cannot integrate directly. From the equation

$$e^x = 1 + x + \frac{x^2}{2!} + \frac{x^3}{3!} + \cdots$$

we get

$$e^{x^2} = 1 + x^2 + \frac{x^4}{2!} + \frac{x^6}{3!} + \cdots$$

Integrating term by term,

$$\int_0^{0.1} e^{x^2} \, dx = \int_0^{0.1} \left(1 + x^2 + \frac{x^4}{2!} + \cdots \right) dx$$

$$= x + \frac{x^3}{3} + \frac{x^5}{5 \cdot 2!} + \cdots \Big|_0^{0.1}$$

$$= 0.1 + \frac{(0.1)^3}{3} + \frac{(0.1)^5}{10} + \cdots$$

$$\approx 0.100334 \quad ∎$$

EXAMPLE 5. Approximate $\int_{0.5}^{1} \dfrac{\sin x}{x}\, dx$.

Using the series

$$\sin x = x - \frac{x^3}{3!} + \frac{x^5}{5!} - \cdots$$

we get

$$\frac{\sin x}{x} = 1 - \frac{x^2}{3!} + \frac{x^4}{5!} - \cdots$$

and thus

$$
\begin{aligned}
\int_{0.5}^{1} \frac{\sin x}{x} &= \int_{0.5}^{1} \left(1 - \frac{x^2}{3!} + \frac{x^4}{5!} - \cdots\right) dx \\
&= x - \frac{x^3}{3 \cdot 3!} + \frac{x^5}{5 \cdot 5!} - \cdots \Big|_{0.5}^{1} \\
&= \left(1 - \frac{1}{18} + \frac{1}{600} - \cdots\right) - \left(0.5 - \frac{(0.5)^3}{18} + \frac{(0.5)^5}{600} - \cdots\right) \\
&\approx (1 - 0.05556 + 0.00166) - (0.5 - 0.00694 + 0.000052) \\
&= 0.452988 \quad \blacksquare
\end{aligned}
$$

The table of integrals in the back of the book gives the following integration rules:

59. $\displaystyle\int \frac{\sin u}{u^n}\, du = -\frac{\sin u}{(n-1)u^{n-1}} + \frac{1}{n-1}\int \frac{\cos u}{u^{n-1}}\, du \qquad n \neq 1$

60. $\displaystyle\int \frac{\cos u}{u^n}\, du = -\frac{\cos u}{(n-1)u^{n-1}} - \frac{1}{n-1}\int \frac{\sin u}{u^{n-1}}\, du \qquad n \neq 1$

These are reduction formulas, which reduce the given integral to another, simpler integral. In these cases, the power in the denominator is decreased. Repeated applications of these formulas will eventually result in an integral of the form

$$\int \frac{\sin u}{u}\, du$$

or

$$\int \frac{\cos u}{u}\, du$$

The formulas themselves do not apply to these integrals, which must be solved by a series approximation, as in the example above.

Practice 3. Use three terms of the appropriate series to approximate $\int_{0.2}^{1} \dfrac{\cos x}{x}\, dx$.

EXAMPLE 6. Approximate the area bounded by $y = e^{-x^3}$, $y = 0$, $x = 0$, and $x = 1$ by using three terms of the appropriate series.

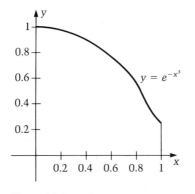

Figure 10-1

We want to approximate the value of $\int_0^1 e^{-x^3}\, dx$. The curve is shown in Figure 10-1. From the Maclaurin series expansion,

$$e^x = 1 + x + \frac{x^2}{2!} + \cdots$$

we get

$$e^{-x^3} = 1 - x^3 + \frac{x^6}{2} + \cdots$$

Thus

$$\int_0^1 e^{-x^3}\, dx = \int_0^1 \left(1 - x^3 + \frac{x^6}{2} + \cdots\right) dx$$

$$= \left(x - \frac{x^4}{4} + \frac{x^7}{14} + \cdots\right)\Big|_0^1$$

$$= \left(1 - \frac{1}{4} + \frac{1}{14} + \cdots\right)$$

$$\approx 1 - 0.25 + 0.07143 = 0.82143 \quad \blacksquare$$

As we have said, a Maclaurin series expansion of a function $f(x)$ is a representation of that function in a polynomial-like form. How good is the representation? There are two issues involved. If

$$f(x) = a_0 + a_1 x + a_2 x^2 + \cdots + a_n x^n + \cdots$$

we have approximated the value $f(c)$ by computing the sum of a finite number of terms in the series. Thus

$$f(c) \approx a_0 + a_1 c + a_2 c^2$$

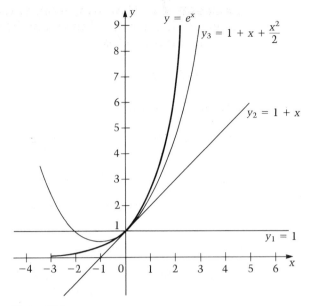

Figure 10-2

if we have used three terms in the series. The two issues are as follows:

1. For a fixed value of c, the approximation will improve as we use more terms.
2. For a fixed number of terms, the approximation will improve as c becomes closer to zero.

Figure 10-2 shows the graph of the curve $y = e^x$, along with the graphs of $y_1 = 1$, $y_2 = 1 + x$, and $y_3 = 1 + x + x/2$. The graph of y_1, representing only the first term in the Maclaurin series expansion of $y = e^x$, has the correct value for $x = 0$ and is not even close to the shape of the curve anywhere else. Using two terms of the series, y_2, we still have a straight line, but at least it is slanted to agree a little more closely with the general direction of y. Using three terms of the series, we see that y_3 agrees fairly closely with y for the positive range of values shown, although it bends away for negative values of x. All three curves have values that agree more closely with those of $y = e^x$ when x is close to zero than when x is farther from zero.

When we drop the remaining terms in the series and use only a finite sum to compute $f(c)$, we introduce some error into the calculation. There are methods available to tell us the maximum amount of error introduced, so that we can determine the accuracy of our approximation. In particular, if the series we are evaluating is an **alternating series** (one whose terms alternate in sign between positive and negative), then the maximum amount of error is the value of the first term omitted.

EXAMPLE 7. How many terms in the series should be taken to compute the value of e^{-1} to within three decimal places?

The series for e^x is

$$e^x = 1 + x + \frac{x^2}{2!} + \frac{x^3}{3!} + \cdots$$

Evaluating when $x = -1$, we get

$$1 - 1 + \frac{(-1)^2}{2!} + \frac{(-1)^3}{3!} + \frac{(-1)^4}{4!} + \frac{(-1)^5}{5!} + \cdots$$

$$= \frac{1}{2!} - \frac{1}{3!} + \frac{1}{4!} - \frac{1}{5!} + \frac{1}{6!} - \frac{1}{7!} + \cdots$$

which is an alternating series. Its values are

$$0.500000 - 0.166667 + 0.041667 - 0.008333 + 0.001389$$
$$- 0.000198 + \cdots$$

The last term computed does not affect the third decimal place; therefore, we can stop with the term before this one. ∎

The problem of finding a polynomial-like representation of a function that will be a good representation for values not close to zero will be dealt with in the next section. A nonpolynomial series representation, useful for certain functions, will be discussed in Section 10-6.

Exercises / Section **10-4**

Exercises 1–8: Use the appropriate Maclaurin series with the number of terms shown in order to approximate the value of the expression.

1. e; 6

2. $\frac{1}{e}$; 6

3. $\ln 1.1$; 4

4. $\ln(0.98)$; 2

5. $\sin 1°$; 2 (Do not forget to work in radian measure.)

6. $\cos 2°$; 2

7. $\sqrt{e}$; 5

8. $\frac{1}{1.1}$; 5

Exercises 9–12: Find Maclaurin series expansions for $f(x) = \sqrt{1 + x}$ and $f(x) = \sqrt[3]{1 + x}$, and use the number of terms shown to approximate the value of the expression.

9. $\sqrt{0.98}$; 3

10. $\sqrt{1.1}$; 4

11. $\sqrt[3]{1.2}$; 4

12. $\sqrt[3]{0.97}$; 3

Exercises 13–18: Use three terms of the appropriate series in order to approximate the integral.

13. $\int_{0.1}^{0.3} e^{\sqrt{x}} \, dx$

14. $\int_{0.01}^{0.02} \frac{e^{-x}}{x} \, dx$

15. $\int_0^1 \cos x^2 \, dx$

16. $\int_0^1 \sin \sqrt{x} \, dx$

17. $\int_0^{0.1} \ln(1 + x^2) \, dx$

18. $\int_0^{0.5} x \cos x^3 \, dx$

19. Find an approximation to the area bounded by the curves $y = e^x/x$, $y = 0$, $x = 0.1$, and $x = 0.3$. Use four terms of a series.

20. Find an approximation to the area bounded by the curves $y = \sin x^2$, $y = 0$, $x = 0$, and $x = 0.8$. Use three terms of a series.

21. Use three terms of a series to approximate the volume formed by revolving the area bounded by $y = \ln(1 + x)$, $y = 0$, $x = 0$, and $x = 0.5$ about the y-axis.

22. Use three terms of a series to approximate the volume formed by revolving the area bounded by $y = \cos x/x^2$, $y = 0$, $x = 0.1$, and $x = 0.4$ about the y-axis.

23. Find an approximation for the first moment with respect to the y-axis of the area bounded by $y = e^{x^3}$, $y = 0$, $x = 0$, and $x = 0.2$. Use two terms of a series and assume a constant density k.

24. Find an approximation for the length of arc of the curve $y = 1/x$ from $x = 10$ to $x = 100$. Use two terms of a series. (*Hint:* See Exercise 8 of Section 10-2.)

25. The distance from the equilibrium position of a fixed point on a vibrating string is given by

$$x = \tfrac{1}{3} \cos 8t$$

where x is given in centimeters and t in seconds. Use three terms of a series to find an approximate value for x when $t = 0.1$ s.

26. Fifty milligrams of a certain medication are administered orally. The amount of the medication present in the blood stream at time t hours later is given by

$$A = 63(e^{-0.25t} - e^{-1.25t})$$

Use four terms of a series to find an approximate value for A when $t = 1$ h.

27. On the same coordinate system, sketch graphs of the curves $y = 1/(1 - x)$, $y_1 = 1$, $y_2 = 1 + x$, and $y_3 = 1 + x + x^2$.

28. On the same coordinate system, sketch graphs of the curves $y = \ln(1 + x)$, $y_1 = x$, $y_2 = x - x^2/2$, and $y_3 = x - x^2/2 + x^3/3$.

10-5 Taylor Series

A Maclaurin series expansion of $f(x)$ is a power series in x and has the form

$$f(x) = a_0 + a_1x + a_2x^2 + \cdots + a_nx^n + \cdots$$

Values of x close to zero result in the values of successively higher powers of x getting rapidly smaller, and the series converges rapidly close to zero.

Suppose we want to use a series representation of $f(x)$ to approximate its value at some x close to a nonzero value a. A series representation of the form

$$f(x) = c_0 + c_1(x - a) + c_2(x - a)^2 + \cdots + c_n(x - a)^n + \cdots \qquad (1)$$

would converge rapidly because $|x - a|$ would be small, and the values of $(x - a)^2$, $(x - a)^3$, and so on would get smaller rapidly.

As we did for the Maclaurin series, let us simply assume that such a series representation exists for $f(x)$ and see what the coefficients c_i must be in that case.

We have assumed Equation (1) to be true; in particular, it is true for $x = a$. Thus

$$f(a) = c_0$$

Beginning with Equation (1), we do successive differentiations and evaluations with $x = a$.

$$f'(x) = c_1 + 2c_2(x - a) + 3c_3(x - a)^2 + \cdots$$
$$f'(a) = c_1$$
$$f''(x) = 2c_2 + 2 \cdot 3c_3(x - a) + \cdots$$
$$f''(a) = 2c_2$$
$$f'''(x) = 2 \cdot 3c_3 + \cdots$$
$$f'''(a) = 2 \cdot 3c_3$$

From these results we see that

$$c_0 = f(a)$$
$$c_1 = \frac{f'(a)}{1!}$$
$$c_2 = \frac{f''(a)}{2!}$$
$$c_3 = \frac{f'''(a)}{3!}$$
$$\vdots$$

and, in general,

$$c_n = \frac{f^{(n)}(a)}{n!}$$

Equation (1) thus has the form

$$f(x) = f(a) + \frac{f'(a)}{1!}(x - a) + \frac{f''(a)}{2!}(x - a)^2$$
$$+ \cdots + \frac{f^{(n)}(a)}{n!}(x - a)^n + \cdots \tag{2}$$

The right side of Equation (2) is called the **Taylor series expansion of** $f(x)$ **about** $x = a$. (One condition for a Taylor series representation of a function $f(x)$ about $x = a$ is that the function and all its derivatives must be defined at a.)

A Taylor series for $f(x)$ is just a generalization of the Maclaurin series, or rather the Maclaurin series is the special case of the Taylor series where $a = 0$.

EXAMPLE 1. Approximate $\sqrt{16.2}$.

The function involved here is $f(x) = \sqrt{x}$, and we want to evaluate $f(x)$ at $x = 16.2$. Here x is not close to zero, so we want to use a Taylor series with $a \neq 0$ rather than a Maclaurin series. (Besides, the Maclaurin series for $f(x) = \sqrt{x}$ does not exist; see Exercise 28 of Section 10-2.) Two conditions govern our choice of a:

1. The x-value of 16.2 should be close to the value of a so that $|x - a|$ is small and convergence will be rapid.
2. The value of a should be such that the coefficients in the Taylor series, $f(a)$, $f'(a)$, . . . , are easy to determine.

We choose $a = 16$ because it satisfies both of the above requirements. We next compute the coefficients in the Taylor series expansion about $a = 16$.

$$f(x) = \sqrt{x} \qquad\qquad f(16) = 4$$

$$f'(x) = \frac{1}{2}x^{-1/2} \qquad\qquad f'(16) = \frac{1}{2\sqrt{16}} = \frac{1}{8}$$

$$f''(x) = -\frac{1}{4}x^{-3/2} \qquad f''(16) = -\frac{1}{4(\sqrt{16})^3} = -\frac{1}{4 \cdot 64} = -\frac{1}{256}$$

$$\vdots$$

Therefore,

$$\sqrt{x} = 4 + \frac{1}{8}(x - 16) - \frac{1}{2 \cdot 256}(x - 16)^2 + \cdots$$

and

$$\sqrt{16.2} = 4 + \frac{1}{8}(16.2 - 16) - \frac{1}{512}(16.2 - 16)^2 + \cdots$$

$$= 4 + \frac{1}{8}(0.2) - \frac{(0.2)^2}{512} + \cdots$$

$$\approx 4 + 0.025 - 0.000078$$

$$= 4.0249$$

Note that our answer is reasonable; $\sqrt{16.2}$ should be a little bigger than 4. Rough estimates can sometimes detect errors.   ■

EXAMPLE 2.   Approximate sin 86°.
We want to evaluate $f(x) = \sin x$ at $x = 86°$. Keeping in mind the two conditions a must satisfy, we choose $a = 90°$. Next we do a Taylor expansion of $\sin x$ about $x = 90°$.

$$f(x) = \sin x \qquad\qquad f(90°) = 1$$
$$f'(x) = \cos x \qquad\qquad f'(90°) = 0$$
$$f''(x) = -\sin x \qquad\qquad f''(90°) = -1$$
$$f'''(x) = -\cos x \qquad\qquad f'''(90°) = 0$$
$$f^{(4)}(x) = \sin x \qquad\qquad f^{(4)}(90°) = 1$$

$$\vdots$$

Using the first three nonzero terms of the Taylor series expansion,

$$\sin x = 1 - \frac{1}{2!}(x - 90°)^2 + \frac{1}{4!}(x - 90°)^4 - \cdots$$

For $x = 86°$, $x - 90° = -4°$, but we must change this to radian measure. From the equation

$$180° = \pi$$

we get

$$-4° = -\frac{\pi}{180} \cdot 4 = -\frac{\pi}{45}$$

Now we evaluate:

$$\sin 86° \approx 1 - \frac{1}{2!}\left(-\frac{\pi}{45}\right)^2 + \frac{1}{4!}\left(-\frac{\pi}{45}\right)^4$$

$$= 1 - 0.0024369 + 0.0000010$$

$$= 0.99756$$

Once again the answer seems reasonable. ■

Practice 1. Find the first three nonzero terms of the Taylor series expansion for cos x about $a = 45°$.

Practice 2. Use the answer to Practice 1 to approximate cos 48°.

Exercises / Section **10-5**

1. Use three terms of the Taylor series of Example 1 of this section to approximate $\sqrt{15.6}$.

2. Use three nonzero terms of the Taylor series of Example 2 of this section to approximate sin 91°.

Exercises 3–12: Use three terms of the appropriate Taylor series in order to approximate the value shown.

3. $\sqrt{101}$

4. $\sqrt{24}$

5. sin 62°

6. cos 95°

7. ln 5.1   (Use $a = 5$ with ln 5 = 1.6094.)

8. $e^{1.1}$ (Use $e = 2.71828$.)

9. $\dfrac{1}{10.2}$

10. $\dfrac{1}{98}$

11. $\sqrt[3]{8.4}$

12. $\sqrt[3]{62}$

10-6 Fourier Series

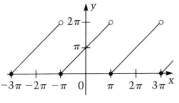

Figure 10-3

A *periodic function* $f(x)$ is one whose pattern of values repeats itself in cycles, as in Figure 10-3. The length of the cycle is called the *period* of the function. The period of the function in Figure 10-3 is 2π. Periodic functions arise in vibration or oscillation problems (vibrating strings or bodies on springs), as well as in electronics problems where a circuit is supplied with a periodic voltage source (as in alternating currents). In fact, you may have seen periodic

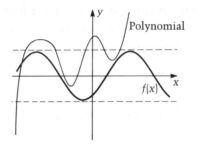

Figure 10-4

patterns or waves on the screen of an oscilloscope; the pattern can be made to change by adjusting the signal.

Suppose that $f(x)$ is a periodic function that we have represented in a Taylor series expansion about $x = a$. We know that for values of x close to a, we do not have to take many terms of the series to approximate the value of $f(x)$. Let us say we choose to use n terms. Then

$$f(x) \approx f(a) + \frac{f'(a)}{1!}(x - a) + \frac{f''(a)}{2!}(x - 2)^2 + \cdots + \frac{f^{(n)}(a)}{n!}(x - a)^n$$

We thus approximate $f(x)$ by a polynomial of degree n. As we get farther away from a, a polynomial of degree n will have values going to $\pm\infty$, whereas the values of our periodic function will in general stay within some fixed range (see Figure 10-4). Therefore, the approximation of a periodic function by a Taylor series can be very poor as x moves away from a.

Instead of trying to approximate a periodic function $f(x)$ by a sum of polynomial terms, we use a sum of functions which are themselves periodic but simple, namely, sines and cosines. Figure 10-5 shows a very simple case, $f(x) = \cos x + \sin x$. The sum of these two periodic functions is a periodic function with period 2π.

We assume that a periodic function $f(x)$ (with period 2π) has a series representation of the form

$$f(x) = a_0 + a_1 \cos x + a_2 \cos 2x + \cdots + a_n \cos nx + \cdots$$
$$+ b_1 \sin x + b_2 \sin 2x + \cdots + b_n \sin nx + \cdots \qquad (1)$$

and we determine what the coefficients a_i and b_i must be. This is the

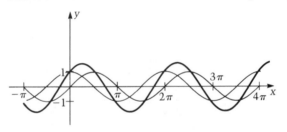

Figure 10-5

same sort of thing we did for the Maclaurin series and the Taylor series. Here, though, instead of differentiating to find the coefficients, we integrate.

$$\int_{-\pi}^{\pi} f(x)\, dx = \int_{-\pi}^{\pi} a_0\, dx$$

$$\left[+ \int_{-\pi}^{\pi} (a_1 \cos x + a_2 \cos 2x + \cdots + a_n \cos nx + \cdots)\, dx \right]$$

$$\left[+ \int_{-\pi}^{\pi} (b_1 \sin x + b_2 \sin 2x + \cdots + b_n \sin nx + \cdots)\, dx \right]$$

$$= a_0 x \Big|_{-\pi}^{\pi} \Bigg]$$

$$\left[+ (a_1 \sin x + \frac{a_2}{2} \sin 2x + \cdots + \frac{a_n}{n} \sin nx + \cdots) \Big|_{-\pi}^{\pi} \right]$$

$$\left[- (b_1 \cos x + \frac{b_2}{2} \cos 2x + \cdots + \frac{b_n}{n} \cos nx + \cdots) \Big|_{-\pi}^{\pi} \right]$$

$$= a_0[\pi - (-\pi)] + 0 \qquad \text{Because the sine of any multiple of } \pi \text{ is zero.}$$

$$- \left[b_1[-1 - (-1)] + \frac{b_2}{2}[1 - 1] + \cdots + \frac{b_n}{n}[-1 - (-1)] + \cdots \right]$$

$$= 2\pi a_0$$

Therefore,

$$a_0 = \frac{1}{2\pi} \int_{-\pi}^{\pi} f(x)\, dx$$

To find the remaining a_i coefficients, we multiply both sides of Equation (1) by $\cos mx$, where m is a fixed integer, and integrate from $-\pi$ to π. The result is

$$\int_{-\pi}^{\pi} f(x) \cos mx = \int_{-\pi}^{\pi} a_0 \cos \overset{\text{①}}{mx}\, dx + \int_{-\pi}^{\pi} (a_1 \cos x \cos mx$$

$$+ \cdots \overset{\text{②}}{+} a_n \cos nx \cos mx + \cdots)\, dx$$

$$+ \int_{-\pi}^{\pi} (b_1 \sin x \cos mx + \overset{\text{③}}{\cdots} +$$

$$b_n \sin nx \cos mx + \cdots)\, dx \qquad (2)$$

We consider each of expressions ①, ②, and ③, in turn.

① $\displaystyle \int_{-\pi}^{\pi} a_0 \cos mx\, dx = \frac{a_0}{m} \sin mx \Big|_{-\pi}^{\pi} = 0$ For any value of m.

② The general term is $\displaystyle \int_{-\pi}^{\pi} a_n \cos nx \cos mx\, dx.$

For the case where $n \neq m$, this matches the pattern of Rule 53 in the table of integrals at the back of the book.

$$\int_{-\pi}^{\pi} a_n \cos nx \cos mx \, dx = a_n \left(\frac{\sin(n-m)x}{2(n-m)} + \frac{\sin(n+m)x}{2(n+m)} \right) \Big|_{-\pi}^{\pi}$$

$$= 0 \qquad \text{Again, the sine of any multiple of } \pi \text{ is zero.}$$

For the case where $n = m$, the term becomes

$$\int_{-\pi}^{\pi} a_n \cos^2 nx \, dx$$

which matches the pattern of Rule 44 in the table of integrals.

$$\int_{-\pi}^{\pi} a_n \cos^2 nx \, dx = \frac{a_n}{n} \left(\frac{nx}{2} + \frac{1}{4} \sin 2nx \right) \Big|_{-\pi}^{\pi}$$

$$= \frac{a_n}{n} \left(\frac{n\pi}{2} - \frac{n(-\pi)}{2} \right) = a_n \pi$$

③   The general term is $\int_{-\pi}^{\pi} b_n \sin nx \cos mx \, dx$.

For the case where $n \neq m$, this matches the pattern of Rule 51.

$$\int_{-\pi}^{\pi} b_n \sin nx \cos mx \, dx = b_n \left(-\frac{\cos(n-m)x}{2(n-m)} - \frac{\cos(n+m)x}{2(n+m)} \right) \Big|_{-\pi}^{\pi}$$

$$= b_n \left[-\frac{\cos(n-m)\pi}{2(n-m)} - \frac{\cos(n+m)\pi}{2(n+m)} \right.$$

$$\Big\downarrow \qquad\qquad \Big\downarrow$$

$$\left. + \frac{\cos(n-m)(-\pi)}{2(n-m)} + \frac{\cos(n+m)(-\pi)}{2(n+m)} \right]$$

The paired terms have a sum of zero because $\cos(\theta) = \cos(-\theta)$. For the case where $n = m$, the term becomes

$$\int_{-\pi}^{\pi} b_n \sin nx \cos nx = \frac{b_n}{n} \frac{\sin^2 nx}{2} \Big|_{-\pi}^{\pi} = 0$$

The only nonzero term in the integration of (2) occurs when $n = m$. Equation (2) then becomes

$$\int_{-\pi}^{\pi} f(x) \cos nx \, dx = a_n \pi$$

from which

$$a_n = \frac{1}{\pi} \int_{-\pi}^{\pi} f(x) \cos nx \, dx$$

In order to find the b_i coefficients, we multiply both sides of Equation (1) by $\sin mx$ and integrate from $-\pi$ to π. Following the same procedures we used in finding the a_i's, we get

$$b_n = \frac{1}{\pi} \int_{-\pi}^{\pi} f(x) \sin nx \, dx$$

Finally, Equation (1) can now be written as

$$f(x) = a_0 + a_1 \cos x + a_2 \cos 2x + \cdots + a_n \cos nx + \cdots$$
$$+ b_1 \sin x + b_2 \sin 2x + \cdots + b_n \sin nx + \cdots$$

where

$$a_0 = \frac{1}{2\pi} \int_{-\pi}^{\pi} f(x) \, dx$$

$$a_n = \frac{1}{\pi} \int_{-\pi}^{\pi} f(x) \cos nx \, dx \qquad n \geq 1$$

$$b_n = \frac{1}{\pi} \int_{-\pi}^{\pi} f(x) \sin nx \, dx \qquad n \geq 1$$

This series is called the **Fourier series** of $f(x)$, named after the French mathematician J. B. J. Fourier (1768–1830).

EXAMPLE 1. Find the Fourier series of the function

$$f(x) = x + \pi, \qquad -\pi \leq x < \pi$$

with $f(x + 2\pi) = f(x)$.

The condition $f(x + 2\pi) = f(x)$ implies that $f(x)$ is a periodic function with period 2π. The graph of $f(x)$ is shown in Figure 10-3. We now find the coefficients in the Fourier series.

$$a_0 = \frac{1}{2\pi} \int_{-\pi}^{\pi} (x + \pi) \, dx = \frac{1}{2\pi} \left(\frac{x^2}{2} + \pi x \right) \Big|_{-\pi}^{\pi}$$

$$= \frac{1}{2\pi} \left[\frac{\pi^2}{2} + \pi^2 - \left(\frac{\pi^2}{2} - \pi^2 \right) \right]$$

$$= \pi$$

$$a_n = \frac{1}{\pi} \int_{-\pi}^{\pi} (x + \pi) \cos nx \, dx$$

$$= \frac{1}{\pi} \left[\int_{-\pi}^{\pi} x \cos nx \, dx + \pi \int_{-\pi}^{\pi} \cos nx \, dx \right]$$

$$= \frac{1}{\pi} \left[\frac{1}{n^2} (\cos nx + nx \sin nx) \Big|_{-\pi}^{\pi} + \frac{\pi}{n} \sin nx \Big|_{-\pi}^{\pi} \right]$$

$$= \frac{1}{\pi} \left\{ \frac{1}{n^2} [\cos n\pi + 0 - \cos(-n\pi) - 0] + 0 \right\}$$

$$= \frac{1}{\pi n^2} (\cos n\pi - \cos n\pi)$$

$$= 0$$

$$b_n = \frac{1}{\pi} \int_{-\pi}^{\pi} (x + \pi) \sin nx \, dx$$

$$= \frac{1}{\pi} \left[\int_{-\pi}^{\pi} x \sin nx \, dx + \pi \int_{-\pi}^{\pi} \sin nx \, dx \right]$$

$$= \frac{1}{\pi} \left[\frac{1}{n^2} (\sin nx - nx \cos nx) \Big|_{-\pi}^{\pi} - \frac{\pi}{n} \cos nx \Big|_{-\pi}^{\pi} \right]$$

$$= \frac{1}{\pi} \left\{ \frac{1}{n^2} [0 - n\pi \cos n\pi - 0 - n\pi \cos(-n\pi)] \right.$$

$$\left. - \frac{\pi}{n} [\cos n\pi - \cos(-n\pi)] \right\}$$

$$= \frac{1}{\pi} \left[\frac{1}{n^2} (-n\pi \cos n\pi - n\pi \cos n\pi) \right.$$

$$\left. - \frac{\pi}{n} (\cos n\pi - \cos n\pi) \right]$$

$$= \frac{1}{\pi n^2} (-2n\pi \cos n\pi)$$

$$= -\frac{2}{n} \cos n\pi$$

Therefore,

$$b_1 = -2 \cos \pi = 2$$
$$b_2 = -\cos 2\pi = -1$$
$$b_3 = -\tfrac{2}{3} \cos 3\pi = \tfrac{2}{3}$$
$$b_4 = -\tfrac{1}{2} \cos 4\pi = -\tfrac{1}{2}$$
$$\vdots$$

and the Fourier series expansion is

$$f(x) = \pi + 2 \sin x - \sin 2x + \tfrac{2}{3} \sin 3x - \tfrac{1}{2} \sin 4x + \cdots$$

Figure 10-6 shows the graph of $f(x)$, together with y_1, y_2, y_3, and

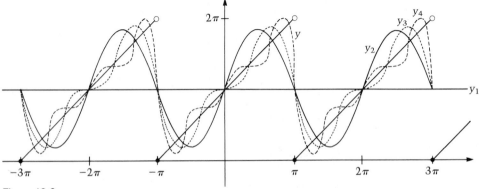

Figure 10-6

y_4, which represent the first, first two, first three, and first four terms, respectively, of the Fourier series for $f(x)$. The more terms of the series used, the closer the resulting curve is to the shape of $f(x)$. ∎

EXAMPLE 2. Find the Fourier series of the "square wave" function

$$f(x) = \begin{cases} 2 & \text{for } -\pi \le x < 0 \\ 0 & \text{for } 0 \le x < \pi \end{cases}$$

with $f(x + 2\pi) = f(x)$.

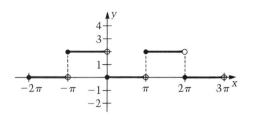

Figure 10-7

The graph of $f(x)$ is shown in Figure 10-7. As we compute the integrals for the Fourier series coefficients, we must break the interval of integration in half, because the function is defined differently in each half.

$$a_0 = \frac{1}{2\pi}\left(\int_{-\pi}^{0} 2\,dx + \int_{0}^{\pi} 0\,dx\right)$$

$$= \frac{1}{2\pi}2x \Big|_{-\pi}^{0} \qquad -\frac{1}{2\pi}(-2\pi) =$$

$$= 1$$

$$a_n = \frac{1}{\pi}\left(\int_{-\pi}^{0} 2\cos nx\,dx + \int_{0}^{\pi} 0\cos nx\,dx\right)$$

$$= \frac{1}{\pi}\left(\frac{2}{n}\sin nx \Big|_{-\pi}^{0}\right) = 0$$

$$b_n = \frac{1}{\pi}\left(\int_{-\pi}^{0} 2\sin nx\,dx + \int_{0}^{\pi} 0\sin nx\,dx\right)$$

$$= \frac{1}{\pi}\left(-\frac{2}{n}\cos nx \Big|_{-\pi}^{0}\right)$$

$$= \frac{1}{\pi}\left(-\frac{2}{n} + \frac{2}{n}\cos n\pi\right)$$

$$= -\frac{2}{n\pi} + \frac{2}{n\pi}\cos n\pi$$

Therefore,

$$b_1 = -\frac{4}{\pi}$$

$$b_2 = 0$$

$$b_3 = -\frac{4}{3\pi}$$

$$b_4 = 0$$

$$b_5 = -\frac{4}{5\pi}$$

The Fourier series expansion is

$$f(x) = 1 - \frac{4}{\pi} \sin x - \frac{4}{3\pi} \sin 3x - \frac{4}{5\pi} \sin 5x - \cdots \quad \blacksquare$$

Practice 1.   Find the Fourier series of the function

$$f(x) = \begin{cases} 0 & \text{for } -\pi \le x < 0 \\ 1 & \text{for } 0 \le x < \pi \end{cases}$$

with $f(x + 2\pi) = f(x)$

In the Fourier series for the functions we have seen so far, the a_n coefficients, $n \ge 1$, have all turned out to be zero. This does not always happen. Also, in all cases seen so far, the function itself has been relatively simple, so that we would not use a Fourier series expansion of the function as a tool for evaluating the function, as we used Maclaurin and Taylor series. What then are the applications of Fourier series? Many problems involving vibrations and heat transfer lead to differential equations, whose solutions involve expansion of functions in Fourier series. Differential equations are discussed in Chapter 11, although the differential equations we see there are not quite that complex. The concept of Fourier series also has applications in electronics.

EXAMPLE 3.   A certain electric circuit has a variable voltage source $E(t)$ given by

$$E(t) = t^2, \qquad -\pi \le t < \pi,$$

with $E(t + 2\pi) = E(t)$. In order to find the steady-state current of this circuit, $E(t)$ must be expanded in a Fourier series. Find some of the terms in this series.

We find the Fourier coefficients, making use of Rules 54–57 in the table of integrals.

$$a_0 = \frac{1}{2\pi} \int_{-\pi}^{\pi} t^2 \, dt = \frac{1}{2\pi} \frac{t^3}{3}\bigg|_{-\pi}^{\pi} = \frac{1}{2\pi}\left(\frac{\pi^3}{3} + \frac{\pi^3}{3}\right) = \frac{\pi^2}{3}$$

$$a_n = \frac{1}{\pi} \int_{-\pi}^{\pi} t^2 \cos nt \, dt = \frac{1}{\pi n^3} \int_{-\pi}^{\pi} (nt)^2 \cos nt \, d(nt)$$

$$= \frac{1}{\pi n^3}\left[(nt)^2 \sin nt \bigg|_{-\pi}^{\pi} - 2 \int_{-\pi}^{\pi} nt \sin nt \, d(nt)\right]$$

$$= -\frac{2}{\pi n^3}\left(\sin nt \bigg|_{-\pi}^{\pi} - nt \cos nt \bigg|_{-\pi}^{\pi}\right)$$

$$= -\frac{2}{\pi n^3}(-n\pi \cos n\pi + n(-\pi) \cos n\pi)$$

$$= \frac{4}{n^2} \cos n\pi$$

Therefore,

$$a_1 = -4$$

$$a_2 = 1$$

$$a_3 = -\frac{4}{9}$$

$$a_4 = \frac{1}{4}$$

$$\vdots$$

$$b_n = \frac{1}{\pi} \int_{-\pi}^{\pi} t^2 \sin nt \, dt = \frac{1}{\pi n^3} \int_{-\pi}^{\pi} (nt)^2 \sin nt \, d(nt)$$

$$= \frac{1}{\pi n^3}\left[-(nt)^2 \cos nt \bigg|_{-\pi}^{\pi} + 2 \int_{-\pi}^{\pi} nt \cos nt \, d(nt)\right]$$

$$= \frac{1}{\pi n^3}\left[-(nt)^2(\cos n\pi - \cos n\pi) + 2(\cos nt + nt \sin nt)\bigg|_{-\pi}^{\pi}\right]$$

$$= \frac{2}{\pi n^3}(\cos n\pi - \cos n\pi + 0)$$

$$= 0$$

The Fourier series is

$$E(t) = \frac{\pi^2}{3} - 4 \cos t + \cos 2t - \frac{4}{9} \cos 3t + \frac{1}{4} \cos 4t + \cdots \quad \blacksquare$$

The equations for the Fourier series coefficients can be modified so that a periodic function with a period different from 2π can be expanded in a series involving sines and cosines.

Exercises / Section 10-6

Exercises 1–8: Find some terms of the Fourier series for the function. Assume that $f(x + 2\pi) = f(x)$.

1. $f(x) = \begin{cases} 1 & -\pi \leq x < 0 \\ 0 & 0 \leq x \leq \pi \end{cases}$

2. $f(x) = \begin{cases} 1 & -\pi \leq x < 0 \\ 2 & 0 \leq x < \pi \end{cases}$

3. $f(x) = \begin{cases} 0 & -\pi \leq x < 0 \\ x & 0 \leq x < \pi \end{cases}$

4. $f(x) = \begin{cases} 1 & -\pi \leq x < -\pi/2 \\ 0 & -\pi/2 \leq x \leq \pi/2 \\ 1 & \pi/2 < x < \pi \end{cases}$

5. $f(x) = x, \qquad -\pi \leq x < \pi$

6. $f(x) = \begin{cases} -x & -\pi \leq x < 0 \\ x & 0 \leq x < \pi \end{cases}$

7. $f(x) = \begin{cases} x & -\pi \leq x < 0 \\ \pi - x & 0 \leq x < \pi \end{cases}$

8. $f(x) = \begin{cases} x & -\pi \leq x < 0 \\ x^2 & 0 \leq x < \pi \end{cases}$

Exercises 9–10: An electronic device called a *half-wave rectifier* removes the negative portion of the wave from an alternating current. Find some terms of the Fourier series expansion for the resulting periodic function and graph several cycles of the function.

9. $f(x) = \begin{cases} 0 & -\pi \leq x < 0 \\ \sin x & 0 \leq x < \pi \end{cases}$

10. $f(x) = \begin{cases} 0 & -\pi \leq x < -\dfrac{\pi}{2} \\ \cos x & -\dfrac{\pi}{2} \leq x \leq \dfrac{\pi}{2} \\ 0 & \dfrac{\pi}{2} < x < \pi \end{cases}$

STATUS CHECK

Now that you are at the end of Chapter 10, you should be able to:

section **10-1**   Given an infinite series, decide whether it seems to converge or diverge, and if it converges estimate the sum.

section **10-2**   Find the first few terms of the Maclaurin series expansion of a given function from the definition of the Maclaurin series.

section **10-3**   Find the first few terms of the Maclaurin series expansion of a given function by operating on known Maclaurin series for other functions.

section **10-4**   Approximate the value of a function at a given value of its independent variable by using terms in the Maclaurin series expansion for the function.
Approximate the value of a definite integral by using terms in the Maclaurin series expansion of the integrand.
Approximate the value of the definite integral expression for the area bounded by given curves or the volume of a solid of revolution by using the Maclaurin series expansion for the integrand.

section **10-5**   Approximate the value of a function at a given value of its independent variable by using terms in the appropriate Taylor series expansion for the function.

section **10-6**   Find some terms of the Fourier series expansion of a given periodic function.

10-7 More Exercises for Chapter **10**

Exercises 1–4: Decide if the series seems to converge or diverge. If it converges estimate the sum.

1. $\frac{2}{3} - \frac{2}{4} + \frac{6}{16} - \frac{18}{64} + \cdots$

2. $1 + 1.1 + (1.1)^2 + (1.1)^3 + \cdots$

3. $\frac{e}{1} + \frac{e^2}{2} + \frac{e^3}{3} + \frac{e^4}{4} + \cdots$

4. $\frac{\sin^2 1}{1^3} + \frac{\sin^2 2}{2^3} + \frac{\sin^2 3}{3^3} + \frac{\sin^2 4}{4^3} + \cdots$

Exercises 5–6: Use the definition of the Maclaurin series to find the first three nonzero terms of the Maclaurin series expansion of the given function.

5. $f(x) = e^{-2x}$

6. $f(x) = e^x \sin x$

Exercises 7–12: Find the first three nonzero terms of the Maclaurin series expansion by operating on known series.

7. $f(x) = e^{-2x}$ (See Exercise 5.)

8. $f(x) = e^x \sin x$ (See Exercise 6.)

9. $f(x) = e^{3x}$

10. $f(x) = \sin \frac{x}{3}$

11. $f(x) = \cos 2x^2$

12. $f(x) = \dfrac{1}{1 - x^2}$

Exercises 13–14: The Maclaurin series expansion for $f(x) = \tan x$ is

$$\tan x = x + \tfrac{1}{3}x^3 + \tfrac{3}{10}x^5 + \cdots$$

(see Example 3, Section 10-3). Use this to find the first three nonzero terms in the Maclaurin series expansion for the function shown.

13. $f(x) = \sec^2 x$ 14. $f(x) = \ln \cos x$

Exercises 15–16: Find the first three nonzero terms of the Taylor series expansion about the value of *a* shown.

15. $f(x) = \sin x;$ $a = 30°$

16. $f(x) = e^{-x};$ $a = 1$

Exercises 17–22: Use three terms of the appropriate series to evaluate the expression.

17. $\cos 1°$ 18. $\ln 1.2$

19. $\cos 58°$ 20. $\sqrt{65}$

21. $\ln 0.9$ 22. $\tan 47°$

Exercises 23–24: Use three terms of the appropriate series to evaluate the integral.

23. $\displaystyle\int_0^1 \sin x^2 \, dx$ 24. $\displaystyle\int_{0.01}^{0.02} e^{1/x} \, dx$

Exercises 25–26: Find some terms of the Fourier series for the function. Assume that $f(x + 2\pi) = f(x)$.

25. $f(x) = \begin{cases} -2 & -\pi \le x < 0 \\ 1 & 0 \le x < \pi \end{cases}$

26. $f(x) = \begin{cases} x & -\pi \le x < 0 \\ 1 & 0 \le x < \pi \end{cases}$

27. Use three terms of an appropriate series to approximate the area bounded by the curves $y = \cos \sqrt{x}$, $y = 0$, $x = 0$, and $x = 1$.

28. Use two terms of an appropriate series to approximate the area bounded by the curves $y = 1/(1 - x^3)$, $y = 0$, $x = 0$, and $x = 0.2$.

chapter **11**

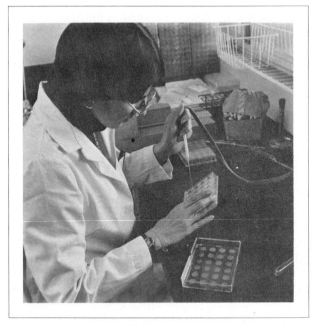

Differential
Equations

*A differential equation can be used to model
the growth rate of certain bacteria.*

11-1 **Differential Equations and Their Solutions**

Any equation that contains a derivative or differential is a **differential equation**. Differential equations arise in many applications because they express relationships between variables and the rates of change of those variables. In fact, differential equations are fundamental tools used by applied mathematicians to model or describe physical situations.

We have already seen differential equations being used, although we did not call them by that name.

EXAMPLE 1. We know from Section 7-1 that a falling body moving only under the influence of gravity experiences a constant acceleration $a = -9.8$ m/s^2. But we also know that acceleration is the time rate of change of velocity, which is in turn the time rate of change of distance.

$$a = \frac{dv}{dt} = \frac{d(ds/dt)}{dt} = \frac{d^2s}{dt^2}$$

Thus the motion of the falling body is described by the differential equation

$$\frac{d^2s}{dt^2} = -9.8 \qquad (1)$$

To find an expression for s as a function of t, we integrate Equation (1) twice with respect to t, each integration introducing an arbitrary constant of integration. Integrating once,

$$\frac{ds}{dt} = -9.8t + C_1$$

and then again

$$s = -\frac{9.8}{2}t^2 + C_1t + C_2 \qquad (2)$$

The relationship between s and t given by Equation (2) is a solution of the differential equation (1) because if we substitute from (2) into (1), making appropriate differentiations, the resulting equation is true. Actually, Equation (2), due to the presence of C_1 and C_2, represents a whole family of functions s. It is therefore called a *general* solution of (1); a *particular* solution of (1) is found by assigning specific values to C_1 and C_2. In order to determine meaningful specific values for C_1 and C_2, we need two further pieces of information, such as the velocity and position of the object at a specific value of t. ■

A differential equation that involves only derivatives with respect to a single independent variable is called an **ordinary differential equation**. If there is more than one independent variable, then the derivatives are partial derivatives and the equation is a **partial differential equation**. The **order** of a differential equation is the order of the highest-order derivative in the equation; thus if the equation contains a second derivative, but no higher-order derivatives, then it is a second-order differential equation. The **degree** of a differential equation is the power of the highest-order derivative.

EXAMPLE 2. Equation (1) is a second-order, ordinary differential equation of degree 1. (The highest-order derivative is a second derivative, and it appears to the first power.) ■

EXAMPLE 3. The equation $dy/dx = x^2 + 2y$ is an ordinary differential equation of order one and degree one. The equation

$$\frac{d^2y}{dx^2} = -4y\left(\frac{dy}{dx}\right)^3$$

is an ordinary differential equation of order two and degree one. (Remember that the degree is the power of the *highest-order deriv-*

ative; here the highest-order derivative is d^2y/dx^2, which appears only to the first power, even though dy/dx appears to the third power.) The equation

$$\frac{\partial z}{\partial x} = 2\frac{\partial^2 z}{\partial y^2} + \frac{\partial z}{\partial w}$$

is a partial differential equation of order two and degree one.   ■

In this chapter we learn some methods for solving differential equations; we consider only ordinary differential equations of the first degree, thereby simplifying the task considerably. You can appreciate why entire books are written on how to solve differential equations.

A **solution** of a differential equation is a relationship between the variables that satisfies the equation; that is, if we substitute from the relationship into the differential equation, the result is a true equation. As in Example 1, the solution may be expressed in terms of arbitrary constants. If the solution of an *n*th-order equation contains *n* arbitrary constants, it is called the **general solution**. If at least one of the constants is specified, the solution is a **particular solution**. A particular solution is thus an instance of the general solution with an extra condition or conditions that it must satisfy. If there are *n* arbitrary constants in the general solution, it requires *n* conditions to determine them; such conditions are called **boundary conditions**. An *n*th-order differential equation with *n* boundary conditions may not have a general solution at all, or it may have a general solution but no particular solution satisfying all *n* boundary conditions. In either case, we are unable to solve the problem as given.

We have really been solving simple differential equations in all our work on integration.

EXAMPLE 4.   Find $\int x^2 \, dx$.

This integration problem translates into "Find a function y such that $dy/dx = x^2$," which means "solve the differential equation $dy/dx = x^2$." The general solution of this first-order, first-degree equation, obtained by integration, is

$$y = \frac{x^3}{3} + C$$

A boundary condition might be

$$y = 5 \quad \text{when} \quad x = 2$$

Then

$$5 = \frac{8}{3} + C$$

$$C = \frac{7}{3}$$

The particular solution satisfying the given equation with the given boundary condition is

$$y = \frac{x^3}{3} + \frac{7}{3} \quad \blacksquare$$

The arbitrary constants in the general solution of a differential equation must be *independent*. We could not claim, for instance, that

$$y = 3x + c_1 + c_2$$

has two arbitrary constants, since c_1 and c_2 can be combined and the equation written as

$$y = 3x + c$$

with a single constant. On the other hand, the form of the general solution can sometimes be simplified by writing the arbitrary constants in a peculiar way—for instance, as $\ln c$ instead of c.

Although we do not yet know how to obtain solutions for any but the simplest differential equations, we can easily use substitution to determine whether a proposed solution is correct.

EXAMPLE 5. Is $y = e^x(c_1 + c_2 x)$ a general solution of the differential equation

$$y'' - 2y' + y = 0 \tag{3}$$

The differential equation is of order two, and the proposed solution does have two independent arbitrary constants. We need to substitute from the proposed solution into the given differential equation. To do this, we find y' and y''.

$$y = e^x(c_1 + c_2 x)$$
$$y' = c_2 e^x + (c_1 + c_2 x)e^x$$
$$y'' = c_2 e^x + c_2 e^x + (c_1 + c_2 x)e^x = 2c_2 e^x + (c_1 + c_2 x)e^x$$

Substituting into (3), we get

$$y'' - 2y' + y = 2c_2 e^x + (c_1 + c_2 x)e^x - 2[c_2 e^x + (c_1 + c_2 x)e^x]$$
$$+ e^x(c_1 + c_2 x) = 0$$

or

$$0 = 0$$

The differential equation is satisfied. $\quad \blacksquare$

EXAMPLE 6. Show that $y = e^x(5 - 3x)$ is a solution of the differential equation

$$y'' - 2y' + y = 0$$

subject to the boundary conditions

$$y = 5 \quad \text{when} \quad x = 0$$
$$y' = 2 \quad \text{when} \quad x = 0$$

From Example 5, we already know that $y = e^x(c_1 + c_2x)$ is a general solution of the differential equation. We need only to make sure that this particular solution satisfies the given boundary conditions.

$$y = e^x(5 - 3x) \qquad\qquad y(0) = e^0(5) = 5$$
$$y' = -3e^x + (5 - 3x)e^x \qquad y'(0) = -3e^0 + 5e^0 = -3 + 5 = 2$$

Both the boundary conditions are satisfied. ■

EXAMPLE 7. Show that $y = 3x^2 + 1$ is a solution of the differential equation

$$\frac{1}{2}xy' - y = -1$$

subject to the boundary condition

$$y = 1 \quad \text{when} \quad x = 0$$

To show that $y = 3x^2 + 1$ satisfies the differential equation, we find dy/dx and substitute.

$$y = 3x^2 + 1$$
$$y' = 6x$$
$$\frac{1}{2}xy' - y = \frac{1}{2}x(6x) - (3x^2 + 1) = -1$$
$$3x^2 - 3x^2 - 1 = -1$$
$$0 = 0$$

The differential equation is satisfied; next we check the boundary condition.

$$y = 3x^2 + 1$$
$$y(0) = 3(0) + 1 = 1$$

The given solution does satisfy the boundary condition. ■

Practice 1. Show that $y = e^{2x} + 1$ is a solution of the differential equation

$$y'' - 2y' = 0$$

subject to the boundary condition

$$y = 2 \quad \text{when} \quad x = 0$$

Exercises / Section **11-1**

Exercises 1–16: Show that the function is a general solution of the given differential equation.

1. $y = x + c; \quad \dfrac{dy}{dx} = 1$

2. $y = x^3 - c; \quad \dfrac{dy}{dx} = 3x^2$

3. $y = ce^x; \quad \dfrac{dy}{dx} = y$

4. $y = cx; \quad \dfrac{dy}{dx} = \dfrac{y}{x}$

5. $y = cx^2; \quad xy' - 2y = 0$

6. $y = \dfrac{c}{x}; \quad xy' = -y$

7. $y = \dfrac{c - \sin x}{x}; \quad xy' + y = -\cos x$

8. $y = c \ln x; \quad xy' \ln x = y$

9. $y = c_1 \sin x + c_2 \cos x; \quad \dfrac{d^2y}{dx^2} + y = 0$

10. $y = c_1 e^x + c_2 e^{2x}; \quad \dfrac{d^2y}{dx^2} - 3\dfrac{dy}{dx} + 2y = 0$

11. $y = c_1 + c_2 e^{4x} - \tfrac{3}{4}x; \quad y'' - 4y' = 3$

12. $y = \ln c_1 x + c_2; \quad x^2 y'' + y' - \dfrac{1}{x} = -1$

13. $xy = x^2 + c; \quad xy' + y = 2x$

14. $x^2 + y^2 = c; \quad x + yy' = 0$

15. $x^2 + y^2 = cx; \quad 2xyy' = y^2 - x^2$

16. $xy - x^2 = cx; \quad x^2 y' - x^2 = 0$

Exercises 17–22: Show that the function is a solution of the given differential equation subject to the boundary conditions.

17. $y = 2e^{x^2}; \quad \dfrac{dy}{dx} = 2xy; \quad y = 2$ when $x = 0$.

18. $y = 1 - 2e^{2x}; \quad -\dfrac{dy}{dx} + 2y = 2; \quad y = -1$ when $x = 0$.

19. $y = x^3 + 2x; \quad y'' - xy' + 3y = 10x; \quad y = 3$ when $x = 1$, $y' = 5$ when $x = 1$.

20. $y = \sin 2x + 3; \quad y'' + 4y = 12; \quad y = 3$ when $x = 0$, $y = 3$ when $x = \pi$.

21. $(1 - x)(1 + y) = 3; \quad (1 - x)y' = 1 + y; \quad y = 2$ when $x = 0$.

22. $y^3 + 3x^2 = 4x^{3/2}; \quad 2xy' + \dfrac{x^2}{y^2} = y; \quad y = 1$ when $x = 1$.

Exercises 23–28: Is the function a solution to the given differential equation and boundary conditions?

23. $y = x^4 + 2x^2; \quad xy' - 2(x^4 + y) = 0; \quad y = 2$ when $x = 1$.

24. $y \ln x = x; \quad x \ln x \dfrac{dy}{dx} + y - x = 0; \quad y = 1$ when $x = e$.

25. $y = \dfrac{3}{1 + x^2}; \quad (1 + x^2)y' + 2xy = 0; \quad y = \dfrac{3}{5}$ when $x = 2$.

26. $y = 4e^{4x} + 2; \quad y'' = 4y'; \quad y = 6$ when $x = 0$, $y' = 16$ when $x = 0$.

27. $(x^2 + 1)y^2 = 3$; $x^2y + (x^2 + 1)y' = 0$;
     $y = \sqrt{3}$ when $x = 0$.

28. $y = xe^x$; $y' - y = xe^x$; $y = e$ when $x = 1$.

Exercises 29–32: Find a value for c such that the general solution to the differential equation satisfies the boundary condition.

29. $y + 1 = c \sin x$; $y' \tan x = 1 + y$; $y = 0$
     when $x = \pi/2$.

30. $y = ce^{3x}$; $y' = 3y$; $y = 1$ when $x = 2$.

31. $xy^2 - y^3 = c$; $y + (2x - 3y)y' = 0$; $y = 2$
     when $x = 0$.

32. $x^3 = 3y^3 \ln cy$; $x^2y - (x^3 + y^3)y' = 0$;
     $y = e/2$ when $x = \sqrt[3]{\frac{3}{8}}\, e$.

33. The decay of a radioactive material can be expressed by the differential equation

$$\frac{dA}{dt} = -kA$$

where A is the amount of material present at time t and k is a constant. Show that the function

$$A = 2000e^{-kt}$$

is a solution to this differential equation satisfying the boundary condition $A = 2000$ when $t = 0$.

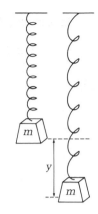

Figure 11-1

34. A body of mass m is hung on the end of a vertical spring of spring constant k; the body is pulled down and then released (see Figure 11-1). In the resulting *harmonic motion*, the displacement y of the body from its resting position is given by the differential equation

$$m \frac{d^2y}{dt^2} + ky = 0$$

Show that the function

$$y = \cos \sqrt{\frac{k}{m}}\, t$$

is a solution to this differential equation satisfying the boundary conditions $y = 1$ when $t = 0$ and $y' = 0$ when $t = 0$.

11-2   Separation of Variables

We said that we shall consider solution methods only for ordinary differential equations of the first degree. Until Section 11-6, we limit ourselves still further and consider only first-order, first-degree equations. Such an equation contains the first derivative to the first power and has the general form

$$\frac{dy}{dx} = f(x, y) \tag{1}$$

where $f(x, y)$ is some function of x and y. If $f(x, y)$ is a function of x alone, we have

$$\frac{dy}{dx} = g(x)$$

and the solution is found by integration,

$$y = \int g(x)\, dx$$

In general, however, $f(x, y)$ is more complex; for example,

$$\frac{dy}{dx} = \frac{\sin x + y}{x^2 + y^2}$$

This equation could be written in differential form as

$$(\sin x + y)\, dx + (-x^2 - y^2)\, dy = 0$$

Similarly, an equivalent form for Equation (1) is

$$M(x, y)\, dx + N(x, y)\, dy = 0 \tag{2}$$

A solution to (2) must also involve integration. If the function $M(x, y)$ involves both x and y, we cannot find $\int M(x, y)\, dx$, but if M involves x alone, perhaps we can do the integration. Similarly, we can hope to find $\int N(x, y)\, dy$ only if $N(x, y)$ involves y alone.

EXAMPLE 1. Solve the differential equation $x^2\, dx + y\, dy = 0$.
Here M, the coefficient of dx, is a function only of x, and N, the coefficient of dy, is a function only of y. Integrating, we get

$$\frac{x^3}{3} + \frac{y^2}{2} = c$$

where c is the constant of integration. This gives us the general solution to the original equation, which can be checked by differentiating (implicitly) and substituting into that equation. ∎

Our goal, then, is to try to "separate the variables," and write the given first-order, first-degree equation in the form

$$M(x)\, dx + N(y)\, dy = 0$$

Then we solve by integrating each term.*

Practice 1. Solve the differential equation $\dfrac{dx}{x} + e^y\, dy = 0$.

---

*Formally, we can write the equation $M(x)\, dx + N(y)\, dy = 0$ as $M(x) + N(y)\, dy/dx = 0$ and then, because y is a function of x, we can integrate throughout with respect to x:

$$\int M(x)\, dx + \int N(y) \frac{dy}{dx}\, dx = 0$$

However, the same result is obtained by writing

$$\int M(x)\, dx + \int N(y)\, dy = 0$$

and integrating with respect to x and with respect to y, in turn.

EXAMPLE 2.   Solve the differential equation $dx + \sqrt{1 - x^2}\, y\, dy = 0$.

In order to separate the variables, we divide both sides of the equation by $\sqrt{1 - x^2}$, giving $dx/\sqrt{1 - x^2} + y\, dy = 0$. Integrating gives

$$\text{Arcsin } x + \frac{y^2}{2} = c \quad \blacksquare$$

Word of Advice

To use separation of variables to solve a first-order, first-degree equation, you must be able to write the equation so that the coefficient of dx is a function of x alone and the coefficient of dy is a function of y alone.

EXAMPLE 3.   Solve the differential equation $(y - 1)\, dx + \cot x\, dy = 0$.

We separate the variables by dividing through by $(y - 1) \cot x$ and write

$$\frac{dx}{\cot x} + \frac{dy}{y - 1} = 0$$

or

$$\tan x\, dx + \frac{dy}{y - 1} = 0$$

Integrating gives

$$-\ln \cos x + \ln(y - 1) = c \tag{3}$$

Because c is an arbitrary constant, we can also write it as $\ln c_1$. The equation then simplifies using the rules of logarithms.

$$-\ln \cos x + \ln(y - 1) = \ln c_1$$

$$\ln \frac{y - 1}{\cos x} = \ln c_1$$

or

$$\frac{y - 1}{\cos x} = c_1$$

$$y = c_1 \cos x + 1 \tag{4}$$

While Equations (3) and (4) do not look similar, both represent general solutions to the given differential equation; Equation (4) looks a bit more tidy. $\blacksquare$

EXAMPLE 4. Solve the differential equation $(y - 1)\, dx + \cot x\, dy = 0$ subject to the boundary condition $y = 3$ when $x = \pi/4$.

We already have (3) as a general solution; we substitute the boundary condition values in (3) in order to determine c.

$$-\ln \cos\left(\frac{\pi}{4}\right) + \ln(3 - 1) = c$$

$$-\ln \frac{1}{\sqrt{2}} + \ln 2 = c$$

$$\ln 2\sqrt{2} = c$$

The particular solution is

$$-\ln \cos x + \ln(y - 1) = \ln 2\sqrt{2}$$

which can be written

$$\ln \frac{y - 1}{\cos x} = \ln 2\sqrt{2}$$

or

$$\frac{y - 1}{\cos x} = 2\sqrt{2}$$

$$y = 2\sqrt{2} \cos x + 1$$

But (4) is also a general solution to the given equation. Substituting the boundary condition values in (4) in order to determine c_1, we get

$$3 = c_1 \cos\left(\frac{\pi}{4}\right) + 1$$

$$3 = \frac{c_1}{\sqrt{2}} + 1$$

$$c_1 = 2\sqrt{2}$$

and the particular solution is still

$$y = 2\sqrt{2} \cos x + 1$$

We see, then, that the choice of how to express the arbitrary constant in the general solution does not affect the correctness of a particular solution. ∎

Practice 2. Solve the differential equation $xy^2\, dx + (1 - x^2)\, dy = 0$.

EXAMPLE 5. Solve the differential equation $\sin x \cos^2 y\, dx + \cos^2 x\, dy = 0$ subject to the boundary condition $y = \pi/4$ when $x = 0$.

We separate the variables by dividing through by $\cos^2 x \cos^2 y$

$$\frac{\sin x}{\cos^2 x}\, dx + \frac{1}{\cos^2 y}\, dy = 0$$

or, using trigonometric identities,

$$\tan x \sec x \, dx + \sec^2 y \, dy = 0$$

Integrating each term,

$$\sec x + \tan y = c$$

This is the general solution; we find the particular solution by substituting values from the boundary condition in order to evaluate c.

$$\sec 0 + \tan \frac{\pi}{4} = c$$

$$1 + 1 = c$$

$$2 = c$$

The particular solution is

$$\sec x + \tan y = 2 \quad \blacksquare$$

Practice 3.   Solve the differential equation $e^x \, dx + y^2 e^{2x} \, dy = 0$ subject to the boundary condition $y = 2$ when $x = 0$.

Exercises / Section 11-2

Exercises 1–20: Solve the given differential equation.

1. $\dfrac{dx}{x} + y^3 \, dy = 0$

2. $(x^2 + 2) \, dx - (1 + y) \, dy = 0$

3. $\dfrac{dx}{1 + x^2} - \sec^2 y \, dy = 0$

4. $\sin^2 x \cos x \, dx + y e^y \, dy = 0$

5. $yy' + xy^3 = 0$

6. $x^2 y' + (1 + y^2) = 0$

7. $(1 + x) \, dy - (1 - y) \, dx = 0$

8. $(x^2 + 1) y \, dy + (y^2 - 3) x \, dx = 0$

9. $\dfrac{e^{x^2}}{y^2} \, dx + \dfrac{dy}{x} = 0$ 10. $e^{x+y} \, dx + dy = 0$

11. $(xy^2 + x) \, dx + (x^2 y - y) \, dy = 0$

12. $(x + x \sin y) \, dx + x^2 \cos y \, dy = 0$

13. $y \, dx + x \, dy = xy(dy - dx)$

14. $x^2 \, dx - y \, dy = 2 \, dx - e^y \, dy$

15. $e^{(\sin x) - y} \, dx + \dfrac{dy}{\cos x} = 0$

16. $\sec y (\sin^2 x - 1) \, dx + \sin y \sin^2 x \, dy = 0$

17. $3e^x \tan y \, dx = (1 - e^x) \sec^2 y \, dy$

18. $(y^2 + 1) \, dx + xy(x + 1) \, dy = 0$

19. $\dfrac{dx}{e^{y^2}} + y\sqrt{1 - x^2} \, dy = 0$

20. $x \cos y \, dx + \sec y \sec x \, dy = 0$

Exercises 21–26: Solve the differential equation subject to the boundary condition shown.

21. $x^2 + (y + 1) y' = 0$; $y = 4$ when $x = 0$

22. $e^{x+y} = y'$; $y = 2$ when $x = 0$

23. $dy = \dfrac{4y}{x(y - 1)} \, dx$; $y = 1$ when $x = e$

24. $(1 + 3y) \, dx + (4 + x^2) \, dy = 0$; $y = 0$ when $x = 2$

25. $\sin x \, dx + \cos x \, dy = y \sin x \, dx$; $y = 2$ when $x = \pi$

26. $1 - xy' = 2(y + y')$; $y = 0$ when $x = 4$

27. The inner surface of a wall is kept at one temperature and the outer surface at another. The temperature T within the wall varies as a function of the distance x from the inner surface. If k is the constant of thermal conductivity of the material of which the wall is made and Q is the constant heat loss per hour through 1 m^2 of the wall, then T and x satisfy the differential equation

$$-k\frac{dT}{dx} = Q$$

Find an equation relating T and x if $T = 30°C$ when $x = 0$ cm.

28. A person's circulatory system is sometimes tested by injecting a harmless dye into the bloodstream near the heart. If the volume of blood the heart contains is a constant V and the heart pumps out blood at a constant rate r, then the amount y of dye contained in the heart cavity at time t satisfies the differential equation

$$dy = -\frac{r}{V}\, y\, dt$$

Find an equation relating y and t if the amount of dye in the heart cavity is k cubic centimeters when $t = 0$ min.

11-3 Regrouping to Advantage

Not every differential equation can be solved by separation of variables, and we need to develop other solution techniques. The technique of this section cannot be systematically applied, however; in fact, we might call this the "luck-out" method. It involves recognizing certain combinations of terms as the differentials of certain expressions and hence being able to integrate those terms directly. Of course, in a broad sense, this is all that is involved in solving any differential equation. We expect to be able to recognize such combinations only in a limited number of cases. For example, we should be able to recognize the left side in each equation below as the differential of the expression shown on the right.

$$x\, dy + y\, dx = d(xy) \tag{1}$$
$$2(x\, dx + y\, dy) = d(x^2 + y^2) \tag{2}$$
$$\frac{x\, dy - y\, dx}{x^2} = d\left(\frac{y}{x}\right) \tag{3}$$
$$\frac{y\, dx - x\, dy}{y^2} = d\left(\frac{x}{y}\right) \tag{4}$$

Therefore, $x\, dy + y\, dx$ is du for some u (namely, $u = xy$), and when this expression appears we would look for some integrable combination, such as du/u, $u^n\, du$, or $\sin u\, du$. The same applies if any of the other expressions appear.

EXAMPLE 1. Solve the differential equation $x^2 y\, dy + xy^2\, dx + dx = 0$.

If we write the equation in the form $x^2 y\, dy + (xy^2 + 1)\, dx = 0$, we see that we cannot separate the variables. However, if we rewrite it in the form

$$xy(x \, dy + y \, dx) + dx = 0$$

we see that the first term has the form $u \, du$. Integrating, we get

$$\frac{(xy)^2}{2} + x = c \quad ∎$$

EXAMPLE 2. Solve the differential equation $x \, dx + y \, dy + x^2 y^2 \, dy + y^4 \, dy = 0$.

Again, we first note that we cannot separate the variables. But we recognize the first two terms as part of Equation (2) above—that is, it is $d(x^2 + y^2)$ except for the 2. We therefore look for $x^2 + y^2$ elsewhere in the equation. Regrouping, we get

$$(x \, dx + y \, dy) + y^2(x^2 + y^2) \, dy$$

Dividing by $x^2 + y^2$ results in

$$\frac{x \, dx + y \, dy}{x^2 + y^2} + y^2 \, dy = 0$$

The first term is now of the form du/u (except for a constant), and the second term is also integrable. The solution is

$$\frac{1}{2} \ln(x^2 + y^2) + \frac{y^3}{3} = c_1$$

or, multiplying by 6,

$$3 \ln(x^2 + y^2) + 2y^3 = c$$

Here $c = 6c_1$, although both are arbitrary constants.   ∎

Practice 1. Solve the differential equation $(y - 2x) \, dx + x \, dy = 0$.

We shall develop a list of solution techniques for solving differential equations, much as we had a list of integration rules. When confronted with a differential equation to solve, we start at the top of the list and run down it to look for an appropriate method. Some methods apply only to certain differential equation types; thus if you have a second-order equation, a solution method for first-order equations will not apply, just as the trigonometric integration patterns cannot be applied to integrals that do not contain trigonometric functions.

Separation of variables should be first on your list of solution methods and should always be tried on a first-order, first-degree equation before going on to other methods.

Word of Advice

On a first-order, first-degree differential equation, always try separation of variables first.

EXAMPLE 3. Solve the differential equation $x\,dy + y\,dx + xy^2\,dy = 0$.

We note the presence of $x\,dy + y\,dx$ and look for a function involving xy. Dividing through by xy, we get

$$\frac{x\,dy + y\,dx}{xy} + y\,dy = 0$$

The first term is of the form du/u, and the second term is also integrable. The solution is

$$\ln xy + \frac{y^2}{2} = c$$

However, if we rewrite the original equation as

$$x(1 + y^2)\,dy + y\,dx = 0$$

we see that we can separate the variables. Doing this gives us

$$\left(\frac{1 + y^2}{y}\right)dy + \frac{dx}{x} = 0$$

or

$$\left(\frac{1}{y} + y\right)dy + \frac{dx}{x} = 0$$

Integrating each term,

$$\ln y + \frac{y^2}{2} + \ln x = c$$

or

$$\ln xy + \frac{y^2}{2} = c \quad\blacksquare$$

EXAMPLE 4. Solve the differential equation $y\,dx + (y^3 - x)\,dy = 0$.

This equation cannot be solved by separation of variables. However, we do detect the presence of $y\,dx - x\,dy$ in the equation. This may suggest use of Equation (4), or—except for a negative sign—Equation (3). After a bit of thought, we notice that we can rewrite the original equation as

$$\frac{y\,dx - x\,dy}{y^2} + y\,dy = 0$$

We can now integrate and get

$$\frac{x}{y} + \frac{y^2}{2} = c \quad\blacksquare$$

Practice 2. Solve the differential equation $x\,dx + y\,dy = \dfrac{x}{x^2+y^2}\,dx$.

Clearly it takes practice and perhaps trial and error to locate an integrable combination in a given differential equation. The method of the next section, when it applies, is more foolproof and dependable.

Exercises / Section **11-3**

Exercises 1–18: Solve the given differential equation.

1. $xy\,dy + y^2\,dx + dy = 0$

2. $\dfrac{y\,dx + x\,dy}{y^2} = dy$

3. $x\,dy - y\,dx = x^2\,dx$

4. $x^3\,dx + x^2y\,dy + y^2x\,dx + y^3\,dy = 0$

5. $y^2x\,dy + x^6\,dx - y^3\,dx = 0$

6. $(x^2 + y^2 + x)\,dx + y\,dy = 0$

7. $y^2x\,dx + x\,dx + x^2y\,dy + y\,dy = 0$

8. $y\,dy - x\,dx + (x^2 - y^2)\,dy = 0$

9. $x^3y\,dx + y^3x\,dx + y^2(x^2 + y^2)\,dy = dy$

10. $y\,dx - x\,dy = xy^2\,dy + y^3\,dx$

11. $e^{xy}(x\,dy + y\,dx) = dx$

12. $\sqrt{\dfrac{y}{x}}(x\,dy - y\,dx) = x^4\,dx$

13. $\sec^2(x^2 + y^2)(x\,dx + y\,dy) = x\,dy + y\,dx$

14. $x^2dy + xy\,dx = \csc(xy)\,dx$

15. $xy\,dx + x^2y\,dy - 2x\,dx + y\,dy = 0$

16. $y\,dy + x\,dx + xy(x\,dy + y\,dx) = 0$

17. $x\,dy + 2y\,dx + x(dx + dy) = 0$

18. $y\,dx - 2x\,dy = y(y^2\,dy - dx)$

Exercises 19–20: Solve the differential equation subject to the boundary condition shown.

19. $(x^2 + y^2 + x)\,dx + (x^2 + y^2 + y)\,dy = 0;\quad y = 1$ when $x = 1$.

20. $e^{x/y}(y\,dx - x\,dy) = y^2x^2\,dx + y^4\,dy;\quad y = 1$ when $x = 0$.

21. Find the equation of the curve whose slope is given by $1 - y/x$ and that passes through the point $(2, 2)$.

22. Find the equation of the curve whose slope is given by $y/(x + y^2)$ and that passes through the point $(0, 1)$.

11-4   **First-Order Linear Differential Equations**

Many first-order, first-degree differential equations that arise in practice can be written in the form

$$dy + Py\,dx = Q\,dx \tag{1}$$

where P and Q denote functions of x only. Such an equation is called a **linear differential equation of the first order**.

We multiply both sides of Equation (1) by the expression $e^{\int P\,dx}$. It is far from obvious why we should do this, but bear with us. The result is

$$e^{\int P\,dx}\,dy + e^{\int P\,dx}\,Py\,dx = e^{\int P\,dx}\,Q\,dx \tag{2}$$

The left side of Equation (2) is the differential of the expression $ye^{\int P\,dx}$. To see this, we take the derivative

$$\frac{d}{dx}\left(ye^{\int P\,dx}\right)$$

By the product rule, we get

$$y\,\frac{d}{dx}\left(e^{\int P\,dx}\right) + e^{\int P\,dx}\,\frac{dy}{dx}$$

By the exponential form,

$$ye^{\int P\,dx}\,\frac{d}{dx}\left(\int P\,dx\right) + e^{\int P\,dx}\,\frac{dy}{dx} \tag{3}$$

The expression $(d/dx)\left(\int P\,dx\right)$ says to take P (a function of x), integrate it with respect to x, and then differentiate the result with respect to x. This will get us back to P again, so (3) can be written

$$ye^{\int P\,dx}P + e^{\int P\,dx}\,\frac{dy}{dx}$$

Thus

$$\frac{d}{dx}\left(ye^{\int P\,dx}\right) = ye^{\int P\,dx}P + e^{\int P\,dx}\,\frac{dy}{dx}$$

or

$$d\left(ye^{\int P\,dx}\right) = ye^{\int P\,dx}P\,dx + e^{\int P\,dx}\,dy \tag{4}$$

By using Equation (4), we can rewrite Equation (2) as

$$d\left(ye^{\int P\,dx}\right) = e^{\int P\,dx}Q\,dx \tag{5}$$

Now we integrate both sides of Equation (5) with respect to x, giving

$$ye^{\int P\,dx} = \int e^{\int P\,dx}Q\,dx \tag{6}$$

Equation (6) represents the solution to the differential equation (1). In order to use this solution, however, we must be able to integrate both $\int P\,dx$ and $\int e^{\int P\,dx}Q\,dx$.

EXAMPLE 1. Solve the differential equation

$$dy + 3y \, dx = 20 \, dx$$

This is a linear differential equation because it matches the form of Equation (1), where $P = 3$ and $Q = 20$. We know that Equation (6) represents the solution; we first have to find $e^{\int P \, dx}$. Here

$$e^{\int P \, dx} = e^{\int 3 \, dx} = e^{3x}$$

Then Equation (6) becomes

$$ye^{3x} = \int e^{3x} \, 20 \, dx$$

or

$$ye^{3x} = \frac{20}{3}e^{3x} + c_1$$

which can be rewritten as

$$3y = 20 + ce^{-3x} \quad \blacksquare$$

EXAMPLE 2. Because of the fact that Q in the differential equation of Example 1 is a constant, that equation can also be solved by separation of variables. Separating the variables in

$$dy + 3y \, dx = 20 \, dx$$

we get

$$\frac{dy}{3y - 20} + dx = 0$$

Integrating gives

$$\frac{1}{3}\ln(3y - 20) + x = \ln c_1$$

or

$$\ln(3y - 20) + 3x - \ln c = 0$$

Rewriting, we have

$$\ln \frac{3y - 20}{c} = -3x$$

$$e^{-3x} = \frac{3y - 20}{c_1}$$

$$3y = 20 + ce^{-3x}$$

which is the same solution as before. $\quad \blacksquare$

In working with the expression $e^{\int P \, dx}$, we may run into something of the form $e^{\ln w}$. This can be simplified to w. To see this, let

$$v = e^{\ln w}$$

and take the logarithm of each side.

$$\ln v = \ln (e^{\ln w}) = \ln w \ln e = \ln w$$

From

$$\ln v = \ln w$$

we conclude that $v = w$, or $e^{\ln w} = w$.*

EXAMPLE 3. Solve the differential equation

$$dy + \frac{y}{x} \, dx = x^3 \, dx$$

This is a linear equation with $P = 1/x$ and $Q = x^3$. We find $e^{\int P \, dx}$.

$$e^{\int P \, dx} = e^{\int 1/x \, dx} = e^{\ln x} = x$$

The solution (from Equation 6) is

$$yx = \int x \cdot x^3 \, dx = \int x^4 \, dx = \frac{x^5}{5} + c_1$$

or

$$5yx = x^5 + c \quad \blacksquare$$

Practice 1. Solve the differential equation $dy + \frac{y}{x} \, dx = x \, dx$.

EXAMPLE 4. Solve the differential equation

$$x \, dy - 2y \, dx = x^3 e^{3x} \, dx$$

By dividing through by x, we can put the equation in the linear form of Equation (1).

$$dy - \frac{2}{x} y \, dx = x^2 e^{3x} \, dx$$

Here $P = -2/x$ and $Q = x^2 e^{3x}$.

$$e^{\int P \, dx} = e^{\int -2/x \, dx} = e^{-2 \ln x} = e^{\ln x^{-2}} = x^{-2}$$

Thus the solution is

$$yx^{-2} = \int x^{-2} \cdot x^2 e^{3x} \, dx$$

or

$$\frac{y}{x^2} = \int e^{3x} \, dx = \frac{1}{3} e^{3x} + c_1$$

which can be written

$$3y = x^2 e^{3x} + cx^2 \quad \blacksquare$$

*It is also true that $e^{m \ln w} = w^m$ and that $e^{n + (m \ln w)} = e^n w^m$. Can you prove these facts?

Practice 2.   Solve the differential equation $x \, dy + 3y \, dx = \dfrac{dx}{x}$.

EXAMPLE 5.   Solve the differential equation

$$y' + 2y = \sin x$$

Rewriting the equation as

$$\frac{dy}{dx} + 2y = \sin x$$

or

$$dy + 2y \, dx = \sin x \, dx$$

we see that it is linear, with $P = 2$ and $Q = \sin x$.

$$e^{\int P \, dx} = e^{\int 2 \, dx} = e^{2x}$$

The solution is

$$ye^{2x} = \int e^{2x} \sin x \, dx$$

The above integral is not a basic form. Consulting a table of integrals (in the table at the back of this book, Rule 63), we find

$$\int e^{2x} \sin x \, dx = \frac{e^{2x}}{5}(2 \sin x - \cos x) + c$$

(This integration can also be done by two applications of integration by parts.)

The solution to the differential equation is, therefore,

$$ye^{2x} = \frac{e^{2x}}{5}(2 \sin x - \cos x) + c \quad \blacksquare$$

Practice 3.   Solve the differential equation $dy = x \, dx - 3y \, dx$.

Word of Advice

To solve a linear differential equation of the first order, first write it in the exact form

$$dy + Py \, dx = Q \, dx$$

and then write P and Q specifically.

There are three conditions that must be met in order to find the solution of a differential equation by the method of this section.

1. The equation must fit the form $dy + Py\, dx = Q\, dx$.
2. The expression $\int P\, dx$ must be one we can integrate.
3. The expression $\int e^{\int P\, dx}\, Q\, dx$ must be one we can integrate.

EXAMPLE 6. Solve the differential equation

$$y' + 2y = e^{-x}$$

subject to the boundary condition $y = 3$ when $x = 0$.
Rewriting the equation in linear form,

$$dy + 2y\, dx = e^{-x}\, dx$$

Here $P = 2$ and $Q = e^{-x}$.

$$e^{\int P\, dx} = e^{\int 2\, dx} = e^{2x}$$

The general solution is

$$ye^{2x} = \int e^{2x} \cdot e^{-x}\, dx = \int e^x\, dx = e^x + c$$

or

$$ye^{2x} = e^x + c$$

We find the particular solution by using the values from the boundary condition.

$$3 \cdot e^0 = e^0 + c$$
$$3 = 1 + c$$
$$c = 2$$

The solution is

$$ye^{2x} = e^x + 2$$

or

$$y = e^{-x} + 2e^{-2x} \quad \blacksquare$$

Exercises / Section **11-4**

Exercises 1–18: Solve the given differential equation.

1. $dy + \dfrac{2y}{x}\, dx = x^2\, dx$

2. $dy - \dfrac{y}{x}\, dx = xe^x\, dx$

3. $x\, dy = (5y + x + 1)\, dx$

4. $x^2\, dy + (3xy - e^{x^2})\, dx = 0$

5. $dy + x\, dx + y \cot x\, dx = 0$

6. $dy + y \tan x\, dx - \cos^2 x\, dx = 0$

7. $xy' = 8xe^{2x} + y(2x - 1)$

8. $x^2 y' = 1 - y(2x + x^2)$

9. $x\, dy + y\, dx = xy\, dx$

10. $dy + yx\,dx = yx^2\,dx$

11. $y' = e^{-x} - y$

12. $y' = e^{-2x} - 2y$

13. $y' = e^x - y$

14. $e^x\,dy + 3ye^x\,dx = dx$

15. $y' = x^2(1 - 3y)$

16. $y' = x(1 - y)$

17. $dy + (y - x)\,dx = 0$

18. $dy + (2xy - x^3)\,dx = 0$

Exercises 19–22: Solve the differential equation subject to the boundary condition shown.

19. $x\,dy - y\,dx = 2x\,dx$; $y = 4$ when $x = 1$.

20. $2x\,dy + y\,dx = 6x^2\,dx$; $y = 1$ when $x = 4$.

21. $y' - 3y - e^x = 0$; $y = 2$ when $x = 0$.

22. $y' + 2yx - x = 0$; $y = e$ when $x = 1$.

Exercises 23–26: Any equation of the form of Equation (1) can be solved by the method of this section, even if the variable names are different. For example, an equation of the form $dx + Px\,dy = Q\,dy$, where P and Q are functions of y alone, has the solution $xe^{\int P\,dy} = \int e^{\int P\,dy} Q\,dy$. Solve the given differential equation.

23. $\dfrac{dx}{dy} + \dfrac{2x}{y} = \dfrac{y - 1}{y}$

24. $\dfrac{dx}{dy} + x - e^{-y}\sin y = 0$

25. $(r^4 + 2t)\,dr - r\,dt = 0$

26. $2\theta\,dz = (\theta^4 + z)\,d\theta$

27. The current i in a certain electric circuit behaves according to the differential equation $di/dt + 3i = 4e^{-2t}$. Find an equation relating i and t if $i = 0$ A when $t = 0$ ms.

28. A tank contains 50 gal of a mixture of water and 10 lb of dissolved granular herbicide. Then a mixture of 2 lb of herbicide per gallon of water is poured into the tank at the rate of 2 gal/min, while 1 gal/min of fluid is drained from the tank. The amount y of pounds of herbicide in the tank at time t satisfies the differential equation

$$dy = 4\,dt - \frac{y}{50 + t}\,dt$$

Find an equation for y as a function of t. (Remember that $y = 10$ when $t = 0$.)

11-5 Applications

First-order, first-degree differential equations arise in many areas of technology, as well as in fields such as physical and biological sciences, psychology, medicine, and economics. When we consider that such an equation basically expresses the rate at which one quantity changes with respect to a second quantity, we see that any two related, nonconstant quantities can be expected to satisfy such an equation. The tricky part is establishing just how the quantities are related—that is, finding just what the appropriate differential equation looks like.

We classify the applications that arise into three or four general types and then a catchall, "other" type.

TYPE 1

In the first type of application, we have a geometric situation in which we are, directly or indirectly, given the slope of a curve and asked to find its equation.

EXAMPLE 1. A given curve has slope everywhere equal to x/y and passes through the point $(2, 4)$. Find its equation.

The slope of the curve is dy/dx, and we therefore know that the curve satisfies the differential equation

$$\frac{dy}{dx} = \frac{x}{y}$$

We can separate the variables and integrate (choosing a convenient expression for the arbitrary constant).

$$y \, dy - x \, dx = 0$$
$$\frac{y^2}{2} - \frac{x^2}{2} = \frac{c}{2}$$

or

$$y^2 - x^2 = c$$

Using the boundary condition that $y = 4$ when $x = 2$, we can find the particular solution by evaluating c

$$16 - 4 = c$$
$$12 = c$$

The equation is

$$y^2 - x^2 = 12 \quad \blacksquare$$

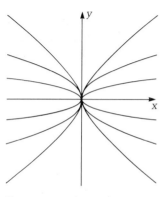

Figure 11-2

Practice 1. Find the equation of the curve whose slope is everywhere $2xy$ and that passes through the point $(2, 1)$.

There are many applications in which we are given a family of curves and we want to know the equation of a second family of curves such that every member of the second family intersects every member of the first family at right angles. The two families are called **orthogonal trajectories** of each other. For example, the orthogonal trajectories of the concentric circles of Figure 11-2 are straight lines radiating from the origin, some of which are shown in Figure 11-2. (See Exercise 7 at the end of this section.)

EXAMPLE 2. In a certain electric field, suppose that the *lines of force* are given by the family of parabolas $y^2 = cx$, shown in Figure 11-3(a). Find the equation of the *lines of equal potential*, which are the orthogonal trajectories of the lines of force.

From the equation $y^2 = cx$, we differentiate and get

$$2yy' = c$$
$$y' = \frac{c}{2y} \tag{1}$$

Figure 11-3   (a)

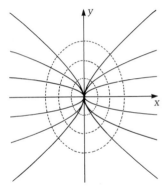

Figure 11-3   (b)

Using the original equation, we solve for c, getting $c = y^2/x$, and put this expression for c in Equation (1), so that

$$y' = \frac{y^2/x}{2y} = \frac{y}{2x}$$

This expression represents the slopes of the lines of force. The slopes of the orthogonal trajectories must be the negative reciprocal of this expression. Therefore the orthogonal trajectories satisfy the differential equation

$$\frac{dy}{dx} = -\frac{2x}{y}$$

Solving this equation by separation of variables, we get

$$y\,dy + 2x\,dx = 0$$

$$\frac{y^2}{2} + x^2 = \frac{c}{2}$$

$$y^2 + 2x^2 = c \tag{2}$$

Equation (2) represents the family of ellipses shown by dashed lines in Figure 11-3(b). Notice that each ellipse intersects each parabola at right angles.   ■

TYPE **2**

The second type of application concerns cases involving growth or decay of a substance. The assumption here is that the rate of the total growth (or loss) is proportional to the amount present; the more there is, the faster the total amount grows. This assumption can be used to model population growth, growth in a money supply invested at a compound rate, decay of radioactive material, and many other situations.

The basic differential equation here is of the form

$$\frac{dy}{dt} = ky$$

where k is a constant of proportionality. This equation is easily solved by separation of variables.

EXAMPLE 3.   A certain colony of bacteria grows at a rate proportional to the number present. If the number of bacteria doubles in 40 min, how long will it take to reach 20 times the original amount?

We begin with the differential equation

$$\frac{dy}{dt} = ky$$

and solve by separation of variables.

$$\frac{dy}{y} = k \, dt$$

$$\ln y = kt + \ln c$$

$$\ln y - \ln c = kt$$

$$\ln \frac{y}{c} = kt$$

$$\frac{y}{c} = e^{kt}$$

$$y = ce^{kt}$$

We denote the initial amount present by y_0; thus $y = y_0$ when $t = 0$. Putting this in the above equation, we get

$$y_0 = ce^0$$

so $c = y_0$ and the solution is

$$y = y_0 e^{kt}$$

Into this equation we put the condition that $y = 2y_0$ when $t = 40$.

$$2y_0 = y_0 e^{k40} = y_0(e^k)^{40}$$

or

$$2 = (e^k)^{40}$$

$$e^k = (2)^{1/40}$$

Thus the solution to the differential equation is

$$y = y_0(2^{1/40})^t = y_0 2^{t/40}$$

The original question asked for the value of t when $y = 20y_0$. Using $y = 20y_0$ in the last equation, we get

$$20y_0 = y_0 2^{t/40}$$

$$20 = 2^{t/40}$$

$$\ln 20 = \left(\frac{t}{40}\right) \ln 2$$

$$t = 40 \frac{\ln 20}{\ln 2} = 172.88 \text{ min} = 2.88 \text{ h} \quad \blacksquare$$

The equation $y = y_0 e^{kt}$ is always the form of the solution to the Type 2 differential equation.

Practice 2. Money invested at an annual interest rate of r percent and compounded continuously satisfies the differential equation

$$\frac{dy}{dt} = \frac{ry}{100}$$

If \$500 is invested at 11% compounded continuously, what will be the balance at the end of four years?

TYPE 3

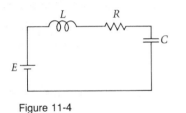

Figure 11-4

The third application results from applying Kirchhoff's law, which says that the voltage supplied to a closed electric circuit equals the sum of all the voltage drops around the circuit. In the circuit of Figure 11-4, which contains a voltage supply of E volts, an inductor with inductance of L henrys, a resistor with resistance of R ohms, and a capacitor with capacitance of C farads, application of Kirchhoff's law results in the differential equation

$$L\frac{di}{dt} + Ri + \frac{q}{C} = E \tag{3}$$

where i is the current and q is the charge on the capacitor. There are a number of variations on Equation (3). In a circuit with no capacitor (an RL circuit), Equation (3) reduces to

$$L\frac{di}{dt} + Ri = E$$

which is a linear differential equation. In a circuit with no inductor (an RC circuit), Equation (3) becomes

$$Ri + \frac{q}{C} = E$$

and, because $i = dq/dt$, this can be written

$$R\frac{dq}{dt} + \frac{q}{C} = E$$

which is linear in q. Using $i = dq/dt$ in Equation (3) results in

$$L\frac{d^2q}{dt^2} + R\frac{dq}{dt} + \frac{q}{C} = E$$

which is a second-order differential equation. We consider it again in Section 11-10.

EXAMPLE 4. In an RL circuit with $E = 20$ V, $R = 20\ \Omega$, and $L = 2$ H, the current is initially zero. Find the value of the current at $t = 0.1$ s.

The equation for an RL circuit is

$$L\frac{di}{dt} + Ri = E$$

For this particular circuit, the equation becomes

$$2\frac{di}{dt} + 20i = 20$$

or

$$di + 10i\ dt = 10\ dt$$

The solution to this linear equation is

$$ie^{\int P \, dt} = \int e^{\int P \, dt} Q \, dt$$

where $P = Q = 10$. We get

$$ie^{10t} = \int 10e^{10t} \, dt = e^{10t} + c$$

From the condition $i = 0$ when $t = 0$, we can evaluate c.

$$0e^0 = e^0 + c$$
$$c = -1$$

The particular solution is

$$ie^{10t} = e^{10t} - 1$$

We now find the value for i when $t = 0.1$ s.

$$ie^{10(.1)} = e^{10(.1)} - 1$$
$$ie = e - 1$$
$$i = \frac{e - 1}{e} = 0.632 \text{ A} \quad \blacksquare$$

Practice 3. In a particular RC circuit, $R = 10 \ \Omega$, $C = 0.003$ F, $E = 50$ V. Assuming that the initial charge on the capacitor in 0.5 C, find an expression for the charge as a function of time.

TYPE 4

Type 4 applications are those for which two opposing forces are at work. Examples of this situation occur with moving bodies subject to a propelling force and a resisting force, mixtures where substances flow into, as well as out of, the mixture, and in general, changes in some quantity y where one condition exists to increase y and another exists to decrease it, so that dy/dt is a difference of two expressions.

EXAMPLE 5. According to Newton's laws of motion, a body of mass m moving under the influence of a force F obeys the equation

$$ma = F \tag{4}$$

where $a = dv/dt$, the acceleration of the body. For a falling body moving only under the influence of gravity, the sole force acting upon the body is its weight W, where $W = mg$, g being the gravitational constant. If we take the positive direction to be upward, then $g = -9.8$ m/s^2. Equation (4) becomes

$$m \frac{dv}{dt} = m(-9.8)$$

or

$$\frac{dv}{dt} = -9.8$$

This is the equation we used in Section 7-1, but it is a special case of the situation with which we deal now. Suppose that we consider air resistance acting upon the falling body. This force acts in the opposite direction from the weight of the object and is generally proportional to the velocity of the object. The force F in Equation (4) now has two terms, and Equation (4) becomes

$$m \frac{dv}{dt} = -9.8m + kv \tag{5}$$

If a body of mass 2 kg is dropped from rest and meets a resisting force equal to half the velocity, find an expression for velocity as a function of time.

Using Equation (5), we write

$$2 \frac{dv}{dt} = -9.8(2) + 0.5v$$

Separating variables and integrating,

$$\frac{dv}{-9.8 + 0.25v} = dt$$

$$\frac{1}{0.25} \ln(-9.8 + 0.25v) = t + c_1$$

$$\ln\left(\frac{-9.8 + 0.25v}{c}\right) = 0.25t$$

$$-9.8 + 0.25v = ce^{0.25t}$$

Using $v(0) = 0$, we evaluate c.

$$-9.8 = ce^0$$

$$c = -9.8$$

Finally,

$$-9.8 + 0.25v = -9.8e^{0.25t}$$

$$0.25v = 9.8(1 - e^{0.25t})$$

$$v = 39.2(1 - e^{0.25t}) \text{ m/s} \quad \blacksquare$$

Practice 4.   An object of mass 5 kg is thrown upward with an initial velocity of 25 m/s. It encounters air resistance equal to $0.8v$. Find an expression for velocity as a function of time.

EXAMPLE 6. A tank contains 50 gal of a mixture of water and 10 lb of dissolved granular herbicide. Then a mixture of 2 lb of herbicide per gallon of water is poured into the tank at the rate of 2 gal/min, while 1 gal/min of fluid is drained from the tank (this is Exercise 28, Section 11-4). Find an expression for the amount y of herbicide in the tank as a function of time.

Here dy/dt is influenced by two opposing "forces," the influx of more herbicide, which tends to increase y, and the draining of fluid, which tends to decrease y. The influx of herbicide occurs at the rate of (2 lb/gal)(2 gal/min) = 4 lb/min. The amount y of herbicide in the tank decreases as the 1 gal/min of fluid is drained from the tank. The amount of herbicide in each gallon drained is given by the ratio

$$\frac{\text{amount of herbicide in tank}}{\text{number of gallons in tank}}$$

The amount of herbicide in the tank is y. The number of gallons in the tank, originally 50, changes at the rate of 2 gal put in minus 1 gal taken out per minute. Therefore the rate of decrease of herbicide due to draining is

$$\left[\frac{y}{50 + (2 - 1)t} \text{ lb/gal}\right](1 \text{ gal/min}) = \frac{y}{50 + t} \text{ lb/min}$$

Finally, the rate of change of y is given by the difference of these two rates, so the differential equation is

$$\frac{dy}{dt} = 4 - \frac{y}{50 + t}$$

This can be written as

$$dy + \frac{1}{50 + t} y \, dt = 4 \, dt$$

which is a linear differential equation. The general solution is

$$y e^{\int 1/(50+t) \, dt} = \int e^{\int 1/(50+t) \, dt} 4 \, dt$$

or

$$y(50 + t) = 200t + 2t^2 + c$$

Using the condition that $y = 10$ when $t = 0$, the particular solution is

$$y(50 + t) = 200t + 2t^2 + 500$$

or

$$y = \frac{200t + 2t^2 + 500}{50 + t} \text{ lb} \quad \blacksquare$$

There are numerous other differential equations that model physical and biological events.

EXAMPLE 7. A tank with constant cross-sectional area A contains a small hole in the bottom with cross-sectional area B. Liquid in the tank flows out through the hole, causing the height h of the liquid in the tank to decrease. The rate of change of h is given by the differential equation

$$\frac{dh}{dt} = -\frac{B}{A}\sqrt{2gh}$$

where g is the absolute value of the gravitational constant. For a tank with $A = 0.1$ m^2, $B = 0.0005$ m^2, and initial height of 0.25 m, find the height of the liquid after 30 s.

For our particular case, the differential equation is

$$\frac{dh}{dt} = -0.005\sqrt{19.6h}$$

or, in separated form,

$$\frac{dh}{\sqrt{19.6h}} = -0.005\ dt$$

Integrating gives

$$\frac{1}{19.6}(19.6h)^{1/2} \cdot 2 = -0.005t + c_1$$

or

$$h^{1/2} = -0.011t + c$$

Using the condition that $h = 0.25$ when $t = 0$, the value of c is 0.5. The solution is thus

$$h^{1/2} = -0.011t + 0.5$$

and the value of h when $t = 30$ is 0.029 m.   ■

Exercises / Section 11-5

1. Find the equation of the curve whose slope is everywhere $1 - x$ and that passes through the point $(1, 1)$.

2. Find the equation of the curve whose slope is everywhere $(1 + y)/x$ and that passes through the point $(1, 2)$.

3. Find the equation of the curve whose slope is everywhere $1 + y/x$ and that passes through the point $(1, 4)$.

4. Find the equation of the curve whose slope is everywhere $e^{-x} - y$ and that passes through the point $(1, 1/e)$.

5. Find the equation of the curve whose slope at any point equals the sum of the coordinates of the point and that passes through the point (0, 2).

6. Find the equation of the curve whose slope at any point equals the ratio of the ordinate of the point to one more than the abscissa of the point and that passes through (1, 2).

7. Find the equation of the orthogonal trajectories of the curves $x^2 + y^2 = c$, which are concentric circles with centers at the origin.

8. Find the equation of the orthogonal trajectories of the curves $y = ce^x$.

9. In a certain electric field the lines of force are given by the curves $x^2 - y^2 = c$. Find the equation of the lines of equal potential, which are the orthogonal trajectories of the lines of force.

10. In a certain two-dimensional fluid flow, the *streamlines* are given by the curves $y = x^3 + c$. Find the equation of the *equipotential lines*, which are the orthogonal trajectories of the streamlines.

11. On a particular weather map, the *isobars* (lines connecting points of equal barometric pressure) are given by the equation $y = \tan x + c$. Find the equation of the orthogonal trajectories, which indicate the direction of wind from high to low pressure areas.

12. On a particular weather map, the *isotherms* (lines connecting points of equal temperature) are given by the equation $y = ce^{-2x} + x$. Find the equation of the orthogonal trajectories, which indicate the direction of heat flow.

13. A bacteria culture grows at a rate proportional to the amount of bacteria present. If there are 100,000 bacteria present at the start of an experiment and there are three times that many at the end of 2 h, how many are there at the end of 7 h?

14. A certain amount of money is invested at an annual rate of 9% compounded continuously. At the end of 5 years, $1600 is in the account. To the nearest dollar, what was the amount of the original investment?

15. Radioactive substances decay at a rate proportional to the amount present. The half-life (the time required for half of the original amount to decay) for radium is 1600 years.

What percentage of the original amount remains after 2500 years?

16. A wet material loses moisture by evaporation at a rate proportional to the amount of moisture present. If a material loses one-third of its moisture content in 1 h, how long will it take to lose one-half of its moisture content?

17. The electrical charge on a spherical surface leaks off at a rate proportional to the charge present. If 4 C of charge is present after 15 min and 3 C is present after 30 min, what was the amount of the original charge?

18. Penicillin leaves a patient's body at a rate proportional to the amount present. If three-quarters of the original dosage is still in the patient's body after 4 h, how long will it take before the original dosage is half gone?

19. Show that in any *RL* circuit with constant E, $\lim_{t \to \infty} i = E/R$.

20. In an *RL* circuit with $E = 100$ V, $R = 40\ \Omega$, $L = 0.03$ H, and $i(0) = 0$, find the time required for the current to reach 99% of its maximum value.

21. In an *RL* circuit with $E = 110$ V, $R = 10\ \Omega$, $L = 3.2$ H and $i(0) = 0$, find an expression for the current as a function of time.

22. In an *RL* circuit with $E = 4$ V, $R = 2\ \Omega$, $L = 0.06$ H, and $i(0) = 6$, find the value of the current at $t = 0.01$ s.

23. If the initial charge on the capacitor in an *RC* circuit is zero, find an expression for the charge as a function of time if the voltage is 6 V, the resistance is $20\ \Omega$, and the capacitance is 0.01 F.

24. If the initial charge on the capacitor in an *RC* circuit is 0.25 C, find the charge at $t = 0.2$ s if $E = 110$ V, $R = 2\ \Omega$, and $C = 0.04$ F.

25. In a particular *RC* circuit, $R = 10\ \Omega$, $C = 0.01$ F, and $E = 200e^{-2t}$ V. If the initial charge on the capacitor is 0.25 C, find the charge at $t = 1$ s.

26. In a particular *RL* circuit, $R = 40\ \Omega$, $L = 2$ H, and $E = 50 \cos 2t$ V. Find an expression for the current as a function of time if the current is initially zero.

27. An object of mass 30 kg is dropped with an initial velocity of -20 m/s. It encounters air resistance equal to $0.2v$. Find an expression for velocity as a function of time.

28. An object of mass 20 kg is dropped from rest. It experiences air resistance equal to the velocity. Find the velocity at $t = 3$ s.

29. A boat of mass 200 kg is at rest when its engine is started. The engine supplies a force of 50 N to propel the boat forward, but the water exerts a resisting force equal to twice the velocity of the boat. Find the velocity of the boat after 5 min. (Note that 1 N is the force required to give a mass of 1 kg an acceleration of 1 m/s^2, so the unit of time is seconds.)

30. An object of mass 2 kg is being propelled at a velocity of 0.3 m/s by a force of 5 N when it encounters frictional resistance equal to $0.8v$. Find an expression for velocity as a function of time.

31. A tank contains 100 gal of brine, in which 40 lb of salt is dissolved. Pure water flows into the tank at the rate of 2 gal/min, while mixture drains from the tank at the same rate. Find the amount of salt in the tank after 10 min.

32. A vessel holds 3 L of a mixture of water and 50 g of sugar. A mixture of 20 g of sugar per liter of water flows into the vessel at the rate of 0.03 L/s, while fluid is drained from the vessel at the rate of 0.01 L/s. What amount of sugar is in the vessel after 3 min?

33. Newton's law of cooling says that the rate of change of temperature T of an object is proportional to the difference between T and the temperature of the surrounding medium. Water is heated to the boiling point, 100°C. When taken off the heat and left at room temperature of 24°C, it cools to 90°C in 5 min. Find the temperature of the water after 15 min.

34. The price P of a certain product changes at a rate proportional to the difference between the demand and the supply. Suppose that the demand is given by the expression $200 - 0.1\,P$ and that the supply is a constant 500. If the price of the product is originally \$20 and the price at the end of one month is \$24, find the price (to the nearest dollar) at the end of 5 months.

35. In a chemical reaction where one molecule of substance Z is formed from one molecule each of substances X and Y, the rate of the reaction is proportional to the product of the amounts of X and Y present. If z denotes the amount of substance Z, then

$$\frac{dz}{dt} = k(x_0 - z)(y_0 - z)$$

where x_0 and y_0 are the initial amounts of X and Y. If x_0 and y_0 are, respectively, 40 g and 70 g and 20 g of Z is formed in 30 min, what quantity of Z has been formed in 1 h?

36. The volume V of a melting snowball changes at a rate proportional to $V^{2/3}$. If the snowball's original volume was 0.002 m^3 and after 10 min, one-third of the volume has melted, find the time required for the snowball to melt away completely.

37. Assume that the concentration y of dissolved oxygen in a moving stream is governed by the differential equation

$$\frac{dy}{dt} = k_1 L - k_2 y$$

where k_1 and k_2 are constants and $L = L_0 e^{-k_1 t}$ is the biochemical oxygen demand at time t. Find an expression for y as a function of time, given that $y(0) = y_0$.

38. The Volterra model of the relationship between a predator and a prey leads to the differential equation

$$\frac{dy}{dx} = \frac{y(k_1 + k_2 x)}{x(s_1 + s_2 y)}$$

where k_1, k_2, s_1, and s_2 are constants. Find the general solution.

39. In a diffraction grating, used to measure wave lengths, the differential equation

$$\frac{d\theta}{d\lambda} = \frac{m}{k \cos \theta}$$

arises, where m and k are constants. Find the general solution.

40. Under certain conditions, the rate of change of pressure p of a gas with respect to the volume v is directly proportional to the pressure and inversely proportional to the volume. Find an equation relating p and v.

11-6 Approximation Techniques

Finding the solution to a differential equation involves integration in one way or another. We already know that not every function is integrable. Therefore it should not surprise us that there are differential equations that cannot be solved "in closed form," meaning that we cannot come up with an equation describing a function that is the solution. In such cases we must resort to approximation techniques. Some of these techniques produce functions that approximate the solution function over some interval. Other techniques do not produce an actual function, but they do produce values that are approximations to the values of the solution function. A characteristic of approximation techniques is that some operation or series of steps must be repeated many times; generally, the more repetitions, the closer the approximation. This makes approximation techniques in which arithmetic computations are involved very suitable to be carried out by computer or calculator. We study three approximation techniques in this section.

The first technique is known as **Picard's method,** after the French mathematician Charles Picard (1856–1941). Suppose that we have a first-order, first-degree differential equation of the form

$$\frac{dy}{dx} = f(x, y) \tag{1}$$

with boundary condition $y(x_0) = y_0$. Picard's method generates a sequence of functions, each a better approximation to the solution function itself in an interval around x_0. The first approximation is $y = y_0$, a constant function. (We already know that y takes on the value y_0 when $x = x_0$, but in this first approximation, we are assuming that y takes on this value everywhere.) Letting $y = y_0$ in the differential equation (1), we get

$$\frac{dy}{dx} = f(x, y_0) \tag{2}$$

Because y_0 is a constant, $f(x, y_0)$ in Equation (2) is a function only of x. If we integrate both sides of (2) and solve for the constant of integration by using the given boundary condition, we get a new expression for y, which we call y_1.

Now the process repeats. We substitute y_1 for y in the original differential equation (1), integrate, and get still another function y_2. Each successive function, $y_1, y_2, y_3, \ldots$, more closely approximates the exact solution. The closer we wish the approximation to be, the more times we repeat the process.

EXAMPLE 1. Use Picard's method to find four approximations to the solution of the differential equation $dy/dx = -xy$ subject to the boundary condition $y = 1$ when $x = 0$.

The first approximation is $y_0 = 1$. Putting this value into the differential equation results in

$$\frac{dy}{dx} = -x$$

Integrating, we get

$$y_1 = -\frac{x^2}{2} + c$$

Using the boundary condition $y = 1$ when $x = 0$, we evaluate c and get

$$y_1 = -\frac{x^2}{2} + 1$$

We now substitute this expression for y_1 in place of y in the original differential equation, which then becomes

$$\frac{dy}{dx} = -x\left(-\frac{x^2}{2} + 1\right) = \frac{x^3}{2} - x$$

Integrating,

$$y_2 = \frac{x^4}{8} - \frac{x^2}{2} + 1$$

where 1 is the constant of integration evaluated by using the boundary condition.

Repeating the process one more time,

$$\frac{dy}{dx} = -x\left(\frac{x^4}{8} - \frac{x^2}{2} + 1\right) = -\frac{x^5}{8} + \frac{x^3}{2} - x$$

from which

$$y_3 = -\frac{x^6}{48} + \frac{x^4}{8} - \frac{x^2}{2} + 1$$

where 1, again, is the constant of integration.

The four approximations are

$$y_0 = 1$$

$$y_1 = -\frac{x^2}{2} + 1$$

$$y_2 = \frac{x^4}{8} - \frac{x^2}{2} + 1$$

$$y_3 = -\frac{x^6}{48} + \frac{x^4}{8} - \frac{x^2}{2} + 1$$

The differential equation $dy/dx = -xy$ with boundary condition $y(0) = 1$ happens to be one for which we can find the exact solution by separation of variables. The solution is

$$y = e^{-x^2/2}$$

From Chapter 10, we recall that the Maclaurin series expansion for e^x is

$$e^x = 1 + x + \frac{x^2}{2!} + \frac{x^3}{3!} + \cdots$$

The series expansion for $e^{-x^2/2}$ can be found by substituting $-x^2/2$ for x in the series for e^x, resulting in

$$e^{-x^2/2} = 1 - \frac{x^2}{2} + \frac{(-x^2/2)^2}{2!} + \frac{(-x^2/2)^3}{3!} + \cdots$$

$$= 1 - \frac{x^2}{2} + \frac{x^4}{8} - \frac{x^6}{48} + \cdots$$

The four approximations, y_0, y_1, y_2, and y_3, are thus successively longer initial segments of the Maclaurin series expansion for the exact solution. The exact solution and the four approximations are shown for an interval around $x = 0$ in Figure 11-5. Notice that each successive approximation is closer to the exact solution throughout the entire interval, but that any particular approximation is closer to the exact solution the closer x is to 0. ■

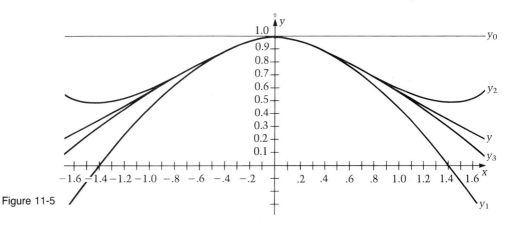

Figure 11-5

EXAMPLE 2.   Use Picard's method to find four approximations to the solution of the differential equation $dy/dx = ye^x$ with $y(0) = 3$. Also find the exact solution, and compare $y(0.5)$ with $y_3(0.5)$.

We begin as before, putting $y_0 = 3$ into the differential equation, integrating, and evaluating the constant of integration by means of the condition $y = 3$ when $x = 0$.

$$\frac{dy}{dx} = 3e^x$$

$$y_1 = 3e^x + c$$

$$3 = 3 \cdot 1 + c$$

$$c = 0$$

$$y_1 = 3e^x$$

Repeating the process, we put $y_1 = 3e^x$ in for y in the original differential equation.

$$\frac{dy}{dx} = (3e^x)e^x = 3e^{2x}$$

$$y_2 = \frac{3}{2}e^{2x} + c$$

$$3 = \frac{3}{2} \cdot 1 + c$$

$$c = \frac{3}{2}$$

$$y_2 = \frac{3}{2}e^{2x} + \frac{3}{2}$$

Finally,

$$\frac{dy}{dx} = \left(\frac{3}{2}e^{2x} + \frac{3}{2}\right)e^x = \frac{3}{2}e^{3x} + \frac{3}{2}e^x$$

$$y_3 = \frac{3}{2} \cdot \frac{1}{3}e^{3x} + \frac{3}{2}e^x + c$$

$$3 = \frac{1}{2} \cdot 1 + \frac{3}{2} \cdot 1 + c$$

$$c = 1$$

$$y_3 = \frac{1}{2}e^{3x} + \frac{3}{2}e^x + 1$$

The exact solution, found by separation of variables, is $y = 3e^{e^x - 1}$. Evaluating each function at $x = 0.5$, we get

$$y_3(0.5) = 5.714$$

and

$$y(0.5) = 5.739 \quad \blacksquare$$

Practice 1. Find the fourth approximation to the solution of the differential equation $dy/dx = x - y$, $y(0) = 0$.

To use Picard's method, we must know a boundary condition that the solution satisfies. We must also be able to integrate the successive functions of x that arise after substitution for y. The second approximation technique avoids these disadvantages, but it has some of its own.

The second method is solution by the use of **series.** We assume that the solution y to a differential equation has a power series expansion

$$y = a_0 + a_1x + a_2x^2 + a_3x^3 + a_4x^4 + \cdots \tag{3}$$

Just as we did in Chapter 10, we differentiate the infinite series term by term. Thus

$$y' = a_1 + 2a_2x + 3a_3x^2 + 4a_4x^3 + \cdots \tag{4}$$

We then substitute from Equations (3) and (4) into the original differential equation and equate coefficients of like terms. This allows us to solve for the a_i's.

EXAMPLE **3.** Use the series method to solve the differential equation $dy/dx = -xy$, $y(0) = 1$.

The differential equation has the form

$$y' = -xy$$

and we substitute from Equations (3) and (4) to get

$$a_1 + 2a_2x + 3a_3x^2 + 4a_4x^3 + \cdots$$
$$= -x(a_0 + a_1x + a_2x^2 + a_3x^3 + \cdots) \tag{5}$$

Equating coefficients of like powers of x on each side of Equation (5) results in the series of equations

$$a_1 = 0$$
$$2a_2 = -a_0$$
$$3a_3 = -a_1$$
$$4a_4 = -a_2$$
$$\vdots$$

The solutions are

$$a_1 = 0$$
$$a_2 = -\frac{a_0}{2}$$
$$a_3 = -\frac{a_1}{3} = 0$$
$$a_4 = -\frac{a_2}{4} = \frac{-(-a_0/2)}{4} = \frac{a_0}{8}$$
$$\vdots$$

Using this information in Equation (3), y becomes

$$y = a_0 - \frac{a_0}{2}x^2 + \frac{a_0}{8}x^4 + \cdots$$

or

$$y = a_0\left(1 - \frac{x^2}{2} + \frac{x^4}{8} + \cdots\right)$$

Successive terms involve higher powers of x, so that when we apply the boundary condition $y(0) = 1$, we get

$$1 = a_0(1)$$
$$a_0 = 1$$

The solution is thus

$$y = 1 - \frac{x^2}{2} + \frac{x^4}{8} + \cdots$$

This is the differential equation we did in Example 1. From that example, we know that the exact solution is $y = e^{-x^2/2}$ and that our solution here is indeed the series representation of the exact solution. ∎

Solution by the use of series is not limited to first-order differential equations. Also, if we do not know any boundary conditions that the differential equation satisfies, then the series solution will contain one or more arbitrary constants a_i, just as we require a general solution to have.

EXAMPLE 4. Use the series method to solve the differential equation

$$y'' + y = 0$$

From the assumption that

$$y = a_0 + a_1x + a_2x^2 + a_3x^3 + a_4x^4 + \cdots$$

we get

$$y' = a_1 + 2a_2x + 3a_3x^2 + 4a_4x^3 + \cdots$$
$$y'' = 2a_2 + 6a_3x + 12a_4x^2 + \cdots$$

Substituting into the original differential equation results in

$$(2a_2 + 6a_3x + 12a_4x^2 + \cdots) + (a_0 + a_1x + a_2x^2 + a_3x^3 + \cdots) = 0$$

Equating coefficients of like powers of x gives us

$$2a_2 + a_0 = 0, \quad \text{so} \quad a_2 = \frac{-a_0}{2}$$

$$6a_3 + a_1 = 0, \quad \text{so} \quad a_3 = \frac{-a_1}{6}$$

$$12a_4 + a_2 = 0, \quad \text{so} \quad a_4 = \frac{-a_2}{12} = \frac{-(-a_0/2)}{12} = \frac{a_0}{24}$$

from which

$$y = a_0 + a_1x - \frac{a_0}{2}x^2 - \frac{a_1}{6}x^3 + \frac{a_0}{24}x^4 + \cdots$$

$$= a_0\left(1 - \frac{x^2}{2} + \frac{x^4}{24} + \cdots\right) + a_1\left(x - \frac{x^3}{6} + \cdots\right)$$

Recalling Chapter 10, we may recognize the parenthetical expressions above as the Maclaurin series for cos x and sin x, so that the exact general solution seems to be

$$y = a_0 \cos x + a_1 \sin x$$

Direct substitution into the original differential equation shows that this is indeed the general solution. ∎

Practice 2. Solve $y' = y$ by the use of series. From the series expression, write the solution function; then verify by using separation of variables to find the exact solution.

The series method may, however, fail to work. In using series, we have assumed that the solution to the differential equation has a Maclaurin series expansion, and we know that some functions exist that cannot be expanded as a Maclaurin series. It is easy to detect when the method fails.

EXAMPLE 5. Consider the equation $x^3 y'' = 1$. By direct integration, we find that the exact solution is

$$y = \frac{1}{2x} + c_1 x + c_2$$

This function does not have a Maclaurin series expansion because it is undefined at $x = 0$. An attempt to use the series method results in the equation

$$x^3 (2a_2 + 6a_3 x + 12a_4 x^2 + \cdots) = 1$$

Equating coefficients of like powers of x from x^3 on up results in $a_2 = 0$, $a_3 = 0$, $a_4 = 0$, . . . , so that the equation becomes

$$0 = 1$$

which cannot have a solution. ∎

When the series method does work, we may not recognize the resulting series solution as the Maclaurin expansion of a particular function, which is why it is an approximation method. We can still approximate the value of the solution y for a particular value x_0 by taking terms of the series and evaluating them at x_0.

The third approximation technique is called **Euler's method,** named for the Swiss mathematician Leonhard Euler (pronounced *Oiler*), 1707–1783. It applies to first-order, first-degree differential equations and requires that one boundary condition of the form

$y(x_0) = y_0$ be known. Unlike the previous two methods, Euler's method will not yield a function that approximates the solution function y. Instead, starting from the known point (x_0, y_0) on the solution, the method generates a series of points (x_i, y_i), which approximate points on the solution curve. Thus if we were to connect these points graphically, the resulting curve should approximate the shape of the solution function itself in some interval around x_0.

We write the differential equation in the form

$$\frac{dy}{dx} = f(x, y)$$

or

$$dy = f(x, y) \, dx$$

From our discussion of differentials in Section 5-6, we recall that for a small change dx or Δx in x, dy is a close approximation to the change in y, Δy. Thus for an x-value of $x_1 = x_0 + \Delta x$ where Δx is small, the corresponding y-value, $y_0 + \Delta y$, is approximated by

$$y_1 = y_0 + f(x_0, y_0)\Delta x$$

Assuming that we know a value for Δx, this equation allows us to write a new point (x_1, y_1) that is close to being on the solution curve itself.

Now we repeat the process, beginning with the point (x_1, y_1) and computing a new point (x_2, y_2) by

$$x_2 = x_1 + \Delta x$$
$$y_2 = y_1 + f(x_1, y_1)\Delta x$$

We may continue this process to get a whole series of points. The smaller the increment Δx used, the more accurate the results.

EXAMPLE 6. Use Euler's method to approximate the solution to the differential equation $dy/dx = -xy$, $y(0) = 1$. Use $\Delta x = 0.1$. (This again is the problem of Examples 1 and 3.)

We compute a series of (x, y)-values, beginning with $x_0 = 0$ and $y_0 = 1$. Here $f(x, y) = -xy$.

$$x_1 = x_0 + \Delta x = 0 + 0.1 = 0.1$$
$$y_1 = y_0 + f(x_0, y_0)\Delta x = 1 - 0(1)(0.1) = 1$$

A table, such as the one shown, can help in organizing our results. The first row begins with x_0 and y_0 on the left and shows the

x_i	y_i	$f(x_i, y_i)$	x_{i+1}	y_{i+1}
0	1	0	0.1	1
0.1	1			

computed value of x_1 and y_1 on the right. This pair of values then becomes the start of the next row.

$$x_2 = x_1 + \Delta x = 0.1 + 0.1 = 0.2$$
$$y_2 = y_1 + f(x_1, y_1)\Delta x = 1 - (0.1)1(0.1) = 0.99$$

If we continue our calculations to $x_i = 1.4$, we get the results shown in the completed table.

x_i	y_i	$f(x_i, y_i)$	x_{i+1}	y_{i+1}
0	1	0	0.1	1
0.1	1	−0.100	0.2	0.99
0.2	0.99	−0.198	0.3	0.97
0.3	0.97	−0.291	0.4	0.94
0.4	0.94	−0.376	0.5	0.90
0.5	0.90	−0.450	0.6	0.86
0.6	0.86	−0.516	0.7	0.81
0.7	0.81	−0.567	0.8	0.75
0.8	0.75	−0.600	0.9	0.69
0.9	0.69	−0.621	1.0	0.63
1.0	0.63	−0.630	1.1	0.57
1.1	0.57	−0.627	1.2	0.51
1.2	0.51	−0.612	1.3	0.45
1.3	0.45	−0.585	1.4	0.39
1.4	0.39	−0.546	1.5	0.33

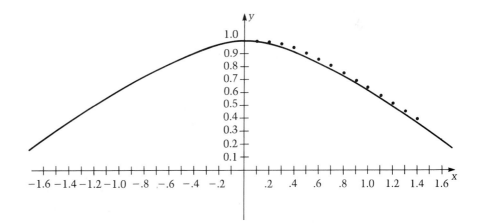

Figure 11-6

These points are plotted against the exact solution in Figure 11-6. Once again, we know that the exact solution is $y = e^{-x^2/2}$, and we can do some comparisons.

	Exact	Approximation
$x = 0$	1	1
$x = 0.3$	0.96	0.97
$x = 0.5$	0.88	0.90
$x = 1.2$	0.49	0.51

In general, the approximation to the exact value for a fixed Δx is better the closer we are to x_0. Also, a better set of approximate points is obtained when Δx is smaller. ■

Word of Advice

Your calculator is ideal for Euler's method. It is helpful to organize your results in tabular form to keep track of what you are doing.

EXAMPLE 7. Use Euler's method to find five points approximating the solution to $y' = \sin y + x$, $y(0) = 0$. Take Δx to be 0.1.

Here $f(x, y) = \sin y + x$, $x_0 = 0$, and $y_0 = 0$. Working as in the previous example, we use the formulas

$$x_{i+1} = x_i + \Delta x$$
$$y_{i+1} = y_i + f(x_i, y_i)\Delta x$$

and construct a table.

x_i	y_i	$f(x_i, y_i)$	x_{i+1}	y_{i+1}
0	0	0	0.1	0
0.1	0	0.100	0.2	0.010
0.2	0.010	0.210	0.3	0.031
0.3	0.031	0.331	0.4	0.064

The five points are $(0, 0)$, $(0.1, 0)$, $(0.2, 0.01)$, $(0.3, 0.031)$, and $(0.4, 0.064)$. ■

Practice 3. Use Euler's method to find five points approximating the solution to $y' = y$, $y(0) = 1$. Use $\Delta x = 0.2$. Also find the exact solution and compare the corresponding values.

Exercises / Section **11-6**

Exercises 1–4: Find y_3 using Picard's method; also find the exact solution. Compute both y_3 and y for the value of x shown.

1. $y' = x + y$; $y(0) = 2$; $x = 0.5$
2. $y' = 2xy$; $y(1) = 1$; $x = 0.8$
3. $y' = 2y$; $y(0) = 1$; $x = 1$
4. $y' = x^2 - y$; $y(1) = 3$; $x = 0.5$

Exercises 5–8: Use Picard's method to find the indicated approximation to the solution.

5. $y' = 1 + xy^2$; $y(0) = 1$; y_2
6. $y' = x - y^2$; $y(0) = 0$; y_3
7. $y' = x + y \sin x$; $y(0) = 1$; y_2
8. $y' = y^2 - \cos x$; $y(0) = 1$; y_2

Exercises 9–10: Solve by the use of series. Try to find the solution function from the series expression; then verify by solving the differential equation for the exact solution.

9. $y' = y - x$
10. $y' = 4xy$

Exercises 11–16: Find a series solution.

11. $y'' + y' = 0$
12. $y'' = y$
13. $(x + 1)y' = xy$; $y(0) = 1$
14. $y' = (x - 1)y$; $y(0) = 2$
15. $y'' + (2x - 1)y' = 0$
16. $y''' = xy$

Exercises 17–18: Use series to approximate the solution for the value of x shown.

17. $y'' + 2y' + 3y = 0$; $y(0) = 0$, $y'(0) = 1$; $x = 1$
18. $(x + 1)y' + (3x - 2)y = 0$; $y(0) = 1$; $x = 0.2$

Exercises 19–24: Use Euler's method to find five points approximating the solution function; the initial point and the value of Δx are given.

19. $y' = x + y$; $y(0) = 0$; $\Delta x = 0.2$
20. $y' = x^2 + y^2$; $y(0) = 1$; $\Delta x = 0.1$
21. $y' = e^{xy}$; $y(0) = 2$; $\Delta x = 0.1$
22. $y' = \sqrt{x + y}$; $y(1) = 3$; $\Delta x = 0.2$
23. $y' = \dfrac{x}{y}$; $y(1) = 2$; $\Delta x = 0.3$
24. $y' = \cos x + 2y$; $y(0) = 1$; $\Delta x = 0.1$

25. The position x of a particular moving body as a function of time t is given by the differential equation $dx/dt = t^2 + x$, with $x = 0$ km when $t = 0$ s. Find the exact solution and evaluate it when $t = 1$ s.

26. For the problem of Exercise 25, use Picard's method, and evaluate x_3 when $t = 1$ s.

27. For the problem of Exercise 25, use series to approximate the value of x when $t = 1$ s.

28. For the problem of Exercise 25, use Euler's method with $\Delta t = 0.1$ to approximate the value of x when $t = 1$ s.

11-7 Second-Order Linear Homogeneous Equations

In Section 11-4, we defined a linear differential equation of first order to be one of the form

$$dy + Py\, dx = Q\, dx$$

where P and Q are functions only of x. We can rewrite this in the form

$$\frac{dy}{dx} + P(x)y = Q(x)$$

Here y and its derivative appear only to the first power and are arranged in a polynomial-like format in which the coefficients are functions only of x.

The linear differential equation of first order is a special case of the **linear differential equation of order n,** which has the form

$$P_0(x) \frac{d^n y}{dx^n} + P_1(x) \frac{d^{n-1} y}{dx^{n-1}} + \cdots + P_{n-1}(x) \frac{dy}{dx} + P_n(x)y = Q(x)$$

Notice again that y and all its derivatives appear only to the first power and are arranged in a polynomial-like format, with co-efficients that are functions of x.

Before dealing with such equations, we make some simplifying assumptions. First, we assume that all the functions $P_i(x)$ are in fact constant functions. Second, we deal only with second-order equations, although the methods we develop do carry over to higher-order linear equations. Our equations thus have the form

$$P_0 \frac{d^2 y}{dx^2} + P_1 \frac{dy}{dx} + P_2 y = Q(x) \tag{1}$$

A final simplification in notation occurs if we write D^n to stand for the operation of taking the nth derivative with respect to x. With this notation, we can write (1) as

$$P_0 D^2 y + P_1 D y + P_2 y = Q(x) \tag{2}$$

or even

$$(P_0 D^2 + P_1 D + P_2)y = Q(x) \tag{3}$$

which means that each term in the expression is to operate upon y and the results are to be added.

To find the solution to Equation (3), we must be able to solve the corresponding **homogeneous equation**—that is, Equation (3) where $Q(x) = 0$,

$$(P_0 D^2 + P_1 D + P_2)y = 0 \tag{4}$$

In this section and the next, we concentrate on solving equations of the form shown in (4); in Section 11-9, we solve the more general **nonhomogeneous** equation—that is, Equation (3) with $Q(x) \neq 0$.

A clue to the form of a solution for (4) may be found in looking at the first-order, linear, homogeneous equation

$$\frac{dy}{dx} + ky = 0$$

whose solution (by separation of variables) is

$$y = ce^{-kx}$$

We might assume that an exponential function of the form $y = e^{mx}$ is a solution to Equation (4). What would have to happen for this to be possible? Substitution of $y = e^{mx}$ into Equation (4) would have to satisfy the equation. Making this substitution, we get

$$(P_0 D^2 + P_1 D + P_0)e^{mx} = 0$$

which becomes, after applying the derivatives,

$$P_0 m^2 e^{mx} + P_1 m e^{mx} + P_0 e^{mx} = 0$$

or

$$(P_0 m^2 + P_1 m + P_0) e^{mx} = 0$$

The factor e^{mx} can never be zero, so this last equation can be true if and only if

$$P_0 m^2 + P_1 m + P_0 = 0 \tag{5}$$

Equation (5) is called the **auxiliary equation** of Equation (4), and it can be found merely by looking at (4). Notice that (5) is a quadratic equation. If its roots are m_1 and m_2, then $y = e^{m_1 x}$ and $y = e^{m_2 x}$ will both be solutions of the homogeneous equation (4). In fact, because the derivative of a sum is the sum of the derivatives and because a constant can be pulled through a derivative, the expression

$$y = c_1 e^{m_1 x} + c_2 e^{m_2 x}$$

is also a solution of (4). This is the general solution because of the two arbitrary constants c_1 and c_2.

EXAMPLE 1.   Solve the differential equation $y'' - 2y' - 3y = 0$. We first write this equation in operator notation.

$$(D^2 - 2D - 3)y = 0$$

The auxiliary equation is

$$m^2 - 2m - 3 = 0$$

which can be factored into $(m - 3)(m + 1) = 0$ and has roots $m_1 = 3$ and $m_2 = -1$. The general solution is

$$y = c_1 e^{3x} + c_2 e^{-x} \quad \blacksquare$$

Practice 1.   Verify by substitution that $y = c_1 e^{3x} + c_2 e^{-x}$ is a solution of the differential equation $y'' - 2y' - 3y = 0$.

The steps to solve a second-order, linear, homogeneous equation with constant coefficients are as follows:

1. Write the equation with the D operator notation.
2. Write the auxiliary equation.
3. Solve the auxiliary equation for its roots m_1 and m_2; because the auxiliary equation is quadratic, the quadratic formula can be used.
4. The general solution is

$$y = c_1 e^{m_1 x} + c_2 e^{m_2 x}$$

EXAMPLE 2.   Solve the differential equation $2y'' + 3y' - y = 0$.
In operator notation this equation becomes

$$(2D^2 + 3D - 1)y = 0$$

The auxiliary equation, $2m^2 + 3m - 1 = 0$, is not easily factored. We use the quadratic formula to find its roots.

$$m = \frac{-3 \pm \sqrt{9 - 4(2)(-1)}}{4} = \frac{-3 \pm \sqrt{17}}{4}$$

The general solution to the differential equation is

$$y = c_1 e^{(-3+\sqrt{17})x/4} + c_2 e^{(-3-\sqrt{17})x/4}$$
$$= e^{-3x/4}(c_1 e^{\sqrt{17}x/4} + c_2 e^{-\sqrt{17}x/4}) \quad \blacksquare$$

Practice 2.   Solve the differential equation $3y'' + 2y' = y$.

EXAMPLE 3.   Solve the differential equation $y'' - y' = 12y$ subject to the boundary conditions $y(0) = \frac{5}{3}$ and $y'(0) = -\frac{22}{3}$.
Because we have a second-order equation, there will be two arbitrary constants, and we need the two boundary conditions to determine them.
We can write

$$(D^2 - D - 12)y = 0$$

from which the auxiliary equation is

$$m^2 - m - 12 = 0$$

or, factoring,

$$(m - 4)(m + 3) = 0$$

The general solution is

$$y = c_1 e^{4x} + c_2 e^{-3x}$$

Now applying the boundary conditions, we get

$$y = c_1 e^{4x} + c_2 e^{-3x} \qquad\qquad y(0) = c_1 + c_2 = \tfrac{5}{3}$$
$$y' = 4c_1 e^{4x} - 3c_2 e^{-3x} \qquad\qquad y'(0) = 4c_1 - 3c_2 = -\tfrac{22}{3}$$

This gives us the system of equations

$$c_1 + c_2 = \tfrac{5}{3}$$
$$4c_1 - 3c_2 = -\tfrac{22}{3}$$

The solution to this system is $c_1 = -\frac{1}{3}$ and $c_2 = 2$, and the particular solution to the differential equation is

$$y = -\tfrac{1}{3}e^{4x} + 2e^{-3x} \quad \blacksquare$$

Exercises / Section 11-7

Exercises 1–22: Solve the given differential equation.

1. $(D^2 - D)y = 0$

2. $(D^2 + 2D)y = 0$

3. $y'' - y = 0$

4. $y'' + 2y' - 15y = 0$

5. $(D^2 - 6D + 8)y = 0$

6. $(D^2 - D - 2)y = 0$

7. $\dfrac{d^2y}{dx^2} + 5\dfrac{dy}{dx} + 6y = 0$

8. $\dfrac{d^2y}{dx^2} + 4\dfrac{dy}{dx} - 12y = 0$

9. $2y'' + y' = 0$

10. $3y'' + 2y' = 0$

11. $(2D^2 - 5D - 12)y = 0$

12. $(3D^2 + 5D - 2)y = 0$

13. $6\dfrac{d^2y}{dx^2} + 11\dfrac{dy}{dx} - 10y = 0$

14. $6\dfrac{d^2y}{dx^2} - 19\dfrac{dy}{dx} + 15y = 0$

15. $y'' + y' - 3y = 0$

16. $y'' - 2y' - 5y = 0$

17. $2y'' + 4y' - 3y = 0$

18. $2y'' + 4y' - y = 0$

19. $3\dfrac{d^2y}{dx^2} - 5\dfrac{dy}{dx} + y = 0$

20. $3\dfrac{d^2y}{dx^2} + 3\dfrac{dy}{dx} - 2y = 0$

21. $y''' + y'' - 2y' = 0$ (Note that this is a third-order equation.)

22. $y''' - y'' - 4y' + 4y = 0$

Exercises 23–24: Solve the differential equation subject to the boundary conditions shown.

23. $2y'' - y' - y = 0$; $y(0) = \frac{5}{2}$, $y'(0) = \frac{7}{4}$

24. $y'' - y' - 6y = 0$; $y(0) = 2$, $y(1) = \dfrac{3 - e^5}{e^2}$

11-8 Repeated or Complex Roots

In the previous section, we learned how to write the general solution to a second-order, linear, homogeneous equation by finding the roots of the auxiliary equation. If the auxiliary equation happens to have repeated roots, however, the form of the solution must be slightly modified. For example, suppose we want to solve the differential equation

$$y'' + 2y' + y = 0$$

In operator notation, the equation becomes

$$(D^2 + 2D + 1)y = 0$$

and the auxiliary equation is

$$m^2 + 2m + 1 = (m + 1)^2 = 0$$

with a double root of -1. If we follow the instructions of the last section, we write the solution as

$$y = c_1 e^{-x} + c_2 e^{-x}$$

But this can be written as

$$y = (c_1 + c_2)e^{-x} = ce^{-x}$$

This is not the general solution, because there is only one arbitrary constant. The following example shows the form of the general solution when there are repeated roots.

EXAMPLE 1.   Solve the differential equation $y'' + 2y' + y = 0$.
Again, using operator notation, the equation becomes

$$(D^2 + 2D + 1)y = 0 \tag{1}$$

If we interpret D times D to mean that we apply the D operator twice, thus getting the second derivative, we can write Equation (1) in the factored form

$$(D + 1)(D + 1)y = 0 \tag{2}$$

Now $(D + 1)y$ is some function of x; we call this function w. Equation (2) is, therefore,

$$(D + 1)w = 0$$

or

$$\frac{dw}{dx} + w = 0$$

This equation can be solved by separation of variables.

$$dw = -w \, dx$$

$$\frac{dw}{w} = -dx$$

$$\ln w = -x + \ln c_1$$

$$\ln \frac{w}{c_1} = -x$$

$$w = c_1 e^{-x}$$

But because $(D + 1)y = w$, we can write

$$(D + 1)y = c_1 e^{-x}$$

or

$$\frac{dy}{dx} + y = c_1 e^{-x}$$

This equation is a linear, first-order equation and its solution is given by

$$ye^{\int dx} = \int e^{\int dx} c_1 e^{-x} \, dx$$

$$ye^x = \int e^x c_1 e^{-x} \, dx$$

$$ye^x = \int c_1 \, dx$$

$$ye^x = c_1 x + c_2$$

$$y = c_1 x e^{-x} + c_2 e^{-x} = e^{-x}(c_1 x + c_2) \tag{3}$$

Equation (3) gives a general solution, with two independent arbitrary constants, to the original differential equation.   ∎

We can see from Example 1 that the solution in which there is a repeated root m consists of two terms involving e^{mx} as before, but one of the terms has an extra x factor thrown in. Specifically, the solution is

$$y = e^{mx}(c_1 x + c_2)$$

None of the equations in the previous section led to repeated roots.

EXAMPLE 2. Solve the differential equation $4\dfrac{d^2 y}{dx^2} - 4\dfrac{dy}{dx} + y = 0$.

In operator notation, the equation becomes

$$(4D^2 - 4D + 1)y = 0$$

and the auxiliary equation is

$$4m^2 - 4m + 1 = 0$$

or

$$(2m - 1)^2 = 0$$

The roots are $m = \frac{1}{2}, \frac{1}{2}$. Because of the repeated root, we use the new solution form, and the solution is

$$y = e^{x/2}(c_1 x + c_2) \quad \blacksquare$$

Practice 1. Solve the differential equation $y'' - 6y' + 9y = 0$.

For the second-order, linear, homogeneous equation, the auxiliary equation is quadratic. The possibility exists that the roots may be complex numbers. This will occur if, in applying the quadratic formula,

$$m = \frac{-b \pm \sqrt{b^2 - 4ac}}{2a}$$

the discriminant $b^2 - 4ac$ is negative. In this case, the roots are the complex conjugate numbers

$$m = \alpha \pm j\beta$$

where $j = \sqrt{-1}$. (Mathematicians use i to represent $\sqrt{-1}$, while engineers prefer j because i is also used as the symbol for electric current.) These are not repeated roots, but two distinct values; by the previous section, the solution to the differential equation is

$$y = c_1 e^{(\alpha + j\beta)x} + c_2 e^{(\alpha - j\beta)x}$$

or

$$y = e^{\alpha x}(c_1 e^{j\beta x} + c_2 e^{-j\beta x}) \tag{4}$$

This expression can be written in another form by making use of **Euler's formulas**

$$e^{j\theta} = \cos\theta + j\sin\theta$$
$$e^{-j\theta} = \cos\theta - j\sin\theta$$

(To verify Euler's formulas, we expand each side of the equation in a Maclaurin series expansion; see Exercises 27 and 28.) Making use of Euler's formulas, we rewrite the solution as follows.

$$
\begin{aligned}
y &= e^{\alpha x}(c_1 e^{j\beta x} + c_2 e^{-j\beta x}) \\
&= e^x[c_1(\cos\beta x + j\sin\beta x) + c_2(\cos\beta x - j\sin\beta x)] \\
&= e^{\alpha x}[(c_1 + c_2)\cos\beta x + (c_1 j - c_2 j)\sin\beta x]
\end{aligned}
$$

or

$$y = e^{\alpha x}(c_3 \cos\beta x + c_4 \sin\beta x)$$

where $c_3 = c_1 + c_2$ and $c_4 = c_1 j - c_2 j$. This form of the solution looks less complicated than Equation (4) because j no longer explicitly appears.

EXAMPLE 3.   Solve the differential equation

$$(D^2 + 2D + 5)y = 0$$

The auxiliary equation is

$$m^2 + 2m + 5 = 0$$

and the roots, by the quadratic formula, are

$$m = \frac{-2 \pm \sqrt{4 - 4(5)}}{2} = \frac{-2 \pm 4j}{2} = -1 \pm 2j$$

Here $\alpha = -1$ and $\beta = 2$. The solution to the differential equation is

$$y = e^{-x}(c_3 \cos 2x + c_4 \sin 2x) \quad \blacksquare$$

Practice 2.   Solve the differential equation $y'' - 4y' + 8y = 0$.

EXAMPLE 4.   Solve the differential equation

$$\frac{d^2y}{dx^2} + 3\frac{dy}{dx} + 3y = 0$$

subject to the boundary conditions $y(0) = 1$ and $y'(0) = 4$.

From the operator form

$$(D^2 + 3D + 3)y = 0$$

we get the auxiliary equation

$$m^2 + 3m + 3 = 0$$

which has roots

$$m = \frac{-3 \pm \sqrt{9 - 4(3)}}{2} = \frac{-3 \pm \sqrt{3}j}{2} = -\frac{3}{2} \pm \frac{\sqrt{3}}{2}j$$

The general solution is

$$y = e^{-3x/2}\left(c_3 \cos \frac{\sqrt{3}}{2}x + c_4 \sin \frac{\sqrt{3}}{2}x\right)$$

Applying the condition $y(0) = 1$ results in

$$1 = e^0(c_3 \cos 0 + c_4 \sin 0)$$
$$1 = c_3$$

In order to use the condition $y'(0) = 4$, we must differentiate the general solution to find y'.

$$y' = e^{-3x/2}\left(-c_3\frac{\sqrt{3}}{2} \sin \frac{\sqrt{3}}{2}x + c_4\frac{\sqrt{3}}{2} \cos \frac{\sqrt{3}}{2}x\right)$$

$$-\frac{3}{2}e^{-3x/2}\left(c_3 \cos \frac{\sqrt{3}}{2}x + c_4 \sin \frac{\sqrt{3}}{2}x\right)$$

$$y'(0) = 4 = e^0\left(-c_3\frac{\sqrt{3}}{2} \sin 0 + c_4\frac{\sqrt{3}}{2} \cos 0\right)$$

$$-\frac{3}{2}e^0(c_3 \cos 0 + c_4 \sin 0)$$

$$4 = c_4\frac{\sqrt{3}}{2} - \frac{3}{2}c_3$$

$$4 = c_4\frac{\sqrt{3}}{2} - \frac{3}{2}$$

$$c_4 = \frac{11}{\sqrt{3}}$$

Finally, the particular solution is

$$y = e^{-3x/2}\left(\cos \frac{\sqrt{3}}{2}x + \frac{11}{\sqrt{3}} \sin \frac{\sqrt{3}}{2}x\right) \quad \blacksquare$$

Word of Advice

Remember that the form of the general solution to a second-order, linear, homogeneous equation depends upon the solutions to the auxiliary equation.

1. Distinct real roots m_1 and m_2:

$$y = c_1 e^{m_1 x} + c_2 e^{m_2 x}$$

2. Repeated real root m:

$$y = e^{mx}(c_1 x + c_2)$$

3. Complex roots $\alpha \pm j\beta$

$$y = e^{\alpha x}(c_3 \cos \beta x + c_4 \sin \beta x)$$

Exercises / Section 11-8

Exercises 1–22: Solve the given differential equation.

1. $(D^2 + 4D + 4)y = 0$
2. $(D^2 - 2D + 1)y = 0$
3. $y'' = 0$
4. $y'' - 8y' + 16y = 0$
5. $\dfrac{d^2y}{dx^2} + 6\dfrac{dy}{dx} + 9y = 0$
6. $\dfrac{d^2y}{dx^2} + 10\dfrac{dy}{dx} + 25y = 0$
7. $4y'' + 12y' + 9y = 0$
8. $16y'' - 40y' + 25y = 0$
9. $(9D^2 - 6D + 1)y = 0$
10. $(4D^2 - 20D + 25)y = 0$
11. $16\dfrac{d^2y}{dx^2} + 72\dfrac{dy}{dx} + 81y = 0$
12. $9\dfrac{d^2y}{dx^2} + 30\dfrac{dy}{dx} + 25y = 0$
13. $y'' + 2y' + 2y = 0$
14. $y'' - 4y' + 5y = 0$
15. $2y'' + 5y = 0$
16. $3y'' + 4y = 0$

17. $(D^2 - 4D + 7)y = 0$
18. $(D^2 + 5D + 7)y = 0$
19. $2\dfrac{d^2y}{dx^2} - 4\dfrac{dy}{dx} + 5y = 0$
20. $3\dfrac{d^2y}{dx^2} + 2\dfrac{dy}{dx} + 4y = 0$
21. $0.1y'' - 0.5y' + 0.8y = 0$
22. $1.6y'' + 0.3y' + 1.2y = 0$

Exercises 23–26: Solve the differential equation subject to the boundary conditions shown.

23. $y'' - 4y' + 4y = 0$; $y(0) = 3$, $y'(0) = 4$
24. $9y'' - 12y' + 4y = 0$; $y(0) = 1$, $y(1) = e$
25. $(D^2 + 4)y = 0$; $y(0) = 2$, $y(\pi/4) = 3$
26. $2\dfrac{d^2y}{dx^2} + 3\dfrac{dy}{dx} + 5y = 0$; $y(0) = 2$, $y'(0) = 1$

27. Use the Maclaurin series expansions for e^x, $\sin x$, and $\cos x$ (Section 10-3) to verify the Euler formula $e^{j\theta} = \cos \theta + j \sin \theta$. (*Hint:* Because $j^2 = -1$, it follows that $j^3 = -j$, $j^4 = 1$, and so on.)

28. Use the Maclaurin series expansions for e^x, $\sin x$, and $\cos x$ (Section 10-3) to verify the Euler formula $e^{-j\theta} = \cos \theta - j \sin \theta$.

11-9   Nonhomogeneous Equations

In Sections 11-7 and 11-8, we learned how to solve second-order, linear, homogeneous equations with constant coefficients. Such equations have the form

$$(P_0 D^2 + P_1 D + P_2)y = 0$$

Now we want to solve the nonhomogeneous case, in which the right side of the equation is not zero but some function of x. Therefore, our equation has the form

$$(P_0 D^2 + P_1 D + P_2)y = Q(x) \tag{1}$$

In order for y to be the general solution to Equation (1), two things must happen:

1. y must contain two independent arbitrary constants.
2. When y and its derivatives are substituted into the left side of Equation (1), the result must be the expression on the right side of Equation (1).

We accomplish these two tasks by constructing y as the sum of two expressions, each of which does one of the two tasks. Thus we let

$$y = y_c + y_p$$

where y_c is called the **complementary solution** and y_p is called the **particular solution**. The complementary solution y_c is the general solution to the corresponding homogeneous equation

$$(P_0 D^2 + P_1 D + P_2)y = 0$$

The complementary solution supplies the two arbitrary constants, but gives a value of zero when substituted into the left side of Equation (1). The particular solution contains no arbitrary constants, but it does result in the value $Q(x)$ after substitution in the left side of Equation (1). Therefore, the sum $y = y_c + y_p$ contains the two required constants and results in $0 + Q(x) = Q(x)$ when substituted into the left side of Equation (1).

EXAMPLE 1.   The solution of the differential equation $y'' - 4y' + 4y = 8x^2$ is $y = y_c + y_p$, where $y_c = e^{2x}(c_1 x + c_2)$ and $y_p = 2x^2 + 4x + 3$. The complementary solution y_c is obtained by solving the homogeneous equation $y'' - 4y' + 4y = 0$, which in turn depends upon solving the auxiliary equation $m^2 - 4m + 4 = (m - 2)^2 = 0$. The auxiliary equation has 2 as a repeated root, so by the last section, we know that $y_c = e^{2x}(c_1 x + c_2)$. We shall see presently how to find y_p. For now, we simply note that

$$
\begin{aligned}
y'' - 4y' + 4y &= D^2(y_c + y_p) - 4D(y_c + y_p) + 4(y_c + y_p) \\
&= (D^2 - 4D + 4)y_c + (D^2 - 4D + 4)y_p \\
&= 0 + (D^2 - 4D + 4)(2x^2 + 4x + 3) \\
&= D^2(2x^2 + 4x + 3) - 4D(2x^2 + 4x + 3) \\
&\quad + 4(2x^2 + 4x + 3) \\
&= 4 - 4(4x + 4) + 4(2x^2 + 4x + 3) \\
&= 8x^2
\end{aligned}
$$

Therefore, y contains two arbitrary constants and satisfies the differential equation. ∎

To find y_p, we note that y_p probably involves the expression $Q(x)$ of Equation (1). Because we must also differentiate y_p, the derivatives of $Q(x)$ will appear. In order to be able to cancel out such derivative terms, we include them in the original y_p as well. We therefore assume that y_p consists of the sum of terms containing $Q(x)$ and its derivatives but whose coefficients are unknown.* By substituting into Equation (1), we can find what the coefficients must be. (This technique is called the **method of undetermined coefficients**.)

EXAMPLE 2.   Again consider the equation $y'' - 4y' + 4y = 8x^2$. We assume that y_p contains an x^2-term, plus its derivatives, which are of the form x and a constant. Thus we let $y_p = Ax^2 + Bx + C$, where A, B, and C are yet to be determined. Then $y_p' = 2Ax + B$ and $y_p'' = 2A$. We substitute into $y'' - 4y' + 4y = 8x^2$ and get

$$
2A - 4(2Ax + B) + 4(Ax^2 + Bx + C) = 8x^2
$$

Coefficients of like powers of x must be the same on each side of the equation, giving us the system of equations

$$
\begin{array}{lrl}
x^2\text{:} & 4A & = 8 \\
x\text{:} & -8A + 4B & = 0 \\
\text{constant:} & 2A - 4B + 4C & = 0
\end{array}
$$

This system has the solution $A = 2$, $B = 4$, and $C = 3$. Therefore,

$$
y_p = 2x^2 + 4x + 3 \quad ∎
$$

---

*If any of the terms of y_p obtained by this method are also terms of y_c, then we must multiply those terms by the least power of x such that the result is not a term in y_c. No such examples occur in this book.

EXAMPLE 3. Solve the differential equation $y'' + 5y = e^{2x} + 2x^2$.

To find the complementary solution, we look at the auxiliary equation $m^2 + 5 = 0$, which has complex roots $\pm\sqrt{5}j$. Therefore, $y_c = c_3 \cos \sqrt{5}x + c_4 \sin \sqrt{5}x$. For y_p, we include terms of the form e^{2x} and x^2, as well as any terms that arise from their derivatives. The derivative of e^{2x} does not introduce any new kind of term, but the derivatives of x^2, as in Example 2, introduce x and constant terms. Therefore, we let $y_p = Ae^{2x} + Bx^2 + Cx + D$. We find the derivatives of y_p and substitute into the differential equation.

$$y'_p = 2Ae^{2x} + 2Bx + C$$
$$y''_p = 4Ae^{2x} + 2B$$

$$4Ae^{2x} + 2B + 5(Ae^{2x} + Bx^2 + Cx + D) = e^{2x} + 2x^2$$

Equating the coefficients of like terms gives us the system of equations

$$
\begin{aligned}
e^{2x}: \quad & 4A + 5A = 1 \\
x^2: \quad & 5B = 2 \\
x: \quad & 5C = 0 \\
\text{constant}: \quad & 2B + 5D = 0
\end{aligned}
$$

for which the solution is $A = \frac{1}{9}$, $B = \frac{2}{5}$, $C = 0$, and $D = -\frac{4}{25}$. Therefore, $y_p = \frac{1}{9}e^{2x} + \frac{2}{5}x^2 - \frac{4}{25}$ and the complete general solution is

$$y = c_3 \cos \sqrt{5}x + c_4 \sin \sqrt{5}x + \frac{1}{9}e^{2x} + \frac{2}{5}x^2 - \frac{4}{25} \quad \blacksquare$$

Word of Advice

Do not forget to do *both* parts of the solution to a nonhomogeneous second-order linear equation,

$$y_c \quad \text{and} \quad y_p$$

EXAMPLE 4. Solve the differential equation $(D^2 - 3D - 4)y = 3 \sin 2x$.

From the auxiliary equation $m^2 - 3m - 4 = (m - 4)(m + 1) = 0$, we get $y_c = c_1 e^{-x} + c_2 e^{4x}$. We let $y_p = A \sin 2x + B \cos 2x$ where $\cos 2x$ is introduced because of differentiating $\sin 2x$. Then

$$y'_p = 2A \cos 2x - 2B \sin 2x$$
$$y''_p = -4A \sin 2x - 4B \cos 2x$$

and

$$-4A \sin 2x - 4B \cos 2x - 3(2A \cos 2x - 2B \sin 2x)$$
$$- 4(A \sin 2x + B \cos 2x) = 3 \sin 2x$$

from which

$$-4A + 6B - 4A = 3$$
$$-4B - 6A - 4B = 0$$

The solution to this system of equations is $A = -\frac{6}{25}$ and $B = \frac{9}{50}$. Therefore,

$$y_p = -\frac{6}{25} \sin 2x + \frac{9}{50} \cos 2x$$

and

$$y = c_1 e^{-x} + c_2 e^{4x} - \frac{6}{25} \sin 2x + \frac{9}{50} \cos 2x. \quad \blacksquare$$

Practice 1. Solve the differential equation $y'' + 3y' + 2y = -4 \cos x$.

EXAMPLE 5. Solve the differential equation

$$\frac{d^2 y}{dx^2} + 2 \frac{dy}{dx} + y = 2xe^x$$

subject to the boundary conditions $y(0) = 4$, $y'(0) = 5$.

From the auxiliary equation $m^2 + 2m + 1 = (m + 1)^2 = 0$, we get the complementary solution $y_c = e^{-x}(c_1 x + c_2)$. The form we assume for y_p is $y_p = Axe^x + Be^x$. Then

$$y'_p = A(xe^x + e^x) + Be^x$$
$$y''_p = A(xe^x + e^x + e^x) + Be^x$$

and

$$A(xe^x + 2e^x) + Be^x + 2A(xe^x + e^x) + 2Be^x + Axe^x + Be^x = 2xe^x$$

from which

$$A + 2A + A = 2$$
$$2A + B + 2A + 2B + B = 0$$

The solution is $A = \frac{1}{2}$ and $B = -\frac{1}{2}$. Therefore, $y_p = \frac{1}{2}xe^x - \frac{1}{2}e^x$, and

$$y = e^{-x}(c_1 x + c_2) + \frac{1}{2}xe^x - \frac{1}{2}e^x$$

Applying the boundary conditions,

$$y(0) = c_2 - \frac{1}{2} = 4, \quad \text{so} \quad c_2 = \frac{9}{2}$$

and

$$y' = e^{-x}(c_1) - e^{-x}(c_1 x + c_2) + \frac{1}{2}(xe^x + e^x) - \frac{1}{2}e^x$$
$$y'(0) = c_1 - c_2 + \frac{1}{2} - \frac{1}{2} = 5, \quad \text{so} \quad c_1 = \frac{19}{2}$$

Finally, the solution is

$$y = e^{-x}(\tfrac{19}{2}x + \tfrac{9}{2}) + \tfrac{1}{2}xe^x - \tfrac{1}{2}e^x \quad \blacksquare$$

Exercises / Section **11-9**

Exercises 1–20: Solve the differential equation.

1. $(D^2 - D - 2)y = 14$

2. $(D^2 - 7D + 12)y = \frac{5}{6}$

3. $y'' - 2y' + y = 4x$

4. $y'' - 4y = 8x^2 + 2$

5. $(D^2 + 2D + 10)y = x^2$

6. $(D^2 + D + 2)y = 6x^3$

7. $2y'' + 7y' + 6y = 5 \sin x$

8. $2y'' + 7y' - 4y = 10 \cos 2x$

9. $\dfrac{d^2y}{dx^2} + 4\dfrac{dy}{dx} + 3y = 12e^{3x}$

10. $\dfrac{d^2y}{dx^2} - 6\dfrac{dy}{dx} + 8y = 5e^{-x} + 8x$

11. $(D^2 - 4D + 5)y = 20 \cos 3x - x$

12. $(D^2 - 4D + 4)y = 25 \sin x + 2x^2$

13. $y'' - 9y = 10e^{2x} + 3 \sin x$

14. $y'' + 2y = 4 \cos 2x + 6e^{-x}$

15. $y'' + 6y' + 9y = 8e^{-x} - 5e^{2x}$

16. $4y'' + 12y' + 9y = 9e^{3x} - 7e^{2x}$

17. $(2D^2 - 9D + 4)y = 3xe^x + 2x$

18. $(D^2 - 5D - 14)y = 9xe^x + 25 \sin x$

19. $\dfrac{d^2y}{dx^2} - 2y = 3x \sin x$

20. $\dfrac{d^2y}{dx^2} + 2y = 5x \cos x + x$

Exercises 21–22: Solve the differential equation subject to the boundary conditions shown.

21. $(3D^2 - 4D - 4)y = 22e^{3x}; \quad y(0) = 4, \, y'(0) = 3$

22. $(4D^2 + 4D + 1)y = 5x^3 + 40x; \quad y(0) = -20, \, y'(0) = 0$

11-10 Applications

One type of problem in which second-order, linear differential equations arise is that of oscillating mechanical motion. A weight hung on the end of a spring is the classic example. Figure 11-7(a) shows a spring of a given natural length. When an object of weight W is hung on the end of the spring, the spring is stretched some distance s beyond its natural length (Figure 11-7(b)), coming to rest in its **equilibrium position**. If the spring is then pulled down some further distance x and released, the weight will bob up and down about the equilibrium position. We can describe this oscillating motion by finding an expression for x (the distance of the weight from the equilibrium position) as a function of time t.

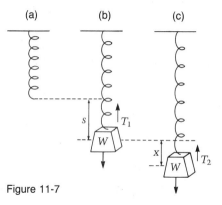

Figure 11-7

Hooke's law tells us that the force required to stretch a spring beyond its natural length is proportional to the distance it is stretched. Thus for the equilibrium case, the weight (force) W has stretched the spring a distance s, and

$$W = ks \tag{1}$$

Here we are taking the downward direction to be positive; W and s are positive, and k is a positive constant. Because the spring is in equilibrium, the downward force of the weight is balanced by an upward restoring force, or tension, T_1, which has the same magnitude but the opposite sign. Thus $T_1 = -ks$, a force acting upward.

For the case of Figure 11-7(c), additional force has been applied to stretch the spring a total distance of $s + x$. The restoring force, or tension, again acting upward, is $T_2 = -k(s + x)$. Therefore, when the spring is released, the resulting force acting on the object is the sum of the downward and upward forces:

$$\text{force} = W + T_2 = ks + [-k(s + x)] = ks - ks - kx = -kx$$
$$\downarrow \qquad \uparrow$$

When x is positive (the spring is below its equilibrium position), the resulting force acts upward; when x is negative (the spring is above its equilibrium position), the resulting force acts downward. This causes the oscillating motion. Newton's second law of motion states that the total force on a body equals the mass of the body times its acceleration. We can therefore write

$$ma = -kx$$

or

$$m\frac{d^2x}{dt^2} = -kx$$

Rewriting this equation results in

$$m\frac{d^2x}{dt^2} + kx = 0 \tag{2}$$

which is the general differential equation for what is called **simple harmonic motion**. Simple harmonic motion occurs not only for weight on a spring, but also for a simple pendulum or an object bobbing in water. Equation (2) is a second-order, linear, homogeneous equation.

EXAMPLE 1. A weight of 8 N is required to stretch a certain spring 0.2 m. The weight is then pulled an additional 0.1 m below the equilibrium position and released. Find the equation of the resulting motion.

The **spring constant** k is obtained from Equation (1).

$$8 = k(0.2)$$
$$k = 40 \text{ N/m}$$

The mass of an object is its weight divided by the gravitational constant, 9.8 (remember we are taking the downward direction to be positive). Therefore, Equation (2) becomes

$$\frac{8}{9.8}\frac{d^2x}{dt^2} + 40x = 0$$

or

$$\frac{d^2x}{dt^2} + 49x = 0$$

The general solution to this differential equation is

$$x = c_3 \cos 7t + c_4 \sin 7t$$

However, we also know that at $t = 0$, $x = 0.1$ and $dx/dt = 0$ (since initially the weight is 0.1 m below the equilibrium position, and it is not moving). From these two conditions, we evaluate the constants c_3 and c_4 and arrive at the solution

$$x = 0.1 \cos 7t$$

The graph of position x as a function of time t is given by Figure 11-8; the curve has amplitude 0.1 and period $2\pi/7$. The oscillating motion of the weight on the end of the spring is easy to see. ■

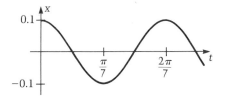

Figure 11-8

In the idealized spring system of Example 1, the oscillations would go on forever; no retarding force, such as that due to air resistance or internal friction within the spring, was incorporated into the differential equation. Any retarding force, or **damping force**, can be assumed to be proportional to the velocity of the moving object. The differential equation when a damping force is incorporated becomes

$$m\frac{d^2x}{dt^2} + b\frac{dx}{dt} + kx = 0 \tag{3}$$

The roots of the auxiliary equation for (3) are

$$\frac{-b \pm \sqrt{b^2 - 4mk}}{2m}$$

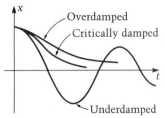

Figure 11-9

Depending on the value of the discriminant $b^2 - 4mk$, the solution to Equation (3) takes one of the following forms:

1. $b^2 - 4mk < 0$ $x = e^{\alpha t}(c_3 \cos \beta t + c_4 \sin \beta t)$ underdamped
2. $b^2 - 4mk > 0$ $x = c_1 e^{m_1 t} + c_2 e^{m_2 t}$ overdamped
3. $b^2 - 4mk = 0$ $x = e^{m_1 t}(c_1 t + c_2)$ critically damped

In the underdamped case, the sine and cosine terms mean that oscillations are present, although the factor $e^{\alpha t}$ will eventually "damp out" the oscillations. In the overdamped case, no oscillations are present. The critically damped case is on the verge of, but does not have, oscillations. Figure 11-9 shows the general behavior for each of these three cases.

EXAMPLE 2. For the spring of Example 1, a retarding force of 4 times the velocity results in $b^2 - 4mk = (4)^2 - 4(8/9.8)(40)$, which is negative. This is an underdamped system. But a retarding force of 15 times the velocity results in $b^2 - 4mk = (15)^2 - 4(8/9.8)(40)$, which is positive and produces an overdamped system. ∎

Finally, an external force F can be applied to the spring system—for instance, by tapping up the weight whenever it hits bottom. This will result in the nonhomogeneous differential equation

$$m \frac{d^2x}{dt^2} + b \frac{dx}{dt} + kx = F \tag{4}$$

EXAMPLE 3. A certain spring stretches 2 m under a weight of 20 N. The weight is then pulled down an additional 0.5 m and released. Find the equation of the resulting motion, assuming a damping force equal to 2.5 times the velocity and an external force $F = 6 \sin 2t$ N.

From Equation (1), $20 = k(2)$ and $k = 10$. The differential equation, from Equation (4), is

$$\frac{20}{9.8} \frac{d^2x}{dt^2} + 2.5 \frac{dx}{dt} + 10x = 6 \sin 2t$$

with boundary conditions $x = 0.5$ and $dx/dt = 0$ when $t = 0$. The auxiliary equation is $(20/9.8)m^2 + 2.5m + 10 = 0$, which has solutions

$$\frac{-2.5 \pm \sqrt{(2.5)^2 - 4(20/9.8)10}}{2(20/9.8)} = -0.61 \pm 2.13j$$

Thus $x_c = e^{-0.61t}(c_3 \cos 2.13t + c_4 \sin 2.13t)$ is the complementary solution. The form we assume for x_p is $x_p = A \sin 2t + B \cos 2t$.

Then

$$x'_p = 2A \cos 2t - 2B \sin 2t$$
$$x''_p = -4A \sin 2t - 4B \cos 2t$$

and

$$\frac{20}{9.8}(-4A \sin 2t - 4B \cos 2t) + 2.5(2A \cos 2t - 2B \sin 2t)$$
$$+ 10(A \sin 2t + B \cos 2t) = 6 \sin 2t$$

Equating coefficients of like terms results in the system of equations

$$1.84A - 5B = 6$$
$$5A + 1.84B = 0$$

The solution is $A = 0.39$ and $B = -1.06$. Therefore, $x_p = 0.39 \sin 2t - 1.06 \cos 2t$ and

$$x = e^{-0.61t}(c_3 \cos 2.13t + c_4 \sin 2.13t) + 0.39 \sin 2t - 1.06 \cos 2t$$

Applying the boundary conditions, we get the system of equations

$$0.5 = 1(c_3) - 1.06$$
$$0 = 1(2.13c_4) + c_3(-0.61) + 2(0.39)$$

from which $c_3 = 1.56$ and $c_4 = 0.08$.

Therefore, the final equation of motion is

$$x = e^{-0.61t}(1.56 \cos 2.13t + 0.08 \sin 2.13t) + 0.39 \sin 2t - 1.06 \cos 2t$$

∎

The factor $e^{-0.61t}$ in the x_c part of the solution to Example 3 approaches zero for large values of t (this is the factor that tends to damp out oscillations), and eventually only the x_p-component of the solution is significant. The x_c-component is called the **transient solution**, and the x_p-component is called the **steady-state solution**; after some period of time the transient terms become negligible and the steady-state solution prevails.

Practice 1. A weight of 25 N is required to stretch a certain spring a length of 0.8 m. When the weight has reached equilibrium position, it is then raised 0.2 m and released. Assume a damping force equal to three times the velocity, and (a) write the differential equation describing the system, (b) tell whether the system is underdamped, overdamped, or critically damped, and (c) find the equation of motion.

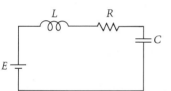

Figure 11-10

In the closed electric circuit of Figure 11-10 containing a voltage source of E volts, an inductor with inductance of L henrys, a resistor with resistance of R ohms, and a capacitor with capacitance of C farads, Kirchhoff's law results in the differential equation

$$L\frac{d^2q}{dt^2} + R\frac{dq}{dt} + \frac{q}{C} = E \qquad (5)$$

where q is the charge on the capacitor. Equation (5) looks exactly like Equation (4), where charge q corresponds to distance x, inductance L corresponds to mass m, and so on. Because the same differential equation describes both types of systems, an electrical circuit can be built to simulate a mechanical system. An electrical circuit can be easier and cheaper to build than a cumbersome mechanical system, and the behavior of the circuit parallels the behavior of the corresponding mechanical system.

If the circuit shown in Figure 11-10 is underdamped, then the transient solution will show that q oscillates; in this case the current i ($i = dq/dt$) is an alternating current. If the voltage E supplied to the system is alternating (a sine or cosine function), then the steady-state current will be alternating.

EXAMPLE 4. In a given electric circuit, $L = 2$ H, $R = 10$ Ω, $C = 0.02$ F, and $E = 0$ V. Find the equation for charge as a function of time, given that $q = 0$ C and $i = 2$ A at $t = 0$ ms.

The differential equation (5) becomes

$$2\frac{d^2q}{dt^2} + 10\frac{dq}{dt} + \frac{q}{0.002} = 0$$

or

$$2\frac{d^2q}{dt^2} + 10\frac{dq}{dt} + 500q = 0$$

The auxiliary equation is $2m^2 + 10m + 500 = 0$, with solutions $2.5 \pm 15.6j$. Therefore, the general solution to the differential equation is

$$q = e^{-2.5t}(c_3 \cos 15.6t + c_4 \sin 15.6t)$$

Applying the conditions that $q = 0$ when $t = 0$ and $i = dq/dt = 2$ when $t = 0$ results in the final solution

$$q = e^{-2.5t}(0.128 \sin 15.6t) \quad \blacksquare$$

EXAMPLE 5. Find the steady-state solution for the current in a circuit with $L = 10$ H, $R = 20$ Ω, $C = 10^{-3}$ F, and $E = 100 \sin t$ V.

Because $i = dq/dt$, we can differentiate the steady-state solution for q to get the steady-state solution for i. The steady-state solution for q is the particular solution q_p to the differential equation

$$10\frac{d^2q}{dt^2} + 20\frac{dq}{dt} + \frac{q}{10^{-3}} = 100 \sin t$$

or, dividing through by 10,

$$\frac{d^2q}{dt^2} + 2\frac{dq}{dt} + 100q = 10 \sin t$$

We let

$$q = A \sin t + B \cos t$$

Then

$$\frac{dq}{dt} = A \cos t - B \sin t$$

$$\frac{d^2q}{dt^2} = -A \sin t - B \cos t$$

Substituting into the differential equation, we have

$$-A \sin t - B \cos t + 2(A \cos t - B \sin t) + 100(A \sin t + B \cos t)$$
$$= 10 \sin t$$

from which

$$-A - 2B + 100A = 10$$
$$-B + 2A + 100B = 0$$

The solution to this system is $A = 0.099$ and $B = -0.002$. Therefore,

$$q_p = 0.099 \sin t - 0.002 \cos t$$

Differentiating with respect to t,

$$i = 0.099 \cos t + 0.002 \sin t \quad \blacksquare$$

Practice 2. Find the steady-state solution for the charge in a circuit with $L = 1$ H, $R = 3$ Ω, $C = 0.25$ F, and $E = 30 \cos t$ V.

Exercises / Section 11-10

1. The end of a tuning fork, neglecting air resistance, vibrates with simple harmonic motion determined by the differential equation

$$\frac{d^2x}{dt^2} + 12x = 0$$

Find the equation of motion.

2. The vertical motion of an automobile may be considered to arise from balancing the automobile on a single spring. Suppose a certain automobile, weighing 13,200 N, is visualized as balanced on a spring with a spring constant of 44,000 N/m. Friction exerts a damping force $b \, dx/dt$. Find the approximate value of b required for critical damping.

3. A weight of 2 N stretches a certain spring 0.2 m. The weight is then pulled down an additional 0.08 m below the equilibrium position and released. No damping is present.
   a. Find the equation of the resulting motion.
   b. Find the velocity of the weight when it first passes through the equilibrium position.

4. A weight of 140 N stretches a certain spring 2 m. When the weight has reached equilibrium position, it is then raised 0.5 m and given an initial upward velocity of 1.2 m/s. No damping is present.
   a. Find the equation of the resulting motion.
   b. Find the distance of the weight from the equilibrium position at $t = 2$ s.

5. For the spring of Exercise 3, find the equation of motion, assuming a damping force equal to twice the velocity.

6. For the spring of Exercise 4, find the equation of motion, assuming a damping force equal to 10 times the velocity.

7. For the spring of Exercise 3, assume that no damping force is present but that an external force of 8 cos 6t N acts on the system. Find the equation of motion.

8. For the spring of Exercise 4, assume that no damping is present but that an external force of 300 sin 3t N acts on the system. Find the equation of motion.

9. Find the equation of motion for the spring of Exercise 3 with the damping force of Exercise 5 and the external force of Exercise 7.

10. Find the equation of motion for the spring of Exercise 4 with the damping force of Exercise 6 and the external force of Exercise 8.

11. In a given electric circuit, $L = 0.5$ H, $R = 6$ Ω, $C = 0.003$ F, and $E = 0$ V. Find the equation for charge as a function of time, given that $q = 0.05$ C and $i = 0$ A at $t = 0$ ms.

12. For the circuit of Exercise 11, find the equation for current as a function of time.

13. Find the equation for charge as a function of time in a circuit with $L = 2$ H, $R = 3$ Ω, $C = 0.05$ F, and $E = 100$ V. Assume that $q = 0$ C and $i = 0$ A at $t = 0$ ms.

14. For the circuit of Exercise 13, find the equation for current as a function of time.

15. In a given electric circuit, $L = 0.5$ H, $R = 4$ Ω, $C = 0.05$ F, and $E = 20 \sin 2t$ V. Find the equation for charge as a function of time, given that $q = 0$ C and $i = 0$ A at $t = 0$ ms.

16. For the circuit of Exercise 15, find the equation for current as a function of time.

17. Find the steady-state solution for the charge in a circuit with $L = 2$ H, $R = 10$ Ω, $C = 0.004$ F, and $E = 10 \cos t$ V.

18. Find the steady-state solution for the current in a circuit with $L = 10$ H, $R = 50$ Ω, $C = 0.06$ F, and $E = 56 \sin 2t$ V.

19. Find the steady-state solution for the current in a circuit with $L = 1$ H, $R = 8$ Ω, $C = 0.5$ F, and $E = 120 \cos 4t$ V.

20. Find the steady-state solution for the charge in a circuit with $L = 5$ H, $R = 4$ Ω, $C = 0.08$ F, and $E = 80 \sin 3t$ V.

21. A box of weight 800 N, which is 0.5 m on a side, is floating in water (Figure 11-11(a)).

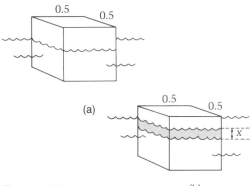

Figure 11-11   (b)

When the box is pushed a distance x m below its equilibrium position and released (Figure 11-11(b)), it bobs up and down in simple harmonic motion. Using Newton's second law, the fact that the restoring force equals the weight of the water displaced, and the fact that water weighs 9928 N/m³, the differential equation for the motion is

$$\frac{800}{9.8}\frac{d^2x}{dt^2} = -(0.5)(0.5)(9928)x$$

or

$$81.63\frac{d^2x}{dt^2} + 2482x = 0$$

Find the general solution for the displacement x as a function of time t.

22. The differential equation describing the motion of a simple pendulum is

$$\frac{d^2\theta}{dt^2} = -\frac{g}{L}\theta$$

where θ is the angle (in radians) that the pen-

dulum arm makes with the vertical, and L is the length of the pendulum arm.
a. Find the general solution for θ as a function of t.
b. Show that the period T of the pendulum is given by $T = 2\pi\sqrt{L/g}$.

11·11 Laplace Transforms

There is one more technique that allows us to solve for the particular solution of a linear differential equation subject to boundary conditions that are **initial conditions**—that is, values when the independent variable is zero. This technique involves applying an operation called a Laplace transform. The Laplace transform, as we shall see, transforms the differential equation into an algebraic equation, which can be easily solved. Then the Laplace transform is reversed by taking an inverse Laplace transform. The result is the solution to the original differential equation.

We use L to denote the process of applying the Laplace transform and L^{-1} to denote finding the inverse Laplace transform. With this notation, we can summarize the above paragraph by Figure 11-12. Previous sections of this chapter have dealt with other methods to solve linear differential equations (the path shown as a dotted line on Figure 11-12). We now suggest using the path shown as a solid line in Figure 11-12. This new approach has two advantages:

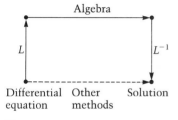

Figure 11-12

1. Instead of solving a differential equation, we have only to solve an algebraic equation.
2. Instead of finding a general solution and having to use the boundary conditions to evaluate constants and find the particular solution, we get the complete solution directly.

In this section we concentrate on how to apply L and L^{-1}; in the next section, we solve differential equations using Laplace transforms.

The Laplace transform is an operation applied to a function. Let $f(x)$ be a function, and let s be a positive constant. The **Laplace transform** of f, $L(f)$, is defined by

$$L(f) = \int_0^\infty e^{-sx} f(x)\, dx \tag{1}$$

Because the upper limit of the integral is not an actual number, evaluation of the integral in (1) cannot be done in the usual way. Instead, we let the upper limit be some variable, and we look at the limit of the integral as that variable gets larger without bound.

Therefore (1) may be rewritten as

$$L(f) = \lim_{c \to \infty} \int_0^c e^{-sx} f(x)\, dx \qquad (2)$$

EXAMPLE 1. Find the Laplace transform of the function $f(x) = x$.
By Equation (2),

$$L(f) = \lim_{c \to \infty} \int_0^c e^{-sx} x\, dx$$

Using integration by parts (or Rule 61 in the table of integrals at the back of the book), we get

$$L(f) = \lim_{c \to \infty} \int_0^c e^{-sx} x\, dx = \lim_{c \to \infty} \frac{1}{s^2}(-sx - 1)e^{-sx}\Big|_0^c$$

$$= \lim_{c \to \infty} \frac{(-sc - 1)e^{-sc}}{s^2} + \frac{1}{s^2}$$

In evaluating the limit, we note that $e^{-sc} \to 0$ as $c \to \infty$, but that $-sc \to -\infty$ as $c \to \infty$. The two factors in the product, $-sc - 1$ and e^{-sc}, seem to go in opposite directions. However, because the exponential factor approaches zero much more rapidly than the polynomial term goes to negative infinity, the limit of the product is zero. Thus

$$L(f) = \frac{1}{s^2} \qquad \blacksquare$$

In Example (1), the Laplace transform of f was a function of s. This will, in general, be true whatever the function f.

Practice 1. Find the Laplace transform of the constant function $f(x) = 1$.

By using Equation (2), we could continue to find the Laplace transform $L(f)$ for many different functions f. (But we will not.) Laplace transforms for a number of common functions are shown in the table of Figure 11-13.

Because a constant can be moved through an integral sign and because the integral of a sum is the sum of the integrals, the Laplace transform obeys the following rule, where a and b are constants and $f(x)$ and $g(x)$ are functions

$$L[af(x) + bg(x)] = aL(f) + bL(g) \qquad (3)$$

By using (3), together with the table of Laplace transforms, we can find the Laplace transforms of more complex functions.

	f	$L(f)$
1.	1	$\dfrac{1}{s}$
2.	x	$\dfrac{1}{s^2}$
3.	$x^n, \quad n = 1, 2, 3, \ldots$	$\dfrac{n!}{s^{n+1}}$
4.	e^{ax}	$\dfrac{1}{s - a}$
5.	xe^{ax}	$\dfrac{1}{(s - a)^2}$
6.	$x^n e^{ax}, \quad n = 1, 2, 3 \ldots$	$\dfrac{n!}{(s - a)^{n+1}}$
7.	$\sin ax$	$\dfrac{a}{s^2 + a^2}$
8.	$\cos ax$	$\dfrac{s}{s^2 + a^2}$
9.	$x \sin ax$	$\dfrac{2as}{(s^2 + a^2)^2}$
10.	$x \cos ax$	$\dfrac{s^2 - a^2}{(s^2 + a^2)^2}$
11.	$e^{ax} \sin bx$	$\dfrac{b}{(s - a)^2 + b^2}$
12.	$e^{ax} \cos bx$	$\dfrac{s - a}{(s - a)^2 + b^2}$
13.	$1 - \cos ax$	$\dfrac{a^2}{s(s^2 + a^2)}$
14.	$ax - \sin ax$	$\dfrac{a^3}{s^2(s^2 + a^2)}$

Figure 11-13

EXAMPLE 2. Find the Laplace transform of $f(x) = x^2 + 2x \cos 4x$.

$$
\begin{aligned}
L(f) &= L(x^2 + 2x \cos 4x) \\
&= L(x^2) + 2L(x \cos 4x) \qquad \text{By (3).} \\
&= \frac{2!}{s^3} + 2\,\frac{s^2 - 16}{(s^2 + 16)^2} \qquad \text{By Transforms 3 and 10 of} \\
&\qquad\qquad\qquad\qquad\qquad\qquad \text{Figure 11-13.} \qquad \blacksquare
\end{aligned}
$$

We will be taking the Laplace transform of a function $f(x)$. We also need to be able to take the Laplace transform of the derivative function $f'(x)$. By Equation (2),

$$
L(f') = \lim_{c \to \infty} \int_0^c e^{-sx} f'(x)\, dx
$$

We can use integration by parts on the integral, letting $u = e^{-sx}$ and $dv = f'(x)\, dx$. Then $du = -se^{-sx}$, $v = f(x)$, and

$$L(f') = \lim_{c \to \infty} e^{-sx} f(x) \Big|_0^c - \lim_{c \to \infty} \int_0^c -se^{-sx} f(x)\, dx$$

$$= 0 - f(0) + s \lim_{c \to \infty} \int_0^c e^{-sx} f(x)\, dx$$

$$= -f(0) + sL(f)$$

Therefore,

$$\boxed{L(f') = sL(f) - f(0)} \qquad (4)$$

By the same sort of process, it is also true that

$$\boxed{L(f'') = s^2 L(f) - sf(0) - f'(0)} \qquad (5)$$

EXAMPLE 3.   Given a function y with $y(0) = 0$ and $y'(0) = 2$, find the Laplace transform of $y'' + 4y'$ in terms of s and $L(y)$.

$$
\begin{aligned}
L(y'' + 4y') &= L(y'') + 4L(y') &&\text{By (3).}\\
&= s^2 L(y) - sy(0) - y'(0) + 4[sL(y) - y(0)] &&\text{By (4)}\\
&&&\text{and (5).}\\
&= s^2 L(y) - 0 - 2 + 4[sL(y) - 0]\\
&= s^2 L(y) - 2 + 4sL(y)\\
&= (s^2 + 4s)L(y) - 2 \quad \blacksquare
\end{aligned}
$$

Practice 2.   Given $y(0) = 1$ and $y'(0) = -1$, find the Laplace transform of $y'' - 3y'$ in terms of s and $L(y)$.

If we are given an expression that is the Laplace transform of a function, we may be able to determine the original function, a process called taking the **inverse Laplace transform** and denoted by L^{-1}. The table of Laplace transforms in Figure 11-13 is useful here.

EXAMPLE 4.   Find $L^{-1}\left(\dfrac{1}{(s - a)^2}\right)$.

From Transform 5 of the table, it is clear that

$$L^{-1}\left(\frac{1}{(s - a)^2}\right) = xe^{ax} \quad \blacksquare$$

EXAMPLE 5. Find $L^{-1}\left(\dfrac{1}{s^2 - 4s}\right)$.

The expression $1/(s^2 - 4s)$ does not appear in the table as $L(f)$ for any f. However, if we write

$$\frac{1}{s^2 - 4s} = \frac{1}{s(s - 4)}$$

we see that the expression could have been obtained as the sum of two fractions of the form A/s and $B/(s - 4)$. We write

$$\frac{1}{s(s - 4)} = \frac{A}{s} + \frac{B}{s - 4} = \frac{A(s - 4) + B(s)}{s(s - 4)}$$

The two numerators must be equal, so

$$1 = A(s - 4) + B(s)$$

Equating coefficients of like terms, we get

$$1 = -4A$$
$$0 = A + B$$

Thus $A = -\frac{1}{4}$ and $B = \frac{1}{4}$. (This is just what we did in using partial fractions for integration in Section 9-9.) Thus

$$L^{-1}\left(\frac{1}{s^2 - 4s}\right) = L^{-1}\left(\frac{A}{s} + \frac{B}{s - 4}\right)$$

$$= L^{-1}\left[\frac{-1}{4s} + \frac{1}{4(s - 4)}\right]$$

$$= -\frac{1}{4}L^{-1}\left(\frac{1}{s}\right) + \frac{1}{4}L^{-1}\left(\frac{1}{s - 4}\right)$$

$$= -\frac{1}{4}(1) + \frac{1}{4}e^{4x}$$

$$= \frac{1}{4}(e^{4x} - 1) \quad \blacksquare$$

Practice 3. Find $L^{-1}\left(\dfrac{1}{s^2 - 6s + 25}\right)$. (*Hint:* Complete the square on the denominator.)

Exercises / Section **11-11**

1. Use the definition of the Laplace transform to verify Transform 7 in Figure 11-13.

2. Find $L(\cos ax)$ by using $L(\sin ax)$ and Equation (4) of this section.

3. Use the definition of the Laplace transform to verify Transform 11 in Figure 11-13.

4. Find $L(e^{ax} \cos bx)$ by using $L(e^{ax} \sin bx)$ and Equation (4) of this section.

Exercises 5–12: Find the Laplace transforms of the given function.

5. $f(x) = x^4$

6. $f(x) = e^{3x}$

7. $f(x) = x^2 e^{7x}$

8. $f(x) = \cos 4x$

9. $f(x) = xe^{2x} + 2x^3$

10. $f(x) = 3x \sin 5x - e^{4x}$

11. $f(x) = 6 \sin \pi x - 2x^2$

12. $f(x) = 4e^x \sin 2x - 3x^2 e^{3x}$

Exercises 13–18: Find the Laplace transform of the given expression in terms of s and $L(y)$.

13. $y'' - 2y'$; $y(0) = 0$, $y'(0) = 4$

14. $2y'' - 6y'$; $y(0) = 1$, $y'(0) = 7$

15. $2y'' + 7y$; $y(0) = -1$, $y'(0) = 2$

16. $3y' + 4y$; $y(0) = 5$

17. $4y'' - 6y' + 3y$; $y(0) = 2$, $y'(0) = 0$

18. $2y'' + 4y' - 7y$; $y(0) = 0$, $y'(0) = \frac{1}{2}$

Exercises 19–26: Find the inverse Laplace transform of the given expression.

19. $\dfrac{1}{s - 3}$

20. $\dfrac{6}{s^2 + 36}$

21. $\dfrac{1}{s^2 - 2s + 1}$

22. $\dfrac{s}{s^2 + 5}$

23. $\dfrac{s}{(s^2 + 9)^2}$

24. $\dfrac{4}{s(s + 2)}$

25. $\dfrac{s - 3}{s^2 + 16}$

26. $\dfrac{s + 1}{s^2 - 4s + 13}$

11·12   Differential Equations and Laplace Transforms

Now that we can take Laplace transforms and inverse Laplace transforms, we can use these techniques to find solutions of linear differential equations subject to initial conditions. The process essentially involves three steps, corresponding to the three sides in Figure 11-12 of the last section.

EXAMPLE 1.   Solve the differential equation $y' + 2y = 0$ with $y(0) = 3$, using Laplace transforms.

We apply the Laplace transform to both sides of the equation.

$$L(y' + 2y) = L(0) \tag{1}$$

By the definition of the Laplace transform, $L(0) = 0$. Using the properties of the Laplace transform from the last section, Equation (1) becomes

$$L(y') + L(2y) = 0$$

or

$$sL(y) - 3 + 2L(y) = 0 \tag{2}$$

Equation (2) is an algebraic equation, which can easily be solved for $L(y)$.

$$L(y) = \frac{3}{s + 2}$$

Remember that y is the solution to the differential equation, and we now know $L(y)$. If we apply the inverse Laplace transform, that will give us y.

$$y = L^{-1}\left(\frac{3}{s + 2}\right) = 3L^{-1}\left(\frac{1}{s + 2}\right) = 3e^{-2x}$$

It can easily be checked that $y = 3e^{-2x}$ is the solution to the differential equation and satisfies the boundary condition. ■

The three steps used to solve a differential equation by Laplace transforms are the following:

1. Take the Laplace transform of both sides of the differential equation. This results in an algebraic equation involving s and $L(y)$.
2. Solve this algebraic equation for $L(y)$ in terms of s.
3. Take the inverse Laplace transform of $L(y)$; this gives you y, the solution to the differential equation.

EXAMPLE 2. Solve the differential equation $y' - y = 6e^x$ with $y(0) = 1$.

Applying the Laplace transform to the equation and solving for $L(y)$, we get

$$L(y' - y) = L(6e^x)$$

$$L(y') - L(y) = \frac{6}{s - 1}$$

$$sL(y) - 1 - L(y) = \frac{6}{s - 1}$$

$$(s - 1)L(y) = \frac{6}{s - 1} + 1$$

$$L(y) = \frac{6}{(s - 1)^2} + \frac{1}{s - 1}$$

Now we take the inverse transform.

$$y = L^{-1}\left(\frac{6}{(s - 1)^2} + \frac{1}{s - 1}\right) = 6xe^x + e^x \quad ■$$

Word of Advice

The Laplace transform must be applied to all terms of the differential equation. In particular, if the equation is not homogeneous, do not forget to take the transform of the right side of the equation.

By the way, the mysterious constant s that occurs due to the Laplace transform disappears when the inverse transform is taken.

Practice 1. Solve the differential equation $y' + 3y = 5$ with $y(0) = 0$.

EXAMPLE 3. Solve the differential equation $y'' + 9y = x$ with $y(0) = 0$ and $y'(0) = 2$.
  Taking the transform of the equation, we have

$$L(y'' + 9y) = L(y'') + 9L(y) = L(x)$$

By Equation (5) of the last section, we can write

$$s^2 L(y) - s \cdot 0 - 2 + 9L(y) = \frac{1}{s^2}$$

Solving for $L(y)$,

$$L(y) = \frac{1}{s^2(s^2 + 9)} + \frac{2}{s^2 + 9}$$

Finally

$$y = L^{-1}\left[\frac{1}{s^2(s^2 + 9)} + \frac{2}{s^2 + 9}\right]$$

$$= \frac{1}{27}(3x - \sin 3x) + \frac{2}{3}\sin 3x$$

$$= \frac{1}{9}x + \frac{17}{27}\sin 3x$$

The solution is $y = \frac{1}{9}x + \frac{17}{27}\sin 3x$. ■

Practice 2. Solve the differential equation $y'' + 4y = 1$ with $y(0) = 0$ and $y'(0) = 1$.

EXAMPLE 4. In a given electric circuit, $L = 1$ H, $R = 10$ Ω, $C = 0.008$ F, and $E = 0$ V. Find the equation for charge as a function of time, given that $q = 0$ C and $i = 10$ A at $t = 0$ ms.
  The differential equation describing this circuit is

$$q'' + 10q' + \frac{q}{0.008} = 0$$

or

$$q'' + 10q' + 125q = 0$$

Following the three-step process, we write

$$L(q'') + 10L(q') + 125L(q) = 0$$
$$s^2 L(q) - s \cdot 0 - 10 + 10[sL(q) - 0] + 125L(q) = 0$$
$$(s^2 + 10s + 125)L(q) = 10$$

$$L(q) = \frac{10}{s^2 + 10s + 125}$$

$$= \frac{10}{(s + 5)^2 + 10^2}$$

$$q = e^{-5t} \sin 10t \quad \blacksquare$$

Exercises / Section 11-12

Exercises 1–16: Use Laplace transforms to solve the differential equation subject to the given boundary conditions.

1. $y' - 3y = 0;$ $y(0) = -1$

2. $2y' + y = 0;$ $y(0) = 2$

3. $y' - 4y = 2;$ $y(0) = -3$

4. $3y' - 2y = 4;$ $y(0) = 1$

5. $y' - 3y = e^{3x};$ $y(0) = 1$

6. $y' + y = 2e^{3x};$ $y(0) = 2$

7. $y'' + 2y = 0;$ $y(0) = 1, y'(0) = 0$

8. $2y'' + y = 0;$ $y(0) = 2, y'(0) = 3$

9. $y'' - 3y' = 0;$ $y(0) = 0, y'(0) = -2$

10. $2y'' + 5y' = 0;$ $y(0) = 6, y'(0) = 1$

11. $y'' - 4y' + 5y = 0;$ $y(0) = 0, y'(0) = 3$

12. $y'' + 2y' + y = 0;$ $y(0) = 0, y'(0) = 4$

13. $y'' + 9y = x;$ $y(0) = 1, y'(0) = 0$

14. $y'' - y' = e^{2x};$ $y(0) = -1, y'(0) = 1$

15. $y'' - 3y' + 2y = 5;$ $y(0) = 0, y'(0) = 0$

16. $y'' - 2y' + y = xe^x;$ $y(0) = 0, y'(0) = 0$

17. A colony of bacteria grows at a rate equal to 0.2 times the amount present. If there are 30,000 present at $t = 0$ h, find an expression for the amount present at any time t (use Laplace transforms).

18. In an *RL* circuit with $E = 8$ V, $R = 10$ Ω, and $L = 1$ H, the current is initially zero. Find an expression for the current as a function of time (use Laplace transforms).

19. A weight of 60 N stretches a certain spring 0.5 m. The weight is then pulled down an additional 0.2 m below the equilibrium position and released. Find the equation of the resulting motion, assuming no damping force (use Laplace transforms).

20. In a given electric circuit, $L = 1$ H, $R = 6$ Ω, $C = 0.02$ F, and $E = 0$ V. Find the equation for charge as a function of time, given that $q = 0$ C and $i = 40$ A at $t = 0$ ms (use Laplace transforms).

STATUS CHECK

Now that you are at the end of Chapter 11, you should be able to:

section **11-1** Show that a given function is a general solution of a given differential equation.

Show that a given function is a solution of a given differential equation subject to specific boundary conditions.

section **11-2**   Use the method of separation of variables to find the general solution of a given differential equation.

Use the method of separation of variables to solve a given differential equation subject to a specific boundary condition.

section **11-3**   Find the general solution to a given differential equation by regrouping terms to isolate integrable combinations.

Find the solution to a given differential equation subject to a specific boundary condition by regrouping terms to isolate integrable combinations.

section **11-4**   Find the general solution to a given linear differential equation of the first order.

Find the solution to a given linear differential equation of the first order subject to a specific boundary condition.

section **11-5**   Find the equation of a curve, given an expression for its slope at any point and the coordinates of a point through which it passes.

Given the equation for a family of curves, find the equation for the family of orthogonal trajectories.

Given a situation where growth or decay of a substance is proportional to the amount present, solve the corresponding differential equation.

For a given closed *RL* electric circuit, find an expression for the current as a function of time.

For a given closed *RC* electric circuit, find an expression for the charge on the capacitor as a function of time.

Find an expression for velocity as a function of time for a falling body encountering air resistance proportional to the velocity.

Find an expression for the amount of substance in a mixture as a function of time, given that solution containing the substance flows into and out of the mixture.

Solve given differential equations modeling other physical situations.

section **11-6**   Approximate the solution to a first-order, first-degree differential equation with a given boundary condition by using Picard's method.

Find a series solution to a given differential equation.

Use Euler's method to generate a series of points approximating the solution curve to a given first-order, first-degree differential equation with a specific boundary condition.

section **11-7**   Find the general solution to a given second-order, linear, homogeneous differential equation with constant coefficients.

Find the solution to a given second-order, linear, homogeneous differential equation with constant coefficients subject to specific boundary conditions.

section **11-8**   Find the general solution to a given second-order, linear, homogeneous differential equation with constant coefficients, where the auxiliary equation has repeated roots.

Find the general solution to a given second-order, linear, homogeneous differential equation with constant coefficients, where the auxiliary equation has complex roots.

Find the solution to a given second-order, linear, homogeneous differential equation with constant coefficients subject to specific boundary conditions, where the auxiliary equation has repeated or complex roots.

section **11-9** Find the general solution to a given second-order, linear, non-homogeneous differential equation with constant coefficients.

Find the solution to a given second-order, linear, nonhomogeneous differential equation with constant coefficients subject to specific boundary conditions.

section **11-10** Write and solve the second-order, linear differential equation corresponding to oscillating motion, such as a weight on a spring.

Write and solve the second-order, linear differential equation corresponding to a closed electric circuit containing a voltage source, inductor, resistor, and capacitor.

section **11-11** Find the Laplace transform of a given function, using a table of basic Laplace transforms.

Given an expression containing a function y and its derivatives together with initial conditions, find the Laplace transform of the expression in terms of s and $L(y)$.

Find the inverse Laplace transform of a given expression in s, using a table of basic Laplace transforms.

section **11-12** Use Laplace transforms to solve a given linear differential equation subject to specific initial conditions.

11-13 More Exercises for Chapter **11**

Exercises 1–22: Solve the given differential equation.

1. $(x^2 - 1)(y^2 + 1)\, dx = x^2 y\, dy$

2. $y\, dx - x\, dy = y^3\, dy$

3. $xy' - y = x^4$

4. $e^{2x-y}\, dx + e^{2y-x}\, dy = 0$

5. $(1 + y)y\,' = x\sqrt{y}$

6. $x\, dy + (2y + 1 - xy)\, dx = 0$

7. $x\, dx + y\, dy = (x^2 + y^2)\, dx$

8. $(y + 1)x\, dx + e^{-x}y\, dy = 0$

9. $y' + y = \cos x$

10. $y' = \cos x \sec y$

11. $xy\, dx - (x + 2)\, dy = 0$

12. $(3 - x)y' + y = (3 - x)^3 x^2$

13. $y'' + 2y' - 3y = 0$

14. $y'' - 2y' - 8y = 0$

15. $y'' + 8y' + 16y = 0$

16. $25y'' - 30y' + 9y = 0$

17. $2\, d^2y/dx^2 + dy/dx - 15y = 0$

18. $6\, d^2y/dx^2 - dy/dx - 2y = 0$

19. $d^2y/dx^2 + 3y = 0$

20. $3\, d^2y/dx^2 - 2\, dy/dx + 2y = 0$

21. $y'' + 2y' - 3y = 9x$

22. $y'' - y' + y = \sin x + x$

Exercises 23–28: Solve the differential equation subject to the boundary conditions shown.

23. $(y + 3)x\,dx + x^{-1}\,dy = 0$; $y(0) = 1$

24. $dy/x = y^2\,dx + [(y^2 \cos x)/x]\,dx$; $y(0) = \frac{1}{2}$

25. $y' = y + e^x$; $y(0) = 2$

26. $x\,dy/dx = 6xe^{2x} + y(2x - 1)$; $y(1) = 1$

27. $y'' - 4y = 6e^x$; $y(0) = -1$, $y'(0) = 0$

28. $y'' - 2y' + y = 5e^{2x} + 3x^2$; $y(0) = 10$, $y'(0) = 0$

Exercises 29–32: Use Laplace transforms to solve the differential equation with the given boundary conditions.

29. $y' + 2y = 1$; $y(0) = 0$

30. $y'' + 3y' = 0$; $y(0) = 0$, $y'(0) = 1$

31. $y'' + 9y = 3x$; $y(0) = 1$, $y'(0) = -1$

32. $y'' - 6y' + 10y = 0$; $y(0) = 1$, $y'(0) = 0$

Exercises 33–34: Using Picard's method, find the indicated approximation to the solution.

33. $y' = xy - 1$, $y(0) = 0$; y_3

34. $y' = x^2 + y$, $y(1) = 2$; y_2

Exercises 35–36: Find a series solution.

35. $y' = xy$

36. $y' = x + 2y$

Exercises 37–38: Use Euler's method to find five points approximating the solution function.

37. $y' = x + 3y$, $y(0) = 0$;   use $\Delta x = 0.2$.

38. $y' = \sin y$, $y(0) = 0.3$;   use $\Delta x = 0.1$.

39. Find the equation of the orthogonal trajectories of the curves $y = cx^4$.

40. The isobars (lines connecting points of equal barometric pressure) on a certain weather map are given by the equation $y = c/x^2$. Find the equation of the orthogonal trajectories that indicate the wind direction from high- to low-pressure areas.

41. A chemical reaction converts substance A into substance B at a rate proportional to the amount of A left unconverted. The reaction begins with 200 g of substance A present; at the end of an hour, 120 g of A remain. How many grams of A will remain at the end of 4 h?

42. A radioactive substance decays at a rate proportional to the amount present. If 10% of a certain isotope has decayed in 300 years, find the half-life of the substance.

43. An electric circuit contains a 10-V voltage source, a resistance of 60 Ω, and an inductance of 0.5 H. Find an expression for the current i (in amperes) as a function of time t (in milliseconds), given that $i(0) = 0$.

44. In an RC circuit with $E = 200$ V, $R = 80\,\Omega$, and $C = 0.01$ F, find the charge on the capacitor at $t = 0.5$ s if the initial charge is 1 C.

45. An object of mass 5 kg is thrown into the air with an initial velocity of 10 m/s. Resistance due to air pressure is equal to half the velocity. Find the velocity at $t = 7$ s.

46. A tank contains 40 gal of brine in which 10 lb of salt has been dissolved. Brine with a concentration of 2 lb/gal of salt is poured into the tank at the rate of 7 gal/min, while 5 gal/min of mixture is drained from the tank. Find an expression for the amount y of salt in the tank as a function of time.

47. A weight of 24 N stretches a certain spring 0.3 m. When the weight has reached equilibrium position, it is then raised 0.1 m and released. Assuming a damping force equal to five times the velocity, what is the equation of motion?

48. A certain spring stretches 0.4 m under a weight of 40 N. The weight is then pulled down an additional 0.1 m and released. Find the equation of the resulting motion, assuming a damping force equal to the velocity and an external force $F = 10 \sin 2t$ N.

49. In a given electric circuit, $L = 2$ H, $R = 20\,\Omega$, $C = 0.005$ F, and $E = 10 \cos t$ V. Find the equation for charge as a function of time, given that $q = 0$ C and $i = 0$ A at $t = 0$ ms.

50. Find the steady-state solution for the current in a circuit with $L = 10$ H, $R = 4\,\Omega$, $C = 0.16$ F, and $E = 100 \sin 2t$ V.

Glossary

Abscissa   The x-coordinate of a point in the plane (page 15).

Absolute error   The actual error in the measurement of a quantity (page 169).

Absolute value   The size of a number independent of its sign (page 11).

Acceleration   The rate of change of velocity with respect to time, and the second derivative of distance with respect to time (page 123).

Algebraic function   A function involving only the operations of addition, subtraction, multiplication, division, and taking roots (page 7).

Alternating series   An infinite series whose terms alternate signs between positive and negative (page 358).

Analytic geometry   A system for visualizing algebraic equations geometrically (page 14).

Antiderivative   Another name for *integral* (page 182).

Antidifferentiation   Another name for *integration* (page 182).

Arithmetic sequence   An infinite sequence of numbers in which each item is obtained from the preceding one by adding a fixed number (page 340).

Asymptote   A line to which a curve gets closer and closer; may be used as a guideline to graph the curve (page 21).

Auxiliary equation   The polynomial equation obtained from a homogeneous linear differential equation (page 417).

Axis   The line through the focus of a parabola perpendicular to the directrix (page 40).

Boundary conditions   The n independent conditions the variables must satisfy in an nth-order differential equation; the boundary conditions allow evaluation of the n constants in the general solution (page 376).

Cardioid   The graph of a type of polar coordinate equation (page 76).

Center of mass   A point in a system of point masses in the plane such that if the entire mass were concentrated at that point and the first moments with respect to the axes were recomputed, they would be the same as when computed for the original system (page 224).

Centroid   Another name for *center of mass*.

Chain rule A procedure used to differentiate the composition of two functions (page 114).

Circle The path traced by a point which moves so that its distance from a fixed point (the center) is a constant (the radius) (page 48).

Circular paraboloid A type of three-dimensional surface (page 69).

Closed circle A dot representing a point on a graph (page 18).

Common logarithms Logarithms to the base 10 (page 275).

Complementary solution For a nonhomogeneous, linear differential equation, the general solution to the corresponding homogeneous equation (page 425).

Components Vectors whose resultant, or sum, equals a given vector (page 156).

Composition of functions The process of merging two existing functions into a new function; if y is a function of u and u is a function of x, then y is also a function of x, the composition function (page 5).

Conic section A general name for parabolas, circles, ellipses, and hyperbolas, all of which can be obtained as the intersection of a plane with a right circular cone (page 56).

Conjugate axis A line segment through the center of a hyperbola perpendicular to the transverse axis (page 52).

Constant A quantity whose value does not change, but remains fixed, or the symbol used to represent such a quantity (page 1).

Constant of integration An arbitrary constant that appears in the process of finding an indefinite integral (page 183).

Continuous A function is everywhere continuous if there are no breaks or gaps in its graph (page 86).

Converge An infinite series converges to a sum S if the sums obtained by successively adding more terms get closer and closer to S (page 341).

Critical point A point on a curve where the first derivative is zero (page 141).

Curvilinear motion Two-dimensional motion in a plane (page 155).

Cylindrical surface A three-dimensional surface whose equation contains only two variables (page 70).

Damping force A retarding force in simple harmonic motion (page 431).

Definite integral The definite integral of a function $f(x)$ from a to b is the limit, as the number of subintervals between a and b becomes infinite, of the sum of the products obtained by multiplying the function evaluated in each subinterval times the width of the subinterval (page 194).

Degree A unit of angle measure; one complete revolution equals 360 degrees (page 12).

Degree of a differential equation The power of the highest-order derivative in the equation (page 375).

Degree of a polynomial The highest power of x that appears in the polynomial (page 8).

Dependent variable In a function, a variable whose unique value is computed by the function rule from the value(s) of the independent variable(s) (page 2).

Derivative The derivative of a function is a mathematical expression involving the function and a limit which allows for measurement of the rate of change of the function (page 94).

Differential For a function $y = f(x)$, the differential of x, dx, is Δx and the differential of y, dy, is $f'(x)\,dx$; dy can be used to approximate $y\Delta$ (page 167).

Differential equation An equation containing a derivative or differential (page 374).

Differentiation The process of finding the derivative of a function (page 91).

Direct proportion In a function of the form $y = kx$, where k is a constant, y is directly proportional to x (page 10).

Directed distance The directed distance of a point from the y-axis is the x-coordinate of the point; the directed distance of a point from the x-axis is the y-coordinate of the point (page 223).

Directrix See *parabola* (page 40).

Discontinuous A function is discontinuous at a point if there is a break or gap in its graph at that point (page 86).

Distance formula A rule for computing the distance between two points in the plane (page 25).

Diverge An infinite series diverges if there is no sum to which it converges (page 342).

Double integral For a function of two variables, the result of integrating first with respect to one variable while holding the other fixed and then integrating with respect to the other variable (page 241).

Ellipse The path traced by a point that moves so that the sum of its distances from two fixed points (the foci) is a constant (page 44).

Equilibrium position The at-rest position of a weighted spring (page 429).

Euler's formulas Identities involving e, j, the sine function, and the cosine function (page 422). Namely,

$$e^{j\theta} = \cos\theta + j\sin\theta,$$
$$e^{-j\theta} = \cos\theta - j\sin\theta$$

Explicit function A function y of x whose equation has the form $y = f(x)$ (page 125).

Exponential function A transcendental function of the form $y = b^x$, where b is a constant; b is called the **base**, and x is called the **exponent** of the function (page 271).

Factorial For a positive integer n, n factorial is the product of whole numbers from 1 to n (page 348).

First derivative The derivative of a function (page 121).

First derivative test A method to help locate maximum or minimum points on a curve (page 142).

First moment The first moment of a point mass with respect to an axis is the product of the mass and its directed distance from the axis (page 223).

Focus A fixed point in the definition of a conic section; see *parabola, ellipse,* and *hyperbola* (pages 40, 44, 50).

Fourier series An infinite series whose terms involve the sine and cosine functions, useful to approximate periodic functions (page 367).

Function A relationship between two variables such that a value for one variable (the independent variable) determines a unique value for the other variable (the dependent variable) (page 2).

Function of several variables A relationship between variables in which the value of the dependent variable is uniquely determined by the values of two or more independent variables (page 11).

General solution For an nth-order differential equation, a general solution is a solution containing n independent arbitrary constants (page 376).

Geometric sequence An infinite sequence of numbers in which each item in the sequence is obtained from the preceding one by multiplying by a fixed number (page 340).

Graph of a point A dot on a rectangular coordinate system corresponding to the coordinates of the point (page 16).

Graph of an equation The collection of all graphs of points whose coordinates satisfy the equation (page 16).

Homogeneous equation A linear differential equation in which the function of x on the right side of the equation is zero (page 416).

Hyperbola The path traced by a point that moves so that the difference of its distances from two fixed points (the foci) is a constant (page 50).

Hyperbolic functions Transcendental functions related to the exponential function (page 276).

Hyperboloid of one sheet A type of three-dimensional surface (page 69).

Identity An equation true for all values of the variables (page 247).

Implicit differentiation The process of differentiating an implicit function (page 125).

Implicit function A function y of x in which y is not isolated (page 125).

Indefinite integral See *integral* (page 183).

Independent variable In a function, a variable whose value may be chosen and then used to compute the value of the dependent variable (page 2).

Infinite sequence An infinite list of numbers (page 340).

Infinite series The indicated sum of terms in an infinite sequence (page 341).

Inflection point A point on a curve where the concavity changes from upward to downward, or vice versa (page 141).

Initial conditions Boundary conditions for a differential equation, where the independent variable has the value zero (page 437).

Integer A number such as 0, 1, 2 ... or -1, -2, -3, ... (page 16).

Integrable A function is integrable if its indefinite integral exists (page 189).

Integral The (indefinite) integral of a function $f(x)$ is any function whose derivative equals $f(x)$ (page 182).

Integral sign The symbol used to denote integration (page 183).

Integrand A function to be integrated (page 183).

Integration The process of finding an integral (page 182).

Interval of convergence The interval of values around a fixed point within which a power series converges (page 344).

Inverse Laplace transform Given an expression that is the Laplace transform of a function, the inverse Laplace transform of the expression is the function (page 440).

Inverse proportion In a function of the form $y = k/x$, where k is a constant, y is inversely proportional to x (page 10).

Irrational number A real number with a decimal representation that does not repeat and does not terminate (page 16).

Joint variation In a function of the form $y = kx_1 x_2$, where k is a constant, y varies jointly as x_1 and x_2 (page 10).

Laplace transform An operation involving an integral that is applied to a function (page 437).

Latus rectum A line segment from one side of an ellipse to the other that is perpendicular to the major axis and goes through a focus (page 46).

Lemniscate The graph of a type of polar coordinate equation (page 76).

Limaçon The graph of a type of polar coordinate equation (page 76).

Limit A number L is the limit of a function $f(x)$ as x approaches a number a if the values of $f(x)$ get closer to L as the values of x get closer to a (page 85).

Linear differential equation A differential equation with a specific polynomial-like format (pages 388, 416).

Linear equation An equation of the form $Ax + By + C = 0$, representing a straight line (page 37).

Logarithmic function A transcendental function related to the exponential function; $y = \log_b x$ means $b^y = x$ (page 272).

Maclaurin series A power series expansion of a function whose interval of convergence centers around 0 (page 348).

Major axis The line segment between the two vertices of an ellipse (page 45).

Maximum A (relative) maximum point on a curve is higher than any surrounding point (page 141).

Method of least squares A technique for fitting a straight line to a collection of data points (page 174).

Method of undetermined coefficients A process of using unspecified constants as coefficients in an expression and then evaluating those constants by putting the expression in an equation and doing algebraic manipulations (pages 322, 441).

Minimum A (relative) minimum point on a curve is lower than any surrounding point (page 141).

Minor axis The line segment from one side of an ellipse to the other that is perpendicular to the major axis and goes through the center (page 46).

Moment of inertia The moment of inertia of a point mass with respect to an axis is the product of the mass and the square of its distance from the axis (page 229).

Natural logarithms Logarithms to the base e, where e is a specific irrational number (page 275).

Nonhomogeneous equation A linear differential equation in which the function of x on the right side of the equation is not zero (page 416).

Normal line A line perpendicular to the tangent line to a curve at a point (page 138).

Open circle A small ring denoting the end of a line segment, but not a point on a graph (page 18).

Order The order of a differential equation is the order of the highest-order derivative in the equation (page 375).

Ordered pair A first and second number, used to locate points on a plane in a rectangular coordinate system (page 15).

Ordinary differential equation A differential equation containing only derivatives with respect to a single independent variable (page 375).

Ordinate The y-coordinate of a point in the plane (page 15).

Origin The point at which the axes of a rectangular coordinate system cross (page 15).

Orthogonal trajectories Two families of curves that intersect everywhere at right angles (page 395).

Parabola The path traced by a point always equidistant from a fixed point (the focus) and a fixed line (the directrix) (page 40).

Parametric equations A set of equations in which

both x and y are given as functions of a third variable, called a **parameter** (page 158).

Partial derivative The derivative of a function of several variables obtained by differentiating with respect to one of the independent variables, holding the remaining independent variables constant (page 129).

Partial differential equation An equation containing partial derivatives (page 375).

Particular solution A solution to a differential equation where at least one arbitrary constant has been specified (pages 376, 425).

Period For a periodic function $f(x)$, the smallest cycle within which the function repeats; that is, the smallest positive number k such that $f(x + k) = f(x)$ for all x (pages 247, 363).

Periodic function A function whose pattern of values repeats itself in cycles (page 246).

Point-slope form The equation of a straight line when a point on the line and the slope of the line are known (page 35).

Polar axis The fixed line in a polar coordinate system (page 71).

Polar coordinate system A fixed point and a fixed line used as references to locate points in the plane (page 71).

Pole The fixed point in a polar coordinate system (page 71).

Polynomial function A function of the form $f(x) = a_n x^n + a_{n-1} x^{n-1} + \cdots + a_1, \quad x + a_0,$ where n is a nonnegative integer and the a's are constants (page 8).

Power rule A formula to differentiate a function raised to a power (page 115).

Power series expansion A power series expansion of a function $f(x)$ is a power series that sums to the value of $f(x)$ for all x in some interval (page 347).

Power series in x An infinite series in the form of an "infinite polynomial" in x (page 344).

Product rule A formula for differentiating the product of two functions (page 109).

Quadrant One of four sections into which the plane is divided by a rectangular coordinate system (page 15).

Quotient rule A formula for differentiating the quotient of two functions (page 111).

Radian A unit of angle measure; one complete revolution equals 2π radians (page 13).

Radius The fixed distance from the center of a circle to any point on the circle (page 48).

Radius of gyration The radius of gyration of an area or volume with respect to an axis is a distance such that if the entire mass were placed at that distance from the axis and the moment of inertia with respect to that axis recomputed, it would be the same as when computed for the original area or volume (page 232).

Radius vector The positive distance from the origin of a point on the terminal side of an angle in standard position (page 12).

Rate of change For y a function of x, the relative change in y as x changes (page 83).

Rational number A real number that can be represented as a terminating or repeating decimal; it can be obtained by dividing one integer by another (page 16).

Real number A number that has a decimal representation (page 16).

Rectangular coordinate system A set of scaled perpendicular axes allowing representation of points in a plane (page 14).

Rectilinear motion Motion along a straight line (page 155).

Relative error The ratio of the absolute error in measuring a quantity to the size of the quantity itself (page 169).

Resultant The resultant of two or more vectors is a single vector, which alone has the same effect as the original vectors acting together (page 156).

Riemann sum The sum over n subintervals of the products obtained by multiplying a function evaluated on each subinterval times the width of the subinterval; it appears in the definition of the definite integral (page 194).

Rose The graph of a type of polar equation (page 76).

Saddle A type of three-dimensional surface (page 70).

Scalar quantity A quantity with magnitude but no direction (page 156).

Second derivative The derivative of the first derivative of a function (page 121).

Second derivative test A method to help find maximum or minimum points on a curve (page 144).

Second moment Another name for *moment of inertia* (page 229).

Section The curve formed by the intersection of a three-dimensional surface with a plane parallel to one of the coordinate planes (page 68).

Semimajor axis   Half of the major axis of an ellipse (page 46).

Semiminor axis   Half of the minor axis of an ellipse (page 46).

Simple harmonic motion   Oscillatory motion described by one type of second-order, linear differential equation (page 430).

Simpson's rule   A formula to approximate the value of a definite integral (page 335).

Slope   The ratio of the vertical distance to the horizontal distance between two points on a line, given by $\dfrac{y_2 - y_1}{x_2 - x_1}$ (page 27).

Slope-intercept form   The equation of a straight line when its slope and y-intercept are known (page 36).

Solution of a differential equation   A relationship between the variables that satisfies the equation (pages 376).

Spring constant   The constant of proportionality in Hooke's law (pages 235, 430).

Standard form   The simplest form for the equation of a conic section, occurring when the axes are horizontal and vertical and the center (or vertex, in the case of a parabola) is at the origin (page 56).

Standard position   The position of an angle on a rectangular coordinate system in which the vertex is at the origin and the initial side is along the positive x-axis (page 12).

Steady-state solution   A component in the solution to a differential equation for damped oscillatory motion (page 433).

Symmetry about the origin   A curve is symmetric about the origin if $(-x, -y)$ is on the graph whenever (x, y) is (page 21).

Symmetry about the x-axis   A curve is symmetric about the x-axis if $(x, -y)$ is on the graph whenever (x, y) is (page 19).

Symmetry about the y-axis   A curve is symmetric about the y-axis if $(-x, y)$ is on the graph whenever (x, y) is (page 19).

Taylor series   A power series expansion of a function whose interval of convergence centers around a point $x = a$ (page 361).

Total differential   The analogy, for a function of several variables, of the differential (page 171).

Trace   The curve formed by the intersection of a three-dimensional surface with a coordinate plane (page 68).

Transcendental function   A nonalgebraic function, such as a trigonometric function (page 8).

Transient solution   A component in the solution to a differential equation for damped oscillatory motion (page 423).

Translation of axes   The process of introducing a new coordinate system parallel to the original one (page 433).

Transverse axis   The line segment with the two vertices of a hyperbola as endpoints (page 52).

Trapezoid rule   A formula to approximate the value of a definite integral (page 332).

Trigonometric functions   Six functions obtained from ratios associated with an angle (page 12).

Variable   A quantity whose value can change, or vary, or the symbol used to represent such a quantity (page 1).

Vector   A directed line segment representing a vector quantity, where the length of the line segment is proportional to the magnitude of the vector quantity, and the line segment points in the direction of the vector quantity (page 156).

Vector quantity   A quantity with both magnitude and direction (page 156).

Velocity   Rate of change of position with respect to time (page 123).

Vertex   The fixed endpoint about which a half-line is rotated to generate an angle; also, the intersection of the axis with a parabola, an endpoint of the major axis in an ellipse, or an endpoint of the transverse axis in a hyperbola (pages 12, 40, 46, 52).

Work   The work done in moving an object a distance d by applying a constant force F is the product Fd (page 234).

x-axis   The horizontal axis on a rectangular coordinate system (page 15).

x-coordinate   The first of the two numbers in the ordered pair of numbers representing a point on the plane in a rectangular coordinate system (page 15).

y-axis   The vertical axis on a rectangular coordinate system (page 15).

y-coordinate   The second of the two numbers in the ordered pair of numbers representing a point on the plane in a rectangular coordinate system (page 15).

y-intercept   The point at which a line crosses the y-axis (page 36).

appendix **1**

Table of Geometric Formulas

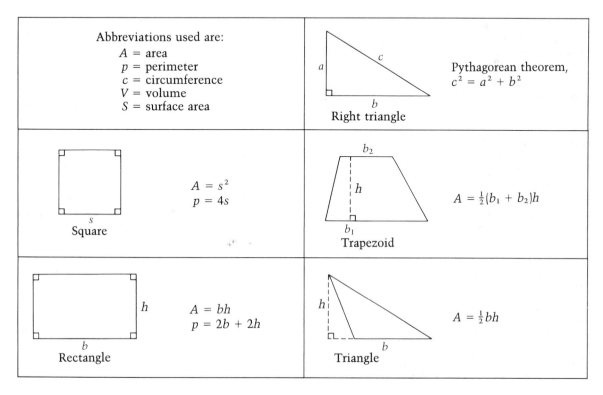

Abbreviations used are:
A = area
p = perimeter
c = circumference
V = volume
S = surface area

Pythagorean theorem,
$c^2 = a^2 + b^2$

Right triangle

$A = s^2$
$p = 4s$

Square

$A = \frac{1}{2}(b_1 + b_2)h$

Trapezoid

$A = bh$
$p = 2b + 2h$

Rectangle

$A = \frac{1}{2}bh$

Triangle

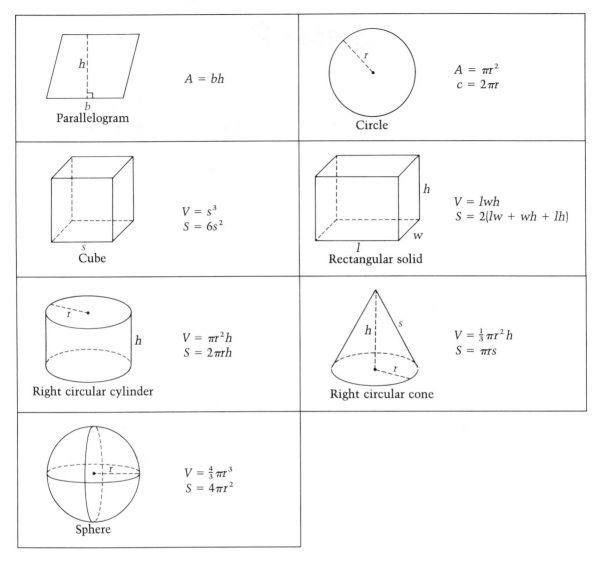

$$A = bh$$

Parallelogram

$$A = \pi r^2$$
$$c = 2\pi r$$

Circle

$$V = s^3$$
$$S = 6s^2$$

Cube

$$V = lwh$$
$$S = 2(lw + wh + lh)$$

Rectangular solid

$$V = \pi r^2 h$$
$$S = 2\pi rh$$

Right circular cylinder

$$V = \tfrac{1}{3}\pi r^2 h$$
$$S = \pi rs$$

Right circular cone

$$V = \tfrac{4}{3}\pi r^3$$
$$S = 4\pi r^2$$

Sphere

appendix 2

International System of Units

(SI, or metric, units)

Common metric prefixes

Name	Symbol	Meaning
kilo	k	1000 times
deci	d	1/10 of
centi	c	1/100 of
milli	m	1/1000 of
micro	μ	1/1,000,000 of

Quantity	Initial	Unit	Symbol	Quantity	Initial	Unit	Symbol
Capacitance	C	farad	F	Luminous Intensity	I	candela	cd
Capacity		liter	L	Mass	m	kilogram	kg
Charge	Q, q	coulomb	C	Pressure	p	pascal	Pa
Current	I, i	ampere	A	Power	P	watt	W
Distance	s	meter	m	Resistance	R	ohm	Ω
Energy (work)	KE, W	joule	J	Temperature	T	°C	°C
Force	F	newton	N	Time	t	second,	s
Illuminance	E	lux	lx			minute,	min
Inductance	L	henry	H			hour	h
				Voltage, potential	V, E	volt	V

appendix 3

Table of Integrals

Basic forms

1. $\int u^n \, du = \dfrac{u^{n+1}}{n+1} + C \quad n \neq -1$

2. $\int \dfrac{1}{u} \, du = \ln|u| + C$

3. $\int e^u \, du = e^u + C$

4. $\int \sin u \, du = -\cos u + C$

5. $\int \cos u \, du = \sin u + C$

6. $\int \sec^2 u \, du = \tan u + C$

7. $\int \csc^2 u \, du = -\cot u + C$

8. $\int \sec u \tan u \, du = \sec u + C$

9. $\int \csc u \cot u \, du = -\csc u + C$

10. $\int \tan u \, du = -\ln|\cos u| + C$

11. $\int \cot u \, du = \ln|\sin u| + C$

12. $\int \sec u \, du = \ln|\sec u + \tan u| + C$

13. $\int \csc u \, du = -\ln|\csc u + \cot u| + C$

14. $\int \dfrac{1}{\sqrt{a^2 - u^2}} \, du = \operatorname{Arcsin} \dfrac{u}{a} + C$

15. $\int \dfrac{1}{a^2 + u^2} \, du = \dfrac{1}{a} \operatorname{Arctan} \dfrac{u}{a} + C$

Forms containing a + bu

16. $\int \dfrac{u}{a + bu} \, du = \dfrac{1}{b^2} \left(a + bu - a \ln|a + bu|\right) + C$

17. $\int \dfrac{u}{(a + bu)^2} \, du = \dfrac{1}{b^2} \left(\ln|a + bu| + \dfrac{a}{a + bu}\right) + C$

18. $\int \dfrac{u}{(a + bu)^3} \, du = \dfrac{1}{b^2} \left(\dfrac{a}{2(a + bu)^2} - \dfrac{1}{a + bu}\right) + C$

458

19. $\int \dfrac{u^2}{a + bu}\, du = \dfrac{1}{b^3}\left[\dfrac{1}{2}(a + bu)^2 - 2a(a + bu) + a^2 \ln|a + bu|\right] + C$

20. $\int \dfrac{du}{u(a + bu)} = -\dfrac{1}{a}\ln\left|\dfrac{a + bu}{u}\right| + C$

21. $\int \dfrac{du}{u(a + bu)^2} = \dfrac{1}{a(a + bu)} - \dfrac{1}{a^2}\ln\left|\dfrac{a + bu}{u}\right| + C$

22. $\int \dfrac{du}{u^2(a + bu)} = -\dfrac{1}{au} + \dfrac{b}{a^2}\ln\left|\dfrac{a + bu}{u}\right| + C$

Forms containing $\sqrt{a + bu}$

23. $\int u\sqrt{a + bu}\, du = \dfrac{2}{15b^2}(3bu - 2a)(a + bu)^{3/2} + C$

24. $\int \dfrac{u}{\sqrt{a + bu}}\, du = \dfrac{2}{3b^2}(bu - 2a)\sqrt{a + bu} + C$

25. $\int \dfrac{u^2}{\sqrt{a + bu}}\, du = \dfrac{2}{15b^3}(8a^2 + 3b^2u^2 - 4abu)\sqrt{a + bu} + C$

26. $\int \dfrac{\sqrt{a + bu}}{u}\, du = 2\sqrt{a + bu} + a\int \dfrac{du}{u\sqrt{a + bu}}$

27. $\int \dfrac{du}{u\sqrt{a + bu}} = \dfrac{1}{\sqrt{a}}\ln\left|\dfrac{\sqrt{a + bu} - \sqrt{a}}{\sqrt{a + bu} + \sqrt{a}}\right| + C \qquad a > 0$

Forms containing $\sqrt{a^2 + u^2}$

28. $\int \sqrt{a^2 + u^2}\, du = \dfrac{u}{2}\sqrt{a^2 + u^2} + \dfrac{a^2}{2}\ln|u + \sqrt{a^2 + u^2}| + C$

29. $\int \dfrac{du}{\sqrt{a^2 + u^2}} = \ln|u + \sqrt{a^2 + u^2}| + C$

30. $\int \dfrac{du}{u\sqrt{a^2 + u^2}} = -\dfrac{1}{a}\ln\left|\dfrac{\sqrt{a^2 + u^2} + a}{u}\right| + C$

31. $\int \dfrac{\sqrt{a^2 + u^2}}{u}\, du = \sqrt{a^2 + u^2} - a\ln\left|\dfrac{a + \sqrt{a^2 + u^2}}{u}\right| + C$

32. $\int \dfrac{u^2}{\sqrt{a^2 + u^2}}\, du = \dfrac{u}{2}\sqrt{a^2 + u^2} - \dfrac{a^2}{2}\ln|u + \sqrt{a^2 + u^2}| + C$

Forms containing $\sqrt{u^2 - a^2}$

33. $\int \sqrt{u^2 - a^2}\, du = \dfrac{u}{2}\sqrt{u^2 - a^2} - \dfrac{a^2}{2}\ln|u + \sqrt{u^2 - a^2}| + C$

34. $\int \dfrac{du}{\sqrt{u^2 - a^2}} = \ln|u + \sqrt{u^2 - a^2}| + C$

35. $\int \dfrac{du}{u\sqrt{u^2 - a^2}} = \dfrac{1}{a}\,\text{Arcsec}\,\dfrac{u}{a} + C$

36. $\int \dfrac{\sqrt{u^2 - a^2}}{u} \, du = \sqrt{u^2 - a^2} - a \text{ Arccos } \dfrac{a}{u} + C$

37. $\int \dfrac{u^2}{\sqrt{u^2 - a^2}} \, du = \dfrac{u}{2} \sqrt{u^2 - a^2} + \dfrac{a^2}{2} \ln|u + \sqrt{u^2 - a^2}| + C$

Forms containing $\sqrt{a^2 - u^2}$

38. $\int \sqrt{a^2 - u^2} \, du = \dfrac{u}{2} \sqrt{a^2 - u^2} + \dfrac{a^2}{2} \text{ Arcsin } \dfrac{u}{a} + C$

39. $\int \dfrac{du}{u\sqrt{a^2 - u^2}} = -\dfrac{1}{a} \ln \left| \dfrac{a + \sqrt{a^2 - u^2}}{u} \right| + C$

40. $\int \dfrac{\sqrt{a^2 - u^2}}{u} \, du = \sqrt{a^2 - u^2} - a \ln \left| \dfrac{a + \sqrt{a^2 - u^2}}{u} \right| + C$

41. $\int \dfrac{u^2}{\sqrt{a^2 - u^2}} \, du = -\dfrac{u}{2} \sqrt{a^2 - u^2} + \dfrac{a^2}{2} \text{ Arcsin } \dfrac{u}{a} + C$

42. $\int \dfrac{du}{u^2\sqrt{a^2 - u^2}} = -\dfrac{1}{a^2 u} \sqrt{a^2 - u^2} + C$

Trigonometric forms

43. $\int \sin^2 u \, du = \dfrac{u}{2} - \dfrac{1}{4} \sin 2u + C$

44. $\int \cos^2 u \, du = \dfrac{u}{2} + \dfrac{1}{4} \sin 2u + C$

45. $\int \tan^2 u \, du = \tan u - u + C$

46. $\int \cot^2 u \, du = -\cot u - u + C$

47. $\int \sin^n u \, du = -\dfrac{1}{n} \sin^{n-1} u \cos u + \dfrac{n-1}{n} \int \sin^{n-2} u \, du$

48. $\int \cos^n u \, du = \dfrac{1}{n} \cos^{n-1} u \sin u + \dfrac{n-1}{n} \int \cos^{n-2} u \, du$

49. $\int \tan^n u \, du = \dfrac{1}{n-1} \tan^{n-1} u - \int \tan^{n-2} u \, du \qquad n \neq 1$

50. $\int \cot^n u \, du = -\dfrac{1}{n-1} \cot^{n-1} u - \int \cot^{n-2} u \, du \qquad n \neq 1$

51. $\int \sin au \cos bu \, du = -\dfrac{\cos(a-b)u}{2(a-b)} - \dfrac{\cos(a+b)u}{2(a+b)} + C \qquad a \neq b$

52. $\int \sin au \sin bu \, du = \dfrac{\sin(a-b)u}{2(a-b)} - \dfrac{\sin(a+b)u}{2(a+b)} + C \qquad a \neq b$

53. $\int \cos au \cos bu \, du = \dfrac{\sin(a-b)u}{2(a-b)} + \dfrac{\sin(a+b)u}{2(a+b)} + C \qquad a \neq b$

54. $\int u \sin u \, du = \sin u - u \cos u + C$

55. $\int u \cos u \, du = \cos u + u \sin u + C$

56. $\int u^n \sin u \, du = -u^n \cos u + n \int u^{n-1} \cos u \, du$

57. $\int u^n \cos u \, du = u^n \sin u - n \int u^{n-1} \sin u \, du$

58. $\int \dfrac{du}{\sin u \cos u} = \ln|\tan u| + C$

59. $\int \dfrac{\sin u}{u^n} \, du = -\dfrac{\sin u}{(n-1)u^{n-1}} + \dfrac{1}{n-1} \int \dfrac{\cos u}{u^{n-1}} \, du \qquad n \neq 1$

60. $\int \dfrac{\cos u}{u^n} \, du = -\dfrac{\cos u}{(n-1)u^{n-1}} - \dfrac{1}{n-1} \int \dfrac{\sin u}{u^{n-1}} \, du \qquad n \neq 1$

Other forms

61. $\int u e^{au} \, du = \dfrac{1}{a^2} (au - 1)e^{au} + C$

62. $\int u^n e^{au} \, du = \dfrac{1}{a} u^n e^{an} - \dfrac{n}{a} \int u^{n-1} e^{au} \, du$

63. $\int e^{au} \sin bu \, du = \dfrac{e^{au}}{a^2 + b^2} (a \sin bu - b \cos bu) + C$

64. $\int e^{au} \cos bu \, du = \dfrac{e^{au}}{a^2 + b^2} (a \cos bu + b \sin bu) + C$

65. $\int \ln u \, du = u \ln u - u + C$

66. $\int u^n \ln u \, du = \dfrac{u^{n+1}}{(n+1)^2} [(n+1)\ln u - 1] + C$

67. $\int \text{Arcsin } u \, du = u \text{ Arcsin } u + \sqrt{1 - u^2} + C$

68. $\int \text{Arctan } u \, du = u \text{ Arctan } u - \frac{1}{2} \ln(1 + u^2) + C$

Answers to Practices

Chapter 1

Section 1.1 / page 1

1. $V = 0.625I = 0.625(21) = 13.125$ V
2. $C = 2\pi r$
3. $h(3) = 2 \cdot 3^2 - 1 = 2 \cdot 9 - 1 = 17$;
   $h(-1) = 2(-1)^2 - 1 = 2 - 1 = 1$; $h(a) = 2a^2 - 1$;
   $h(h(2)) = 2(h(2))^2 - 1 = 2[2 \cdot 2^2 - 1]^2 - 1$
   $= 2(7)^2 - 1 = 97$
4. $y = 2z = 2x^3$
5. $x \geq 0, x \neq 3$

Section 1.2 / page 7

1. $f(x) = (5x - 2)/(3x - 4)^{1/2}$
2. $E = 11,860/d^2$
3. $f(4) = |4| = 4$; $f(-6) = |-6| = 6$
4. 12
5. $\sin 3 = 0.1411$, $\cos(-1.2) = 0.3624$,
   $\sin(\pi/4) = 0.7071$, $\tan(-0.3) = -0.3093$

Section 1.3 / page 14

1.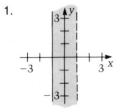

2. (Three points on the graph are $(-1, 6)$, $(0, 2.6)$, and $(1.8, -3.52)$. Of course, you may have computed others.)

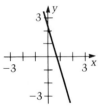

3.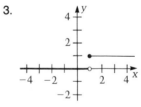

4.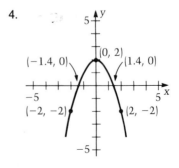

462

5. Horizontal asymptote: $y = 0$; vertical asymptote: $x = 0$; curve is symmetric about the origin.

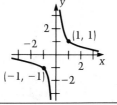

Section 1.4 / page 24

1. $\sqrt{10}$
2. $-\frac{4}{5}$
3. Both lines have slope of 4.5.

Chapter 2

Section 2.2 / page 35

1. $4x + y - 13 = 0$
2. $3x + 10y + 1 = 0$
3. $m = -3$

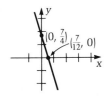

Section 2.3 / page 40

1. $F(0, 3)$, $y = -3$

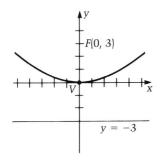

2. $x^2 = 16y$

Section 2.4 / page 44

1. $V(0, 2)$, $V(0, -2)$, $F(0, \sqrt{3})$, $F(0, -\sqrt{3})$

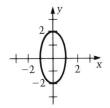

2. $(y^2/13) + (x^2/4) = 1$

Section 2.5 / page 50

1. $V(3, 0)$, $V(-3, 0)$, $F(5, 0)$, $F(-5, 0)$

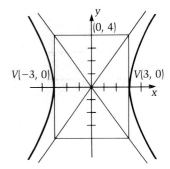

2. $(y^2/16) - (x^2/20) = 1$ 3.

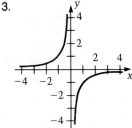

Section 2.6 / page 55

1. $9x^2 + 5y^2 - 54x + 10y + 41 = 0$
2. Hyperbola, center at $(2, -1)$.

$$4(x^2 - 4x) - 5(y^2 + 2y) = -91$$
$$4(x^2 - 4x + 4) - 5(y^2 + 2y + 1) = -91 + 16 - 5$$
$$4(x - 2)^2 - 5(y + 1)^2 = -80$$
$$5(y + 1)^2 - 4(x - 2)^2 = 80$$
$$\frac{(y + 1)^2}{16} - \frac{(x - 2)^2}{20} = 1$$

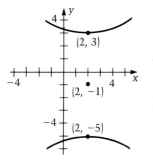

Chapter 3

Section 3.1 / page 66

1.
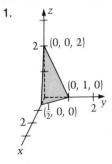

2. $x^2 + y^2 = 4$
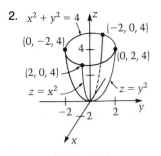

3.
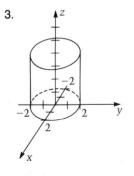

Section 3.2 / page 71

1.
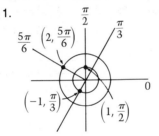

2. $(1.41, -1.41)$

3. $(3.16, 2.82)$

4. $r^2 = \cos^2\theta - \sin^2\theta$, or (using the trigonometric identity $\cos^2\theta - \sin^2\theta = \cos 2\theta$) $r^2 = \cos 2\theta$.

5. $x^2 + y^2 = 3x$

Section 3.3 / page 76

1.
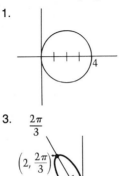

2.
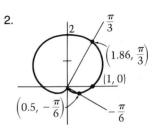

3.
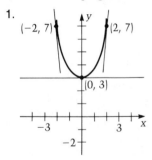

Chapter 4

Section 4.1 / page 81

1.
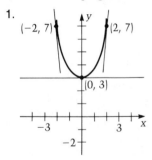

When $x = 0$, slope is zero; when $x = -2$, slope is negative; when $x = 2$, slope is positive.

2. 1, 17

Section 4.2 / page 84

1.

x	2.500	2.800	2.900	2.990	2.999
$f(x)$	-2.500	-3.400	-3.700	-3.970	-3.997

x	3.400	3.200	3.110	3.010	3.001
$f(x)$	-5.200	-4.600	-4.330	-4.030	-4.003

$\lim\limits_{x\to 3} (5 - 3x) = -4$

2. 9.4 3. $\frac{8}{3}$ 4. 2

Section 4.3 / page 91

1. $\lim\limits_{\Delta x \to 0} [f(x + \Delta x) - f(x)]/\Delta x = \lim\limits_{\Delta x \to 0} (2x + \Delta x) = 2x;$
   at (0, 0), slope $= 0$.

2. $\dfrac{ds}{dt} = \lim\limits_{\Delta t \to 0} [f(t + \Delta t) - f(t)]/\Delta t = \lim\limits_{\Delta t \to 0} (6t + \Delta t) = 6t$

Section 4.4 / page 97

1. 3.2 km/h 2. 3.04 J/A

Section 4.5 / page 101

1. $9x^8$ 2. $28x$ 3. $21x^2 + 4x - 6$

Section 4.6 / page 108

1. $36x^2 + 30x - 8$
2. $(-6x^2 + 16x)/4x^4$, which reduces to $(-3x + 8)/2x^3$.
3. $(8x^3 + 6x^2 - 12x + 7)/(8x^2 - 8x + 2)$

Section 4.7 / page 114

1. $3(3x^2 - 2x + 4)^2(6x - 2)$

2. $(6x + 12)(2)(2x^2 + 7x)(4x + 7) + (2x^2 + 7x)^2(6)$
   $= 6(2x^2 + 7x)(10x^2 + 37x + 28)$
3. $-30x^2(2x^3 + 7)^{-6}$ 4. $-\frac{1}{2} x^{-3/2}$
5. $x^3(x^2 - 3x)^{-1/2}(8x - 21)/2$

Section 4.8 / page 121

1. 8 2. 2/27 3. 49.14 m/min²

Section 4.9 / page 125

1. $(2x + 3)/3y^2$ 2. $(2y - 3x^2)/(2y - 2x)$

3. $\dfrac{ds}{dt} = \dfrac{-2t}{3s^2}; \dfrac{ds}{dt}\bigg|_{t=1,\, s=1} = -\dfrac{2}{3}$ m/min;

   $\dfrac{d^2s}{dt^2} = \dfrac{-6s^2 + 12ts\, ds/dt}{9s^4}; \dfrac{d^2s}{dt^2}\bigg|_{t=1,\, s=1} = -\dfrac{14}{9}$ m/min²

4. -1

Section 4.10 / page 128

1. $\partial z/\partial y = 2x; \dfrac{\partial z}{\partial y}\bigg|_{(1,\, 2,\, 8)} = 2$
2. $\partial r/\partial s = 2t^2,\ \partial r/\partial t = 4st - 4t^3$

Chapter 5

Section 5.1 / page 136

1. $\dfrac{dy}{dx} = 9x^2; \dfrac{dy}{dx}\bigg|_{x=-1} = 9;$ the equation is $y + 4$
   $= 9(x + 1)$, or $9x - y + 5 = 0$.
2. Slope $= -1/9$; the equation is $x + 9y + 37 = 0$.
3. $\dfrac{dy}{dx} = \dfrac{1}{y - 1};$ tangent line: slope $= \frac{1}{2}$, equation is
   $x - 2y + 9 = 0$; normal line: slope $= -2$, equation
   is $2x + y + 3 = 0$.

Section 5.2 / page 141

1. $y' = 3x^2 + 6x - 9 = 3(x + 3)(x - 1);$
   $(1, -15), (-3, 17)$
2. $(1, -15)$ is a minimum point, $(-3, 17)$ is a maximum point.
3. $y'' = 6x + 6, y''|_{x=1} = 12 > 0$ and $(1, -15)$ is a minimum; $y''|_{x=-3} = -12 < 0$ and $(-3, 17)$ is a maximum.
4. $(-1, 1)$ is an inflection point.

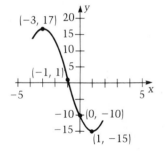

5. $(0, 0)$ is a minimum point by the first derivative test.

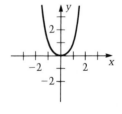

6. $y' = 4x^3 - 4x = 4x(x^2 - 1);$ critical points are
   $(0, 0), (1, -1), (-1, -1); y'' = 12x^2 - 4; y''(0) = -4,$
   $y''(1) = 8$, and $y''(-1) = 8$, so $(0, 0)$ is a maximum
   point and $(1, -1)$ and $(-1, -1)$ are minimum
   points; inflection points are $(\sqrt{\frac{1}{3}}, -\frac{5}{9})$ and
   $(-\sqrt{\frac{1}{3}}, -\frac{5}{9})$

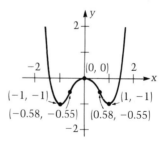

Section 5.3 / page 148

1. $437.50 (the cost for $x = 250$).
2. From the equations for area and perimeter, $A = 30 = xy$ and $p = 2x + 2y$, we get $p = 60/y + 2y$. The solution is $x = y = \sqrt{30} = 5.48$ m.

Section 5.4 / page 155

1. **R** has magnitude 5.695 and direction 43.3°.
2. Velocity has magnitude 10, direction 53.1°; acceleration has magnitude 24.7, direction 76°.
3. $v = 7.2$, $\alpha = 33.7°$
4. 5.7

Section 5.5 / page 162

1. 2700 J/s
2. $x^2 + y^2 = 36$; find dy/dt when $y = 3$, given that $dx/dt = 0.3$. When $y = 3$, $x^2 + 9 = 36$ and $x = \sqrt{27} = 5.196$; differentiating,

$2x \dfrac{dx}{dt} + 2y \dfrac{dy}{dt} = 0$ or $(5.196)(0.3) + (3) \dfrac{dy}{dt} = 0$;

$\dfrac{dy}{dt} = -0.52$. The top is sliding down the wall at 0.52 m/s.

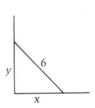

Section 5.6 / page 167

1. $dy = 2(2x - 1)(2)\, dx = (8x - 4)\, dx$
2. $\Delta y = 5.76$, $dy = 5.6$
3. 3.468 cm³
4. $0.088 = 8.8\%$

Section 5.7 / page 170

1. $dA = w\, dl + l\, dw = 32$ cm²
2. $\partial z/\partial x = -2x = 0$ and $\partial z/\partial y = -2y = 0$; solution is $x = 0$, $y = 0$; the point $(0, 0, 6)$ is a maximum point.
3. $\bar{x} = 5$, $\bar{y} = 33.2$, $\overline{xy} = 172$, $\overline{x^2} = 33$; the least-squares line is $y = 0.75x + 29.45$.

Chapter 6

Section 6.1 / page 182

1. $3x^4$ 2. $2x - x^3 + C$ 3. $2\sqrt{x} + C$
4. $\frac{1}{15}(x^5 - 2)^3 + C$ 5. $y = \frac{2}{3}(x + 2)^{3/2} + \frac{2}{3}$

Section 6.2 / page 190

1. $A_3 = 9$, $A_6 = 9.75$ 2. $\int_2^4 x^2\, dx$

Section 6.3 / page 196

1. $\left.\dfrac{3x^2}{2}\right|_0^2 = 6$ 2. $64 - \frac{2}{3}\cdot 8 - (1 - \frac{2}{3}) = \frac{175}{3}$
3. $-\frac{1}{6}[1 - (-3)^{-2}] = -\frac{4}{27}$

Chapter 7

Section 7.1 / page 202

1. $s = 2t^2 + 2t - 1$ 2. 10.9 m   3. 19 A

Section 7.2 / page 209

1. $\int_0^2 (y_U - y_L)\, dx = \int_0^2 [(1 - x^2) - (-2x + 1)]\, dx = \frac{4}{3}$
2. $\int_{-3}^1 (x_R - x_L)\, dy = \int_{-3}^1 [(1 - y)^{1/2} - \frac{1}{2}(1 - y)]\, dy = \frac{4}{3}$
3. $\frac{1}{2}$ (use horizontal elements).
4. $\int_{-1}^2 [0 - (x^2 - 4)]\, dx + \int_2^3 [(x^2 - 4) - 0]\, dx = \frac{34}{3}$

Section 7.3 / page 215

1. $V = \pi \int_0^1 (-2x + 2)^2\, dx = \frac{4}{3}\pi$
2. $V = \pi \int_0^2 (1 - y/2)^2\, dy = \frac{2}{3}\pi$
3. $V = \pi \int_0^2 (8 - 2x^2)^2\, dx = \frac{1024}{15}\pi$

4. $V = 2\pi \int_0^1 x(-2x + 2)\, dx = \frac{2}{3}\pi$
5. $V = 2\pi \int_0^2 y[(-y + 2)/2]\, dy = \frac{4}{3}\pi$
6. $V = 2\pi \int_0^8 (8 - y)(y/2)^{1/2}\, dy = \frac{1024}{15}\pi$
   (It requires some algebraic manipulation to get this answer, which is the same as that of Practice 3.)

Section 7.4 / page 223

1. $(0, -1)$
2. $M = \frac{3}{2}$, $M_y = 1$, $M_x = 3$, $(\bar{x}, \bar{y}) = (\frac{2}{3}, 2)$
3. $M = \pi/3$, $M_{yz} = \pi/12$, centroid is at $(\frac{1}{4}, 0)$
4. $I_y = \frac{3}{4}k$, $I_x = \frac{27}{4}k$
5. $I_x = \int_0^1 y^2 k(2\pi y)x\, dy = \frac{1}{10}\pi k$

Section 7.5 / page 232

1. $F = \int_{-3}^{-2} w(-y)x \, dy = -9800 \int_{-3}^{-2} \frac{1}{2} y \, dy$
   $= 12{,}250$ N

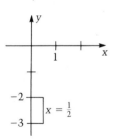

Section 7.6 / page 234

1. $\int_{0}^{0.08} 100x \, dx = 50x^2 \big|_{0}^{0.08} = 0.32$ J
2. $(15)^2 \cdot 8 + \int_{0}^{15} (15 - y)8 \, dy = 2700$ J
3. $s = \int_{0}^{12} \sqrt{1 + (\frac{3}{2} x^{1/2})^2} \, dx = 43.6$
4. $A = 2\pi \int_{0}^{1} x\sqrt{1 + 1} \, dx = \sqrt{2}\pi$

Section 7.7 / page 240

1. $-13/2$
2. $\int_{0}^{4} \int_{0}^{-(x/2)+2} (-x/2 - y + 2) \, dy \, dx = \frac{8}{3}$

Chapter 8

Section 8.1 / page 246

1. $\sin[\pi/2 - (\pi/2 - \beta)] = \cos(\pi/2 - \beta)$, or $\sin \beta = \cos(\pi/2 - \beta)$

Section 8.2 / page 250

1.

θ	0.5	0.1	0.01	-0.5	-0.1	-0.01
$\sin \theta$	0.4794	0.0998	0.0099	-0.4794	-0.9998	-0.0099
$\dfrac{\sin \theta}{\theta}$	0.9589	0.9983	0.99998	0.9589	0.9983	0.99998

$\lim\limits_{\theta \to 0} \dfrac{\sin \theta}{\theta} = 1$

2. $dy/dt = -\cos(2 - x)$ 3. $y' = 12 \sin^2 2x \cos 2x$

Section 8.3 / page 255

1. $\dfrac{d}{dx}\left(\dfrac{\cos u}{\sin u}\right) = \dfrac{\sin u(-\sin u)\,du/dx - \cos u \cos u \,du/dx}{\sin^2 u}$

$= \dfrac{-(\sin^2 u + \cos^2 u)}{\sin^2 u} \dfrac{du}{dx}$

$= -\dfrac{1}{\sin^2 u}\dfrac{du}{dx} = -\csc^2 u \dfrac{du}{dx}$

2. $y' = 2\sec^2 2x - 2\sin x \cos x = 2\sec^2 2x - \sin 2x$

3. $\dfrac{d(\sin u)^{-1}}{dx} = (-1)(\sin u)^{-2}(\cos u)\dfrac{du}{dx} = -\dfrac{\cos u}{\sin^2 u}\dfrac{du}{dx}$

$= -\dfrac{1}{\sin u} \cdot \dfrac{\cos u}{\sin u}\dfrac{du}{dx} = -\csc u \cot u \dfrac{du}{dx}$

4. $\dfrac{dy}{dx} = 2(x - \csc 2x)(1 + 2\csc 2x \cot 2x)$

Section 8.4 / page 259

1. Arcsin $(0.3) = 0.305$; Arcsin $(0.73) = 0.818$; Arcsin $(-0.8) = -0.927$
2. An error occurs; Arcsin$(2) = y$ means $\sin y = 2$, and the sine of any angle cannot exceed 1.

3. $\sin(\text{Arccos } 0.6) = 0.8$;
   Arctan$(\sin 1.2) = 0.7502$
4. $\sin(\text{Arccos } x) = \sqrt{1 - x^2}$
5. $y' = -3/\sqrt{2x - x^2}$

Section 8.6 / page 271

1.

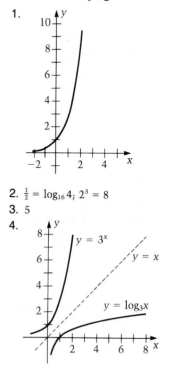

2. $\frac{1}{2} = \log_{16} 4$; $2^3 = 8$
3. 5
4.

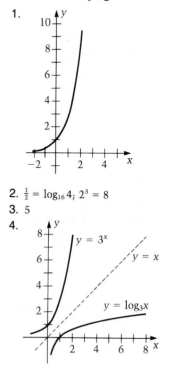

5. $\log_b 10/w$

6.

z	1	0.1	0.01	0.001	0.0001	-0.1	-0.01	-0.001	-0.0001
$(1 + z)^{1/z}$	2	2.5937	2.7048	2.7169	2.7182	2.8680	2.732	2.7196	2.7184

Section 8.7 / page 276

1. $\dfrac{5}{5x + 2}$ 2. $3/x$; $(3 \ln^2 x)/x$ 3. $8xe^{4x^2}$ 4. $2xe^{2x} + e^{2x}$

Chapter 9

Section 9.1 / page 286

1. $-(\cos 2x + 1)^{3/2}/3 + C$

Section 9.2 / page 290

1. $-\ln|1 - x| + C$

Section 9.3 / page 294

1. $\frac{1}{2}e^{2x} + C$ 2. $\frac{1}{6}e^{6x} + C$

Section 9.4 / page 297

1. $\frac{1}{3}\tan(3x + 1) + C$

2. Using the identity $1 + \tan^2 x = \sec^2 x$, we get

$$\frac{\sec^2 x}{2} = \frac{1 + \tan^2 x}{2} = \frac{\tan^2 x}{2} + \left(\frac{1}{2}\right).$$

3. $-\dfrac{1}{2} \ln|\cos x^2| + C$

Section 9.5 / page 302

1. $\int \sin^3 x \, dx = \int \sin x \sin^2 x \, dx$
$= \int \sin x (1 - \cos^2 x) \, dx$
$= \int (\sin x - \sin x \cos^2 x) \, dx$
$= -\cos x + (\cos^3 x)/3 + C$

2. $\int \cos^2 3x \, dx = \int \dfrac{(1 + \cos 6x)}{2} \, dx$
$= \frac{1}{2}\int (1 + \cos 6x) \, dx$
$= \frac{1}{2}(x + \frac{1}{6}\sin 6x) + C$
$= \frac{1}{2}x + \frac{1}{12}\sin 6x + C$

3. $\int \cot^3 x \, dx = \int \cot x \cot^2 x \, dx$
$= \int \cot x (\csc^2 x - 1) \, dx$
$= \int (\cot x \csc^2 x - \cot x) \, dx$
$= -(\cot^2 x)/2 - \ln|\sin x| + C$

Section 9.6 / page 306

1. $\frac{1}{3}\text{Arcsin}\,(3x/2) + C$

2. $\frac{1}{2}\text{Arctan}\,(x/2) + C$
3. $\frac{1}{2}\text{Arctan}\,[(x + 1)/2] + C$

Section 9.7 / page 309

1. $x \sin x - \int \sin x \, dx = x \sin x + \cos x + C$
2. $\frac{1}{8}(\cos 3x \cos x + 3 \sin 3x \sin x) + C$

Section 9.8 / page 315

1. $-\frac{1}{4}\sqrt{4 - x^2}/x + C$
2. $\ln|\sqrt{1 + x^2} + x| + C$
3. $\sqrt{x^2 - 16} - 4 \,\text{Arcsec}\,(x/4) + C$

Section 9.9 / page 321

1. $x + \dfrac{2x^2 - 3x - 2}{x^3 + x^2 - 2x}$

2. $\dfrac{2x^2 - 3x - 2}{x^3 + x^2 - 2x} = \dfrac{2x^2 - 3x - 2}{x(x + 2)(x - 1)}$

3. $\dfrac{2x^2 - 3x - 2}{x(x + 2)(x - 1)} = \dfrac{A}{x} + \dfrac{B}{x + 2} + \dfrac{C}{x - 1}$

4. $A = 1$, $B = 2$, $C = -1$

5. $\int \dfrac{x^4 + x^3 - 3x - 2}{x^3 + x^2 - 2x} \, dx$
$= \int \left(x + \dfrac{1}{x} + \dfrac{2}{x + 2} - \dfrac{1}{x - 1}\right) dx$
$= \dfrac{x^2}{2} + \ln|x| + 2 \ln|x + 2| - \ln|x - 1| + C$
$= \dfrac{x^2}{2} + \ln\left|\dfrac{x(x + 2)^2}{x - 1}\right| + C$

Section 9.10 / page 328

1. $\frac{1}{2}\ln|\tan x^2| + C$

Section 9.11 / page 330

1. $\Delta x = 0.2$; $S = 0.2754$ 2. 1.189

Chapter 10

Section 10.1 / page 340

1. $2.\overline{6}$ 2. $S = a/(1 - r) = 2/(1 - \frac{1}{4}) = 2\frac{2}{3}$
3. $f(0.1) \approx 1.11$; $f(0.2) \approx 1.22$

Section 10.2 / page 347

1. $e^{x/2} = 1 + \dfrac{1}{2}x + \dfrac{1}{4 \cdot 2!}x^2 + \dfrac{1}{8 \cdot 3!}x^3 + \cdots$

2. $1 - 2x^2 + \frac{2}{3}x^4 + \cdots$

Section 10.3 / page 351

1. $e^{x^2} = 1 + x^2 + \dfrac{x^4}{2} + \cdots$

2. $e^x \cos x = 1 + x - \dfrac{x^3}{3} + \cdots$

Section 10.4 / page 353

1. 1.105 2. 1.045 3. 1.379840

Section 10.5 / page 360

1. $\dfrac{1}{\sqrt{2}} - \dfrac{1}{\sqrt{2}}(x - 45°) - \dfrac{1}{2!\sqrt{2}}(x - 45°)^2 + \cdots$

2. 0.66911 (Remember to change to radian measure.)

Section 10.6 / page 363

1. $f(x) = \dfrac{1}{2} + \dfrac{2}{\pi}\sin x + \dfrac{2}{3\pi}\sin 3x$

   $+ \dfrac{2}{5\pi}\sin 5x + \cdots$

Chapter 11

Section 11.1 / page 374

1. $y = e^{2x} + 1$; $y' = 2e^{2x}$; $y'' = 4e^{2x}$. The equation is satisfied. Also $y(0) = e^0 + 1 = 2$, and the boundary condition is satisfied.

Section 11.2 / page 380

1. $\ln|x| + e^y = c$

2. $-\frac{1}{2}\ln(1 - x^2) - 1/y = c$, or $\frac{1}{2}\ln(1 - x^2) + \dfrac{1}{y} = c_1$,

   or $\ln c_2\sqrt{1 - x^2} + 1/y = 0$, and so on.

3. $-e^{-x} + y^3/3 = \frac{5}{3}$, or $-3e^{-x} + y^3 = 5$

Section 11.3 / page 385

1. $xy - x^2 = c$

2. $(x^2 + y^2)^2 = 2x^2 + c$

Section 11.4 / page 388

1. $3yx = x^3 + c$

2. $2yx^3 = x^2 + c$

3. $9y = 3x - 1 + ce^{-3x}$

Section 11.5 / page 394

1. $\ln|y| = x^2 - 4$

2. \$776.36

3. $q = 0.15 + 0.35e^{-100t/3}$

4. $v = 61.25 - 36.25e^{0.16t}$

Section 11.6 / page 405

1. $y_3 = x^2/2 - x^3/6 + x^4/24$

2. $y = a_0(1 + x + x^2/2! + x^3/3! + x^4/4! + \cdots) = a_0e^x$

3.

x_i	y_i	Exact y-value
0	1	1
0.2	1.20	1.22
0.4	1.44	1.49
0.6	1.73	1.82
0.8	2.08	2.23

Section 11.7 / page 415

1. $\dfrac{d^2(c_1e^{3x} + c_2e^{-x})}{dx^2} - 2\dfrac{d(c_1e^{3x} + c_2e^{-x})}{dx} - 3(c_1e^{3x} + c_2e^{-x})$

   $= 9c_1e^{3x} + c_2e^{-x} - 2(3c_1e^{3x} - c_2e^{-x}) - 3(c_1e^{3x} + c_2e^{-x})$

   $= 0$

2. $y = c_1e^{x/3} + c_2e^{-x}$

Section 11.8 / page 419

1. $y = e^{3x}(c_1x + c_2)$

2. $y = e^{2x}(c_3\cos 2x + c_4\sin 2x)$

Section 11.9 / page 425

1. $y = c_1e^{-2x} + c_2e^{-x} - \frac{2}{5}\cos x - \frac{6}{5}\sin x$

Section 11.10 / page 429

1. a. $\dfrac{25}{9.8}\dfrac{d^2x}{dt^2} + 3\dfrac{dx}{dt} + 31.25x = 0$;

   b. $b^2 - 4mk = 9 - 4\left(\dfrac{25}{9.8}\right)(31.25) < 0$,

   so the system is underdamped;

   c. $x = e^{-0.59t}(c_3\cos 3.45t + c_4\sin 3.45t)$; from the conditions $x = -0.2$ and $dx/dt = 0$ at $t = 0$, we can evaluate c_3 and c_4 and get $x = e^{-0.59t}(-0.2\cos 3.45t - 0.034\sin 3.45t)$.

2. $q = 5\cos t + 5\sin t$

Section 11.11 / page 437

1. $L(f) = \lim\limits_{c\to\infty} \int_0^c e^{-sx}\,dx$

   $= \lim\limits_{c\to\infty} -\frac{1}{s}e^{-sx}\Big|_0^c$

   $= \lim\limits_{c\to\infty} \left[-\frac{1}{s}e^{-sc}\right] + \frac{1}{s} = \frac{1}{s}$

2. $(s^2 - 3s)L(y) + 4 - s$

3. $L^{-1}\left[\dfrac{1}{s^2 - 6s + 25}\right] = L^{-1}\left[\dfrac{1}{(s - 3)^2 + 16}\right]$

   $= \frac{1}{4}L^{-1}\left[\dfrac{4}{(s - 3)^2 + 16}\right]$

   $= \frac{1}{4}e^{3x}\sin 4x$,

   from Transform 11 of the table.

Section 11.12 / page 442

1. $L(y) = 5/s(s + 3) = A/s + B/(s + 3)$;

   $A = \frac{5}{3}$, $B = -\frac{5}{3}$; $y = \frac{5}{3}(1 - e^{-3x})$

2. $L(y) = 1/s(s^2 + 4) + 1/(s^2 + 4)$;

   $y = \frac{1}{4}(1 - \cos 2x) + \frac{1}{2}\sin 2x$

Answers to Odd-Numbered Exercises

Chapter 1

Section 1.1 / page 6

1. $A = s^2$

3. $A = 5h$

5. $F = \frac{9}{5}C + 32$

7. $V = x(136 - 2x)^2$; $0 < x < 68$

9. $P = \dfrac{480}{w} + 3w$

11.
$$P = \begin{cases} 20 & 0 < w \le 1 \\ 37 & 1 < w \le 2 \\ 54 & 2 < w \le 3 \\ 71 & 3 < w \le 4 \\ 88 & 4 < w \le 5 \end{cases}$$

13. $x = (c - 600)/2.8$

15. -9; 11

17. 146; 107

19. $a^2/(a - 4)$; $9a^2/(3a - 4)$

21. $(x^2 + 1)/(x^4 + 5)$; $\frac{21}{83}$

23. 4; 25

25. 90

27. $y = 2(x + 1)$

29. $y = x - 1$

31. $x \ne 1$

33. $t > 0$

35. $t \ne 0$, $t \ne 2$, $t \ne -2$

Section 1.2 / page 13

1. $f(x) = (4x^2 - 2)^2$

3. $g(x) = (x^2 + 2)^{1/2}(x^2 + 2x + 1)$

5. $h(x) = (x + 2)^2$

7. $f(x) = (5x + 1)/(1 - x)$

9. $P = 17.78T$

11. $t = 108/r$

13. $C = kr^4/L^2$

15. $V = LWH$

17. $V = \pi r^2 h$

19. $P = 2W + 2L$

21. -6

23. $-\frac{4}{3}$

25. 10

27. $4x^2 + 4x + y + 1$

29. $2x^2 - (x^4 - y^4)/2y^2$ $- 2x + (x^2 - y^2)/2y$

31. $y \ne -1$

33. $x \ne y$, $x \ne -y$

Section 1.3 / page 22

1. $A = (2, -3)$, $B = (-1, 4)$, $C = (-4, 0)$, $D = (-2, -1)$, $E = (3, 2)$

3.

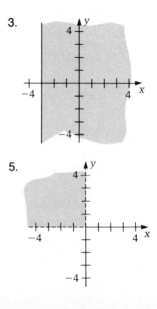

5.

7.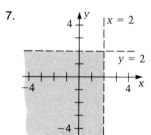

9.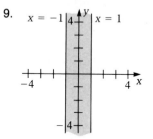

11. All points (x, y) with $x > 1$ and $y > 1$.

13. All points (x, y) with $-3 < x \le -1$.

15. $y = 2$

17.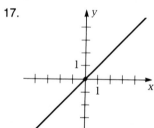

19.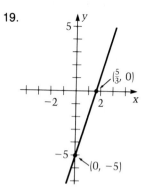

21.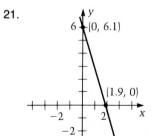

23.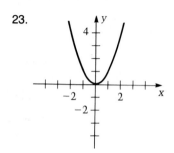

25.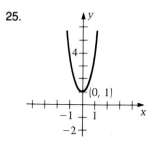

27.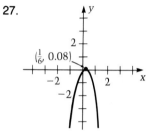

29.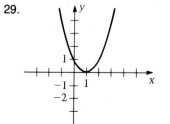

31.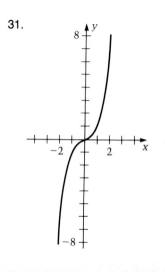

33.

35.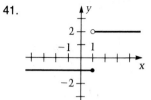

37.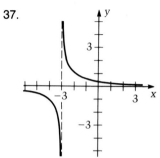

39.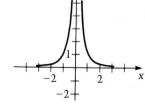

41.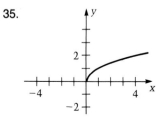

43.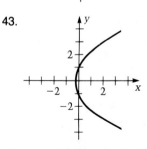

45. Symmetric about the y-axis: Exercises 23, 24, 25, 26, 39, 40, 44; symmetric about the x-axis: Exercises 43, 44.

47. Exercise 37: $x = -3$; Exercise 38: $x = 1$; Exercise 39: $x = 0$; Exercise 40: $x = 3$, $x = -3$.

49.

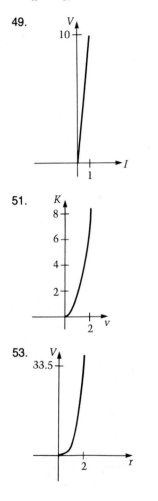

51.

53.

55. $3 = 4 \cdot 0 + 3$; $3 = 2 \cdot 0^2 + 3$; $11 = 4 \cdot 2 + 3$; $11 = 2 \cdot 2^2 + 3$

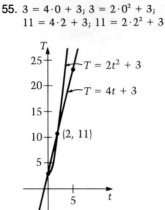

The equation $T = 4t + 3$ is correct. $T(4) \approx 19$.

Section 1.4 / page 30

1. $\sqrt{17}$; 4   **3.** $\sqrt{53}$; $-\frac{2}{7}$

5. $\sqrt{130}$; $\frac{9}{7}$ **7.** $2\sqrt{2}$; 1

9. 10.4; -1.82

11. $m_1 = m_2 = \frac{2}{5}$

13. $m_1 = 4$, $m_2 = -\frac{1}{4}$, $m_1 m_2 = -1$

15. 6, -2

17. -3

19. Slopes of lines between points are all $\frac{2}{3}$.

21. Line between $(1, 3)$ and $(3, 5)$ $(m = 1)$ is perpendicular to line between $(3, 5)$ and $(9, -1)$ $(m = -1)$. Right angle is at $(3, 5)$.

23. Two sides $= \sqrt{65}$.

25. Slope of two sides is $\frac{1}{4}$; slope of other two sides is -4.

Section 1.5 / page 31

1. $f(x) = (x^2 + 1)^{3/4}$

3. -35, $18{,}100$

5. 23

7. $-3 \le x \le 3$

9. $x \ne y$, $y \le 1$

11. $\sqrt{58}$; $-\frac{3}{7}$

13. $A = 6V^{2/3}$

15. $W = 27{,}600t$

17. $a = 2A/b$

19. $A = 5L$

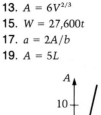

21. $P = 28/V$

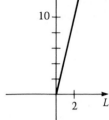

23.

25.

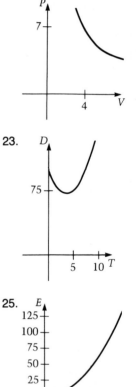

Chapter 2

Section 2.2 / page 39

1. $3x - y - 6 = 0$

3. $x - 2y - 7 = 0$

5. $2x + 7y - 17 = 0$

7. $2x - 11y - 50 = 0$

9. $3x - 2y - 12 = 0$

11. $x - 3y + 6 = 0$

13. $x = 1$

15. $2x - 3y - 4 = 0$

17. $2x + y + 2 = 0$

19. $x - y + 5 = 0$

21. $x + 3y - 11 = 0$

23. $2x - 5y + 3 = 0$

25. $3x - 5y + 36 = 0$

27. $x + 2y + 10 = 0$

29. $m = \frac{2}{3}$

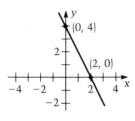

31. $m = -2$

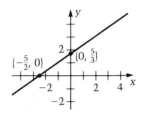

33. $m_1 = m_2 = \frac{1}{2}$

35. $m_1 m_2 = (\frac{1}{2})(-2) = -1$

37. $r = 68.75i$

39. $V = 0.075T - 1.95; 3.3$ V

41. 764.3 mm of mercury

43. $C = 3400 + 0.16x$

Section 2.3 / page 43

1. $F(1, 0), x = -1$

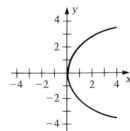

3. $F(-3, 0), x = 3$

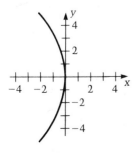

5. $F(0, 2), y = -2$

7. $F(0, -4), y = 4$

9. $F(\frac{3}{4}, 0), x = -\frac{3}{4}$

11. $F(0, -\frac{1}{4}), y = \frac{1}{4}$

13. $F(\frac{1}{6}, 0), x = -\frac{1}{6}$

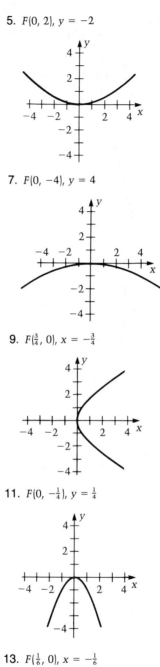

15. $x^2 = -8y$

17. $x^2 = 20y$

19. $x^2 = 12y$

21. $x^2 = -4y$

23. $x^2 = \frac{3}{2}y$

25. 5 units

27. 210 m

29.

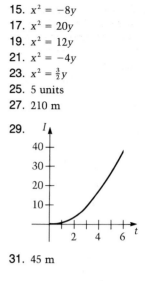

31. 45 m

Section 2.4 / page 49

1. $V(5, 0), V(-5, 0), F(3, 0),$ $F(-3, 0)$

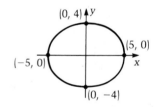

3. $V(0, 6), V(0, -6), F(0, 2\sqrt{5}),$ $F(0, -2\sqrt{5})$

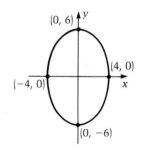

5. $V(2\sqrt{3}, 0), V(-2\sqrt{3}, 0),$ $F(\sqrt{3}, 0), F(-\sqrt{3}, 0)$

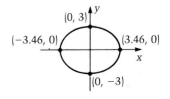

7. $V(0, 4)$, $V(0, -4)$, $F(0, 2\sqrt{2})$, $F(0, -2\sqrt{2})$

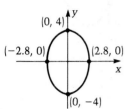

9. $V(2, 0)$, $V(-2, 0)$, $F(\sqrt{3}, 0)$, $F(-\sqrt{3}, 0)$

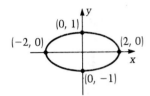

11. $V(0, 5)$, $V(0, -5)$, $F(0, \sqrt{21})$, $F(0, -\sqrt{21})$

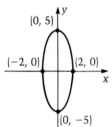

13. $V(\frac{1}{2}, 0)$, $V(-\frac{1}{2}, 0)$, $F\left(\dfrac{\sqrt{5}}{6}, 0\right)$, $F\left(-\dfrac{\sqrt{5}}{6}, 0\right)$

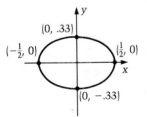

15. $V(0, 1)$, $V(0, -1)$, $F\left(0, \dfrac{1}{\sqrt{2}}\right)$, $F\left(0, -\dfrac{1}{\sqrt{2}}\right)$

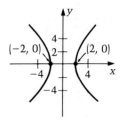

17.

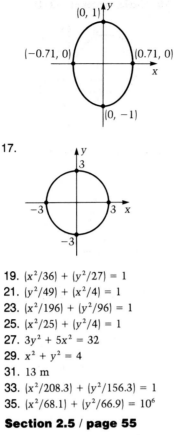

19. $(x^2/36) + (y^2/27) = 1$
21. $(y^2/49) + (x^2/4) = 1$
23. $(x^2/196) + (y^2/96) = 1$
25. $(x^2/25) + (y^2/4) = 1$
27. $3y^2 + 5x^2 = 32$
29. $x^2 + y^2 = 4$
31. 13 m
33. $(x^2/208.3) + (y^2/156.3) = 1$
35. $(x^2/68.1) + (y^2/66.9) = 10^6$

Section 2.5 / page 55

1. $V(4, 0)$, $V(-4, 0)$, $F(\sqrt{41}, 0)$, $F(-\sqrt{41}, 0)$

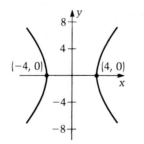

3. $V(0, 2)$, $V(0, -2)$, $F(0, \sqrt{5})$, $F(0, -\sqrt{5})$

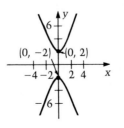

5. $V(2, 0)$, $V(-2, 0)$, $F(\sqrt{11}, 0)$, $F(-\sqrt{11}, 0)$

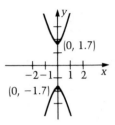

7. $V(0, \sqrt{3})$, $V(0, -\sqrt{3})$, $F(0, 2)$, $F(0, -2)$

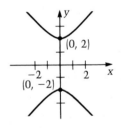

9. $V(0, 2)$, $V(0, -2)$, $F(0, \sqrt{6})$, $F(0, -\sqrt{6})$

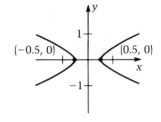

11. $V(\frac{1}{2}, 0)$, $V(-\frac{1}{2}, 0)$, $F(\sqrt{5}/4, 0)$, $F(-\sqrt{5}/4, 0)$

13. $x^2/49 - y^2/51 = 1$
15. $x^2/7 - y^2/9 = 1$
17. $y^2/36 - x^2/9 = 1$
19. $y^2/25 - x^2/9 = 1$

21.

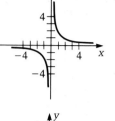

23.

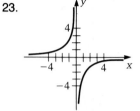

25. $x^2/1.6 - y^2/1.3 = 10^{-3}$

27.

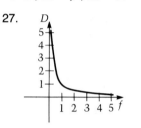

Section 2.6 / page 62

1. $y^2 - 20x - 10y + 65 = 0$
3. $x^2 + 4x + 8y + 44 = 0$
5. $16x^2 + 25y^2 - 64x - 300y + 564 = 0$
7. $64x^2 + 39y^2 - 640x + 156y - 740 = 0$
9. $x^2 + y^2 - 6x - 8y + 21 = 0$
11. $x^2 + y^2 + 10x + 16y + 64 = 0$
13. $8x^2 - y^2 - 64x - 4y + 116 = 0$
15. $16x^2 - 9y^2 + 192x + 72y + 576 = 0$
17. Hyperbola, center at $(-4, 0)$.

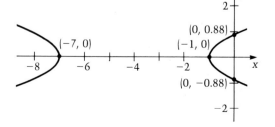

19. Circle, center at $(-3, 1)$.

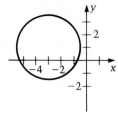

21. Circle, center at $(1, 0)$.

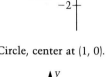

23. Parabola, vertex at $(3, 0)$.

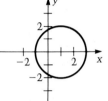

25. Ellipse, center at $(-2, 1)$.

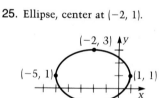

27. Hyperbola, center at $(3, 1)$.

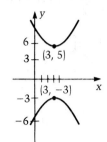

29. Ellipse, center at $(1, 1)$.

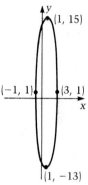

31. Hyperbola, center at $(2, -3)$.

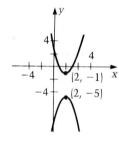

33. Parabola.

35. $4(y')^2 - (x')^2 = 36$

Section 2.7 / page 64

1. $x + 3y - 14 = 0$
3. $2x + 3y - 16 = 0$
5. $x^2 = 20y$
7. $x^2 + y^2 = 9$
9. $x^2/25 + y^2/16 = 1$
11. $y^2/75 - x^2/25 = 1$
13. $x^2 - 8x - 8y - 8 = 0$
15. $16x^2 + 25y^2 + 128x + 150y + 81 = 0$

17. $m = \frac{1}{2}$

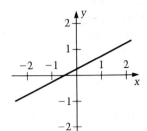

19. $F(3, 0)$, $x = -3$

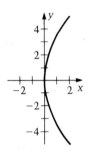

21. $r = 5$

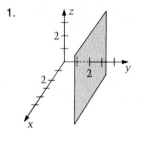

23. $V(0, \pm 9)$, $F(0, \pm 4\sqrt{2})$

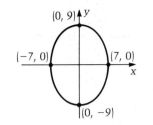

25. $V(\pm 7, 0)$ $F(\pm \sqrt{65}, 0)$

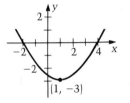

27. Parabola, vertex at $(1, -3)$.

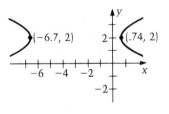

29. Hyperbola, center at $(-3, 2)$.

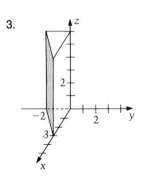

31. Circle, center at $(-2, -3)$.

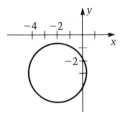

33. Ellipse, center at $(-2, -1)$.

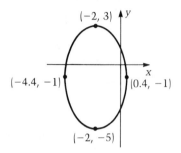

35. 5.12 m
37. $x^2/16 + y^2/9 = 1$
39. $x^2/2.235 + y^2/2.2344 = 10^{16}$

Chapter 3

Section 3.1 / page 70

1.

3.

5.

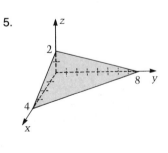

7.

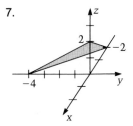

9.

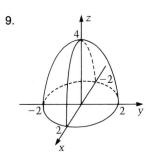

11.

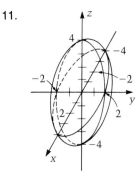

13.

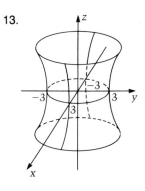

15.

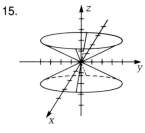

17.

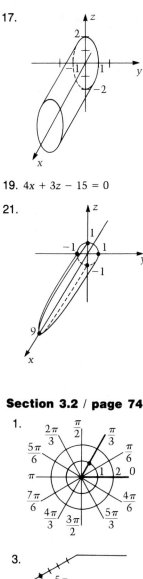

19. $4x + 3z - 15 = 0$

21.

Section 3.2 / page 74

1.

3.

$4, -\dfrac{5\pi}{6}$

5

$-2, \dfrac{\pi}{2}$

7.

$-0.8, -\pi$

9. $(1.62, 1.18)$
11. $(0.5, -0.87)$
13. $(0.45, -1.95)$
15. $(1.07, 1.19)$
17. $(5.39, 1.19)$
19. $(3.05, -0.72)$
21. $(3.61, 2.16)$
23. $(1.43, 4.28)$

(Trigonometric identities have been used to produce the alternate forms of answers for Exercises 25–29.)

25. $r \cos\theta = -4$, or $r = -4 \sec\theta$
27. $\sin\theta = 2\cos\theta$, or $\tan\theta = 2$
29. $r(\cos^2\theta + \sin^2\theta) = 3\sin\theta$, or $r = 3\sin\theta$
31. $x^2 + y^2 = 2y$
33. $x^2 + y^2 = 5x + 3y$
35. $(x^2 + y^2 + y)^2 = x^2 + y^2$
37. $9x^2 + 25y^2 - 72x - 81 = 0$

39. (a) $\dfrac{s}{\theta} = \dfrac{2\pi r}{2\pi}$ (circumference of whole circle) (central angle of whole circle)
so $s = r\theta$.

(b) $\dfrac{a}{\theta} = \dfrac{\pi r^2}{2\pi}$ (area of whole circle) (central angle of whole circle)
so $a = \frac{1}{2}r^2\theta$.

41.

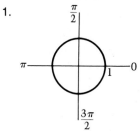

43. $r = \dfrac{1}{\sin\theta - 2\cos\theta}$

Section 3.3 / page 79

1.

3.

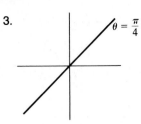

$\theta = \dfrac{\pi}{4}$

5.

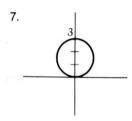

2

7.

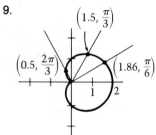

3

9.

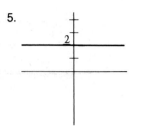

$\left(1.5, \dfrac{\pi}{3}\right)$

$\left(0.5, \dfrac{2\pi}{3}\right)$ $\left(1.86, \dfrac{\pi}{6}\right)$

1 2

11.

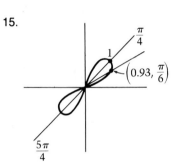

$\left(2, \dfrac{2\pi}{3}\right)$

3 1

$\left(-0.7, -\dfrac{\pi}{6}\right)$

13.

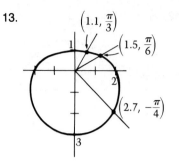

$\left(1.1, \dfrac{\pi}{3}\right)$

$\left(1.5, \dfrac{\pi}{6}\right)$

1

2

$\left(2.7, -\dfrac{\pi}{4}\right)$

3

15.

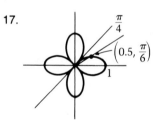

$\dfrac{\pi}{4}$

1

$\left(0.93, \dfrac{\pi}{6}\right)$

$\dfrac{5\pi}{4}$

17.

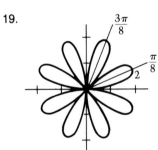

$\dfrac{\pi}{4}$

$\left(0.5, \dfrac{\pi}{6}\right)$

1

19.

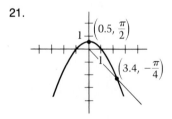

$\dfrac{3\pi}{8}$

$\dfrac{\pi}{8}$

2

21.

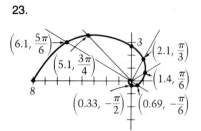

$\left(0.5, \dfrac{\pi}{2}\right)$

1

1

$\left(3.4, -\dfrac{\pi}{4}\right)$

23.

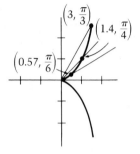

$\left(6.1, \dfrac{5\pi}{6}\right)$ 3

$\left(5.1, \dfrac{3\pi}{4}\right)$ $\left(2.1, \dfrac{\pi}{3}\right)$

$\left(1.4, \dfrac{\pi}{6}\right)$

8

$\left(0.33, -\dfrac{\pi}{2}\right)$ $\left(0.69, -\dfrac{\pi}{6}\right)$

25.

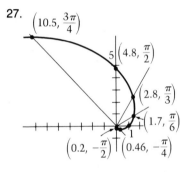

$\left(3, \dfrac{\pi}{3}\right)$ $\left(1.4, \dfrac{\pi}{4}\right)$

$\left(0.57, \dfrac{\pi}{6}\right)$

27.

$\left(10.5, \dfrac{3\pi}{4}\right)$

5 $\left(4.8, \dfrac{\pi}{2}\right)$

$\left(2.8, \dfrac{\pi}{3}\right)$

$\left(1.7, \dfrac{\pi}{6}\right)$

1

$\left(0.2, -\dfrac{\pi}{2}\right)$ $\left(0.46, -\dfrac{\pi}{4}\right)$

Section 3.4 / page 80

1. $\tan \theta = -\dfrac{1}{3}$

3. $r = 5 \cos \theta$

5. $x^2 + y^2 = 16$

7. $15x^2 - y^2 - 24x + 9 = 0$

9.

z

3

2 y

6

x

11.

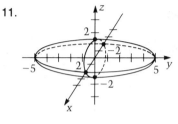

13.

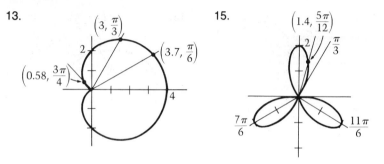

15.

Chapter 4

Section 4.1 / page 83

1. $x = 0$: zero;
$x = 1.2$: positive;
$x = -2.4$: negative.

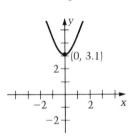

3. $x = 0$: negative;
$x = 2$: zero;
$x = 3$: positive.

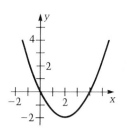

5. 22
7. -24
9. -6
11. -2
13. $\ldots, -3, -2, -1, 1, 2, 3, \ldots$
15. 3
17. 0
19. $-1, 0, 1$

21. Slope $= 2.74$.

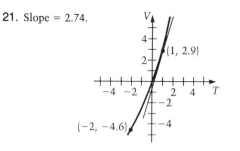

Section 4.2 / page 90

1.

x	1.400	1.450	1.490	1.499
$f(x)$	-3.000	-2.850	-2.730	-2.703

x	1.600	1.550	1.510	1.501
$f(x)$	-2.400	-2.550	-2.670	-2.697

$$\lim_{x \to 1.5} (3x - 7.2) = -2.7$$

3.

x	1.80	1.90	1.99
$f(x)$	3.64	4.31	4.93

x	2.11	2.01	2.001
$f(x)$	5.78	5.07	5.01

$$\lim_{x \to 2} \frac{x^3 + x^2 - 11x + 10}{x - 2} = 5$$

5.

x	10	100	1000	10,000	100,000	1,000,000
$f(x)$	-0.33	0.094	0.25	0.308	0.325	0.331

$$\lim_{x \to \infty} \frac{\sqrt{x} - 7}{2 + 3\sqrt{x}} = 0.333$$

7. None. 9. 4 11. 2 13. 5
15. None. 17. None. 19. −5
21. 66 23. 0.82 25. −0.05
27. 4 29. 7 31. −2 33. 3
35. 0 37. Does not exist.
39. 3 41. $-\frac{3}{4}$ 43. −2 45. 0
47. Not effective. 49. $5
51. 1, 2, 3, 4, 5, 6, 7, . . .

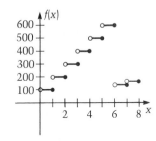

Section 4.3 / page 97

1. 4 3. 3 5. −6 7. 4
9. 0.75 11. $4q$ 13. $-2 + 9n^2$

Section 4.4 / page 100

1. 5 m/min
3. 1775.04 m/min
5. (a) 2 (b) $10 - 2t$ (c) $t = 4$ h

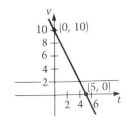

7. 1°C/h

Section 4.5 / page 107

1. 0
3. $2x$
5. $9.3x^2$
7. $20x^4$
9. $30x^2 - 8x$
11. $16x^3 - 6x^2 + 5$
13. $5.4x^2 + 5.8x$
15. $\frac{20}{3}x^9 - \frac{3}{8}x^2 + \frac{3}{4}$
17. $40x^4 - \sqrt{7}$

19. −19 km/min
21. 4.32 km/min
23. 9.97 km/min
25. 120
27. −10,240
29. 13,506
31. 2.58 A/s
33. 1.1 $/unit
35. 19.5 V/°C
37. $k_1(2x - 2k_2)$
39. (a) $-19.8t + 79.2$ (b) 4 (c) 158.4 m
41. $1 - \frac{1}{5}t + \frac{63}{100}t^2 - \frac{1}{25}t^3$

Section 4.6 / page 112

1. $36x^2 - 42x$
3. $16x + 2$
5. $-24x^2 + 24x + 36$
7. $36x^5 - 40x^4$
9. $50x^4 - 56x^3 + 75x^2 - 90x + 14$
11. $\dfrac{2x^2 + 5}{3x^2}$
13. $\dfrac{9x^2 - 6x + 23}{(3x - 1)^2}$
15. $\dfrac{-12x + 14}{(3x^2 - 7x + 4)^2}$
17. 61.6 m/s
19. −0.173 m/s
21. −35
23. 0.473
25. 1.676 A/s
27. $\dfrac{-2x^3 + 300x^2 + 300}{(100 - x)^2}$
29. $\dfrac{-k[4s^3 + 3(a + A)s^2 + 2(aA + b)s + Ab + c]}{[(s + A)(s^3 + as^2 + bs + c)]^2}$

Section 4.7 / page 120

1. $24x^2$
3. $24x^3(2x^4 - 1.9)^2$
5. $\dfrac{-24x}{(x - 3)^3}$
7. $\dfrac{-(x + 1)^{-1/2}(3x^2 + 12x + 4)}{(3x - 2)^3}$
9. $-1/(2x^2)$
11. $-2(2x^3 - 4x + 7)^{-3}(6x^2 - 4)$
13. $-4(x - 1)^{-3}$
15. $(2x - 8)^{-1/2}$

17. $x(x - 1)^{-1/2}\left(\dfrac{5x - 4}{2}\right)$
19. $\dfrac{(x - 1)^{-1/2}(-3x + 4)}{6x^3}$
21. $\left(\dfrac{2x}{x^2 + 1.8}\right)^{-1/2}\dfrac{-x^2 + 1.8}{(x^2 + 1.8)^2}$
23. $\frac{1}{4}x^{-3/4}$
25. $\frac{2}{3}(x^3 - 4x^2)^{-1/3}(3x^2 - 8x)$
27. $\frac{2}{5}(x\sqrt{x - 1})^{-1/5}$

$$(x - 1)^{-1/2}(3x - 2)$$

29. $-6x^{-4} - 35x^{-6}$
31. $-\left(\dfrac{x}{\sqrt{x + 4}}\right)^{-3}\dfrac{x + 8}{(x + 4)^{3/2}}$,

which reduces to $-\dfrac{x + 8}{x^3}$

33. $\frac{2}{5}$
35. 2.311 km/h
37. $-86C^{-3/2}$
39. $\dfrac{12t^{-4} + 2t}{(4t^3 - t^2)^2}k$
41. $20.16t^{-4/5}$ °C/h
43. $-5at(4 - t^2)^{-1/2} + 2t^{-1/3}$

Section 4.8 / page 123

1. $36x^2 - 3.6$
3. $216x - 168$
5. $\dfrac{2}{(x - 1)^3}$
7. $(x - 1)^{-3/2}\left(\dfrac{3x - 4}{4}\right)$

9. $960x^3 - 288x$
11. −820
13. −6.125
15. 0.652
17. $1.\overline{6}$
19. 24.44 km/h^2
21. 76.632 km/h^2
23. 0.0328 km/h^2
25. 0.0549 km/h^2
27. −0.405 km/h^2
29. 6.01 km/h^2

31. $f''(0) = 4$, $f''(1) = -8$,
  $f''(-1) = -8$

33. 145.6 V

35. $-m'(t)v'(t) - v(t)m''(t)$

37. 1

Section 4.9 / page 128

1. $\frac{7}{2}$

3. $2x$

5. $(1 - 4x)/3y^2$

7. $x/(y + 1)$

9. $(y - 2x)/(2y - x)$

11. $\dfrac{2x + 2 - y^2}{2xy + 3y^2}$

13. $\dfrac{5y^3 - 2 - 8xy^2 - 4x^3}{8x^2y - 15xy^2}$

15. $\dfrac{-y}{(x - 1)[2y(x - 1) - 1]}$

17. $3(2x + 1)^2/y$

19. $3(2x - 9)/(y^2 + 1)y$

21. $-\frac{1}{2}$ km/h; $-\frac{1}{8}$ km/h^2

23. -5 km/h; 9 km/h^2

25. $\frac{1}{6}$ km/h; $-\frac{1}{108}$ km/h^2

27. $\frac{1}{2}$

29. -2

31. $7.75/v$

Section 4.10 / page 131

1. $3y^2 - 4xy$; $6xy - 2x^2$

3. $1/(1 + y)$; $-x/(1 + y)^2$

5. $(r + 2)4q$; $2q^2 - 1$

7. $2u\sqrt{1 + v}$; $\dfrac{u}{2}(1 + v)^{-1/2}$

9. $2(wv + w^3)(v + 3w^2)$;
  $2(wv + w^3)w$

11. 12

13. 2

15. $x(3ax^2) + y(3by^2) = 3(ax^3 + by^3)$

17. $x\left[\dfrac{(x^2 + y^2)y - xy(2x)}{(x^2 + y^2)^2}\right]$
  $+ y\left[\dfrac{(x^2 + y^2)x - xy(2y)}{(x^2 + y^2)^2}\right] = 0$

19. $2\pi rh$

21. $2IR$; I^2

23. $kA(3.4t + 4.2x)$

25. $\sqrt{kVS/2P}$

27. $12k$ cm/mL, $36.296k$ cm/°C

Section 4.11 / page 133

1. 9.03

3. 19

5. 6

7. 2

9. 2

11. $\frac{2}{3}$

13. 0

15. $6x^2 - x$

17. $-\frac{1}{2}x^{-2}$

19. $2x + \frac{1}{2}x^{-2}$

21. $0.2/(1 + x)^2$

23. $4x$

25. $2(x^3 - 2x)(3x^2 - 2)$

27. $x(x^2 - 3)^{-1/2}$

29. $(3x^2 - 4)(21x^4 - 12x^2)$

31. $(x - 1)x^{-1/2}\left(\dfrac{5x - 1}{2}\right)$

33. $\dfrac{(x - 1)^{-1/2}(-11x^2 + 12x + 4)}{2(x^2 + 4)^4}$

35. $\dfrac{3[\sqrt{x}(x - 1)]^{1/2}(3x^{1/2} - x^{-1/2})}{4}$

37. $\dfrac{-y}{8y + x}$

39. $\dfrac{3x^2(x^3 - 1)^{-1/2}}{2(3y^2 - 1)}$

41. $24x - 24$

43. $(2 + x^2)(\frac{143}{4}x^{5/2}$
  $+ 23x^{1/2} - x^{-3/2})$

45. $\dfrac{2x^3 - 6x}{(1 + x^2)^3}$

47. $2y(xy + 1)$, $2x(xy + 1)$

49. $3(1 + y^2)x^2$, $2x^3y$

51.

12°C/h, 1500 m^2

53. 4.01 s

55. 0; charges sufficiently far apart experience negligible force.

57. 2.89 m/s, 0.38 m/s^2

59. 12

61. 4

63. -19

65. 0.9832 m^3/°C

67. $\dfrac{9t^4 + 6t^3 - 6t^2 + 6t + 1}{(3t^2 + t)^2}$

69. $-\dfrac{2mgR^2}{x^3}$

71. $2/(3kV^{1/2})$

73. $(33 - T)(5w^{-1/2} - 1)$,
  $-10\sqrt{w} - 10.4 + w$

Chapter 5

Section 5.1 / page 140

1. $3x - 2y + 4 = 0$; $2x + 3y - 19 = 0$

3. $7x + y - 1 = 0$; $x - 7y + 57 = 0$

5. $24x - y - 40 = 0$; $x + 24y - 194 = 0$

7. $x - y + 1 = 0$; $x + y - 3 = 0$

9. $x + 2y - 3 = 0$; $2x - y - 1 = 0$

11. $y = x^2/80$; $x - 2y - 10 = 0$; $x + 2y + 10 = 0$

13. $(1 + \sqrt{5}, 2 + 2\sqrt{5})$; $x + 2y - 5 = 0$

Section 5.2 / page 147

1. Minimum at $(2, -12)$, and intercepts at $(0, 0)$ and $(4, 0)$.

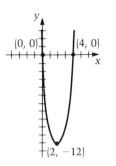

3. Minimum at $(0, 0)$, maximum at $(1, 1)$, inflection point at $(\frac{1}{2}, \frac{1}{2})$, intercepts at $(0, 0)$, and $(\frac{3}{2}, 0)$.

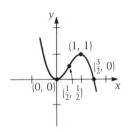

5. Minimum at $(-1, -\frac{2}{15})$, maximum at $(1, \frac{2}{15})$, inflection points at $(0, 0), (1/\sqrt{2}, 0.08)$ and $(-1/\sqrt{2}, -0.08)$ symmetric about the origin.

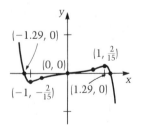

7. Minimum at $(0, 7)$, maxima at $(\sqrt{3}, 16)$ and $(-\sqrt{3}, 16)$, inflection points at $(1, 12)$ and $(-1, 12)$, symmetric about the y-axis.

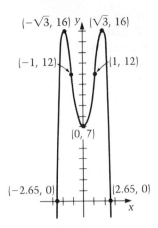

9. Maximum at $(3, 54)$ and inflection points at $(0, 0)$ and $(2, 32)$, intercepts at $(0, 0)$ and $(4, 0)$.

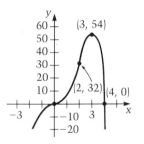

11. Inflection points at $(0, 0)$, $(1, 8)$, and $(-1, -8)$, symmetric about the origin.

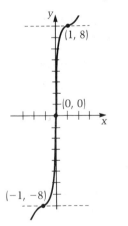

13. Asymptotes $x = 4$ and $y = 1$; function is everywhere decreasing, concave downward for $x < 4$, concave upward for $x > 4$, intercept $(0, 0)$.

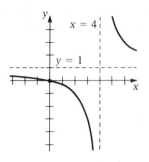

15. Maximum at $(0, 3)$, inflection points at $(1/\sqrt{3}, 9/4)$ and $(-1/\sqrt{3}, 9/4)$, asymptote $y = 0$, symmetric about the y-axis.

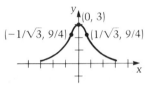

17. Minimum at $(0, 0)$, maximum at $(-2, -4)$, asymptotes $x = -1$ and $y = x - 1$.

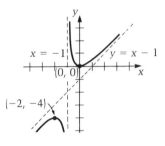

19. Maximum at $(0, -1)$, asymptotes $x = 1$, $x = -1$, and $y = 0$, concave upward for $x < -1$, $x > 1$, concave downward for $-1 < x < 1$, symmetric about the y-axis.

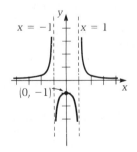

21. $A = 2\pi r^2 + 120/r$, minimum at $(2.12, 84.8)$, asymptote $r = 0$.

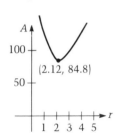

Section 5.3 / page 153

1. 24.1 units
3. 26.7 m
5. $T = \sqrt[3]{1250} = 10.77°C$
7. $\frac{1}{2}$
9. 21.99 cm should be used for the circle, the rest for the square.
11. 2.45 m by 4.9 m
13. 40.15 m by 65.25 m
15. $6234.50
17. 26¢
19. 15.8 cm by 31.6 cm
21. $r = 0.48$ m, $h = 1.12$ m
23. A square with side $= r\sqrt{2}$ units
25. The cable should reach the other side of the canyon at a point 7.07 m away from the point directly opposite A.

Section 5.4 / page 161

1. Magnitude 4.25, direction 47.3°.
3. Magnitude 3.02, direction 245.9°.
5. Velocity has magnitude 74.5, direction 75.2°; acceleration has magnitude 156.2, direction 87.1°.
7. Velocity has magnitude 18.96, direction 171.5°; acceleration has magnitude 93.77, direction 1.2°.
9. Velocity has magnitude 54, direction 89.5°; acceleration has magnitude 36, direction 90.1°.

11. 16.5
13. 2.03
15. 12.2, 80.5°
17. 48, 90°
19. 26.8 m/min, 26.6°
21. 2.01, 84.3°

Section 5.5 / page 165

1. 0.192 Ω/s
3. −27.43 kPa/min
5. −0.26 Ω/min
7. 1.08×10^{-5} cm/min²
9. 0.23 cm/min
11. 942.5 km²/day
13. 0.7 m³/s
15. 0.015 m/s
17. 109.3 km/h
19. 13.3 km/h

Section 5.6 / page 170

1. $(12x^5 - 12x^2 + 10x)\,dx$
3. $\frac{8}{3}x(2x^2 - 1)^{-1/3}\,dx$
5. $\dfrac{x^2 + 1}{2x^2}\,dx$
7. $x/\sqrt{x^2 - 3}\,dx$
9. −0.12, −0.2
11. 4.844, 4.6
13. 0.06121, 0.06
15. 0.010225, 0.01
17. 15.08 cm²
19. 0.075398 m², 0.07%
21. 0.0503 cm³

Section 5.7 / page 176

1. $dz = (3x^2 - 4x)\,dx + 9y^2\,dy$
3. $dz = (3x^2 - 3y + 2xy^2)\,dx$ $+ (-3x + 2x^2y)\,dy$
5. Maximum at $(1, 2, 2)$.
7. Minimum at $(4/3, -2/3, 11/3)$.
9. $y = -3.05x + 95.3$
11. 8.89 W
13. 0.648 Ω, 1.08%
15. 1260 cm³
17. $q = r = 5$, $s = 10$
19. $(0, 0)$
21. $L = H = 12.87$ cm, $W = 19.32$ cm
23. $i = 0.98V + 6.265$

Section 5.8 / page 179

1. $x + y + 1 = 0$, $x - y - 3 = 0$
3. $3x + 2y - 7 = 0$, $2x - 3y + 4 = 0$
5. Maximum at $(\frac{3}{8}, \frac{9}{16})$, intercepts at $(0, 0)$ and $(\frac{3}{4}, 0)$.

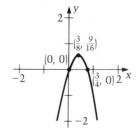

7. Minimum at $(1, 0)$, maximum at $(-1, 8)$, inflection point at $(0, 4)$, intercepts at $(1, 0)$, $(-2, 0)$ and $(0, 4)$.

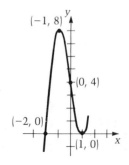

9. Minimum at $(\frac{3}{2}, -\frac{27}{16})$, inflection points at $(0, 0)$ and $(1, -1)$, intercepts at $(0, 0)$ and $(2, 0)$.

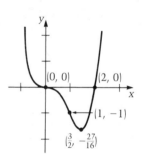

11. Asymptotes $x = 1$ and $y = 1$; function is everywhere decreasing, concave downward for $x < 1$, concave upward for $x > 1$, intercept $(0, 0)$.

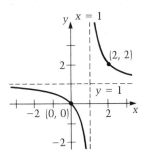

13. 10.77, 158.2°; 22.36, 169.7°
15. 4.03, 97.13°; 8.04, 84.64°
17. 132.19
19. $(21x^2 - 6x + 4)\, dx$
21. $\dfrac{5 - x^2}{(x^2 + 5)^2}\, dx$
23. -0.41, -0.4
25. $(12y^3 - 6y^2)\, dy + 6x\, dx$
27. Minimum at $(0, 0, 4)$.
29. $y = -1.49x + 87.03$
31. $x - 4y + 2 = 0$,
    $x - 4y - 6 = 0$
33. 31.2 cm by 44 cm
35. 4.79 cm by 4.79 cm by 9.58 cm

37. 0.74 units
39. 3 days
41. 12.17, 80.54°
43. 192 W/min
45. 0.29 cm/h
47. 28.8 km/h
49. $-0.46\ \Omega$
51. $-0.00398k$ units
53. W = L = 7.94 cm,
    H = 15.87 cm

Chapter 6

Section 6.1 / page 189

1. $x^4/4 + C$
3. $-x^5/10 + C$
5. $\frac{3}{5}x^{5/3} + C$
7. $-4/x + C$
9. $\frac{5}{6}x^6 - \frac{3}{4}x^4 + x^2 + C$
11. $\frac{4}{3}x^{3/2} - \frac{3}{5}x^5 + C$
13. $\frac{4}{5}x^5 - \frac{4}{3}x^3 + x + C$
15. $\frac{2}{7}x^{7/2} - \frac{12}{5}x^{5/2} + 6x^{3/2} + C$
17. $(3x^2 - 1)^4/4 + C$
19. $(1 - 3x)^5/5 + C$
21. $\frac{2}{3}(x^2 + 1)^{3/2} + C$
23. $(x^2 + 14)^6/12 + C$
25. $(1 - 4x)^{3/2}/(-6) + C$
27. $\frac{3}{10}(4x^2 - 7)^{5/3} + C$
29. $\sqrt{2 + x^2} + C$
31. $2(x^4 - 2x^2 + 5)^5/5 + C$
33. $y = 2x^2 + 2x + 1$

35. $y = (2x^3 - 2x)^3/6 + 12$
37. $y = \frac{2}{9}(3x - 8)^{3/2} - \frac{113}{9}$
39. $y = 5x^2/2 - 3x - 1$

Section 6.2 / page 195

1. $A_8 = 5.25$, $A = 6$
3. $A_4 = 11$, $A = 12$
5. 2.25, 3.094
7. 1.083, 0.9956
9. $\int_1^2 (3x^2 + 1)\, dx$
11. $\int_{-1}^1 (-2x + 6)\, dx$

Section 6.3 / page 200

1. $\frac{15}{4}$
3. 15
5. $-\frac{8}{3}$
7. $\frac{64}{3}$
9. $\frac{1}{2}$
11. $\frac{2}{3}$
13. $\frac{38}{15}$
15. $\frac{19}{6}$

17. $\frac{1}{3}(13\sqrt{13} - 8)$
19. $\frac{1}{3}(5\sqrt{5} - 3\sqrt{3})$
21. $-\frac{1}{6}$
23. $2\sqrt{5} - \sqrt{6}$

Section 6.4 / page 201

1. $x^4/2 - 7x^2/2 + C$
3. $\frac{4}{7}x^{7/4} + C$
5. $\frac{2}{9}(1 + 3x)^{3/2} + C$
7. $-\frac{1}{32}(2 - 4x^2)^4 + C$
9. $-1/[4(x^2 - 4x + 2)^2] + C$
11. $\frac{1}{2}$
13. $\frac{58}{5}$
15. $\frac{9}{40}$
17. $\frac{112}{9}$
19. $2\sqrt{21}$
21. $\int_1^5 \sqrt{1 + x^2}\, dx$
23. $y = \frac{4}{3}(2x - 1)^{3/2} - 1$

Chapter 7

Section 7.1 / page 207

1. $v = \frac{4}{3}t^{3/2} + \frac{28}{3}$
3. 102.5 m
5. $s = \frac{1}{3}t^3 + 2t^2 + \frac{8}{3}t$

7. 8 m/s^2
9. 1.429 s
11. 1.224 s; 1.837 m
13. 24.25 m/s

15. 16
17. 3.99 C
19. $i = 1.25t - 0.5$
21. 4 H

23. $8833.33

25. 225 V

27. 2.458

29. 25°C

31. 15,000 grams per sample

Section 7.2 / page 214

1. $\frac{39}{2}$

3. $\frac{14}{3}$

5. $\frac{2}{3}$

7. $\frac{20}{3}$

9. $\frac{11}{2}$

11. $\frac{14}{3}$

13. $\frac{3}{4}$

15. $\frac{1}{12}$

17. $\frac{9}{2}$

19. $\frac{7}{6}$

21. $\frac{28}{3}$

23. 3.609

25. 2

27. $\frac{1}{2}$

29. 7.54 m²

31. 20.63 m

Section 7.3 / page 222

1. $\frac{32}{3}\pi$

3. $\frac{128}{3}\pi$

5. 8π

7. $\frac{3}{5}\pi$

9. $\frac{256}{15}\pi$

11. 12π

13. $\frac{2}{3}\pi$

15. $\pi/3$

17. $\pi/15$

19. $\frac{124}{5}\pi$

21. $\frac{163}{14}\pi$

23. $\frac{4}{3}\pi a^3$

25. 36π m³

27. $\frac{1280}{3}\pi$

Section 7.4 / page 231

1. $(0, \frac{5}{9})$

3. $(-\frac{27}{13}, \frac{17}{13})$

5. $(\frac{4}{3}, \frac{4}{3})$

7. $(-\frac{4}{5}, -\frac{32}{7})$

9. $(\frac{9}{20}, \frac{9}{20})$

11. $(\frac{2}{3}b, \frac{1}{3}h)$

13. $(0, \frac{9}{2})$

15. $(\frac{1785}{1016}, 0) \approx (1.757, 0)$

17. On the axis of symmetry, $\frac{1}{4}h$ from the base.

19. $I_y = 8k, I_x = \frac{32}{3}k$

21. $I_y = \frac{32}{3}k, I_x = \frac{3}{10}(2)^{10}k$

23. $I_y = 48\pi k/5$

25. $I_x = 2\pi k/45$

27. $R_y = \sqrt{2}, R_x = 2\sqrt{2}/3$

29. $R_y = 2\sqrt{2}/3, R_x = 16/\sqrt{10}$

31. $R_y = 3/\sqrt{35}$

33. $R_y = \sqrt{\frac{6}{5}}$

35. $R_x = 1/\sqrt{3}$

Section 7.5 / page 233

1. 132,300 N

3. 78,400 N

5. $\frac{4900}{3}$ N

7. 48,352 N

9. 111,464.5 N

Section 7.6 / page 238

1. 2.25 J

3. 1.25 J

5. 2400 J

7. 8480 J

9. 2,822,400π J

11. 893.04π J

13. 1.575c J

15. $(3.92 \times 10^{-24})p$ J

17. $3\sqrt{5}$

19. 13.54

21. $48\sqrt{10}\pi$

23. $\frac{13}{3}\pi$

25. $\pi/3$

Section 7.7 / page 243

1. 56/15

3. 48

5. 128/15

7. 21/4

9. 1666/27

11. 1

13. 8

15. $\frac{1}{3}(3)^{3/2} = 1.732$

Section 7.8 / page 244

1. 229.33

3. 15.34 m/s

5. 10 C

7. 18.7 C

9. $c(x) = \frac{0.01}{3}x^3 - \frac{1}{6}x^2 + 1000$

11. $\frac{32}{3}$

13. $\frac{11}{12}$

15. 24π

17. $\pi/15$

19. $(\frac{9}{4}, \frac{27}{10})$

21. $(\frac{5}{2}, 0)$

23. $I_y = \frac{243}{5}k, I_x = \frac{729}{7}k$

25. $2187\pi k/2$

27. $\frac{7}{2}\omega$ N

29. 552.5 J

31. 3675π J

33. 8.35

35. $-\frac{43}{15}$

37. 4

Chapter 8

Section 8.1 / page 250

1. $\cos\theta = x/r = 1/(r/x) = 1/\sec\theta$ 3. $\cot\theta = x/y = (x/r)/(y/r) = \cos\theta/\sin\theta$

5. $\cot^2\theta + 1 = x^2/y^2 + 1 = (x^2 + y^2)/y^2 = r^2/y^2 = (r/y)^2 = \csc^2\theta$

7. From Equation (17) with β replaced by $\alpha + \beta$,

$\cos(\alpha + \beta) = \sin[\pi/2 - (\alpha + \beta)]$

$= \sin[(\pi/2 - \alpha) - \beta]$

$= \sin(\pi/2 - \alpha)\cos\beta - \cos(\pi/2 - \alpha)\sin\beta$ By Equation (16).

$= \cos\alpha\cos\beta - \sin\alpha\sin\beta$ By Equations (17) and (18).

9. From Equation (15) with β replaced by α, $\sin(\alpha + \alpha) = \sin\alpha\cos\alpha + \cos\alpha\sin\alpha$, or $\sin 2\alpha = 2\sin\alpha\cos\alpha$.

11. $\cos 2\alpha = \cos^2\alpha - \sin^2\alpha$ Equation 22

$\qquad\qquad = \cos^2\alpha - (1 - \cos^2\alpha)$ From Equation 12.

$\qquad\qquad = 2\cos^2\alpha - 1$

13. From Equations (19) and (16) with α replaced by $\alpha/2$, β replaced by $\beta/2$,

$$\cos\left(\frac{\alpha}{2} + \frac{\beta}{2}\right)\sin\left(\frac{\alpha}{2} - \frac{\beta}{2}\right) = \left(\cos\frac{\alpha}{2}\cos\frac{\beta}{2} - \sin\frac{\alpha}{2}\sin\frac{\beta}{2}\right)\left(\sin\frac{\alpha}{2}\cos\frac{\beta}{2} - \cos\frac{\alpha}{2}\sin\frac{\beta}{2}\right)$$

$$= \cos\frac{\alpha}{2}\sin\frac{\alpha}{2}\cos^2\frac{\beta}{2} - \sin^2\frac{\alpha}{2}\sin\frac{\beta}{2}\cos\frac{\beta}{2} - \cos^2\frac{\alpha}{2}\cos\frac{\beta}{2}\sin\frac{\beta}{2} + \sin\frac{\alpha}{2}\cos\frac{\alpha}{2}\sin^2\frac{\beta}{2}$$

$$= \sin\frac{\alpha}{2}\cos\frac{\alpha}{2}\left(\cos^2\frac{\beta}{2} + \sin^2\frac{\beta}{2}\right) - \sin\frac{\beta}{2}\cos\frac{\beta}{2}\left(\sin^2\frac{\alpha}{2} + \cos^2\frac{\alpha}{2}\right)$$

$$= \sin\frac{\alpha}{2}\cos\frac{\alpha}{2} - \sin\frac{\beta}{2}\cos\frac{\beta}{2} \quad \text{By Equation (12).}$$

Therefore,

$$\cos\left(\frac{\alpha + \beta}{2}\right)\sin\left(\frac{\alpha - \beta}{2}\right) = \sin\frac{\alpha}{2}\cos\frac{\alpha}{2} - \sin\frac{\beta}{2}\cos\frac{\beta}{2}$$

and

$$2\cos\left(\frac{\alpha + \beta}{2}\right)\sin\left(\frac{\alpha - \beta}{2}\right) = 2\sin\frac{\alpha}{2}\cos\frac{\alpha}{2} - 2\sin\frac{\beta}{2}\cos\frac{\beta}{2}$$

$$= \sin\alpha - \sin\beta \quad \text{By Equation (21).}$$

15. From Equation (23) with α replaced by $\alpha/2$,

$$\cos\alpha = 2\cos^2\frac{\alpha}{2} - 1$$

$$2\cos^2\frac{\alpha}{2} = 1 + \cos\alpha$$

$$\cos^2\frac{\alpha}{2} = \frac{1 + \cos\alpha}{2}$$

$$\cos\frac{\alpha}{2} = \pm\sqrt{\frac{1 + \cos\alpha}{2}}$$

17. $\cos\theta\tan\theta = \cos\theta\dfrac{\sin\theta}{\cos\theta} = \sin\theta$

19. $\tan^2\theta - \sin^2\theta = \dfrac{\sin^2\theta}{\cos^2\theta} - \sin^2\theta = \dfrac{\sin^2\theta - \cos^2\theta\sin^2\theta}{\cos^2\theta} = \dfrac{\sin^2\theta(1 - \cos^2\theta)}{\cos^2\theta}$

$$= \frac{\sin^2\theta}{\cos^2\theta}(1 - \cos^2\theta) = \tan^2\theta\sin^2\theta$$

21. $\dfrac{\cos\theta}{\tan\theta + \sec\theta} - \dfrac{\cos\theta}{\tan\theta - \sec\theta} = \dfrac{\cos\theta(\tan\theta - \sec\theta) - \cos\theta(\tan\theta + \sec\theta)}{(\tan\theta + \sec\theta)(\tan\theta - \sec\theta)}$

$$= \frac{\cos\theta\left(\dfrac{\sin\theta}{\cos\theta} - \dfrac{1}{\cos\theta}\right) - \cos\theta\left(\dfrac{\sin\theta}{\cos\theta} + \dfrac{1}{\cos\theta}\right)}{\tan^2\theta - \sec^2\theta}$$

$$= \frac{\sin\theta - 1 - \sin\theta - 1}{-1}$$

$$= \frac{-2}{-1} = 2$$

23. $\sin(\alpha + \beta)\sin(\alpha - \beta) = (\sin \alpha \cos \beta + \cos \alpha \sin \beta)(\sin \alpha \cos \beta - \cos \alpha \sin \beta)$

$$= \sin^2\alpha \cos^2\beta - \cos^2\alpha \sin^2\beta$$

$$= \sin^2\alpha(1 - \sin^2\beta) - (1 - \sin^2\alpha)\sin^2\beta$$

$$= \sin^2\alpha - \sin^2\alpha \sin^2\beta - \sin^2\beta + \sin^2\alpha \sin^2\beta$$

$$= \sin^2\alpha - \sin^2\beta$$

25. $(\sin \theta + \cos \theta)^2 = \sin^2\theta + 2 \sin \theta \cos \theta + \cos^2\theta$

$$= (\sin^2\theta + \cos^2\theta) + 2 \sin \theta \cos \theta$$

$$= 1 + \sin 2\theta$$

27. $\dfrac{2}{1 - \cos 2\theta} = \dfrac{2}{1 - (\cos^2\theta - \sin^2\theta)} = \dfrac{2}{1 - \cos^2\theta + \sin^2\theta} = \dfrac{2}{\sin^2\theta + \sin^2\theta} = \dfrac{2}{2 \sin^2\theta} = \csc^2\theta$

29. $B = \dfrac{I}{A \cos \theta} = \dfrac{I}{A} \dfrac{1}{\cos \theta} = \dfrac{I}{A} \sec \theta$

Section 8.2 / page 254

1. $2 \cos 2x$

3. $(2x - 2)\cos(x^2 - 2x)$

5. $6x - 2x \cos x^2$

7. $6x \cos(2x + 1) + 3 \sin(2x + 1)$

9. $\sqrt{x} \cos x + (\sin x)/(2\sqrt{x})$

11. $(x \cos x - 2 \sin x)/x^3$

13. $6 \sin 3x \cos 3x$

15. $-[\sin(1 - 2x)]^{-1/2}\cos(1 - 2x)$

17. $6(\sin^2 2x \cos 2x - \sin 3x \cos 3x)$

19. $2 \cos (\sin 2x) \cos 2x$

21. $y' = 2 \cos 2x$, and

$$2 - 2y \tan x = 2 - 2 \sin 2x \tan x$$

$$= 2 - 2(2 \sin x \cos x)(\sin x/\cos x)$$

$$= 2 - 2(2 \sin^2 x)$$

$$= 2(1 - 2 \sin^2 x) = 2 \cos 2x$$

23. -3 **25.** 25.08 m/s **27.** -18.85 A/s

Section 8.3 / page 258

1. $-3 \sin(3x + 2)$

3. $12 \sec^2 4x$

5. $-2x \csc^2 x^2$

7. $3 \sec 3x \tan 3x$

9. $2 \csc(1 - x)\cot(1 - x)$

11. $-4 \cos(2x + 4)\sin(2x + 4)$

13. $8 \tan(4x - 2)\sec^2(4x - 2)$

15. $2 \cot(1 - x)\csc^2(1 - x)$

17. $8 \sec^4 2x \tan 2x$

19. $-6 \csc^2(3x + 1)\cot(3x + 1)$

21. $\csc x^2(1 - 2x^2\cot x^2)$

23. $x^2\sec^2 x(2x \tan x + 3)$

25. $\sec (x - 1)[\cot (x + 1)$
$\tan (x - 1) \csc^2 (x + 1)]$

27. $(-x \csc^2 x - \cot x)/x^2$

29. $\dfrac{2 \cos x \sin 4x + \sin x \sin^2 2x}{\cos^2 x}$

31. $9 \tan^2 3x \sec^2 3x - 2 \sin 4x$

33. $4(\sec^2 x + 1)(\sec^2 x \tan x)$

35. $3[\sec^2 2x - \tan(x - 1)]^2$
$$[4 \sec^2 2x \tan 2x - \sec^2(x + 1)]$$

37. $(1 + \sin^2 x) \sin 4x - 2(1 + \sin^2 x)^2 \sin 2x$

39. $y = \sin x$, $dy/dx = \cos x$, $d^2y/dx^2 = -\sin x$,
$d^3y/dx^3 = -\cos x$, $d^4y/dx^4 = \sin x = y$

41. -5.0488

43. -49.47

Section 8.4 / page 263

1. $\pi/2$ **3.** $\pi/2$ **5.** $\pi/4$ **7.** 0.42 **9.** -0.6435

11. 1.4505 **13.** 1.209 **15.** 0.5547 **17.** 0.7902

19. 0.3145

21. 0.4292

23. $x/\sqrt{1 + x^2}$

25. $\sqrt{1 - x^2}/x$

27. $2x\sqrt{1 - x^2}$

29. $-1/\sqrt{x - x^2}$

31. $64x^3/\sqrt{1 - 16x^8}$

33. $-3/\sqrt{1 - 9x^2}$

35. $1/4(\sqrt{x}\sqrt{1 - x})$

37. $-1/(2 - 2x + x^2)$

39. $4x \text{ Arcsin } x^2/\sqrt{1 - x^4}$

41. $1/(2\sqrt{\text{Arccos}(1 - x)} \sqrt{2x - x^2})$

43. $2x^2/\sqrt{1 - x^4} + \text{Arcsin } x^2$

45. $2(2x - 1)^2/(1 + 4x^2)^2$

47. $(-x - \sqrt{1 - x^2} \text{ Arccos } x)/(x^2\sqrt{1 - x^2})$

49. $y = \text{Arccot } u$, $\cot y = u$; differentiating,

$$-\csc^2 y \frac{dy}{dx} = \frac{du}{dx} \quad \text{or} \quad \frac{dy}{dx} = \frac{-1}{\csc^2 y} \frac{du}{dx}$$

$$= \frac{-1}{1 + \cot^2 y} \frac{du}{dx} = \frac{-1}{1 + u^2} \frac{du}{dx}.$$

51. $y = \text{Arccsc } u$, $\csc y = u$; differentiating,

$$-\csc y \cot y \frac{dy}{dx} = \frac{du}{dx} \quad \text{or} \quad \frac{dy}{dx} = \frac{-1}{\csc y \cot y} \frac{du}{dx}$$

$$= \frac{-1}{\sqrt{\csc^2 y (\csc^2 y - 1)}} \frac{du}{dx}$$

$$= \frac{-1}{\sqrt{u^2 (u^2 - 1)}} \frac{du}{dx}.$$

53. 3.2175

Section 8.5 / page 269

1. $y = 2.219x - 3.269$

3. $y = 3.6255x - 0.819$

5. Minima at $(0, 0)$ and $(2\pi, 0)$; maximum at $(\pi, 1)$; inflection points at $(\pi/2, \frac{1}{2})$, and $(3\pi/2, \frac{1}{2})$.

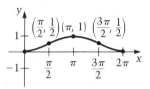

7. Minima at $(0, 1)$, $(\pi, -1)$ and $(2\pi, 1)$; maxima at $(\pi/3, 1.25)$ and $(5\pi/3, 1.25)$; inflection points at $(0.567, 1.13)$, $(2.2, 0.05)$, $(4.08, 0.05)$ and $(5.716, 1.13)$.

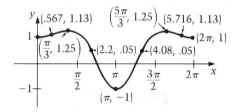

9. $\pi/6$ s

11. $A = 2a^2$

13. $\sqrt{5} = 2.24$ m

15. Magnitude 3.7, direction 2.138.

17. 8 m/s

19. 103.88 m/min

21. 1.32 m/s

23. -0.25 rad/h

25. 0.5676 A

27. 0.016

Section 8.6 / page 275

1. $2 = \log_5 25$

3. $0 = \log_6 1$

5. $\frac{1}{3} = \log_{27} 3$

7. $-2 = \log_4 \frac{1}{16}$

9. $3^4 = 81$

11. $4^{-1} = \frac{1}{4}$

13. $36^{1/2} = 6$

15. $9^{-1/2} = \frac{1}{3}$

17. 16

19. -3

21. 49

23. -1

25. 2

27. $2 \log_b r + \log_b s$

29. $2 \log_b r - \log_b s$

31. $\log_b \frac{3}{7}$

33. $\log_b(x^3 y^2)$

35. $\log_b(y/x^6)$

37. 782 cm/s

39. 0.972 V

41. $\dfrac{(e^x + e^{-x})^2}{4} - \dfrac{(e^x - e^{-x})^2}{4}$

$$= \frac{e^{2x} + 2 + e^{-2x} - e^{2x} + 2 - e^{-2x}}{4} = \frac{4}{4} = 1$$

43. $\dfrac{e^{2x} + e^{-2x}}{2} = \dfrac{2(e^{2x} + 2 + e^{-2x})}{4} - \dfrac{4}{4}$

$$= 2 \left(\frac{e^x + e^{-x}}{2} \right)^2 - 1$$

Section 8.7 / page 279

1. $3/(3x - 4)$ **3.** $1/(2x)$ **5.** $1/(\cos x \sin x)$

7. $2x + 2x \ln x^2$

9. $\dfrac{2(2x - 1)\ln(2x - 1) - 2(2x + 1)\ln(2x + 1)}{(2x + 1)(2x - 1)\ln^2(2x - 1)}$

11. $2 \cot x$ **13.** $\dfrac{2x}{x^2 - 1} - \dfrac{1}{x + 2}$

15. $\dfrac{2 \ln \tan x}{\cos x \sin x}$ **17.** $2e^{2x}$ **19.** $e^{\sqrt{x}}/(2\sqrt{x})$

21. $x(\cos x)e^{\sin x} + e^{\sin x}$ **23.** $\dfrac{xe^x - e^x}{2x^2}$

25. $2xe^{x^2}\sec^2 e^{x^2}\cos(\tan e^{x^2})$ **27.** $3 - \dfrac{1 + e^x}{x + e^x}$

29. $3e^{3x}/\sqrt{1 - e^{6x}}$

31. $y' = -\ln x$, and $xy' + x - y$

$$= -x \ln x + x - (x - x \ln x) = 0$$

33. $y' = e^x + 2e^{2x}$, $y'' = e^x + 4e^{2x}$, and $y'' - 3y' + 2y$

$$= e^x + 4e^{2x} - 3(e^x + 2e^{2x}) + 2(e^x + e^{2x}) = 0$$

35. -0.736 **37.** 0.348   **39.** 5.4366

41. $\dfrac{d[(e^x - e^{-x})/2]}{dx} = \dfrac{e^x + e^{-x}}{2} = \cosh x$

43. $x^x(1 + \ln x)$ **45.** $(\ln x)^x[1/\ln x + \ln\ln x]$

Section 8.8 / page 281

1. $y = -0.4x + 2.45$

3. $y = 2x - e$

5. Minimum at $(0, 1)$; everywhere concave up.

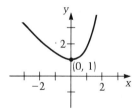

7. Minimum at $(1, 0)$, inflection point at $(e, 1)$, $x > 0$.

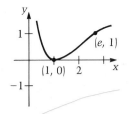

9. 0.2424

11. Magnitude 1.39, direction -0.014.

13. $36e^{-0.3t}$ V/s

15. $a(t) = \dfrac{dv}{dt} = u \cdot \dfrac{1}{m(0)/m(t)} \, m(0)(-1)[m(t)]^{-2}m'(t)$

$= \dfrac{-um'(t)}{m(t)}$ and $a(t)m(t) = -um'(t)$

17. $y' = e^{-bt}\cosh(ct) - \dfrac{b}{c}\sinh(ct)e^{-bt}$;

$y'' = ce^{-bt}\sinh(ct) - 2b\cosh(ct)e^{-bt} + \dfrac{b^2}{c}\sinh(ct)e^{-bt}$

19. 0.0376 cm/s

21. $dz = e^{(x+y^2)}\left[\dfrac{2x}{x^2+1} + \ln(x^2+1)\right]dx$
$+ 2y \ln(x^2+1)e^{(x+y^2)}dy$

Section 8.9 / page 284

1. $\sec\theta(1 - \sin^2\theta) = \sec\theta\cos^2\theta$

$= (1/\cos\theta)\cdot\cos^2\theta = \cos\theta$

3. $\tan^2\theta\cos^2\theta + \cot^2\theta\sin^2\theta$

$= \dfrac{\sin^2\theta}{\cos^2\theta}\cdot\cos^2\theta + \dfrac{\cos^2\theta}{\sin^2\theta}\cdot\sin^2\theta$

$= \sin^2\theta + \cos^2\theta = 1$

5. $2\sin\theta + \sin 2\theta$

$= 2\sin\theta + 2\sin\theta\cos\theta = 2\sin\theta(1 + \cos\theta)$

$= \dfrac{2\sin\theta(1+\cos\theta)(1-\cos\theta)}{1-\cos\theta} = \dfrac{2\sin\theta(1-\cos^2\theta)}{1-\cos\theta}$

$= \dfrac{2\sin^3\theta}{1-\cos\theta}$

7. $-2 = \log_3\frac{1}{9}$ **9.** $4^{-1/2} = \frac{1}{2}$ **11.** 8   **13.** $3x^2\cos x^3$

15. $2x\cos x\sec^2 x^2 - \tan x^2\sin x$

17. $\dfrac{x\sec x\tan x - \sec x}{x^2}$ **19.** $-6/\sqrt{6-9x^2}$

21. $\dfrac{1}{2(2+x)\sqrt{x+1}}$ **23.** $x^2/\sqrt{1-x^2} + 2x\,\text{Arcsin } x$

25. $3x^2e^{x^3}$ **27.** $x^2e^x + 2xe^x$ **29.** $e^{2x}/\sqrt{e^{2x}+4}$

31. $2/(2x - 7)$ **33.** $\dfrac{2x^3}{x^4-2}$ **35.** $\dfrac{1+3\ln x}{x\ln x}$

37. $2e^{\sin 2x}\cos 2x$ **39.** $2xe^{x^2}/(1 + e^{2x^2})$

41. $\cos x/\sin x - 2/x$ **43.** $e^{x^2}(1/x + 2x\ln x)$

45. $\dfrac{2x+\cos x}{x^2+\sin x}$ **47.** $\dfrac{1/x - \cos x\ln x}{e^{\sin x}}$

49. $\dfrac{y(2xy - (\ln y)e^x)}{e^x - x^2y}$ **51.** $\dfrac{y^3\sin x + 2ye^{\sin^2 x}\sin x\cos x}{3y^2\cos x - e^{\sin^2 x}}$

53. $y = -1.474x + 1.5866$; $y = 0.678x + 0.7253$

55. $y = 2e^\pi x + e^\pi(1 - \pi)$

57. Maxima at $(\pi/4, 2)$ and $(5\pi/4, 2)$; minima at $(3\pi/4, -2)$ and $(7\pi/4, -2)$; inflection points at $(\pi/2, 0)$, $(\pi, 0)$, and $(3\pi/2, 0)$.

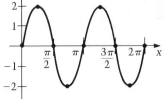

59. No maximum or minimum points; y is everywhere increasing; inflection point at $(0.5, 0.14)$; undefined for $x = 0$; asymptote $y = 1$.

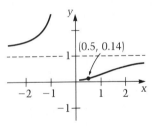

61. 25.4 km

63. $1/e$

65. Magnitude 0.923, direction -1.4.

67. 4.24 m/min

69. 14.11 m/h

71. 4.84 m/min

73. $2.8(-0.02e^{-0.02t} + 0.6e^{-0.6t})$

75. 3×10^{-5}

77. -2.07

Chapter 9

Section 9.1 / page 289

1. $(1 + 2x^3)^{3/2}/9 + C$

3. $(\sin^3 x)/3 + C$

5. $2(\sin x)^{3/2}/3 + C$

7. $2(\tan x)^{3/2}/3 + C$

9. $\frac{9}{128}$

11. $(\text{Arcsin } x)^3/3 + C$

13. $(3 \ln^5 x)/5 + C$

15. $-(1 - \ln x)^3/3 + C$

17. $(e - 1)^3/6$

19. $(\sin^3 e^x)/3 + C$

21. $(\ln \sec x)^2/2 + C$

23. $(e^{2x} + e^{-2x})^{3/2}/3 + C$

25. $(\ln \sin x^2)^2/4 + C$

27. $\frac{1}{2}$

29. $k(\sqrt{a^2 + d^2} - d)$

29. $2 - \pi$

31. 1

33. 4

35. 2π

37. $0.3078k$

39. $T = gka^2/[2(1 - \cos a)]$

Section 9.2 / page 293

1. $\frac{1}{3} \ln|3x + 1| + C$

3. $\frac{1}{2} \ln(x^2 + 1) + C$

5. $\frac{1}{2} \ln \frac{11}{3}$

7. $-\ln \frac{1}{2} = \ln 2$

9. $-1/\sin x + C$

11. $-\ln|\cos x| + C$

13. $\ln|\ln x| + C$

15. $-\ln|1 - e^x| + C$

17. $\frac{1}{2} \ln|2x - e^{2x}| + C$

19. $\ln|e^{\sin x} + x| + C$

21. $2\sqrt{x} + C$

23. $2 \ln(1 + \sqrt{x}) + C$

25. 2

27. 1

29. $\pi \ln \frac{5}{4}$

31. $\ln 1.02$

33. $T = e^{(kt + C)} + 14$

Section 9.3 / page 296

1. $\frac{1}{3} e^{3x} + C$

3. $\frac{1}{3}(e - 1)$

5. $\frac{1}{2} e^{(2x + 3)} + C$

7. $e - 1$

9. $e^{e^x} + C$

11. $2e^{\sqrt{x}} + C$

13. $\frac{1}{2} e^{2x} + \frac{1}{2} e^{-2x} + C$

15. $(e^{2x} - x)^3/3 + C$

17. $\frac{1}{5} e^{5x} + C$

19. $-\frac{1}{2} \ln|2 - e^{2x}| + C$

21. $\frac{1}{2}(e^2 + 1) - e$

23. $(5 - e^{-4})/2$

25. $\pi(e^2 - 1)$

27. $s = \frac{1}{2} e^{\tan^2 t} + \frac{9}{2}$

29. 1114 billion barrels

Section 9.4 / page 301

1. $\frac{1}{2} \cos(1 - 2x) + C$

3. $\frac{1}{5} \tan 5x + C$

5. $\sec(e^x) + C$

7. $\ln|\sin e^x| + C$

9. $-2 \ln|\csc \frac{x}{2} + \cot \frac{x}{2}| + C$

11. 0.2298

13. $-\sin(e^{-x}) + C$

15. $-\frac{1}{2} \cot x^2 + C$

17. $\ln|\sec x + \tan x| - \ln|\csc x + \cot x| + C$

19. $-\cot x + 2 \ln|\csc x + \cot x| + x + C$

21. $x - \cos x + C$ **23.** $e^{\sin x} + C$ **25.** $(-\cot^3 x)/3 + C$

27. $-\cot x + \ln|\csc x + \cot x| + (\sin x)^{-1}$
    $+ \ln|\sin x| + C$

Section 9.5 / page 305

1. $\frac{1}{2} \sin 2x - \frac{1}{6} \sin^3 2x + C$ **3.** $\frac{1}{3} \sin^3 x - \frac{1}{5} \sin^5 x + C$

5. $(\frac{1}{2} x + \frac{1}{8} \sin 4x)|_0^{\pi/4} = \pi/8$ **7.** $\frac{1}{8} x - \frac{1}{64} \sin 8x + C$

9. $-\frac{2}{3} \cos^{3/2} x + \frac{2}{7} \cos^{7/2} x + C$ **11.** $\sin x - \frac{1}{3} \sin^3 x + C$

13. $\frac{1}{2} \sec^2 x + C$ **15.** $\frac{1}{6} \tan^3 2x - \frac{1}{2} \tan 2x + x + C$

17. $\frac{1}{4} \tan^4 x - \frac{1}{2} \tan^2 x - \ln|\cos x| + C$

19. $(\frac{1}{5} \sec^5 x - \frac{1}{3} \sec^3 x)|_0^{\pi/6} = 0.0307$

21. $-\frac{1}{2} \cot^2 x - \frac{1}{4} \cot^4 x + C$ or $-(\csc^4 x)/4 + C$

23. $2\sqrt{\sec x} + C$ **25.** $-\frac{1}{3} \sin^{-3} x + (\sin x)^{-1} + C$

27. $\tan x - \sec x + C$ **29.** $-(1 + \tan x)^{-1} + C$

31. $\pi/2$ **33.** 0.549

Section 9.6 / page 308

1. $\text{Arcsin}(x/3) + C$ **3.** $2 \text{ Arcsin } x + C$

5. $\frac{1}{\sqrt{7}} \text{Arcsin}\left(\frac{\sqrt{7} x^2}{4}\right) + C$

7. $\text{Arcsin}\left(\frac{e^x}{3}\right) + C$

9. $-(1 - e^{2x})^{1/2} + C$ **11.** $\frac{1}{20} \text{Arctan}(5x/4) + C$

13. $\frac{1}{2} \ln(9 + x^2) + C$ **15.** 0.2318

17. $\frac{1}{3} \text{Arctan}[(\sin x)/3] + C$

19. $\frac{1}{2} \text{Arctan}[(x + 3)/2] + C$

21. $4 \text{ Arctan}(x - 3) + C$

23. 0.6797 **25.** 0.3835

27. 11.57 **29.** 0.6995

Section 9.7 / page 315

1. $\frac{1}{2} x e^{2x} - \frac{1}{4} e^{2x} + C$ **3.** $-3\pi/4$

5. $(x^2/2) \ln x - x^2/4 + C$ **7.** $\frac{1}{2} e^{x^2} + C$

9. $x^2 \sqrt{x^2 + 1} - \frac{2}{3}(x^2 + 1)^{3/2} + C$

11. $x \tan x + \ln|\cos x| + C$

13. $\frac{1}{2} x^2 \sin 2x + \frac{1}{2} x \cos 2x - \frac{1}{4} \sin 2x + C$

15. $\frac{1}{2}(e^x \sin x + e^x \cos x) + C$

17. $\frac{1}{3} \cos x \sin 2x - \frac{2}{3} \sin x \cos 2x + C$

19. 0.5708 **21.** 0.5914

23. k **25.** 3348

Section 9.8 / page 320

1. $\ln|(\sqrt{9 + x^2} + x)/3| + C$
3. $\ln|(x + \sqrt{x^2 - 9})/3| + C$
5. $x/\sqrt{1 + x^2} + C$
7. $-\frac{9}{5}(9 - x^2)^{5/2} + (9 - x^2)^{7/2}/7 + C$
9. $\frac{1}{2}\text{Arcsec}(x/2) + C$
11. $\sqrt{16 + x^2} + C$ **13.** 0.2668
15. $\frac{1}{125}\left[\ln|(\sqrt{25x^2 + 16} + 5x)/4|\right.$

$$\left. - 5x/\sqrt{25x^2 + 16}\right] + C$$

17. $\ln|(\sqrt{9 + \sin^2 x} + \sin x)/3| + C$
19. $\ln|\sqrt{x^2 + 2x + 1} + x + 1| + C$
21. 25π **23.** $\pi(\pi + 2)/64$
25. $0.1525k$

Section 9.9 / page 327

1. $3\ln|x| + 2\ln|x - 1| + C = \ln|x^3(x - 1)^2| + C$
3. $2\ln|2x - 3| + \ln|x + 5| + C$
$$= \ln|(2x - 3)^2(x + 5)| + C$$
5. $\ln\left|\dfrac{x^3(x - 1)^4}{x + 1}\right| + C$
7. $\ln|x^3(x - 1)| - 2/x + C$
9. $-2/x - 3/(2x^2) - \ln|x^2(x - 2)^5| + C$
11. $\frac{39}{28} + \ln\sqrt{7} = 2.3658$
13. $x^2/2 - 2x + \ln|(x + 1)^3/(2x + 3)^2| + C$
15. $4\ln|x| + \frac{3}{2}\ln(x^2 + 1) - \text{Arctan } x + C$
17. $5\ln|x| + \frac{3}{2}\ln|x^2 + 6x + 10| + 3\text{Arctan}(x + 3) + C$
19. $3\ln(x^2 + 1) - 9\text{Arctan } x - 2/(x^2 + 1) + C$
21. 0.305 **23.** $203,500$

Section 9.10 / page 330

1. Rule 31; $\sqrt{9 + x^2} - 3\ln|(3 + \sqrt{9 + x^2})/x| + C$
3. Rule 34; $\ln|x + \sqrt{x^2 - 9}| + C$
5. Rule 54; $\sin x - x\cos x + C$
7. Rule 65; 2.545
9. Rule 56 (and 55);
$-x^2\cos x + 2\cos x + 2x\sin x + C$
11. Rule 48 (and 5); $\frac{1}{3}\cos^2 x \sin x + \frac{2}{3}\sin x + C$
13. Rule 29; $\frac{1}{2}\ln|2x + \sqrt{9 + 4x^2}| + C$
15. Rule 35; $\text{Arcsec } 3x + C$
17. Rule 46; $-\frac{1}{2}(\cot x^2 + x^2) + C$
19. Rule 61; $\frac{1}{8}(2x^2 - 1)e^{2x^2} + C$

21. Rule 37;
$\frac{1}{16}(x^2\sqrt{4x^4 - 49} + \frac{49}{2}\ln|2x^2 + \sqrt{4x^4 - 49}|) + C$
23. Rule 66; $(x^9/27)(3\ln x^3 - 1) + C$
25. Rule 17; 0.00589
27. Rule 44; $\frac{1}{5}(5x/2 + \frac{1}{4}\sin 10x)$
29. Rule 25; $\frac{1}{120}(8 + 12x^4 - 8x^2)\sqrt{1 + 2x^2} + C$
31. Rule 30; $-\frac{1}{2}\ln\left|\dfrac{\sqrt{4 + 9x^2} + 2}{3x}\right| + C$
33. Rule 19; $\frac{1}{3456}[\frac{1}{2}(5 + 12x^2)^2 - 10(5 + 12x^2)$
$+ 25\ln(5 + 12x^2)] + C$
35. 0.1556 **37.** 5.182 A

Section 9.11 / page 336

1. Trapezoid rule, 6.03125;   Simpson's rule, 6; 6.
3. Trapezoid rule, -0.1805;   Simpson's rule, -0.14306; -0.11438.
5. Trapezoid rule, 3.26081;   Simpson's rule, 3.19561; 3.19453.
7. Trapezoid rule, 1.50171;   Simpson's rule, 1.49713.
9. Trapezoid rule, 0.01238;   Simpson's rule, 0.01224.
11. Trapezoid rule, 0.64681;   Simpson's rule, 0.64791.
13. 18.7 **15.** 1.1467

Section 9.12 / page 338

1. $\frac{1}{9}\tan^3(3x - 1) + C$ **3.** $\frac{1}{2}(x - \frac{1}{2}\sin 2x) + C$
5. $\sqrt{x^2 - 9} - 3\text{Arcsec}(x/3) + C$
7. $x^2\sin x + 2x\cos x - 2\sin x + C$
9. $e^{\sin x} + C$ **11.** $-\frac{1}{2}\ln|\cos 2x| + C$
13. $-\ln|\sqrt{1 + x^2}/x + 1/x| + \sqrt{1 + x^2} + C$
15. $3x + \ln|(x + 6)^5/x^2| + C$
17. $(\sec^4 x)/4 + C$
19. $\frac{1}{2}(\sec x \tan x + \ln|\sec x + \tan x|) + C$
21. $(\ln \sin x)^2/2 + C$ **23.** $\frac{1}{8}\sin 4x^2 + C$
25. $-\text{Arcsin}(\cos x) + C$ or $x + C$
27. $-\frac{2}{5}(1 + e^{-2x})^{5/4} + C$
29. $\frac{3}{4}\ln(2 + x^{4/3}) + C$ **31.** $1/[2(25 - x^2)] + C$
33. $\ln(1 + \sin^2 x) + C$ **35.** $-\cos(\ln x) + C$
37. $-\frac{1}{3}x^2 e^{-3x} - \frac{2}{9}xe^{-3x} - \frac{2}{27}e^{-3x} + C$
39. 13.372 **41.** $\phi = (k/2\pi)\ln(r_2/r_1)$
43. 9.67 A   **45.** $\pi \tan 1 = 4.893$
47. $3\pi I_m$ **49.** $\frac{1}{2}\text{Arctan } 1 = 0.3927$
51. $198,239$
53. $y = -\frac{1}{2}\ln|(2 + \sqrt{4 - x^2})/x| + 4$ **55.** 0.561

Chapter 10

Section 10.1 / page 346

1. Converges to 2 (geometric series with $r = \frac{2}{3}$).
3. Converges; $S \approx 1.2$.
5. Diverges (geometric series with $r = -\frac{4}{3}$)
7. Converges; $S \approx 0.312$.
9. Converges to 99 (geometric series with $r = 0.99$).
11. Diverges.
13. 0.83

Section 10.2 / page 350

1. $x - \frac{1}{3!}x^3 + \frac{1}{5!}x^5 - \cdots$

3. $1 + x + \frac{x^2}{2!} + \cdots$

5. $1 + x + x^2 + \cdots$

7. $2x - \frac{4}{3}x^3 + \frac{4}{15}x^5 + \cdots$

9. $-2x - 2x^2 - \frac{8}{3}x^3 - \cdots$

11. $1 + x^2 + \frac{1}{2}x^4 + \cdots$

13. $x - \frac{1}{3}x^3 + \cdots$

15. $-\frac{1}{2}x^2 - \frac{1}{12}x^4 - \cdots$

17. $1 + \frac{1}{2}x - \cdots$

19. $e - \frac{e}{2}x^2 + \cdots$

21. $1 - x + x^2 - x^3 + \cdots$

23. $f(0) = 1$, $f''(0) = 4$, all other terms are zero.

25. $f(0)$ is undefined.

27. $f(0)$ is undefined.

Section 10.3 / page 353

1. $1 - x + \frac{x^2}{2!} - \cdots$

3. $-x^2 - \frac{1}{2}x^4 - \frac{1}{3}x^6 - \cdots$

5. $\frac{1}{2}x - \frac{1}{48}x^3 + \frac{1}{3840}x^5 - \cdots$

7. $-5x - \frac{25}{2}x^2 - \frac{125}{3}x^3 - \cdots$

9. $x - x^2 + \frac{1}{3}x^3 + \cdots$

11. $1 + \frac{1}{2}x^2 + \frac{5}{24}x^4 + \cdots$

13. $x + \frac{x^3}{3!} + \frac{x^5}{5!} + \cdots$

15. $x^2 - \frac{1}{3}x^4 + \frac{2}{45}x^6 + \cdots$

17. $1 + x + x^2 + \cdots$

19. $\dfrac{d}{dx}\left(x - \dfrac{x^3}{3!} + \dfrac{x^5}{5!} - \cdots\right) = 1 - \dfrac{3x^2}{3!} + \dfrac{5x^4}{5!} - \cdots$

$$= 1 - \frac{x^2}{2!} + \frac{x^4}{4!} - \cdots$$

21. $\dfrac{d}{dx}\left(x + \dfrac{x^3}{3!} + \dfrac{x^5}{5!} + \cdots\right) = 1 + \dfrac{3x^2}{3!} + \dfrac{5x^4}{5!} + \cdots$

$$= 1 + \frac{x^2}{2!} + \frac{x^4}{4!} + \cdots$$

23. $\int(-1 - x - x^2 - x^3 - \cdots)\,dx$

$$= -x - \tfrac{1}{2}x^2 - \tfrac{1}{3}x^3 - \cdots$$

25. $\dfrac{1}{1-x} = 1 + x + x^2 + x^3 + x^4 + \cdots$;

$\dfrac{d}{dx}\left(\dfrac{1}{1-x}\right) = \dfrac{1}{(1-x)^2} = 1 + 2x + 3x^2 + 4x^3 + \cdots$;

$\dfrac{d}{dx}\left[\dfrac{1}{(1-x)^2}\right] = \dfrac{2}{(1-x)^3}$

$$= 1 \cdot 2 + 2 \cdot 3x + 3 \cdot 4x^2 + \cdots ;$$

The function is $\dfrac{2}{(1-x)^3}$.

Section 10.4 / page 359

1. $2.71\overline{6}$	9. 0.98995	17. 0.000332
3. 0.09536	11. 1.06272	19. 1.32006
5. 0.017454	13. 0.308463	21. 0.22580
7. 1.64844	15. 0.904630	23. 0.02006k
		25. 0.23236 cm

27.

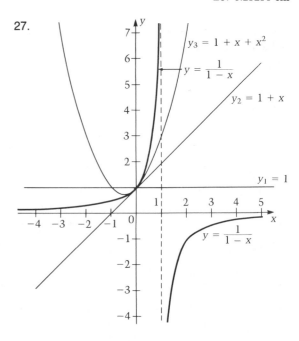

Section 10.5 / page 363

1. 3.949688
3. 10.049875; $(f(x) = \sqrt{x}, a = 100)$
5. 0.882951; $(f(x) = \sin x, a = 60°)$
7. 1.6292; $(f(x) = \ln x, a = 5)$
9. 0.09804; $(f(x) = 1/x, a = 10)$
11. 2.032778; $(f(x) = \sqrt[3]{x}, a = 8)$

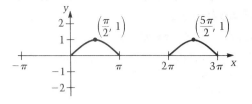

Section 10.6 / page 372

1. $f(x) = \dfrac{1}{2} - \dfrac{2}{\pi}\sin x - \dfrac{2}{3\pi}\sin 3x - \dfrac{2}{5\pi}\sin 5x - \cdots$

3. $f(x) = \dfrac{\pi}{4} - \dfrac{2}{\pi}\cos x - \dfrac{2}{9\pi}\cos 3x - \dfrac{2}{25\pi}\cos 5x - \cdots$

$\qquad + \sin x - \dfrac{1}{2}\sin 2x + \dfrac{1}{3}\sin 3x - \dfrac{1}{4}\sin 4x + \cdots$

5. $f(x) = 2\sin x - \sin 2x + \dfrac{2}{3}\sin 3x - \dfrac{1}{2}\sin 4x + \cdots$

7. $f(x) = \dfrac{4}{\pi}\cos x + \dfrac{4}{9\pi}\cos 3x + \dfrac{4}{25\pi}\cos 5x + \cdots$

$\qquad + 2\sin x + \dfrac{2}{3}\sin 3x + \dfrac{2}{5}\sin 5x + \cdots$

9. $f(x) = \dfrac{1}{\pi} - \dfrac{2}{3\pi}\cos 2x - \dfrac{2}{15\pi}\cos 4x - \cdots + \dfrac{1}{2}\sin x$

Section 10.7 / page 373

1. Converges to $\frac{8}{21}$ (geometric series with $r = -\frac{3}{4}$).
3. Diverges.
5. $1 - 2x + 2x^2 - \cdots$
7. $1 - 2x + 2x^2 - \cdots$
9. $1 + 3x + \frac{9}{2}x^2 + \cdots$
11. $1 - 2x^4 + \frac{2}{3}x^8 - \cdots$
13. $1 + x^2 + \frac{3}{2}x^4 + \cdots$
15. $0.5 + (\sqrt{3}/2)(x - 30°) - (0.5/2!)(x - 30°)^2 - \cdots$
17. 0.99985
19. 0.529925
21. $-0.105\overline{3}$
23. 0.310281
25. $f(x) = -\dfrac{1}{2} + \dfrac{6}{\pi}\sin x + \dfrac{2}{\pi}\sin 3x + \dfrac{6}{5\pi}\sin 5x + \cdots$
27. 0.7639

Chapter 11

Section 11.1 / page 379

1. $dy/dx = 1$

3. $dy/dx = ce^x = y$

5. $y' = 2cx; \; xy' - 2y = x(2cx) - 2cx^2 = 0$

7. $y' = [x(-\cos x) - (c - \sin x)]/x^2; \; xy' + y = (-x\cos x - c + \sin x)/x$

$\qquad + (c - \sin x)/x = -(x\cos x)/x = -\cos x$

9. $y' = c_1\cos x - c_2\sin x; \; y'' = -c_1\sin x - c_2\cos x;$

$\qquad d^2y/dx^2 + y = -c_1\sin x - c_2\cos x + c_1\sin x + c_2\cos x = 0$

11. $y' = 4c_2e^{4x} - \frac{3}{4}; \; y'' = 16c_2e^{4x}; \; y'' - 4y' = 16c_2e^{4x} - 4(4c_2e^{4x} - \frac{3}{4}) = 3$

13. $xy' + y = 2x$

15. $2x + 2yy' = c; \; 2x^2 + 2xyy' = cx; \; 2xyy' = cx - 2x^2 = (x^2 + y^2) - 2x^2 = y^2 - x^2$

17. $dy/dx = 2e^{x^2} \cdot 2x = (y)2x = 2xy;$ from $y = 2e^{x^2}, \, 2 = 2e^0$

19. $y' = 3x^2 + 2; \; y'' = 6x; \; y'' - xy' + 3y = 6x - x(3x^2 + 2) + 3(x^3 + 2x) = 10x;$

$\qquad$ from $y = x^3 + 2x, \, 3 = 1^3 + 2(1);$ from $y' = 3x^2 + 2, \, 5 = 3(1)^2 + 2.$

21. $(1 - x)(y') + (1 + y)(-1) = 0$, so $(1 - x)y' = 1 + y$; from $(1 - x)(1 + y) = 3$, $(1 - 0)(1 + 2) = 3$.

23. No; differential equation is satisfied, but $y = 3$ when $x = 1$.

25. Yes.

27. No; the boundary condition is satisfied, but the differential equation is not.

29. $c = 1$

31. $c = -8$

33. $dA/dt = -k2000e^{-kt} = -kA$; from $A = 2000e^{-kt}$, $2000 = 2000e^{-k(0)}$

Section 11.2 / page 384

1. $\ln|x| + y^4/4 = c$

3. Arctan $x - \tan y = c$

5. $-1/y + x^2/2 = c$

7. $y = c/(1 + x) + 1$

9. $e^{x^2}/2 + y^3/3 = c$

11. $(x^2 - 1)(y^2 + 1) = c$

13. $\ln|xy| + x - y = c$

15. $e^{\sin x} + e^y = c$

17. $\tan y(1 - e^x)^3 = c$

19. Arcsin $x + e^{y^2}/2 = c$

21. $x^3/3 + y^2/2 + y = 12$

23. $y + 3 = \ln|x^4 y|$

25. $(\cos x)(1 - y) = 1$

27. $kT + Qx - 30k = 0$

Section 11.3 / page 388

1. $xy + \ln|y| = c$

3. $y = x^2 + cx$

5. $(y/x)^3 + x^3 = c$

7. $(xy)^2 + x^2 + y^2 = c$

9. $(x^2 + y^2)^2 = 4\ln|y| + c$

11. $e^{xy} = x + c$

13. $\tan(x^2 + y^2) = 2xy + c$

15. $\frac{1}{2}\ln(x^2 + 1) + y + 2\ln|y - 2| = c$

17. $4xy + x^2 = c$

19. $2x + 2y + \ln(x^2 + y^2) = 4 + \ln 2$

21. $xy = x^2/2 + 2$

Section 11.4 / page 393

1. $5yx^2 = x^5 + C$

3. $20y = -5x - 4 + cx^5$

5. $y \sin x = x \cos x - \sin x + c$

7. $yx = (4x^2 + c)e^{2x}$

9. $ce^x = xy$ **11.** $ye^x = x + c$

13. $2e^x y = e^{2x} + c$

15. $3ye^{x^3} = e^{x^3} + c$

17. $y = x - 1 + ce^{-x}$

19. $y = 2x\ln|x| + 4x$

21. $2e^{-3x}y = 5 - e^{-2x}$ **23.** $6xy^2 = 2y^3 - 3y^2 + c$

25. $2t = r^4 + cr^2$ **27.** $ie^{3t} = 4e^t - 4$

Section 11.5 / page 402

1. $y = -(1 - x)^2/2 + 1$ **3.** $y = x\ln|x| + 4x$

5. $y = -x - 1 + 3e^x$ **7.** $y = cx$ **9.** $yx = c$

11. $y = -(\frac{1}{2}x + \frac{1}{4}\sin 2x) + c$ **13.** 4,676,540

15. 34% **17.** $5\frac{1}{3}$ C

19. From the equation $L\,di/dt + Ri = E$, we get the general solution
$ie^{(R/L)t} = (E/R)e^{(R/L)t} + c$ or $i = E/R + ce^{(-R/L)t}$. As $t \to \infty$, $e^{(-R/L)t} \to 0$ and $i \to E/R$.

21. $i = 11(1 - e^{(-10/3.2)t})$ **23.** $q = 0.06(1 - e^{-5t})$

25. 0.3382 C **27.** $v = 150(9.8 - 9.93e^{t/150})$

29. 23.76 m/s **31.** 32.75 lb **33.** 74°C **35.** 28.3 g

37. $y = \dfrac{k_1 L_0}{k_2 - k_1}(e^{-k_1 t} - e^{-k_2 t}) + y_0 e^{-k_2 t}$

39. $(m/k)\lambda + \sin\theta + c = 0$

Section 11.6 / page 415

1. $y_3 = \frac{1}{24}x^4 + \frac{1}{2}x^3 + \frac{3}{2}x^2 + 2x + 2$; $y = -x - 1 + 3e^x$;
$y_3(0.5) = 3.44$; $y(0.5) = 3.45$

3. $y_3 = \frac{4}{3}x^3 + 2x^2 + 2x + 1$; $y = e^{2x}$; $y_3(1) = 6.33$; $y(1) = 7.39$

5. $y_2 = \frac{1}{24}x^6 + \frac{1}{5}x^5 + \frac{1}{2}x^4 + \frac{2}{3}x^3 + \frac{1}{2}x^2 + x + 1$

7. $y_2 = (x^2/2)(1 - \cos x) + x\sin x - \cos x - (\sin^2 x)/2 + 2$

9. $y = a_0(1 + x + \frac{1}{2}x^2 + \frac{1}{6}x^3 + \frac{1}{24}x^4 + \cdots) + (-\frac{1}{2}x^2 - \frac{1}{6}x^3 - \frac{1}{24}x^4 - \cdots)$

$= a_0 e^x + (1 + x - e^x)$

$= (a_0 - 1)e^x + 1 + x$

$= ce^x + 1 + x$

11. $y = a_0 + a_1(x - \frac{1}{2}x^2 + \frac{1}{6}x^3 - \frac{1}{24}x^4 + \cdots)$

13. $y = 1 + \frac{1}{2}x^2 - \frac{1}{3}x^3 + \frac{3}{8}x^4 + \cdots$

15. $y = a_0 + a_1(x + \frac{1}{2}x^2 - \frac{1}{6}x^3 + \frac{1}{8}x^4 + \cdots)$ **17.** $y(1) \approx 0.33$

19.

x_i	y_i
0	0
0.2	0
0.4	0.040
0.6	0.128
0.8	0.274

21.

x_i	y_i
0	2
0.1	2.10
0.2	2.22
0.3	2.38
0.4	2.49

23.

x_i	y_i
1	2
1.3	2.15
1.6	2.33
1.9	2.54
2.2	2.76

25. $x = -t^2 - 2t - 2 + 2e^t$; 0.437 km

27. 0.417 km

Section 11.7 / page 419

1. $y = c_1 + c_2 e^x$ **3.** $y = c_1 e^x + c_2 e^{-x}$

5. $y = c_1 e^{4x} + c_2 e^{2x}$ **7.** $y = c_1 e^{-2x} + c_2 e^{-3x}$

9. $y = c_1 + c_2 e^{-x/2}$ **11.** $y = c_1 e^{4x} + c_2 e^{-3x/2}$

13. $y = c_1 e^{2x/3} + c_2 e^{-5x/2}$

15. $y = c_1 e^{(-1 + \sqrt{13})x/2} + c_2 e^{(-1 - \sqrt{13})x/2}$

17. $y = c_1 e^{(-2 + \sqrt{10})x/2} + c_2 e^{(-2 - \sqrt{10})x/2}$

19. $y = c_1 e^{(5 + \sqrt{13})x/6} + c_2 e^{(5-\sqrt{13})x/6}$

21. $y = c_1 + c_2 e^{-x} + c_3 e^{2x}$

23. $y = 2e^x + \frac{1}{2}e^{-x/2}$

Section 11.8 / page 424

1. $y = 3^{-2x}(c_1 x + c_2)$ **3.** $y = c_1 x + c_2$

5. $y = e^{-3x}(c_1 x + c_2)$ **7.** $y = e^{-3x/2}(c_1 x + c_2)$

9. $y = e^{x/3}(c_1 x + c_2)$ **11.** $y = e^{-9x/4}(c_1 x + c_2)$

13. $y = e^{-x}(c_3 \cos x + c_4 \sin x)$

15. $y = c_3 \cos \dfrac{\sqrt{10}}{2} x + c_4 \sin \dfrac{\sqrt{10}}{2} x$

17. $y = e^{2x}(c_3 \cos \sqrt{3}x + c_4 \sin \sqrt{3}x)$

19. $y = e^x \left(c_3 \cos \dfrac{\sqrt{6}}{2} x + c_4 \sin \dfrac{\sqrt{6}}{2} x \right)$

21. $y = e^{2.5x}(c_3 \cos 1.32x + c_4 \sin 1.32x)$

23. $y = e^{2x}(-2x + 3)$

25. $y = 2 \cos 2x + 3 \sin 2x$

27. $e^{j\theta} = 1 + j\theta + \dfrac{(j\theta)^2}{2!} + \dfrac{(j\theta)^3}{3!} + \dfrac{(j\theta)^4}{4!} + \cdots$

$$= 1 + j\theta - \dfrac{\theta^2}{2!} - j\dfrac{\theta^3}{3!} + \dfrac{\theta^4}{4!} + \cdots$$

$$= 1 - \dfrac{\theta^2}{2!} + \dfrac{\theta^4}{4!} + \cdots + j\theta - j\dfrac{\theta^3}{3!} + \cdots$$

$$= \cos \theta + j \sin \theta$$

Section 11.9 / page 429

1. $y = c_1 e^{-x} + c_2 e^{2x} - 7$

3. $y = e^x(c_1 x + c_2) + 4x + 8$

5. $y = e^{-x}(c_3 \cos 3x + c_4 \sin 3x) + \frac{1}{10}x^2 - \frac{2}{50}x - \frac{3}{250}$

7. $y = c_1 e^{-2x} + c_2 e^{-3x/2} + \frac{4}{13}\sin x - \frac{7}{13}\cos x$

9. $y = c_1 e^{-x} + c_2 e^{-3x} + \frac{1}{2}e^{3x}$

11. $y = e^{2x}(c_3 \cos x + c_4 \sin x)$
$\qquad - \frac{1}{2}\cos 3x - \frac{3}{2}\sin 3x + \frac{1}{5}x + \frac{4}{25}$

13. $y = c_1 e^{3x} + c_2 e^{-3x} - 2e^{2x} - \frac{3}{10}\sin x$

15. $y = e^{-3x}(c_1 x + c_2) + 2e^{-x} - \frac{1}{5}e^{2x}$

17. $y = c_1 e^{x/2} + c_2 e^{4x} - xe^x + \frac{5}{3}e^x + \frac{1}{2}x + \frac{9}{8}$

19. $y = c_1 e^{\sqrt{2}x} + c_2 e^{-\sqrt{2}x} - x \sin x - \frac{2}{3}\cos x$

21. $y = -\frac{5}{8}e^{2x} + \frac{21}{8}e^{-2x/3} + 2e^{3x}$

Section 11.10 / page 435

1. $x = 0.3 \cos 2\sqrt{3}t$

3. a. $x = 0.08 \cos 7t$;

    b. -0.56 m/s (the negative sign shows that the weight is moving upward).

5. $x = e^{-4.9t}(0.08 \cos 5t + 0.078 \sin 5t)$

7. $x = 0.08 \cos 7t + 3.015 \cos 6t$

9. $x = e^{-4.9t}(0.08 \cos 5t + 0.078 \sin 5t)$
$\qquad + 0.14 \cos 6t + 0.64 \sin 6t$

11. $q = e^{-6t}(0.05 \cos 25.1t + 0.012 \sin 25.1t)$

13. $q = e^{-0.75t}(-5 \cos 3.07t - 1.22 \sin 3.07t) + 5$

15. $q = e^{-2t}(0.41 \cos 2.45t - 0.42 \sin 2.45t)$
$\qquad + 0.92 \sin 2t - 0.41 \cos 2t$

17. $q = 0.04 \cos t + (1.6 \times 10^{-3}) \sin t$

19. $i = -8.12 \sin 4t + 18.56 \cos 4t$

21. $x = c_3 \cos 5.5t + c_4 \sin 5.5t$

Section 11.11 / page 441

1. $L(\sin ax) = \lim\limits_{c \to \infty} \int_0^c e^{-sx} \sin ax$

$$= \lim_{c \to \infty} \left[\dfrac{e^{-sx}}{s^2 + a^2} (-s \sin ax - a \cos ax) \right]_0^c \quad \text{Rule 63.}$$

$$= \lim_{c \to \infty} \dfrac{e^{-sc}}{s^2 + a^2} (-s \sin ac - a \cos ac)$$

$$\qquad - \dfrac{1}{s^2 + a^2} (-a)$$

$$= \dfrac{a}{s^2 + a^2}$$

3. $L(e^{ax} \sin bx) = \lim\limits_{c \to \infty} \int_0^c e^{(a-s)x} \sin bx \, dx$

$$= \lim_{c \to \infty} \dfrac{e^{(a-s)x}}{(a-s)^2 + b^2} [(a-s) \sin bx - b \cos bx] \Big|_0^c$$

$$= \lim_{c \to \infty} \dfrac{e^a \cdot e^{-sc}}{(a-s)^2 + b^2} [(a-s) \sin bc - b \cos bc]$$

$$- \dfrac{1}{(a-s)^2 + b^2} (-b) = \dfrac{b}{(a-s)^2 + b^2} = \dfrac{b}{(s-a)^2 + b^2}$$

5. $24/s^5$

7. $2/(s-7)^3$

9. $1/(s-2)^2 + 12/s^4$

11. $6\pi/(s^2 + \pi^2) - 4/s^3$

13. $(s^2 - 2s)L(y) - 4$

15. $(2s^2 + 7)L(y) + 2s - 4$

17. $(4s^2 - 6s + 3)L(y) - 8s + 12$

19. e^{3x}

21. xe^x

23. $\frac{1}{6}x \sin 3x$

25. $\cos 4x - \frac{3}{4} \sin 4x$

Section 11.12 / page 445

1. $y = -e^{3x}$

3. $y = -\frac{1}{2} - \frac{5}{2}e^{4x}$

5. $y = xe^{3x} + e^{3x}$

7. $y = \cos \sqrt{2}x$

9. $y = \frac{2}{3}(1 - e^{3x})$

11. $y = 3e^{2x} \sin x$

13. $y = \frac{1}{27}(3x - \sin 3x) + \cos 3x$

15. $y = \frac{5}{2} + \frac{5}{2}e^{2x} - 5e^x$

17. $y = 30,000e^{0.2t}$

19. $x = 0.2 \cos \sqrt{19.6}t$

Section 11.13 / page 447

1. $x + x^{-1} = \frac{1}{2} \ln(y^2 + 1) + c$

3. $3y = x^4 + cx$

5. $12\sqrt{y} + 4y^{3/2} = 3x^2 + c$

7. $\ln(x^2 + y^2) = 2x + c$

9. $y = (\cos x + \sin x)/2 + ce^{-x}$

11. $\ln |(x + 2)^2 y| = x + c$

13. $y = c_1 e^x + c_2 e^{-3x}$

15. $y = e^{-4x}(c_1 x + c_2)$

17. $y = c_1 e^{5x/2} + c_2 e^{-3x}$

19. $y = c_3 \cos \sqrt{3}x + c_4 \sin \sqrt{3}x$

21. $y = c_1 e^x + c_2 e^{-3x} - 3x - 2$

23. $x^3/3 + \ln|y + 3| = \ln 4$

25. $ye^{-x} = x + 2$

27. $y = e^{2x} - 2e^x$

29. $y = \frac{1}{2}(1 - e^{-2x})$

31. $y = \frac{1}{3}x - \frac{4}{9} \sin 3x + \cos 3x$

33. $y_3 = -\frac{1}{15}x^5 - \frac{1}{3}x^3 - x$

35. $y = a_0(1 + \frac{1}{2}x^2 + \frac{1}{8}x^4 + \cdots)$

37.

x_i	y_i
0	0
0.2	0
0.4	0.04
0.6	1.44
0.8	2.42

39. $4y^2 = -x^2 + c$

41. 26 g

43. $i = \frac{1}{6}(1 - e^{-120t})$

45. -240 m/s

47. $x = e^{-t}(-0.1 \cos 5.6t$
$- 0.02 \sin 5.6t)$

49. $q = e^{-5t}(-0.05 \cos 8.66t$
$- 0.03 \sin 8.66t)$
$+ 0.05 \cos t$
$+ (5.05 \times 10^{-3}) \sin t$

Index